Green Chemistry and Sustainable Technology

Series editors

Aims and Scope

The series *Green Chemistry and Sustainable Technology* aims to present cutting-edge research and important advances in green chemistry, green chemical engineering and sustainable industrial technology. The scope of coverage includes (but is not limited to):

- Environmentally benign chemical synthesis and processes (green catalysis, green solvents and reagents, atom-economy synthetic methods etc.)
- Green chemicals and energy produced from renewable resources (biomass, carbon dioxide etc.)
- Novel materials and technologies for energy production and storage (biofuels and bioenergies, hydrogen, fuel cells, solar cells, lithium-ion batteries etc.)
- Green chemical engineering processes (process integration, materials diversity, energy saving, waste minimization, efficient separation processes etc.)
- Green technologies for environmental sustainability (carbon dioxide capture, waste and harmful chemicals treatment, pollution prevention, environmental redemption etc.)

The series *Green Chemistry and Sustainable Technology* is intended to provide an accessible reference resource for postgraduate students, academic researchers and industrial professionals who are interested in green chemistry and technologies for sustainable development.

More information about this series at http://www.springer.com/series/11661

Peter C.K. Lau
Editor

Quality Living Through Chemurgy and Green Chemistry

Springer

Editor
Peter C.K. Lau
Tianjin Institute of Industrial Biotechnology,
Chinese Academy of Sciences
Tianjin
China

and

McGill University
Montreal
Québec
Canada

ISSN 2196-6982 ISSN 2196-6990 (electronic)
Green Chemistry and Sustainable Technology
ISBN 978-3-662-53702-2 ISBN 978-3-662-53704-6 (eBook)
DOI 10.1007/978-3-662-53704-6

Library of Congress Control Number: 2016954596

Printed on acid-free paper

This Springer imprint is published by Springer Nature
The registered company is Springer-Verlag GmbH Germany
The registered company address is: Heidelberger Platz 3, 14197 Berlin, Germany

To those who care though not necessarily understand

"There does not exist a category of science to which one can give the name applied science. There are science and the applications of science, bound together as the fruit and the tree that bears it."

—Louis Pasteur, 1871

Foreword

As we all know from our school years, George Washington Carver sought to transform the U.S. agricultural sector of his day by paving the way for crop diversity and bioproduct manufacturing, which allowed the American South to move away from cotton monoculture.

Carver and his students developed more than 300 industrial uses for peanuts, sweet potatoes, and other crops that could be grown in rotation with cotton and corn. Carver's inventions included plastics, glues, soaps, paints, dyes for cloth and leather, medicines, and cosmetic ingredients. He was a prime mover in the early twentieth century chemurgy movement, which created ways to replace petrochemicals with farm-derived bioproducts. Industrial biotechnology pioneers restarted the movement in the late twentieth century and continue it today. The discovery of the structure of DNA, the development of modern genetic engineering methods and their application in industrial biotechnology accelerated the innovations and transformed the chemurgy space.

In June 2008, The Economist magazine published an article, "Better living through chemurgy," directly evoking the "Better Things for Better Living" branding campaign of a prominent traditional chemical manufacturer. The article outlined the economic potential for agricultural feedstocks to replace petroleum in everyday consumer products and the breakneck pace of industrial biotechnology developments enabling this substitution. It also highlighted the efforts of a growing breed of entrepreneurs to replace petrochemicals with renewable ones, boldly stating, "[C]hemurgy is back with a vengeance in the shape of modern industrial biotechnology." This book's title, "Quality Living through Chemurgy and Green Chemistry," further evokes that economic and technological potential.

In the 2008 Economist article, one pioneer of industrial biotech marveled that genetic engineering "processes that once took five years now take just one." Since 2008, the pace of technological development has accelerated still further. Synthetic biology and other new genetic engineering techniques have increased the speed and improved the precision of engineering biology. The processes that once took one year can now be accomplished more quickly. In Chap. 6 of this volume, Chun You of the Chinese Academy of Sciences and Y.H. Percival Zhang of Virginia Tech describe a proposed consolidated biomass to ethanol process, in which cellulose is both hydrolyzed and fermented in the same vessel. Advanced enzyme engineering and new methods for pretreating cellulosic biomass would enable this proposed methodology, according to the authors.

Industrial biotech companies continue to accelerate the process of bringing products to market as they expand the universe of renewable products available to replace petroleum. In Chap. 3 of this volume, Liang Wu of the DSM Biotechnology Center describes the successful commercialization of biobased succinic acid, a precursor to polyesters and a replacement for fossil-based polyesters. Commercialization of succinic acid paved the way for proof-of-principle fermentation of adipic acid and caprolactam, nylon intermediates that can replace additional fossil-based polyesters, according to Wu. In Chap. 8, Jerald Lalman and colleagues from the University of Windsor and Aberystwyth University review the feedstocks and biofuels currently being produced and developed and find that additional resources are needed to fully displace petroleum. They describe options for fourth generation feedstocks, which could include genetically engineered aquatic plants that are also designed for CO_2 uptake and sequestration. Biohydrogen may also be a biofuel of the future, they find.

The need for additional biobased resources continues to push companies toward novel feedstocks, even as commercial development of cellulose, algae, solid waste streams, and methane continues. In Chap. 5, Huimin Zhao and colleagues from the University of Illinois at Urbana-Champaign provide an overview of the potential use of macroalgae biomass as a feedstock for biofuels and renewable chemicals. They also detail the advances in metabolic engineering that enable microorganisms to metabolize algal biomass sugars to ethanol and possible chemicals. And in Chap. 9, Jairo Lora of Lora Consulting LLC, describes the various uses of lignin—a component of biomass—as a feedstock for polymers and carbon materials. Companies have made progress in commercializing ingredients for consumer products from lignin. In Chap. 10, Fanny Monteil-Rivera of the National Research Council of Canada evaluates those green technologies including biocatalysis that could possibly be commercialized to convert lignin to valuable aromatic chemicals. She further identifies research gaps in the most promising approaches.

Renewable chemicals have made their way into consumer products, with some brand name manufacturers actively seeking commercial-scale quantities of renewable chemicals. Back in 2008, the global management firm McKinsey & Co. projected that the industrial biotech sector would capture $100 billion in value by 2010, primarily through biofuels and plant extracts. The focus of the sector has now shifted as new technologies, applications and opportunities have come into view. More recently, the business consultancy Lux Research examined the subsector of renewable chemical production and found that it was experiencing the most rapid growth within industrial biotechnology, potentially reaching 13.2 million metric tons of annual capacity by 2017.

In addition to the economic potential of the sector, there is the promise of better living. A 2007 report from EPA, "Bioengineering for Pollution Prevention," noted that industrial biotechnology processes are naturally consistent with the principles of green chemistry and cleaner than petrochemical processes, since they reduce pollution at the source, save energy, and make use of renewable carbon. In the first chapter of this volume, David Constable, director of the American Chemical Society's Green Chemistry Institute, examines in detail how biobased and renewable chemicals can fit into the principles of green chemistry. He notes that chemurgy both influenced and continues to be influenced by the concept of green chemistry. In Chap. 4, Dunming Zhu of the Chinese Academy of Sciences and Ling Hua of DuPont Industrial Sciences expand on the unique benefits of using enzyme catalysis to replace chemical reactions in industry. They provide several examples of the positive environmental impacts—such as reduced chemical waste and production costs as well as improved energy efficiency—achieved through enzyme applications in chemical manufacturing.

There is also the promise of sustainably providing better living to people around the world. The availability and affordability of petroleum-based products and transportation has improved the standard of living of consumers in the United States and many other countries. The imbalance in living standards is also reflected in competition for natural resources. World population is expected to reach 9 billion by 2050, according to UN estimates, with rapid growth in Asia and Africa. Growing populations combined with economic growth will increase demands for food and consumer goods; at the same time, changing climate conditions could exacerbate resource constraints.

In Chap. 2 of this volume, Dongming Xie, Quinn Zhu and colleagues at DuPont Industrial Biosciences describe progress in commercializing an omega-3 fatty acid nutrition and feed supplement. The fatty acid is conventionally derived from fish, which are a scarce natural resource. With industrial biotechnology, the nutritional supplement can be produced in large quantities by yeast in a fermentation vessel, reducing demand for fish. Further, the omega-3 fatty acid can supplement fish food for farm raised salmon, further reducing demand for wild caught fish. And in Chap. 11, Baixue Lin and Yong Tao of the Chinese Academy of Sciences discuss progress in using metabolic engineering for production of sialic acids, which have applications in medicine, including antiviral drugs, and as a nutraceutical. These ingredients are traditionally derived from milk whey and egg yolks with typically low yields, which make them insufficient to meet growing demand.

Jim Lane at Biofuels Digest put the challenge of providing economic opportunity to growing worldwide populations in stark terms, saying, "Either we find a second planet to provide the resources. Or we in the United States find a way to tell the rest of the world that they can't have the material lifestyle we enjoy." The better option is to transition to a biobased economy that can generate the food, material goods and fuels needed to meet the demands of the world's growing population while providing new economic opportunities and creating a cleaner environment. Chemurgy—the technology needed to build the biobased economy—can also become the basis for manufacturing and economic growth, providing a means for people around the world to achieve a quality living standard.

In Chap. 7 of this book, Mingyu Wang and Jin Hou of Shandong University review the economic and technical hurdles that the industrial biotechnology and biorefinery industry must continue to address. Progress has been made in reducing the cost of converting cellulosic biomass to fermentable sugars as feedstock for biofuels and renewable chemicals; metabolic engineering provides tools for further cost reductions as well as the ability to produce new and more cost-effective fine chemicals. And in Chap. 12, Krista Morley of the National Research Council Canada and Peter C.K. Lau of Tianjin Institute of Industrial Biotechnology (Chinese Academy of Sciences) and McGill University map out the value chain for new functional polymers, detailing the steps from biomass to phenolic acids to monomers and finally proof-of-concept polymers. The biotech advances that enabled the development of these new polymers include the discovery of new enzymes.

In Chap. 13 of this volume, Manfred Kircher of CLIB2021 details how traditional chemical manufacturers can work with biomass producers to develop infrastructure and value chains to build state and international bioeconomies. The example of Europe's leading chemical cluster straddling Belgium, the Netherlands and Germany is presented as a bottom up approach to chemurgy. If George Washington Carver were alive today he might be very pleased and gratified to see how far chemurgy has come. Chemurgy has evolved to be modern-day industrial biotechnology. The chemurgy story is far from finished and more innovation and problem solving is in our future.

Brent Erickson

Preface

Control of consciousness determines the quality of life.

Mihaly Csikszentmihalyi

The impetus to editing this book stemmed from the overarching needs to fuel the fledgling bioeconomy or circular economy by realizing industrial and environmental sustainability through knowledge-based solutions, innovation, and cleaner production. The heart of this matter is promoting a greater awareness and deployment of environmentally benign biological processes to achieve various bioconversions by using naturally occurring or genetically improved enzymes, whole cells and associated bioprocesses. Whenever possible, renewable bio-based materials should be the feedstock of choice for the supply of chemicals and alternative energy sources to meet human societal needs.

This book, *Quality Living through Chemurgy and Green Chemistry*, is designed to give a deep appreciation of what the future holds, empowered by the synergy of two biomasses—a renewable feedstock and an infinite microbial biomass and its diversity as "game changers" in value addition and creation. Biocatalysis is integral to the 12 principles of green chemistry, an environmental priority and gateway to sustainability. Putting green chemistry to work on renewable feedstocks besides the conventional biomass is an extension of the original definition of chemurgy—applying "chemistry" (chemi) to "work" (ergon—Greek) on agricultural residues or simply on the farm.

I am deeply honored to have Brent Erickson writing a Foreword. He is Executive Vice President of the Industrial and Environmental Section of the Biotechnology Innovation Organization (BIO, formerly Biotechnology Industry Organization), and recently his name made the upper 10 % list of the Biofuels Digest "Top 100 people in the Advanced Bioeconomy," a premier trade publication for the industrial biotechnology sector. I am equally honored to have Dr. David Constable, Director of the American Chemical Society's Green Chemistry Institute to contribute a key chapter on Green Chemistry and Sustainability.

This book comprises 13 chapters and they have been written by scientists and engineers who are experts in their own fields. I am most grateful to these authors and colleagues who kindly accepted to undertake the charitable task of writing an informative chapter beyond their call of duty. The contents of the book are organized as follows:

- Chapter 1 serves as an introduction to the concept and practice of green chemistry and outlines the opportunities for applying sustainable (green) chemistry to maximize biobased resource efficiency.
- Chapters 2 and 3 highlight the industrial needs and perspectives in the bioproduction of key commodity chemicals, e.g., dicarboxylic acids and omega-3 fatty acids.
- Chapters 4 and 5 are directed toward the development of a potpourri of biocatalysts to meet chemical and pharmaceutical needs, chiral building blocks, in particular; and also specialized enzymes from underexplored feedstock such as macroalgae for the production of biofuels and biochemicals.
- Chapters 6–8 focus on value creation from cellulosic and lignocellulosic materials to produce starch and derivatives thereof, as well as to gain access to a variety of biochemicals and biofuels including hydrogen, methane and butanol. Novel approaches to fermentation, e.g., simultaneous enzymatic biotransformation and microbial fermentation are highlighted. In the latter two chapters, various socioeconomic factors and technical bottlenecks that researchers and industries face in the biorefinery of lignocellulosics are discussed.
- Chapters 9 and 10 are devoted to the depolymerization and valorization of lignins from cleaner production (green approaches) point of view, and an updated review on the wide range of polymeric products that can be derived from this abundant aromatic feedstock. Notably, a great number of patents are cited in Chap. 9.
- Chapter 11 discusses the production of a family of amino sugars, sialic acid and its derivatives, and Chap. 12 presents a value chain in phenolic acids and polymeric production from agricultural feedstocks as well as the development of new enzymes and associated bioprocesses.
- Last but not least, Chap. 13 is considered a lagniappe—baker's dozen. Without an infrastructure and receptors any technology development will be futile. Extrapolations of present day pillars of competitiveness in fossil- based economy to those of chemurgy and green chemistry in the bio- or circular economy are discussed by Dr. Manfred Kircher using a European model.

All in all, this book provides invaluable insights and perspectives in biobased economy and bioproduction from academia, national laboratories, health and environment industries as well as consulting enterprises. Although there is no shortage of books in the general context of biomass and nonbiological green chemistry approaches, *Quality Living through Chemurgy and Green Chemistry* is considered unique while adding knowledge to current advances in the literature. Graduate and undergraduate students in various disciplines, aspiring "sustainability scientists," practitioners of industrial biotechnology and biobased industry, researchers and engineers, granting officers and policy makers, etc., are expected to benefit from this book.

Started in the 1930s, a "chemurgic movement" was said to have taken place, formalized by the 1935 "Declaration of Dependence upon the Soil and of the Right of Self-Maintenance" and initiated by Dr. William J. Hale of Dow Chemical Company, among prominent proponents like Henry Ford and George Washington Carver: "When in the course of the life of a Nation, its people become neglectful of the laws of nature…, necessity impels them to turn to the soil in order to recover the right of self-maintenance." Sustainability in either environmental or industrial setting is the new framework in which we all have a role to play.

I will be remiss if I do not express my gratitude to June Tang of Springer Beijing who first approached me for a project on Sustainable Development but agreed on this alternative title. Support and assistance in contacting the authors, preparing and realizing the final launching of the book came from the unyielding effort and patience on the part of June Tang and her team, Heather Feng, in particular. The interactions and exchanges of information with the Publisher, not forgetting the timely cooperations of the authors, have been a most gratifying experience. Thank you all. I hope readers will enjoy the book. Any suggestions or comments are welcome.

Tianjin, China and Montreal, Canada
September 2016

Peter C.K. Lau

Contents

Contributors

Ahmet Badur Department of Chemical and Biomolecular Engineering, University of Illinois at Urbana-Champaign, Urbana, IL, USA; Illumina, San Diego, CA, USA

David J.C. Constable American Chemical Society, Green Chemistry Institute, Washington, DC, USA

Joe Gallagher Institute of Biological, Environmental and Rural Sciences, Aberystwyth University, Aberystwyth, Ceredigion, UK

Eva Garcia-Ruiz Department of Chemical and Biomolecular Engineering, Carl R. Woese Institute for Genomic Biology, University of Illinois at Urbana-Champaign, Urbana, IL, USA

Jin Hou State Key Laboratory of Microbial Technology, School of Life Sciences, Shandong University, Jinan, China

Ling Hua DuPont Industrial Biosciences, Experimental Station, Wilmington, DE, USA

Ethel N. Jackson Industrial Biosciences, E.I. Du Pont de Nemours and Company, Wilmington, DE, USA

Manfred Kircher KADIB Kircher Advice in Bioeconomy, Frankfurt, Germany; CLIB2021, Düsseldorf, Germany

Jerald A. Lalman Department of Civil and Environmental Engineering, University of Windsor, Windsor, ON, Canada

Peter C.K. Lau Tianjin Institute of Industrial Biotechnology, Chinese Academy of Sciences, Tianjin, China; McGill University, Montreal, QC, Canada

Baixue Lin CAS Key Laboratory of Microbial Physiological and Metabolic Engineering, Institute of Microbiology, Chinese Academy of Sciences, Beijing, China

Jairo H. Lora Lora Consulting LLC, Media, PA, USA

Edward Miller Industrial Biosciences, E.I. Du Pont de Nemours and Company, Wilmington, DE, USA

Fanny Monteil-Rivera National Research Council of Canada, Aquatic and Crop Resource Development, Montreal, QC, Canada

Krista L. Morley National Research Council Canada, Montreal, QC, Canada

Christopher V. Rao Department of Chemical and Biomolecular Engineering, University of Illinois at Urbana-Champaign, Urbana, IL, USA

Sreenivas Ravella Institute of Biological, Environmental and Rural Sciences, Aberystwyth University, Aberystwyth, Ceredigion, UK

Wudneh A. Shewa Department of Civil and Environmental Engineering, University of Windsor, Windsor, ON, Canada

Yong Tao CAS Key Laboratory of Microbial Physiological and Metabolic Engineering, Institute of Microbiology, Chinese Academy of Sciences, Beijing, China

Bjorn Tyreus Industrial Biosciences, E.I. Du Pont de Nemours and Company, Wilmington, DE, USA

Mingyu Wang State Key Laboratory of Microbial Technology, School of Life Sciences, Shandong University, Jinan, China

Liang Wu DSM Biotechnology Center, Delft, The Netherlands

Dongming Xie Industrial Biosciences, E.I. Du Pont de Nemours and Company, Wilmington, DE, USA; Department of Chemical Engineering, University of Massachusetts-Lowell, Lowell, MA, USA

Chun You Tianjin Institute of Industrial Biotechnology, Chinese Academy of Sciences, Tianjin, China; Biological Systems Engineering Department, Virginia Tech, Blacksburg, VA, USA

Y.H. Percival Zhang Tianjin Institute of Industrial Biotechnology, Chinese Academy of Sciences, Tianjin, China; Biological Systems Engineering Department, Virginia Tech, Blacksburg, VA, USA

Huimin Zhao Department of Chemical and Biomolecular Engineering, Carl R. Woese Institute for Genomic Biology, University of Illinois at Urbana-Champaign, Urbana, IL, USA; Departments of Chemistry, Biochemistry, and Bioengineering, University of Illinois at Urbana-Champaign, Urbana, IL, USA

Dunming Zhu National Engineering Laboratory for Industrial Enzymes and Tianjin Engineering Research Center of Biocatalytic Technology, Tianjin Institute of Industrial Biotechnology, Chinese Academy of Sciences, Tianjin, China

Quinn Zhu Industrial Biosciences, E.I. Du Pont de Nemours and Company, Wilmington, DE, USA

Chapter 1
Green Chemistry and Sustainability

David J.C. Constable

Abstract The idea of preparing industrial and commercial products from bio-based and renewable materials is not a new one. Humans have relied on agriculture for food, clothing, and shelter throughout human history. In the 1930s, United States saw a resurgence of interest in producing a greater number of industrial and commercial products from agricultural sources, although this interest did not survive much past the end of the Second World War and the growth of the petrochemical industry. In recent years, there has been another resurgence of interest in obtaining chemical feedstocks from bio-based and renewable resources. This is largely a result of society being confronted by the inherent lack of renewability of petroleum and some of the large environmental impacts such as global climate change that are associated with petroleum, especially for transportation fuels. Turning these bio-based and renewable feedstocks into chemicals should be undertaken in a more sustainable way than what is currently being practiced in the chemical manufacturing enterprise. Sustainable and green chemistry are a way of thinking about and practicing chemistry and should be a hallmark of the transition to a bio-based, renewable and sustainable chemical manufacturing enterprise.

Keywords Green chemistry · Sustainable chemistry · Green engineering · Bio-based · Renewable

1.1 Chemistry and Green Chemistry: The Need for Change

It has been interesting to see recent publications pointing to the need for a change in the application and practice of chemistry and chemical engineering [1, 2]. If one were to compare the global chemistry enterprise (industry, academia, and government)

D.J.C. Constable (✉)
American Chemical Society, Green Chemistry Institute, 1155 Sixteenth Street, NW, Washington, DC 20036, USA
e-mail: D_Chichester-Constable@acs.org

P.C.K. Lau (ed.), *Quality Living Through Chemurgy and Green Chemistry*, Green Chemistry and Sustainable Technology,
DOI 10.1007/978-3-662-53704-6_1

of the post-Second World War era from the mid-1940s to the early 1980s with the global chemistry enterprise that has evolved since the early 1980s into its current form, there are many similarities, but perhaps a greater number of profound differences. As just one example of a profound difference, large chemical companies with active research organizations working on basic or fundamental chemical knowledge and/or the commercialization of transformative chemical technologies are a thing of the past. Instead, companies are pursuing what they believe is greater shareholder value by making small, incremental improvements or minor variations to existing product lines to extract as much market share and profitability as they can from known assets. Chemical and chemical technology innovation of the type routinely undertaken in the post-war era is generally seen as being a too risky and comparatively longer term investment. Most companies are relentlessly driven to reduce risk in all areas of the business, so research, development, and demonstration at scale is seen as being too risky for any one company to undertake.

Toward the end of the 1960s and the beginning of the 1970s, a number of environmental disasters led to the promulgation of comprehensive government regulation. Governments sought to prevent releases of waste to the air, water, and land that were leading to very noticeable decreases in environmental quality; i.e., rivers caught on fire, vehicles belched noxious fumes, cities were enveloped in smog, and drinking water was being contaminated. Following the installation of a considerable amount of costly end-of-pipe controls, and in response to an ever-increasing array of environmental legislation that culminated in the U.S. Pollution Prevention Act of 1990, a variety of people in industry, academia, and government began to think about how pollution might be avoided and not just treated to make it less "bad", or to shift it from air to water or land, or vice versa. By the mid-1990s, this thought process was codified and labeled as green chemistry. Most people who are familiar with the term green chemistry usually understand it through, or associate it with, the 12 principles of green chemistry published by Anastas and Warner [3].

Since the publication of this first set of principles, there have been a variety of attempts to further expand upon the ideas and concepts embodied in green chemistry and these have recently been summarized and simplified into 3 categories; maximize resource efficiency, eliminate and minimize hazards and pollution, and design systems holistically and using life cycle thinking [4]. Most people who talk about green chemistry, as typified by the U.S. EPA, emphasize the elimination or reduction of chemical hazards and pollution [5]. The addition of resource efficiency, design, and systems or life cycle thinking is more readily associated with engineering and less with chemistry. It is assumed that the reader is able to link the concepts contained in the "eliminate and minimize hazards and pollution" category to the bio-based and renewable context without further exposition.

It is worth taking a moment to consider two of these different categories and draw out a subset of those concepts/principles in each area that are most relevant to a sustainable and green chemistry and engineering evaluation of bio-based and renewable chemicals production. It can be argued that as one undertakes life cycle and systems level assessments or considerations, one more naturally is driven toward sustainability considerations and the interplay between economic, environmental and social impacts and benefits. The boxes that follow do not contain all the concepts and principles in these categories, and it is not an exhaustive treatment of the potential opportunities. What follows is offered as a limited illustration of how one might think about the application of these concepts/principles to this sector.

Maximize resource efficiency

Principle	Bio-based and renewable opportunity
Design	
Find alternatives: The use of auxiliary substances (e.g., solvents, separation agents) should be made unnecessary whenever possible and innocuous when used [3]	Investigate nonsolvent separation options (e.g., membrane-based, absorption or other solid mass separating agent) that do not entail distillation or other energy-intensive separations technology
Use catalysis: Catalytic reagents (as selective as possible) are superior to stoichiometric reagents [3]	Actively avoid use of platinum group catalysts (e.g., Pt, Pd, Ru, etc.) that are actively being researched for use in, for example, lignin conversion. Favor use of biocatalysts (i.e., enzymes), base metal organometallic catalysts, or organocatalysts
Measure	
Mass balances: Establish full mass balances for a process [6]	Gate-to-gate and cradle-to-gate mass balances should be considered at all phases of R&D to commercialization
Heat and mass transfer: Anticipate heat and mass transfer limitations [6]	Energy and mass integration across a biorefinery will be essential for making it cost competitive
Conversion: Report conversions, selectivities, and productivities [6]	Conversions: Biomass to sugar or to higher value-added chemicals remains a major challenge Selectivities: The inherent selectivity of genetically modified/evolved organisms to produce enantiomerically pure and complex molecules has yet to be realized to as great extent as possible. As the technology evolves and it becomes easier and less expensive, its use will become more widespread Productivities: Productivities, for example, in fermentation remain challenging for many bio-based chemicals and an opportunity for continuing improvement
By-product formation: Identify and quantify by-products [6]	Harness selectivity of living systems and develop separations capability that will enable continuous fermentations
Utilities: Quantify and minimize the use of utilities [6]	Energy efficiency and use of low grade energy (i.e., low temperature) will be key in a fermentation facility and biorefinery. Consider large volume of water in a fermenter at 36 °C as an energy source

(continued)

(continued)

Be efficient	
Reduce: Separation and purification operations should be designed to minimize energy consumption and materials use [7]	The bio-based and renewable sector has an opportunity to develop and implement alternative separations technologies. The industry is characterized by large, water-based, and dilute systems, all of which are challenging from a separation perspective
Minimize: Energy requirements should be recognized for their environmental and economic impacts and should be minimized. Synthetic methods should be conducted at ambient temperature and pressure [3]	The industry is characterized by processes that may operate at close to ambient temperatures and certainly do not require higher pressures except for chemical processing that takes place after fermentation
Optimize: Products, processes, and systems should be designed to maximize mass, energy, space, and time efficiency [7]	Biggest opportunity here is to move toward continuous fermentation processes, or total recycle/zero waste discharge
INTEGRATE: Design of products, processes, and systems must include integration and interconnectivity with available energy and materials flows [7]	Biorefineries should work toward the same level of heat and mass integration that is found in petrochemical facilities
Be sustainable	
Minimize: Minimize depletion of natural resources [8]	The transition to bio-based and renewable chemicals will require careful attention to how this can be done in a way that protects phosphorus and employs a sustainable process for ammonia/ammonium production
Conserve and improve: Conserve and improve natural ecosystems while protecting human health and well-being [8]	Huge opportunity here to employ conservation and improvement activities in all parts of the biotech industry—from raw material extraction to closing the loop on products/waste

Design systems holistically use life cycle thinking

Principle	Bio-based and renewable opportunity
Design	
Holistically: Engineer processes and products holistically, use systems analysis, and integrate environmental impact assessment tools [8]	Just as the modern petrochemical industry has evolved over the course of the past 100 years, so will the biorefinery. At the moment, biorefineries have not achieved the level of mass and energy integration that will move them toward more sustainable processes
Use life cycle thinking: Use life cycle thinking in all engineering activities [8]	
End of use: Products, processes, and systems should be designed for performance in a commercial "afterlife" [7]	These concepts have more to do with moving toward a closed-loop economy and thinking about how waste can be better utilized rather than disposing in air, water, or a landfill
Durability: Targeted durability, not immortality, should be a design goal [7]	

(continued)

(continued)

Principle	Bio-based and renewable opportunity
Conserve complexity: Embedded entropy and complexity must be viewed as an investment when making design choices on recycle, reuse, or beneficial disposition [7]	This is an area that is not being exploited from a molecular perspective. It is interesting to note that early assessments of bio-based and renewable chemical opportunities were from the perspective of harnessing framework molecules that could be modified into various end uses. Most recent assessments have moved away from this toward delivering drop-in replacements. The reader is referred to Table 1.1 in this chapter
Be sustainable	
Think locally: Develop and apply engineering solutions while being cognizant of local geography, aspirations, and cultures [8]	One of the potential benefits from the more widespread use of bio-based and renewable chemicals is that production can be more distributed; i.e., production is possible in many places and not constrained by transportation of petroleum
Engage: Actively engage communities and stakeholders in development of engineering solutions [8]	
Innovate to achieve: Create engineering solutions beyond current or dominant technologies; improve, innovate and invent (technologies) to achieve sustainability [8]	The bio-based and renewable chemicals industry is in its infancy and will require a significant number of innovations in chemistry, molecular biology, engineering, and many allied disciplines to move the industry to be more sustainable. At this point in time, there are aspects of the industry that are more sustainable than the petrochemical industry and aspects that are not
Use renewables: A raw material feedstock should be renewable rather than depleting whenever technically and economically practical [3] Material and energy inputs should be renewable rather than depleting [7]	The only addition here is that raw materials should be both renewable and sustainable. Once again, there is great potential for the industry to live up to these concepts/principles, but at present, it is not

It could also be argued that in the case of bio-based and renewable chemicals production, there is a great opportunity for businesses to avoid many of the problems of the existing global chemistry enterprise by choosing to develop chemical feedstocks that are not extremely hazardous to humans or the environment, and to consider what can be done to minimize or eliminate adverse impacts across the life cycle of the chemicals that are produced. It is naïve to think, although it is common to hear this, that just because a chemical or product is derived from agriculture it is automatically green, or at least greener, than a chemical or product derived from petroleum or some other form of fossil carbon. The assessment of whether or not the chemical or product is green, greener, sustainable, or more sustainable is beyond the scope of this chapter, but it is important for the assessment to be carried out and there are many examples of how this may be accomplished [9–12]. In general, what one finds in these comparisons of a bio-based renewable or related chemical, or product with the same chemical or product made from a set of petrochemical building blocks is that significant impacts are associated with each approach. The reality is that in a comprehensive multivariate assessment, one set of impacts or another predominate, and the question of whether one or more impacts are better or worse than others ultimately comes down to the values and assumptions of the person or organization doing the assessment.

Focusing more narrowly for a moment on the situation of green chemistry and chemistry, it is generally true that chemists over the past 100 years have been trained in an academic setting to think only about what happens to chemicals in a beaker or flask, with little to no knowledge of, or regard to, where things come from or that waste goes anywhere other than down the sink (and therefore, no longer a problem), up the hood exhaust stack, or in the waste can in the lab. Energy is conveniently delivered in the form of an oil bath, a heating coil, running water, dry ice, and liquid nitrogen; and why should a chemist care where these conveniences originate or at what cost? Green chemistry was an attempt to get chemists to start thinking about these sorts of things, and while there has been progress over the past 20 years, there remains a considerable amount of work to be done to integrate green chemistry concepts and new ways of thinking about chemistry into the practice of chemistry.

For the purpose of this book, it is worth noting that in the original 12 principles of green chemistry there is a strong association with the idea of Chemurgy in principle 7 which is: "A raw material feedstock should be renewable rather than depleting whenever technically and economically practical." Those who read this would assume the principle only refers to a chemical that is bio-based and renewable, although from a sustainability perspective, one should also consider the renewability of elements that are not, nor could they be sourced from agriculture or some other living system. A variety of elements on the periodic table are critical to chemical manufacturing regardless of feedstock source and to enabling modern society. The problem is that the production and use of most of these critical elements is not currently sustainable. One other distinction to make here is that feedstocks may be renewable and bio-based but not sustainably produced. For example, one might argue that feedstocks like a vegetable, plant, or algae-based oil, while bio-based and renewable, is not sustainably produced. It is beyond the scope of this chapter to get into the details of this and is offered here only in recognition that it is a very important issue to consider at some point. It is, however, worth a little time to think about raw material feedstocks given the subject of this book.

1.2 Feedstocks

The broader global chemistry enterprise (industry, academia, and government as funders and regulators) has been in the making over the past 250 years or so, and now delivers an overwhelming number of chemicals with some of these being delivered in very high volumes. There is now a large, complex, heavily capitalized petrochemical enterprise, and it is arguably a triumph of post-war creativity, investment, and technological innovation. The petrochemical industry as it currently exists is very good at producing a relatively few key building blocks from petroleum in very high volumes; e.g., alkanes, alkenes, olefins, aromatics like benzene, aniline, terephthalic acid, etc., on the order of millions of pounds per year. Petroleum-derived molecules are, in turn, used to build a much larger number of compounds that are converted into the products used in all parts of our lives and deemed by world markets to be essential to modern society. For the most part, these molecules are straight carbon chains or aromatic compounds having varying degrees of saturation and mostly composed of carbon and hydrogen with few hetero atoms like nitrogen, sulfur, or oxygen. The carbon atoms in petroleum are in a highly reduced state, and getting them to react usually requires extremes in temperature and pressure in the presence of a large number of catalysts.

In the case of the high volume chemicals, these are produced with considerable mass and energy efficiency with constant, incremental improvements in both mass, and energy efficiency still being implemented. It has been recently reported that there are about 120 molecules that account for a majority of the chemical frameworks or building blocks used by chemists to make the more complex molecules that are associated with industrial sectors like the specialty chemicals, agrichemical, and pharmaceutical industries [13, 14]. This is arguably a surprisingly small number of chemical framework molecules chemists work with, especially when one considers the large number of chemicals that have been discovered in living systems, or what chemical diversity might be available as one considers all the possible chemical isomers as the number of C, H, N, O, S, and N atoms in a given molecule is increased.

Given the availability of a set of simple, relatively inexpensive building blocks, chemists have over many years developed a vast array of reactions to effect a variety of chemical transformations. Synthetic organic chemists, in particular, generally have an encyclopedic recall of named reactions, methodologies, or protocols for activating, functionalizing, and otherwise chemically transforming molecules or basic organic chemical building blocks into more complex molecules. These building block molecules are functionalized (adding hetero atoms like N, S, O, etc.) or coupled with other building blocks in a specific order to produce molecules that are in almost everything that humans need to produce not only the basics like food, shelter, and clothing, but also the other things that make our lives so interesting; e.g., electronics, plastics, advanced composites used in transportation, etc.

1.3 Moving from Petroleum to Sugars, Lignocellulosics, and Proteins: A Different Set of Chemical Building Blocks

It is interesting to note that these simple framework molecules derived from petroleum are not the kinds of simple or complex building blocks that are made by plants and other forms of life. As the global chemistry enterprise begins its slow transition away from the types of petrochemical building blocks used today, a set of building blocks like the ones you will read about in the remainder of this book, the chemistries, methodologies, and protocols chemists use to effect chemical transformations will also have to transition. The difficulty in doing this is that chemists are not, in general, trained in a great degree to effect chemical transformations on chemicals derived from plants or other organisms. They also do not have the experience with the kinds of reactions and synthetic strategies required to construct new molecules based on these types of building blocks, and as is the case with most synthetic chemistry; the reagents used to effect chemical transformations are quite hazardous from an environmental, safety, and health perspective. Moreover, the chemical industry has a dedicated infrastructure for transforming and processing a different set of molecules in different ways; e.g., a fermentation facility is very different compared to a petrochemical facility.

This situation has therefore led start-up biotech companies to adopt commercialization strategies for providing "drop-in" replacements from bio-based or other feedstocks like methane or CO_2. For example, a company identifies a micro organism platform (i.e., a yeast, bacteria, algae, etc.) that may be genetically modified or evolved to produce chemicals from sugar that are at a minimum cost competitive to that same chemical if it was made from petrochemical feedstocks. While this approach is enabling the implementation of a variety of technologies at commercial scale, it is unfortunate that these drop-in replacements have a range of known and well-understood undesirable properties and hazards. The tragedy of the current growing availability of bio-based and renewable feedstocks is that chemists are not rethinking the chemical building blocks they normally use, nor the solvents and the products that are being made. This is an opportunity to design out undesirable effects of chemicals, the chemistries used to undertake chemical transformations, and the processes that are used at scale, instead of just replicating the undesirable chemicals and all their associated baggage.

Commercially, transitioning to biologically based and renewable carbon framework molecules is not going to be easy, and there are a variety of strategies that have been proposed over the past 10–15 years. There has been a tremendous amount of research and development directed toward using many different types of biomass and an even a greater number of chemical, thermal, and biological processes. While many of these are under advanced development or in the early stages of commercialization, it is difficult to develop markets and displace existing supply chains. Table 1.1 contains the U.S. Department of Energy National Renewable Laboratory's top value-added chemicals derived from biomass as assessed in 2004, 2010, and 2016. It is quite interesting to see the changes to this list over time.

The vision for the biorefinery of the future is very similar to what is found in a modern petrochemical refinery which contains a highly integrated infrastructure that is highly mass and energy efficient [18–21]. Sugars, starches, proteins, and other bio-based building block molecules are quite different than those obtained from petroleum as they contain significant amounts of oxygen and in some cases, N, S, and P atoms, but none of them have any aromaticity.

Table 1.1 Bio-based chemical targets put forth by U.S. National Renewable Energy Lab in 2004, 2010 and 2016 [15–17]

2004	2010	2016
1,4-Dicarboxylic acids (succinic, fumaric and malic)	Succinic acid	Succinic acid
2,5-Furan dicarboxylic acid	Furanics	Furfural
3-Hydroxypropionic acid	Hydroxypropionic acid/aldehyde	–
Glycerol	Glycerol and derivatives	Glycerine
Sorbitol	Sorbitol	–
Xylitol/Arabinitol	Xylitol	–
Levulinic acid	Levulinic acid	–
Aspartic acid	–	–
Glucaric acid	–	–
Glutamic acid	–	–
Itaconic acid	–	–
3-Hydroxybutyrolactone	–	–
–	Biohydrocarbons	–
–	Lactic acid	Lactic acid
–	Ethanol	–
–	–	Butadiene
–	–	Butanediol
–	–	Ethyl lactate
–	–	Fatty alcohols
–	–	Isoprene
–	–	Propylene glycol
–	–	*p*-Xylene

As you will see in this book and others like it, there are a number of potential biomass sources to consider and a variety of conversion strategies, but a few sources and associated conversion strategies are more common than others. The first type of biomass to be considered is represented by feedstocks from the forest products industry or other parts of agriculture like sugar, starch, cellulose, hemicellulose, lignin, pine chemicals, etc. The second type of biomass to consider is the development of feedstocks from a marine-based polymer that is chemically similar to lignin, and that is chitin. Chitin is found in many marine organisms like shrimp, lobster, etc. The third is the development of chemical feedstocks or chemicals from waste biomass like food wastes; e.g., citrus peels and pulp, potato skins, etc.

Within the forest products industry there are two general strategies for producing useful chemical products. The first is comparatively recent and that is looking to produce a variety of chemicals and materials from lignocellulosics and other by-products of the paper pulping process other than paper, or burning lignin for its fuel value to make steam for use in the paper-making process. Within the ligno-cellulosic arena, there have been many advances in separating cellulose, hemicellulose, and lignin, with lignin having great potential because of the many aromatic compounds that compose its polymeric backbone.

While there are many research groups throughout the world working on processes that will break lignin down into different chemical components, there is currently no viable commercial process for turning lignin into useful chemical products. Because the chemical composition of lignin is highly aromatic there is great potential to use these lignin sub-structures as chemical building blocks for a variety of end uses. Finding a nonfuel use for lignin is perhaps a similar situation as finding a viable high value-added commercial use for glycerol produced as a part of biodiesel production. Nonetheless, considerable efforts are made to valorize lignin in ways that go beyond an old pulp and paper industry saying: "You can make anything out of lignin except money." See Chaps. 9 and 10 for possibilities. The problem with lignin conversion is a common one for chemist's as they work to discover mechanisms to break and make new bonds precisely where it makes the most sense while doing it in a mass and energy efficient manner.

Another part of the forest products industry that few people are aware of is the pine chemicals industry, but this industry represents a good model for obtaining increasingly higher value-added products from bio-based and renewable feedstocks. While the pine chemicals industry is not new, it is a small part of the specialty chemicals industry. It is interesting to note that useful non-timber products such as turpentine and other materials like rosins and pitch have been produced from pine trees for far longer than the modern chemical industrial era that rapidly expanded throughout the twentieth century. As the pulp and paper industry has grown so has the production of crude tall oil and crude sulfate turpentine. Crude tall oil may be separated into a fatty acid fraction, a tall oil fraction, a tall oil rosin fraction, and a tall oil pitch fraction. The crude sulfate fraction is separated into a variety of terpene monomers that may be converted into a variety of terpene resins. All of these streams can be used as raw materials for coatings, various specialty oil applications, surfactants, adhesives, inks, etc. [22].

While the pine chemicals industry from a chemical engineering perspective is not currently as large, integrated, and technically sophisticated an industry as the petrochemical industry, it does use many similar unit operations such as the use of distillation as the primary means of separating complex mixtures. It does, however, represent a model for the integrated biorefinery of the future where various distillate fractions are further separated and converted into useful framework molecules and other more complex chemicals.

The second area within the forest products sector to mention is those businesses seeking opportunities for higher value-added chemicals derived from the leaves, roots, seeds, etc. of trees, bushes, etc. Substances like turpentine, linseed oil, rosins, pitches, and similar materials have been productively used for hundreds of years, while newer uses for various fatty acid fractions or other materials from Tall oil are being sought. In other cases, structural building blocks for a variety of drugs, an early example of which was taxol, although a genetically modified organism was eventually created to get around the variety of issues of sourcing the molecule from such a large volume of plant material. A second example of this is artemisinin, an important new anti-malarial treatment, where plant extracts were used until engineered yeasts were developed to produce the compound at much lower cost and less environmental impact.

1.4 Sugar: The New Oil?

One company achieving success in converting woody biomass to sugars is Renmatix [23]. Renmatix has developed and is actively licensing a near-critical water process that separates the woody biomass into streams of cellulose, hemicellulose, and lignin, and then further decomposes the cellulose and hemicellulose fractions into their component sugars. Lignin continues to be burned in the forest products industry for its energy value (usually to produce steam for heating), but this is clearly not the best use of this very complex and chemically interesting naturally produced polymer. Once constituent sugar streams are separated, the sugars may be used in different fermentation processes to produce fuels (ethanol or biodiesel) or chemicals, depending on which microbial platform is used. Sugars in their various forms (C5 and C6) are sometimes referred to as the next oil, i.e., a readily available carbon source that can be converted to a variety of chemicals and chemical building blocks through an expanding number of industrial fermentation processes.

While there is great promise in using biotechnology to produce chemicals, there are two recent commercial examples where companies proposing to use sugar as a source for chemicals have struggled to do so because of the low cost of petroleum. Both BioAmber [24] and Myriant [25] identified a market opportunity for succinic acid, a 4-carbon chain molecule that is more cost effectively obtained via fermentation than from petrochemical sources. Succinic acid is a useful building block for a variety of commodity (e.g., 1,4-butanediol) and specialty chemical end use applications, so market demand for succinic acid should be quite high. Sadly, these companies are examples of the difficulty of building a market in the face of severe competition from petrochemicals; Myriant has ceased production and BioAmber is struggling. It is hoped that this is a relatively short phenomena, but it does point to the difficulty of being an early entry into the market, and the uphill battle of competing against an entrenched, well established supply chain.

1.5 Oils and Other Specialty Chemicals

Two additional companies working to develop markets for bio-based and renewable chemical building blocks from sugar are Solazyme and Amyris; two of the winners of the 2014 U.S. Presidential Green Chemistry Challenge Awards [26, 27].

Solazyme has recently split into two different companies, Solazyme Industrials [28] and Terra Via [29] and both are exploiting an algae platform to make either tailored oils for a variety of industrial end use applications or consumer food products. Because the basic genetic coding for making oils exists in algae and is conserved in higher plants that make vegetable oils, it was possible to genetically modify the algae to overexpress different oils containing specific fatty acid ratios. This technology allows Solazyme and TerraVia to exactly replicate vegetable oils like olive oil, sunflower oil, soybean oil, or palm oil, or if they need to, they may develop "tailored" oils having the exact ratio of fatty acids and triglycerides that a chemist may convert to a higher value-added chemical. There are two interesting aspects to this. First, there is a unique ability to create framework molecules that are not currently available and which can be produced without the environmental footprint of a petrochemical feedstock. Second, there are no geographic restrictions to making these oils since one is not dependent on a climate that supports a palm or olive tree, for example.

Amyris is perhaps best known for its work in developing a yeast-based platform to produce artemisinin, a potent anti-malarial drug. However, Amyris is also a great example of a biotech company that has diversified into several different commercial directions [30]. In Amyris' case, a yeast instead of an algae was modified to produce a biodiesel drop-in replacement, although the molecule does not look like a petrochemically or "typical" biodiesel molecule. The molecule in this case, known as Farnesene, is a true drop-in replacement and it has the added advantage of not having emissions like SO_X or excessive NO_X, and it burns more cleanly than traditional biodiesel. Perhaps it is not surprising that Amyris is currently seeing greater demand for a molecule like squalene which is used in cosmetics and personal care products. Competing against petroleum-derived diesel given current prices for a barrel of oil is challenging, although there are a few airlines like United that are interested in using greater amounts of bio-based diesel. However, the low cost of oil is likely to push most companies in the biotech sector into higher value-added compounds like squalene, or into compounds that are not easily synthesized from petroleum. Competing on cost and performance is a must and presents a challenge for anyone trying to enter an existing market. While it is great to have a sustainability benefit it in no way guarantees commercial success in the short-term. This is an important point since the hurdle for sustainability, from a commercial perspective, is higher than the hurdle for a company whose feedstocks come from petrochemicals.

1.6 Plastics: PHA's

Another example of using sugar to make a large volume commodity chemical is the production of polyhydroxyalkanoates (PHA), the most developed example being polylactic acid (PLA). While DuPont discovered this molecule in the 1950s, it was not commercialized except for limited medical applications because it was difficult to produce from petrochemical sources. In the early 2000s, Dow and Cargill entered into a joint venture to produce PLA and created a company known as NatureWorks [31]. While Dow has had a long history of making plastics like polyethylene, polystyrene, polyurethanes, etc., Cargill was a large diversified agricultural company that was looking to expand its corn milling operations beyond food and animal feed into commodity chemicals. Dow eventually withdrew from the joint venture because of the slow progress to realize commercial success, but Cargill proved to be a more patient investor.

By enzymatically converting corn starch to sugar, NatureWorks obtained lactic acid through a fermentation process, followed by polymerization of the lactic acid to PLA. PHAs in general, and PLA in particular, are seen as desirable polymers since they can be obtained directly from a variety of microorganisms or from agriculturally derived sugar sources. PLA is also biodegradable, although the caveat to PLA biodegradability is that it only degrades when the temperature of composting can be maintained above 50 °C for an extended time, i.e., in industrial composting facilities. If industrial composting facilities are not used, PLA will biodegrade at about the same rate as many other plastics. A key to the commercial success of PLA has been in the development of end uses where waste streams can be tightly managed and PLA-containing wastes can be diverted to industrial composting facilities. NatureWorks has carefully developed their PLA market into food service and eating utensils at professional sports arenas where team owners like to be seen as "going green" and the waste streams are largely food product wastes.

Although lactic acid may be reliably obtained via fermentation from sugar and PLA is biodegradable, it is not without a few problems. Polymers like polyvinyl chloride, polyethylene, polycarbonate, polystyrene, or other polyesters have been on the market and under continuous development for decades and have a variety of commercial applications that are valuable to consumers. PLA, from a performance and customer acceptability perspective is not seen as being as good as these other already well established plastics. Despite the environmental benefits, it has still taken over 10 years and the consistent financial backing of Cargill for NatureWorks to make its production profitable and to develop a strong market for PLA.

1.7 Closing the Loop: Chemicals from Waste

In recent years, there has been increased attention given to the need for society to become increasingly less mass and energy intensive; i.e., to make more things from less stuff. This is quite a challenge for chemists since the idea of working with waste to get something of value is not terribly appealing to most chemists. However, the idea of using wastes from the food processing industry to produce a range of value-added chemicals has been discussed as part of a broader discussion of "closing the loop" on materials in our economy. One example of a company that has found a way to make a range of specialty chemicals from food processing industry wastes is Florida Chemical [32]. In addition to obvious products like flavors and fragrances that are extracted from citrus peels, the citrus industry has been looking to expand into chemicals production by working to develop products like extractable solvents, oils, and other chemicals. There are large volumes of citrus wastes from juicing operations throughout the world (over 8 million tons per year in Brazil, alone [33]), and these wastes represent a ready supply of chemicals. About 14 % of the peel is a mixture of oils, terpenes, a variety of oils, waxes, and other alkanes. While these materials may be isolated from citrus peels, the largest marketed chemical is D-limonene. Limonene can be used as a building block for plastics like polyethylene terephthalate (PET), or it may be converted into chemicals like cresol or *p*-α-dimethyl styrene [34]. Citrus peels and other wastes are unlikely to provide a sufficient amount of chemicals to displace all the supplies of these chemicals currently required to support the market, but companies are demonstrating that it is possible to obtain a variety of interesting and useful chemicals from nonpetroleum, bio-based and renewable sources.

1.8 Final Words

Clearly this chapter has merely scratched the surface of what is currently happening in the bio-based and renewable chemicals sector and is discussing briefly only the opportunities for applying sustainable and greener chemistry. In many ways, the resources put into the pursuit of bio-based and renewable fuels overwhelms the resources being put into bio-based and renewable chemicals production and this will inevitably slow the commercialization of bio-based and renewable chemicals. There also seems to be a shift over the past 10 years or so from a focus on producing bio-based and renewable chemicals that are unique and may perhaps be used to make new and interesting materials and products to a focus on replicating commodity chemicals currently produced from petroleum. This is a truly unfortunate development, especially if these chemicals are inherently hazardous. Ideally, we would want to bring to market much less hazardous chemicals.

It is also true that the price of petroleum will continue to dominate the development of the sector until the cost per barrel is more than U.S. $100.00 in 2016 dollars. Some would argue that this represents an inherent weakness in the case for bio-based and renewable chemicals, but I think it is more accurate to say that the commercialization of bio-based and renewable chemicals is no different than the commercialization of any new technology. In this sense, the global chemical manufacturing enterprise has largely moved away from the type of commercial innovation development that characterized the industry of the post-war era up until the 1980s and is more a reflection of Wall Street dictating the supremacy of shareholder value and short-term returns over the need for longer term investments in chemical innovation.

Regardless of these Wall Street dictates, market forces and the technology being developed that characterizes the industrial biotech sector is disruptive and it is only a matter of time before it achieves widespread, large-scale, commercial success. We are poised to see the realization of what was once seen to be the way of the future in the early part of the twentieth century. It is unfortunate that we had to wait so long to see it happen.

References

1. Matlin SA, Mehta G, Hopf H, Krief A (2016) One-world chemistry and systems thinking. Nat Chem 8:393–398
2. Whitesides GM (2015) Reinventing chemistry. Angew Chem Int Ed 54:3196–3209
3. Anastas P, Warner JC (1998) Green chemistry: theory and practice. Oxford University Press, Oxford
4. Jimenez-Gonzalez C, Constable DJC (2011) Green chemistry and engineering, a practical design approach. Wiley, London
5. https://www.epa.gov/greenchemistry
6. Winterton N (2001) Twelve more green chemistry principles? Green Chem 3:G73–G75
7. Anastas PT, Zimmerman JB (2003) Design through the twelve principles of green engineering. Environ Sci Technol 37(5):94A–101A
8. Abraham MA, Nguyen N (2003) "Green engineering: defining the principles"—results from the San Destin Conference. Environ Prog 22(4):233–236
9. Sheldon RA, Sanders JPM, Marinas A (2015) Sustainability metrics of chemicals from renewable biomass. Catal Today 239:1–2
10. Sheldon RA, Sanders JPM (2015) Toward concise metrics for the production of chemicals from renewable biomass. Catal Today 239:3–6
11. Cséfalvay E, Akien GR, Qi L, Horváth IT (2015) Definition and application of ethanol equivalent: sustainability performance metrics for biomass conversion to carbon-based fuels and chemicals. Catal Today 239:50–55
12. Patela AD, Telalovićb S, Bitterb JH, Worrella E, Patela MK (2015) Analysis of sustainability metrics and application to the catalytic production of higher alcohols from ethanol. Catal Today 239:56–79
13. Bishop KJM, Klajn R, Grzybowski BA (2006) The core and most useful molecules in organic chemistry. Angew Chem Int Ed 45:5348–5354
14. Grzybowski BA, Bishop KJM, Kowalczykm B, Wilmer CE (2009) The 'wired' universe of organic chemistry. Nat Chem 1:31–36
15. Werpy T, Petersen G (eds) (2004) Top value added chemicals from biomass, Volume I: Results of screening for potential candidates from sugars and synthesis gas. NREL Report No. NREL/TP-510-35523
16. Bozell JJ, Petersen GR (2010) Technology development for the production of biobased products from biorefinery carbohydrates: the US Department of Energy's "Top 10" revisited. Green Chem 12:539–554
17. Biddy MJ, Scarlata C, Kinchin C (2016) Chemicals from biomass: a market assessment of bioproducts with near-term potential. NREL Report No.: NREL/TP-5100-65509
18. Integrated Biorefinery Research Facility: Partnering with Industry to advance biofuels and bioproducts. 2011, NREL Report No. FS-5100-53429, http://www.nrel.gov/biomass/publications.html. Accessed 31 July 2015
19. Davis R, Tao L, Tan ECD, Biddy MJ, Beckham GT, Scarlata C, Jacobson J, Cafferty K, Ross J, Lukas J, Knorr D, Schoen (2013) Process design and economics for the conversion of lignocellulosic biomass to hydrocarbons: dilute-acid and enzymatic deconstruction of biomass to sugars and biological conversion of sugars to hydrocarbons. NREL Report No. NREL-TP-5100-60223

20. Biorefinery Developments for Europe. http://www.biosynergy.eu/fileadmin/biosynergy/user/docs/Results_of_the_Integrated_Project_BIOSYNERGY_2007-2010.pdf. Accessed 31 July 2015
21. Stuart PR, El-Halwagi MM (2012) Integrated biorefineries: design, analysis, and optimization. CRC Press, Boca Raton
22. http://www.arizonachemical.com/Global/Graphs/Company_Materials_Tree_of_Life.pdf. Accessed 10 Sept 2016
23. http://renmatix.com/. Accessed 10 Sept 2016
24. https://www.bio-amber.com/bioamber/en. Accessed 10 Sept 2016
25. http://www.myriant.com/. Accessed 10 Sept 2016
26. http://solazymeindustrials.com/. Accessed 10 Sept 2016
27. http://terravia.com/. Accessed 10 Sept 2016
28. http://www2.epa.gov/green-chemistry/2014-greener-synthetic-pathways-award. Accessed 11 Sept 2016
29. http://www2.epa.gov/green-chemistry/2014-small-business-award. Accessed 11 Sept 2016
30. https://amyris.com/. Accessed 11 Sept 2016
31. http://www.natureworksllc.com/. Accessed 11 Sept 2016
32. http://www.floridachemical.com/. Accessed 11 Sept 2016
33. Ferreira-Leitão V, Gottschalk LMF, Ferrara MA, Nepomuceno AL, Molinari HBC, Bon EPS (2010) Biomass residues in brazil: availability and potential uses. Waste Biomass Valoriz 1 (1):65–76
34. Clark JH, Pfaltzgraff LA, Budarin VL, Hunt AJ, Gronnow M, Matharu AS, Macquarrie DJ, Sherwood JR (2013) From waste to wealth using green chemistry. Pure Appl Chem 85:1625–1631

Chapter 2
Sustainable Production of Omega-3 Eicosapentaenoic Acid by Fermentation of Metabolically Engineered *Yarrowia lipolytica*

Dongming Xie, Edward Miller, Bjorn Tyreus, Ethel N. Jackson and Quinn Zhu

Abstract The omega-3 fatty acid, *cis*-5,8,11,14,17-eicosapentaenoic acid (C20:5; EPA) has wide-ranging benefits in improving heart health, immune function, and mental health. Currently, the major source for EPA is from fish oil, which is subject to challenges in its availability and sustainability due to the concerns of overfishing and contamination in the ocean environment. DuPont has developed a sustainable source of the omega-3 EPA through fermentation using metabolically engineered strains of *Yarrowia lipolytica*. EPA biosynthesis and supporting pathways have been engineered to accumulate high EPA content in *Yarrowia* biomass under fermentation conditions. Many production strain candidates were generated in the molecular biology group, and fermentation research was conducted to (1) identify the best production strains by high throughput fermentation screening; (2) optimize the fermentation medium and process conditions for the selected production strains to achieve the highest EPA titer, rate, and yield; and (3) scale-up the developed fermentation process for commercialization. This chapter summarizes the major fermentation engineering work that has been accomplished at DuPont to achieve large-scale production of *Yarrowia* biomass with high EPA content. This work led to two commercial products, New Harvest™ EPA oil and Verlasso® sustainably farmed salmon.

Keywords *Yarrowia lipolytica* · Omega-3 fatty acid · Eicosapentaenoic acid (EPA) · Fermentation engineering

D. Xie (✉) · E. Miller · B. Tyreus · E.N. Jackson · Q. Zhu
Industrial Biosciences, E.I. Du Pont de Nemours and Company, Wilmington, DE, USA
e-mail: Dongming_Xie@uml.edu

Q. Zhu
e-mail: Quinn.Zhu@dupont.com

D. Xie
Department of Chemical Engineering, University of Massachusetts-Lowell, One University Avenue, 01854 Lowell, MA, USA

P.C.K. Lau (ed.), *Quality Living Through Chemurgy and Green Chemistry*, Green Chemistry and Sustainable Technology,
DOI 10.1007/978-3-662-53704-6_2

2.1 Introduction

Fatty acids have two ends, the carboxylic acid (−COOH) end, which is considered the beginning (alpha) of the chain, and the methyl ($-CH_3$) end, which is considered the tail (omega) of the chain. Omega-3 fatty acids refer to the long-chain polyunsaturated fatty acids (LCPUFA) that contain the first C=C double bond at the third carbon from the methyl (omega) end of the carbon chain. Many clinical studies have shown a wide range of health benefits from two major omega-3 fatty acids, the eicosapentaenoic acid (C20:5; EPA) and docosahexaenoic acid (C22:6; DHA) [1, 2]. While most omega-3 products on the current market address only the "omega-3" concepts in the labels, there are significant differences in health benefits among different omega-3 fatty acids, even between EPA and DHA. Based on the clinical studies, it is believed that EPA is able to improve cardiovascular health, mental health, and immune function, while DHA tends to be better in improving mental health and cognitive development in infants [2]. The Japanese EPA Lipid Intervention Study (JELIS) suggests that EPA is a promising treatment for prevention of major coronary events for people at risk of heart health issues [3]. In addition, the AMR101 study by Amarin Pharma Inc. showed that pure EPA ethyl ester significantly reduced triglyceride levels in adult patients with severe hypertriglyceridemia [4].

EPA and DHA are essential fatty acids in human health. However, the human body cannot efficiently synthesize EPA and DHA de novo and as such, these fatty acids are typically obtained from protein sources within the diet [2], especially the cold water oceanic fishes [1]. Certain sea fishes (e.g., wild salmon, Pacific sardine) accumulate significant amounts of EPA and DHA by consuming microalgae cells in the ocean, which are capable of synthesizing EPA and DHA de novo.

Fish oil is the main source of EPA and DHA on the market. Farm raised salmon accumulate DHA and EPA in their flesh and muscle when the oil from wild caught sea fishes is included in the salmon feed (Fig. 2.1). Usually, 4 kg of sea fish are required to raise 1 kg of salmon. The future availability and sustainability of current fish oil sources of omega-3 essential oils have been questioned due to potential overfishing and contamination issues in the ocean environment. To overcome these challenges, biotechnology companies such as Martek (now a part of DSM) started to produce DHA-enriched oil directly from microalgae in large-scale fermentation processes [5]. There is still no large-scale land-based EPA production from wild-type organisms due to the challenges in achieving high EPA productivity and low cost targets essential for commercial viability.

To address this need, DuPont initiated a research program to develop a sustainable EPA source by metabolic engineering of *Yarrowia lipolytica*, an industrial fungal work horse (Figs. 2.1, 2.2). The first targeted product for commercialization was EPA due to its unique health benefits and the lack of a land-based sustainable supply. *Yarrowia* cells have been metabolically engineered to accumulate EPA up

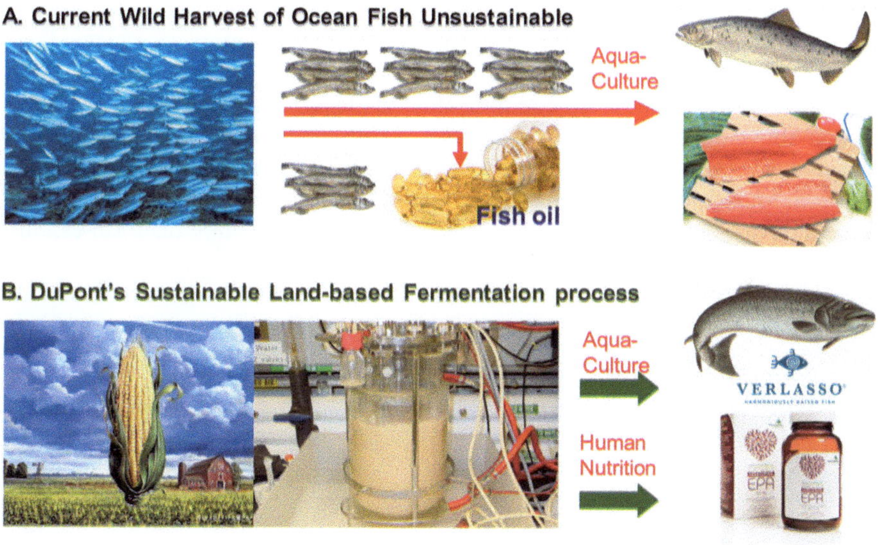

Fig. 2.1 Current major source of omega-3 from wild harvested ocean fishes and DuPont's sustainable land-based source of omega-3 from fermentation of *Yarrowia lipolytica*

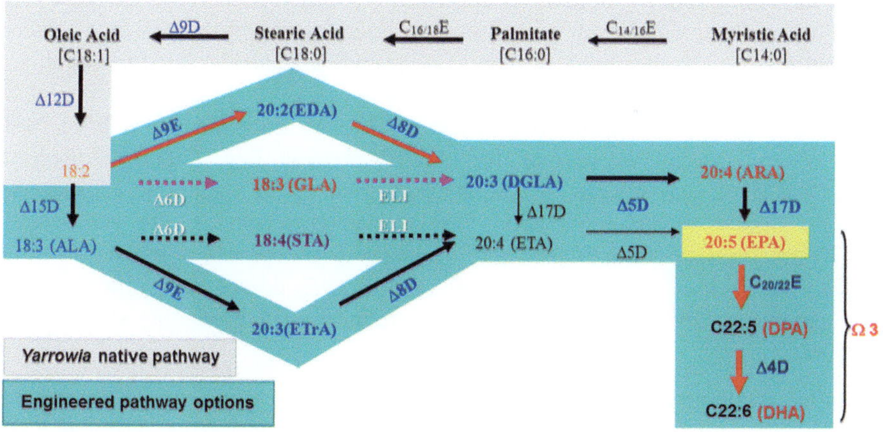

Fig. 2.2 EPA and DHA production by *Yarrowia lipolytica* through both Δ6 (the Δ6-desaturase) and Δ9 (Δ9-elongase and Δ8-desaturase) pathway engineering [7]

to 25 % of total biomass [6–8]. The purified lipids from these EPA producing strains have been used to develop New Harvest™ EPA oil, a commercial human nutritional supplement. The high EPA biomass of *Y. lipolytica* yeast has also been used to raise sustainably farmed salmon, Verlasso®. Sustainability being driven by the fact that only one kilogram sea fish is required to raise one kilogram of salmon due to the use of the yeast biomass in the salmon feed (Fig. 2.1). In addition, the *Y.*

lipolytica metabolic engineering technology has been sufficiently developed such that engineered strains were constructed to produce a variety of omega-3 and omega-6 fatty acids [7, 9].

This chapter summarizes the fermentation engineering research efforts of the *Y. lipolytica* metabolic engineering program to achieve successful commercialization of omega-3 EPA production using this technology. The work has paved the way for further improvement in omega-3 and omega-6 production for more specific applications in future. The advanced *Y. lipolytica* biotechnology platform can also be used for production of other high-value products.

2.2 Metabolic Engineering in *Yarrowia lipolytica* for EPA Production

The nonconventional yeast *Y. lipolytica* was selected as the host for metabolic engineering for a variety of reasons. *Y. lipolytica* is found primarily in foods such as dairy products and meat. *Y. lipolytica* was first used to produce single cell protein using cheap and abundant *n*-paraffins as the sole carbon source for animal feed [10]. It has also been classified as "Generally Recognized as Safe (GRAS)" for commercial production of food grade citric acid (U.S. Food and Drug Administration list of microbial-derived ingredients approved for use in food; Title 21, Part 173, Sec. 165). Other applications include production of erythritol, wax esters, 2-ketoglutaric, 2-hydroxyglutaric, and isopropylmalic acids and secretion of heterologous proteins, including several food enzymes [10, 11]. Forty different *Y. lipolytica* strains from various public depositories all over the world have been collected and screened as part of this effort. American Type Culture Collection (ATCC) #20362 was selected as the starting point of the metabolic engineering program largely based on its good fermentation performance: dry cell weight (DCW) greater than 100 g/L and lipid productivity greater than 1 g/L/h.

Wild type *Y. lipolytica* strains such as ATCC 20362, do not make omega-3 fatty acids naturally. The following strategies to engineer EPA producing strains were employed:

1. Build an efficient EPA biosynthetic pathway by (a) using strong promoters; (b) codon optimization of heterologous genes; (c) increasing/optimizing the copy numbers of structural genes; (d) focusing on limiting steps in the EPA pathway; and (e) pushing and pulling carbon flux.
2. Screen various strain libraries for high lipid and EPA productivities.
3. Generate mutants with increased lipid content by modification of the peroxisome and by disrupting certain genes in the ß-oxidation pathway.
4. Control fatty acid transport.
5. Manipulate global regulators.

The fatty acid profile of the wild-type strain ATCC #20362 shows that it can synthesize the omega-6 fatty acid, linoleic acid (LA, C18:2 n-6), but not any of the omega-3 fatty acids [9, 12, 13]. There are different biosynthetic routes to make EPA, the anaerobic polyketide synthase pathway [14] or an aerobic desaturase and elongase pathway [15]. However the rate, titer, and yield from these organisms could not meet the requirement for commercial production [16]. Recently, the aerobic pathways including both Δ6 (the Δ6-desaturase pathway, found in algae, mosses, fungi, and others) and Δ9 pathways (Δ9-elongase and Δ8-desaturase pathway) [17, 18] have been well studied (Fig. 2.2). In the Δ6 pathway, the LA and/or alpha-linolenic acid (ALA, 18:3n-6) is converted to gamma-linolenic acid (GLA, 18:3 n-6) and/or steridonic acid (STA, 18:4 n-3) by the Δ6-desaturase. Subsequently, the GLA and/or STA is converted to dihomo-gamma-linoleic acid (DGLA, 20:3n-6) and/or eicosatetraenoic acid (ETA, 20:4 n-3) by a $C_{18/20}$-elongase. In the Δ9 pathway, the LA and/or ALA is converted to eicosadienoic acid (EDA, 20:2n-6) and/or eicosatrienoic acid (ETrA, 20:3n-3) by a Δ9 elongase. The EDA and/or EtrA is then converted to DGLA and/or ETA by a Δ8-desaturase. The last two steps leading to EPA are the same for both pathways (Fig. 2.2) employing Δ15- and Δ17-desaturases. The desaturation and elongation enzymes carry out their reactions in the endoplasmic reticulum membrane [12, 15, 18].

Promoters for *Y. lipolytica* genes are no exception among various microbial hosts in controlling the expression of foreign. A set of promoters from genes expressing enzymes involved in glucose central metabolism that are at least as strong as the TEF (translation elongation factor) promoter have been isolated [19]. In screening studies, each individual promoter was placed in front of the ß-glucuronidase (GUS) reporter and the signal strength recorded in quantitative fluorometric assays [20]. Results showed that the *FBA in* promoter was the strongest among the first set of six promoters. The *GPM1* promoter was as strong as the *TEF* promoter, a benchmark comparison; the *GPD1* promoter was 2.5 times stronger than the *GPM1* promoter; and the *FBAin* promoter activity was 5.5 and 2.2 times stronger than the *GPD1* and *GPM1* promoters, respectively [21]. In addition, the YAT1 promoter activity increased by 35-fold when the fermentation was switched from nitrogen-rich to nitrogen-limiting conditions [22], which provides a unique advantage for application in the two-stage omega-3 fermentation process (see Fig. 2.6). The relative strength of the promoters examined under nitrogen-limiting conditions is as follows: FBA_{in} > YAT1 > FBA > GPD, EXP > GPAT > GPM = TEF. In addition, other new *Y. lipolytica* promoters could be generated by random mutagenesis using a specific promoter as template [23].

Codon optimization is also critical for further improving the expression of foreign genes in *Y. lipolytica*. Hence, all genes used in the construction of various strains which include the Δ6-desaturase, $C_{18/20}$-elongase, Δ5-desaturase and Δ17-desaturase, were codon optimized according to the codon usage pattern and GC content of highly expressed genes of *Y. lipolytica*, which typically have a GC content 52–54 % GC. In addition, they contain the consensus sequence (5′-ACCATGG-3′) around the 'ATG' translation initiation codon with a built-in *Nco*I site and a *Not*I site after its stop codon [7, 12] to facilitate cloning in conventional vectors. It was found that the substrate conversion was increased in almost all of the codon-optimized genes except the Δ5-desaturase gene derived from *M. alpine* [12].

Another strategy to further improve EPA pathway activity is by increasing the copy number of the overexpressed genes. In early engineering of the Δ6 pathway, strain Y9027 was generated to produce EPA at 40 % of the total lipid fraction [12] by increasing the copy number of the pathway genes. The "pushing" of carbon flux into the engineered pathway was accomplished by overexpression of the $C_{16/18}$-elongase gene [24] and the Δ12-desaturase gene of *Fusarium moniliforme* [25]. The in vivo substrate conversion catalyzed by the overexpressed Δ6-, Δ5-, and Δ17-desaturases in strain Y9027 were about 86, 90 and 97 %, respectively, of the total fatty acids entering the triglyceride pool. The strategy of increasing the overexpressed gene copy number was further applied in the Δ9 pathway engineering for generating more advanced EPA production strains.

One bottleneck to achieving high carbon flux in the initial engineering of the Δ6 pathway was the low efficiency of the $C_{16/18}$-elongase, which led to high GLA content in the lipid profile. To reduce GLA accumulation, genes encoding the Δ9-elongases [26] and Δ8-desaturases [27] (Fig. 2.2) were isolated and characterized from *Euglena gracilis, E. anabaena,* and *Eutreptiella,* sp. CCMP389. Three Δ9-elongase and Δ8-desaturase bifunctional fusion genes were also created [28], which almost doubled the Δ8-desaturase activity while keeping similar Δ9-elongase activity. In addition, three genes encoding Δ5-desaturases from *E. gracilis, E. anabaena,* and *Eutreptiella,* sp. CCMP626 [29] and three genes encoding Δ17-desaturases from *Pythium aphanidermatum*, *Phytophthora ramorum,* and *Phytophthora sojae* were also isolated and studied for their activities and substrate selectivity [13]. Several genes encoding different acyltransferases were also isolated and used to improve fatty acid traffic in the endoplasmic reticulum [30]. "Pushing" and "pulling" the carbon flux into the engineered Δ-9 pathway was achieved by overexpression of the $C_{16/18}$-elongase gene and the Δ12-desaturase gene [25] and by using multiple copies of Δ17-desaturase genes [13].

One critical finding during engineering of the Δ9 pathway was the mutation of PEX encoding genes that are involved in peroxisome biogenesis and matrix protein import, such as PEX3 or PEX10. Mutation of these genes improved the EPA content in the lipid fraction by two fold [9, 31]. Deletion of the *pex10* gene in engineered strains producing DGLA and ARA were also discovered to double DGLA and ARA titers in the lipid fraction compared to the parent strains with a wild-type *pex*10 gene [9]. In *pex10* mutation strains, ß-oxidation was greatly reduced and unidentified membrane-like structures, possibly deformed nonfunctional peroxisomes, were also observed. By combining the above Δ9 pathway engineering strategies, strain Z5567 [20] was generated that contains 41 copies of 19 different genes. Z5567 produced increased lipid fraction as high as 50 % of dry cell weight (DCW) with an EPA content as high as 25 % of DCW under typical fed-batch fermentation conditions.

2.3 Selection of High EPA Production Strains

Fermentation research plays a critical role in converting the omega-3 metabolic engineering research to commercial application. Both fermentation and strain engineering research were conducted in concert from the beginning of the program. The major fermentation research activities included (1) establishing high throughput strain screening protocols at various scales of fermentation and selecting the top producing strains using these protocols, (2) optimizing fermentation conditions for selected promising candidate strains, and (3) fermentation process scale-up to achieve high EPA production at pilot and commercial scales.

Figure 2.3 summarizes a typical process workflow for the omega-3 strain screening. Tens of thousands of new candidate strains were generated yearly by the metabolic engineering program. Candidate strains were first tested in 24-well plates to identify the top strains based on the EPA content in the lipid fraction as determined by GC analysis [32]. Among them, thousands of high EPA strains were selected for testing under shake flask conditions. Those strains with both high lipid titer in biomass and high EPA content in the lipid fraction were selected for further testing in the Micro-24 bioreactor system [33]. In the Micro-24 bioreactor experiments, the titer and yield of both lipid and EPA were obtained to better evaluate each candidate strain's performance. Dozens of top strains identified from the Micro-24 bioreactor experiments were further evaluated in lab-scale fermentors (2–10 L), where the titer, rate, and yield of cell density (DCW or dry cell weight), lipid, and EPA were carefully compared to identify the very top production strain candidates. Some of the top strains (e.g., Y4305) were described previously in

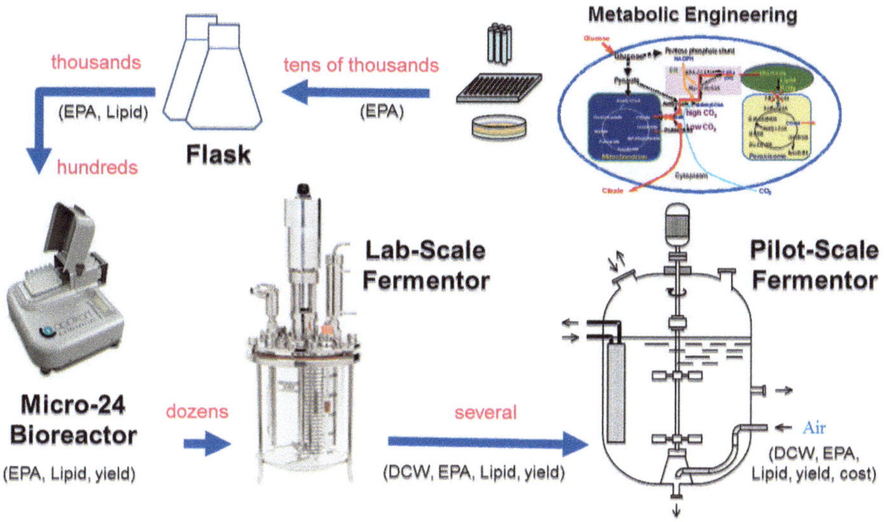

Fig. 2.3 Typical process flow for screening and selection of high EPA production strains under various bioreactor conditions

details [9]. Those top candidates from the lab-scale fermentation experiments were evaluated in pilot-scale fermentors using commercial medium and process conditions. Then the cost to produce EPA by each candidate strain was evaluated to determine the final production strain.

A key step in the strain screening process is evaluation in small scale bioreactors. It is desirable to have a multi-bioreactor system with small working volumes to allow testing of hundreds of candidate strains, yet each small bioreactor system must have high-quality process controls so that the data obtained from these small bioreactors predict the performance in lab- and pilot-scale fermentors. There are many bioreactor tools available for strain screening. The 24-well blocks and test tubes are the simplest bioreactors and can be easily operated. However, the data is much less reliable due to the low controllability and insufficient sample volumes available for both lipid and cell density analysis [34, 35]. Shake flasks are also simple and easy to run at larger working volumes (10–100 mL), but they still do not monitor and control pH values and dissolved oxygen (DO) levels. Besides that, they are more labor intensive, especially for the two-stage omega-3 process [36]. Recently, some Microtiter plates/bioreactors such as the BioLector system have been developed, which have a large number of small reaction wells (1 mL or less) with each well's pH value and DO level monitored [37, 38], but precise control of pH and DO levels are still not available. The small working volume of each well also limited the microtiter bioreactor's application in the omega-3 project due to the large sample size needed for DCW and lipid analysis.

Recently, EPA strain screening and fermentation optimization was performed in a Micro-24 Bioreactor system [33]. The Micro-24's ability to monitor and control each reactor's temperature, dissolved oxygen level, pH values, and potential for real-time data acquisition made it amenable for performing scale-down experiments

Bioreactor	Reactor/ working vol.	Controllability	Experimental Data	Work Capacity
	10~50 flasks/shaker, 25-50 mL	T	Final point – titer, rate, yield	2000 individual experiments /year/person
	24 reactors, 3-7 mL	Individual T, pH, pO_2	Online process- T, pH, pO2 Final point – titer, rate, yield	1000 individual experiments /year/person
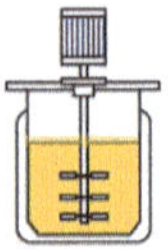	Single reactor, 2-10 L	T, pH, pO_2 feeding	Online process- T, pH, pO2, feeding Time course – titer, rate, yield	40 individual experiments /year/person

Fig. 2.4 Comparison of process controllability, data quality, and work capacity between shake flasks, Micro-24 bioreactor, and lab-scale fermentors

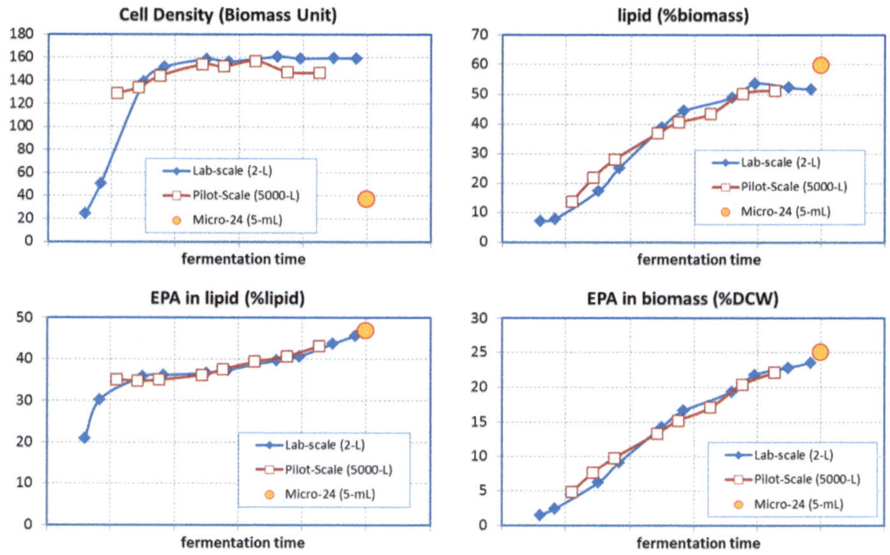

Fig. 2.5 Comparison of omega-3 fermentation data between the Micro-24 Bioreactor, 2-L fermentor, and 5000-L fermentor experiments

of larger lab-scale fermentations in a high throughput manner (Fig. 2.4). Using this system environmental conditions (T, pH, pO_2, medium…) could be varied across a wide range to examine impacts on end of run fermentation performance metrics, including byproduct analysis, DCW, lipid content, EPA content in lipid, and EPA conversion yield. These data significantly increased the predictability of each individual strain's performance in lab- or large-scale fermentation. As can be seen in Fig. 2.5, the lipid and EPA data from the Micro-24 bioreactor experiments agree well to those from the 2-L and 5000-L experiments though were operated at much lower cell densities in the Micro-24 bioreactor experiments.

2.4 Two-Stage Fermentation Process for EPA Production

Optimal fermentation conditions are often different for production strains with different genetic backgrounds. After selection of a top production strain from thousands of candidate strains (Fig. 2.3), fermentation conditions including seed culture, inoculation process, fermentation medium, substrate feeding and control, other process parameters (e.g., T, pH, DO, etc.) were optimized for the given strain. The goal of fermentation optimization is to maximize rate, titer, and yield of EPA production and minimize by-product formation (e.g., organic acids). The omega-3 fatty acids are an intracellular product, thus maintaining both high cell density and high EPA content in biomass is critical for fermentation optimization.

The omega-3 fermentation is a typical two-stage fermentation process (Fig. 2.6) with oil and EPA produced under nitrogen-limiting conditions that is largely non-growth associated. In the first stage of the fermentation, the *Yarrowia* cells are grown on the carbohydrate substrate with nitrogen provided by both an organic source (e.g., yeast extract) and an inorganic source (e.g., $(NH_4)_2SO_4$) in the initial medium. An alternative approach is to supply inorganic nitrogen in the form of NH_4OH used for pH control in the growth phase. After a certain period of time, the residual nitrogen in the fermentation medium is consumed, cell growth stops, and the *Yarrowia* cells start to accumulate lipids by converting the carbohydrate continuously supplied to the nitrogen limited culture.

While flasks and microfermenters gave general guidance around optimization of the fermentation conditions, they are not sufficient for further optimizing the rate, titer, and yield of a complete fermentation run due to limitations in process controllability and low capacities for mixing and mass transfer that do not mimic larger scale fermentor capabilities. For that purpose, the optimization work for a selected production strain was mainly conducted in lab-scale fermenters. The optimized fermentation conditions were further examined in pilot- or commercial scale fermentors to test their scalability. Scale down and re-optimization experiments are sometimes required when significant difference in fermentation performance is seen between lab-scale and large-scale fermentors.

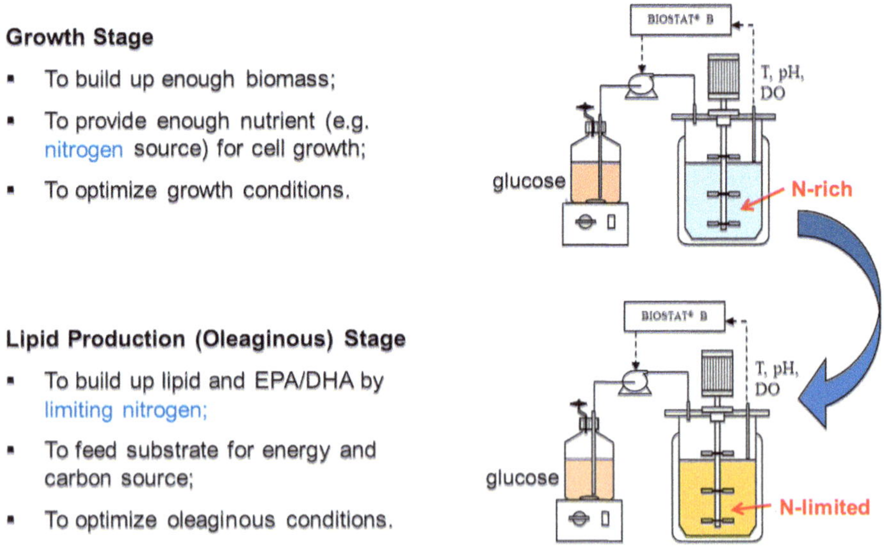

Fig. 2.6 Two-stage (growth and production) fed-batch fermentation process for EPA production by metabolically engineered *Yarrowia lipolytica*

2.5 Using Modeling to Guide Fermentation Optimization and Scale-up

Optimization of lab-scale and large-scale fermentation is critical to achieve target EPA production and guide decisions as to further strain engineering. In practice, it was always a challenge to complete optimization experiments for each strain due to the multitude of strains to be screened opposite time and resource limitations. Our strategy was to use mathematical models to help analyze fermentation performance under different conditions and then to design and predict optimal conditions. Dynamic models were developed to describe fermentation behavior based on the significant amount of fermentation data available and on the understanding of both the strain and the process that had been accumulated. The model was established by fitting to old experimental data, validated and modified by matching with more experimental results under new conditions, and then used to predict new conditions for improved fermentation performance (Fig. 2.7).

A set of unstructured kinetic equations were built from first principles, which included the equations of cell growth, substrate consumption, nitrogen utilization, oxygen uptake, lipid and EPA formation, and byproduct accumulation. The models

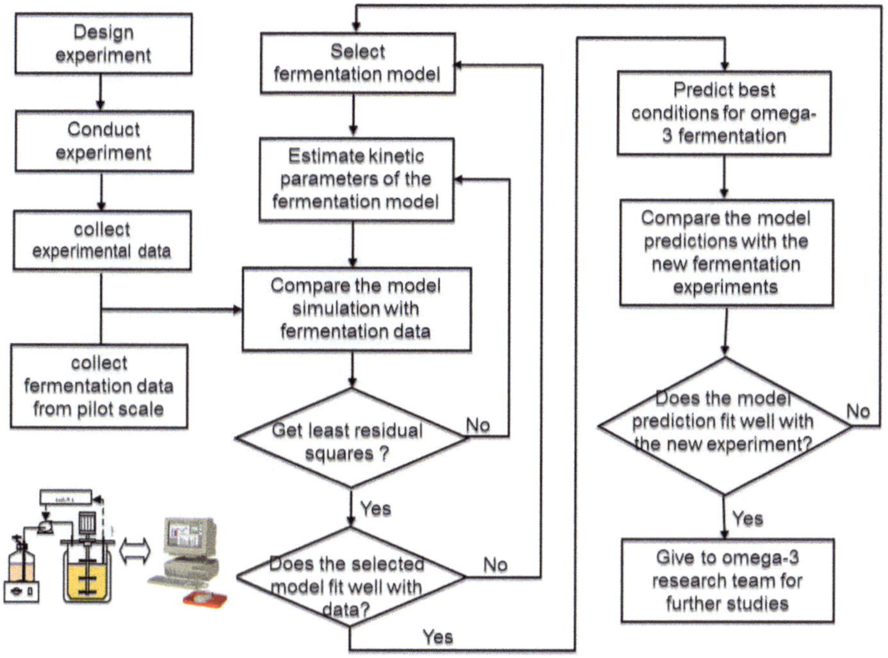

Fig. 2.7 A schematic diagram of establishing a mathematical model for omega-3 fermentation based on experimental data. The model is then used to help guide experiments under new fermentation conditions

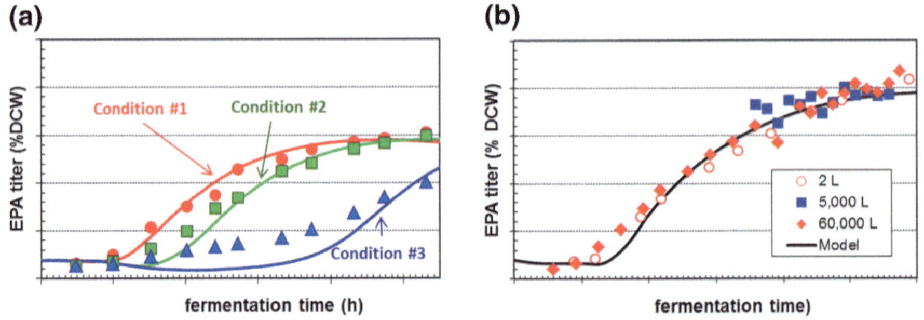

Fig. 2.8 Examples of using mathematical model for process optimization (**a**) and scale-up (**b**). In graph **a**, the model (*solid lines*) guided the fermentation experiments (*symbols*) to improve EPA titers by changing conditions from #3 to #1 and #2. In graph **b**, the model was used to guide the fermentation scale-up from 2L to pilot and commercial scale

could predict cell growth, DCW, DO level, oxygen uptake rate (OUR), CO_2 evolution rate (CER), and other process parameters that were also measured during the fermentation as a function of various medium and process conditions. When the production strain or key process conditions were changed, only a few model parameters needed to be adjusted to maintain the model's predictability. The changes were often associated with rate constants that were used to describe phenotypic changes as the result of certain genetic constructs. The dynamic models were able to predict the key performance parameters (e.g., titer, rate, and yield of a product) before and during the run, and further help to guide the fermentation optimization and process scale-up. Examples of the dynamic model's capability are shown in Fig. 2.8.

2.6 Fermentation Scale-Up for Commercial Production

Achieving high titer, rate, and yield of EPA production in lab-scale fermentation experiments marked the beginning of fermentation research efforts. The ultimate, more challenging goal was to demonstrate similar or better fermentation performance at large scale. Scale-up studies were necessary and became extremely important in the latter stage of the R&D program. The general criteria used for fermentation scale-up included key factors of geometry/size, power input per volume, mass transfer coefficient K_{La}, etc. [39]. Selected production strains were tested in pilot-scale facilities to determine the best criteria for fermentation scale-up, and to gather more information on strain performance, dynamic behavior, and scale-up effects that could influence EPA rate, titer, and yield. During scale-up the set of measured variables expanded beyond temperature, pH, feed rate, and DO levels, and included agitation, aeration rate, and mass transfer characteristics. The latter became specific focal points for scale-up.

The benefits of highly predictive dynamic models became evident as the process moved from lab scale to pilot plant and commercial production. The models were used to study the mass transfer characteristics of the broth as a function of superficial gas velocity and agitation power. The dynamic models were used to predict the commercial scale fermentation performance taking into account the mass transfer characteristics obtained from the pilot trials which were used to set specific agitation and aeration rates. With the aid of the models a successful process scale-up to commercial production was obtained using the least number of pilot trials (Fig. 2.8b).

The major challenges encountered during the scale-up process were limitations in either the medium/feed components or in the achievable process controls in large-scale fermentors. For example, most commercial scale fermentation uses complex media with commercially available raw materials. However, the complex media may have some other minor components affecting the fermentation performance. In addition, fluid dynamic behavior in large-scale fermentors is significantly different from that in lab scale. When significant difference/inconsistency between the fermentation performance in lab-scale and large-scale fermentors was observed, a series of lab-scale "scale-down" experiments were designed to mimic the fermentation medium and/or process conditions at commercial scale [40, 41]. The scale-down studies elucidated a few important factors that affect the scale-up. The identified scale-up issues were fixed by further engineering of the strain or by modifying fermentation protocols to better follow lab-scale and pilot-scale experiments.

2.7 Omega-3 EPA Oil Produced by *Yarrowia lipolytica*

The end result of this effort is the production of an omega-3 oil with a lipid profile that has certain heart health advantages over other omega-3 oils on the market (Fig. 2.9). The *Yarrowia* yeast oil has more than 50 % EPA as a fraction of total lipid and around 5 % saturated fatty acids making the EPA oil unique among omega-3 containing oils that have not been derivatized or fractionated. Saturated fatty acids (C14:0, C16:0 & C18:0) are relatively high in most other oils, which was believed to increase low-density lipoprotein (LDL) cholesterol, a risk factor for heart disease [42]. Fish oil contains roughly 30 % omega-3 content largely as EPA and DHA in an 18:12 ratio and is less desirable from a heart health point of view. Algae oil has close to 40 % DHA content, and is also high in saturated fatty acids. Flax oil has about 40 % ALA, but ALA has very low efficiency (5–10 %) for further conversion to EPA and DHA in the human body. The *Y. lipolytica* technology platform provides an alternative route to high omega-3 content oil with low saturated fatty acids levels that is obtainable without the need for further concentration and purification using conventional distillation technology. The consistent supply and high targeted specific fatty acid content also provides for downstream purification advantages in applications where a single fatty acid is required. In this

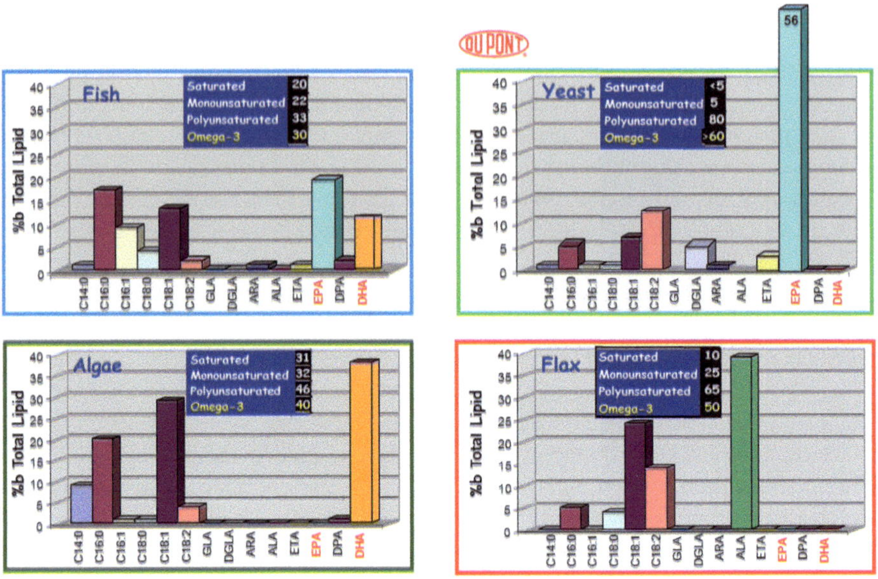

Fig. 2.9 Comparison of the lipid profiles between DuPont's *Yarrowia lipolytica* oil and other major omega-3 oil products on market

latter case, the technology platform developed by DuPont is amenable to produce tailored omega-3 (EPA, DHA) and/or omega-6 (ARA, GLA) fatty acid mixtures in the cellular lipid profiles, which can meet more technical or customer needs in the future [7, 9].

2.8 Conclusion

The nonconventional yeast *Y. lipolytica* was engineered in DuPont to produce a high content of omega-3 EPA (C20:5) in biomass under commercial-scale fermentation conditions. A combination of enzymes were overexpressed to synthesize EPA via the Δ9/Δ8 pathway. The lipid metabolism was optimized to maximize total lipid and EPA production. The expression levels of all other pathway enzymes were also carefully balanced. A key finding during pathway engineering and evaluation was that disruption of peroxisome biogenesis had a major positive impact on the production of EPA and the metabolism of storage lipid, as well as reduction of the major by-products. The result of the metabolic engineering effort produced a high EPA production strain, Z5567, which is capable of producing a single fatty acid, in this case EPA, to levels as high as 25 % of dry biomass.

Fermentation engineering played a critical role in helping to transition the omega-3 research to commercial production. Research in both strain engineering and fermentation process development was initiated at the same time and carried out in parallel. Close collaboration between strain engineering and fermentation research was essential to success. An advanced microfermentor system with well-controlled process parameters significantly improved the efficiency of strain screening and the predictability of the selected strain's performance in larger scale fermentation. The omega-3 fed-batch fermentation process consisted of a growth phase to maximize biomass production and a production (oleaginous) phase to maximize EPA production while minimizing byproduct formation. Mathematical modeling of the fermentation process was developed and used to guide fermentation optimization and scale-up. Two commercial products, NewHarvest™ EPA oil and Verlasso® salmon were developed by using the sustainable EPA source from *Yarrowia lipolytica*. The road to commercial success was made possible by the power of modern biotechnology to combine both fundamental scientific research and industrial engineering.

Acknowledgments We would like to thank Dr. Henry Bryndza and Dr. Michael Saltzberg for their strong support. We also thank all members in the omega-3 technical, manufacturing, and business teams for their collaboration and contributions.

References

1. Martins DA, Custódio L, Barreira L, Pereira H, Ben-Hamadou R, Varela J, Abu-Salah KM (2013) Alternative sources of n-3 long-chain polyunsaturated fatty acids in marine microalgae. Mar Drugs 11:2259–2281
2. Kapoor R, Patil UK (2011) Importance and production of omega-3 fatty acids from natural sources. Int Food Res J 18:493–499
3. Yokoyama M, Origasa H, Matsuzaki M, Matsuzawa Y, Saito Y, Ishikawa Y, Oikawa S, Sasaki J, Hishida H, Itakura H, Kita T, Kitabatake A, Nakaya N, Sakata T, Shimada K, Shirato K (2007) Effects of eicosapentaenoic acid on major coronary events in hypercholesterolaemic patients (JELIS): a randomised open-label, blinded endpoint analysis. Lancet 369:1090–1098
4. Ballantyne CM, Bays HE, Kastelein JJ, Stein E, Isaacsohn JL, Braeckman RA, Soni PN (2012) Efficacy and safety of eicosapentaenoic acid ethyl ester (AMR101) Therapy in statin-treated patients with persistent high triglycerides (from the ANCHOR Study). Am J Cardiol 110:984–992
5. Kyle DJ (2001) The large-scale production and use of single-cell highly enriched oil in docosahexaenoic acid. In: Shahidi F, Finley J (eds) Omega-3 fatty acids: chemistry, nutrition, and health effects. Oxford University Press, Oxford, pp 92–107
6. Hong S-Y, Seip J, Sharpe PL, Xue Z, Yadav NS, Zhang H, Zhu QQ (2011) Recombinant microbial host cells for high eicosapentaenoic acid production. U.S. Patent Application 20120052537
7. Xie D, Jackson E, Zhu Q (2015) Sustainable source of omega-3 eicosapentaenoic acid from metabolically engineered *Yarrowia lipolytica*: from fundamental research to commercial production. Appl Microbiol Biotechnol 99:1599–1610
8. Zhu Q, Jackson E (2015) Metabolic engineering of *Yarrowia lipolytica* for industrial application. Curr Opin Biotechnol 36:65–72

9. Xue Z, Sharpe PL, Hong SP, Yadav NS, Xie D, Short DR, Damude HG, Rupert RA, Seip JE, Wang J, Pollak DW, Bostick MW, Bosak MD, Macool DJ, Hollerbach DH, Zhang H, Arcilla DM, Bledsoe SA, Croker K, McCord EF, Tyreus BD, Jackson EN, Zhu Q (2013) Production of omega-3 eicosapentaenoic acid by metabolic engineering of *Yarrowia lipolytica*. Nat Biotech 31:734–740
10. Ratledge C (2005) Single cell oils for the 21st century. In: Cohen Z, Ratledge C (eds) Single cell oils. AOCS Press, Champaign, pp 1–20
11. Groenewald M, Boekhout T, Neuvéglise C, Gaillardin C, van Dijck PW, Wyss M (2014) *Yarrowia lipolytica*: safety assessment of an oleaginous yeast with a great industrial potential. Crit Rev Microbiol 40:187–206
12. Zhu Q, Xue Z, Yadav N, Damude H, Pollak D, Rupert R, Seip J, Hollerbach D, Macool D, Zhang H (2010) Metabolic engineering of an oleaginous yeast for the production of omega-3 fatty acids. In: Cohen Z, Ratledge C (eds) Single cell oils, 2nd edn. AOCS Press, Urbana, pp 51–73
13. Xue Z, He H, Hollerbach D, Macool DJ, Yadav NS, Zhang H, Szostek B, Zhu Q (2013) Identification and characterization of new Δ-17 fatty acid desaturases. Appl Microbiol Biotechnol 97:1973–1985
14. Metz JG, Roessler P, Facciotti D, Levering C, Dittrich F, Lassner M, Valentine R, Lardizabal K, Domergue F, Yamada A, Yazawa K, Knauf V, Browse J (2001) Production of polyunsaturated fatty acids by polyketide synthases in both prokaryotes and eukaryotes. Science 293:290–293
15. Meesapyodsuk D, Qiu X (2012) The front-end desaturase: structure, function, evolution and biotechnological use. Lipids 47:227–237
16. Napier (2004) The production of long chain polyunsaturated fatty acids in transgenic plants by reverse-engineering. Biochimie 86(11):785–792
17. Wallis JG, Browse J (1999) The Delta8-desaturase of *Euglena gracilis*: an alternate pathway for synthesis of 20-carbon polyunsaturated fatty acids. Arch Biochem Biophys 365:307–316
18. Sayanova O, Napier JA (2004) Eicosapentaenoic acid: biosynthetic routes and the potential for synthesis in transgenic plants. Phytochemical 65:147–158
19. Muller S, Sandal T, Kamp-Hansen P, Dalboge H (1998) Comparison of expression systems in the yeasts *Saccharomyces cerevisiae*, *Hansenula polymorpha*, *Klyveromyces lactis*, *Schizosaccharomyces pombe* and *Yarrowia lipolytica*. Cloning of two novel promoters from *Yarrowia lipolytica*. Yeast 14:1267–1283
20. Jefferson RA, Kavanagh TA, Bevan MW (1987) GUS fusions: beta-glucuronidase as a sensitive and versatile gene fusion marker in higher plants. EMBO J 6:3901–3907
21. Hong S-Y, Seip J, Walters-Pollak D, Rupert R, Jackson R, Xue Z, Zhu Q (2011) Engineering *Yarrowia lipolytica* to express secretory invertase with strong FBA1IN promoter. Yeast 29:59–72
22. Xue Z, Zhu QQ (2012) Ammonium transporter promoter for gene expression in oleaginous yeast. US Patent 8323960
23. Alper H, Fischer C, Nevoigt E, Stephanopoulos G (2005) Turning genetic control through promoter engineering. Proc Natl Acad Sci USA 102:12678–12683
24. Macool, DJ, Xue Z, Zhu Q (2008) A *Mortierellaalpina* C16/18 fatty acid elongase. US Patent 7470532
25. Yadav NS, Zhu Q, Zhang H (2009) Δ12 desaturases suitable for altering levels of polyunsaturated fatty acids in oleaginous yeast. US Patent 7504259
26. Damude HG, Zhu Q (2007) Delta-8 desaturase and its use in making polyunsaturated fatty acids. US Patent US 7256033
27. Damude HG, McGonigle B, Zhu Q, Zhixiong X (2007) Delta-9 elongases and their use in making polyunsaturated fatty acids. US Patent Application US20070118929
28. Damude HG, Kinney AJ, Ripp KG, Zhu QQ (2008) Multizymes and their use in making polyunsaturated fatty acids. US Patent Application 20080254191

29. Pollak DW, Bostick MW, Yoon H, Wang J, Hollerbach DH, He H, Damude HG, Zhang H, Yadav NS, Hong SP, Sharpe P, Xue Z, Zhu Q (2012) Isolation of a delta 5 desaturase gene from *Euglena gracilis* and functional dissection of its HPGG and HDASH motifs. Lipids 47:913–926
30. Zhang H, Damude HG, Yadav NS (2012) Three diacylglycerolacyl transferases contribute to oil biosynthesis and normal growth in *Yarrowia lipolytica*. Yeast 29:25–38
31. Hong SP, Sharpe PL, Xue Z, Yadav NS, Zhu QQ (2009) Peroxisome biogenesis factor protein (pex) disruptions for altering polyunsaturated fatty acids and total lipid content in oleaginous eukaryotic organisms. US Patent Application 20090117253
32. Cahoon EB, Ripp KG, Hall SE, Kinney AL (2001) Formation of conjugated Δ8, Δ10-double bonds by Δ12-oleic acid desaturase-related enzymes: biosynthetic origin of calendic acid. J Biol Chem 276:2637–2643
33. Xie D (2012) Using an advanced microfermentor system for strain screening and fermentation optimization. In: Cheng Q (ed) Methods in molecular biology, 834 (microbial metabolic engineering). Humana Press, New York, pp 217–231
34. Danielson PB, Büchs J, Stockmann C, Fogleman JC (2004) Maximizing cell densities in miniprep-scale cultures with H15 medium and improved oxygen transfer. Biochem Eng J 17:175–180
35. Stockmann C, Losen M, Dahlems U, Knocke C, Gellissen G, Büchs J (2003) Effect of oxygen supply on passaging, stabilizing and screening of recombinant *Hansenula polymorpha* production strains in test tube cultures. FEMS Yeast Res 4:195–205
36. Büchs J (2001) Introduction to advantages and problems of shaken cultures. Biochem Eng J 7:91–98
37. Amanullah A, Otero JM, Mikola M, Hsu A, Zhang J, Aunins J, Schreyer HB, Hope JA, Russo AP (2010) Novel micro-bioreactor high throughput technology for cell culture process development: Reproducibility and scalability assessment of fed-batch CHO cultures. Biotechnol Bioeng 106:57–67
38. Duetz WA, Ruedi L, Hermann R, O'Connor K, Büchs J, Witholt B (2000) Methods for intense aeration, growth, storage and replication of bacterial strains in microtiter plates. Appl Environ Microbiol 66:2641–2646
39. Shuler ML, Kargi F (2002) Bioprocess Engineering, 2nd edn. Prentice Hall PTR, Upper Saddle River, pp 286–306
40. Nienow AW, Nordkvist M, Boulton CA (2011) Scale-down/scale-up studies leading to improved commercial beer fermentation. Biotechnol J 6:911–925
41. Ozbek B (1997) Mathematical modeling and simulation studies of scale-down fermentation systems. Chem Eng Technol 20:259–267
42. Katan M, Zock P, Mensink R (1994) Effects of fats and fatty acids on blood lipids in humans: an overview. Am J Clin Nutr 60(Suppl):1017S–1022S

Chapter 3
Toward Fermentative Production of Succinic Acid, Adipic Acid, and Caprolactam: An Industrial Perspective

Liang Wu

Abstract Over a century, chemical building blocks have been produced from fossil feedstocks by the petrochemical industry. However, there is a growing need for more sustainable production processes based on renewable resources. Here, we provide an industrial perspective on the development of fermentation processes for the production of succinic acid, adipic acid and caprolactam. We describe up-front choices for process configuration, microbial host, metabolic pathway, and strain designs and show that this is critical to reach commercially attractive yield, titer and productivity. For succinic acid, we demonstrate the successful development and commercialization of a low pH yeast-based fermentation process. For adipic acid and caprolactam, we show that systematic exploration of the pathway design space led to the proof-of-principle biosynthesis of these non-natural Nylon-intermediates.

Keywords Bio-based building blocks · Succinic acid · Adipic acid · Caprolactam · 6-Aminocaproic acid · Metabolic engineering · Pathway design · Process design · Low pH fermentation process

3.1 Introduction

Chemical building blocks such as ethylene, propylene, terephthalic acid, adipic acid, and caprolactam can be processed into a wide range of end products like polymers, resins, solvents, and lubricants. These are currently produced mainly from fossil feedstocks like oil, natural gas, and coal by the petrochemical industry. Due to depletion of fossil feedstocks and associated climate change issues, there is a growing need to produce these molecules from renewable feedstocks like biomass and CO_2. This provides a strong driving force to develop and commercialize bio-based, sustainable production processes. In accordance, the production capacity

L. Wu (✉)
DSM Biotechnology Center, A. Fleminglaan 1, 2613 AX Delft, The Netherlands
e-mail: Liang.Wu@dsm.com

P.C.K. Lau (ed.), *Quality Living Through Chemurgy and Green Chemistry*,
Green Chemistry and Sustainable Technology,
DOI 10.1007/978-3-662-53704-6_3

of bio-based building blocks is expected to increase from 5.7 million tons in 2014 to nearly 17 million tons by 2020 [1]. Fermentation processes have been successfully developed and scaled up for a range of bulk chemical building blocks, e.g., 1,3-propandiol [2], farnesene [3], isobutanol [4], isoprene [5] and 1,4-butanediol [6], by various industrial players. Here, we address the direct fermentative production of the chemical building blocks succinic acid, adipic acid, and caprolactam.

Succinic acid (SA) is a versatile building block, potentially giving rise to a wide range of products such as biodegradable polybutylene succinate (PBS), polyurethanes, resins, plasticizers, and 1,4-butanediol (Fig. 3.1). SA was listed as one of the top value-added bio-based chemicals since the publication of the BREW report [7] and its update [8]. The total market size (~40 kt/a in 2011) has so far been limited, since petro-based SA is relatively expensive. Development of large-scale cost-effective bioprocesses will reduce the cost price and open up a large market for potential applications. The bio-based SA production capacity has increased over the years to ~39 kt/a in 2014 and is expected to increase further to ~400 kt/a by 2020 [1]. Companies developing and commercializing bio-based SA include a.o. Reverdia (joint venture between DSM and Roquette), BioAmber, Myriant, and Succinity (joint venture between BASF and Corbion).

Adipic acid (AA) and caprolactam (CL) are among the most important commodity chemical building blocks worldwide, with a global production of 2.3 and 4.0 Mt/a, respectively [9]. AA is mainly used as monomer for polyamide 6,6 (Nylon 6,6) and produced by catalytic oxidation of petro-based KA oil (a mixture of cyclohexanol and cyclohexanone). CL is solely used as the monomer for polyamide

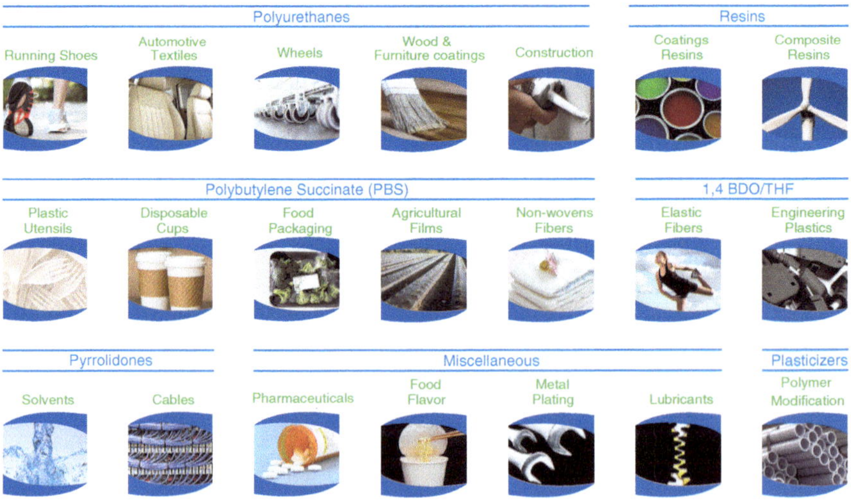

Fig. 3.1 Potential applications of bio-based succinic acid (http://www.reverdia.com/products/markets/) used with permission from Reverdia

6 (Nylon 6) and is produced from petro-based cyclohexanone. Both polymers have a high global warming potential in terms of kg CO_2 equivalent produced per kg product as compared to a wide range of polymers [10]. There is a clear opportunity to reduce the environmental footprint of this industry by replacing petro-derived AA and CL with bio-based drop-in products. Drath and Frost first reported fermentative production of muconic acid from glucose, which can be converted to adipic acid by chemical hydrogenation [11]. Multiple routes have since been developed to convert renewable feedstocks to AA and CL, i.e., by chemocatalytic processes, by fermentation, or a combination of both [9]. Several companies have adopted bio-based AA and/or CL in their R&D programs, e.g., BioAmber, Rennovia, Genomatica, and Verdezyne. However, commercial production of bio-based AA and CL is still in an early stage; so far only Verdezyne has demonstrated pilot-scale production of bio-based AA [12].

For large-scale fermentation process to produce chemical building blocks such as SA, AA, and CL, a set of three key performance indicators (KPI) largely determine the cost price of the product, namely, close to theoretical conversion yields of the product on raw material, volumetric productivity and product titer. Optimization of these KPIs is typically achieved through extensive cycles of microbial strain development and process optimization [13, 14], which involves multi-million dollar R&D expenses and multiple years of development. It is therefore critical to make up-front sound choices w.r.t. process configuration, microbial host, as well as metabolic pathway and strain designs, to maximize the chance of success. In this chapter, we review in detail the design considerations and demonstrate how implementation of the designs led to a proof-of-principle of direct fermentative production of AA and CL, and a cost-effective full-scale commercial fermentation process for SA.

3.2 Succinic Acid

SA occurs naturally in cellular metabolism in the tricarboxylic acid cycle (TCA cycle). Several microbes, in particular rumen bacteria, such as *Actinobacillus succinogenes*, *Anaerobiospirillum succiniciproducens*, *Mannheimia succiniciproducens* [15], and the recently identified *Basfia succiniciproducens* [16], are known to produce significant amounts of SA. In addition, bacteria and fungi that are non-native SA producers have been successfully engineered to produce SA, e.g., *Escherichia coli* [17], *Corynebacterium glutamicum* [18], *Saccharomyces cerevisiae* [19], *Yarrowia lipolytica* [20], *Aspergillus niger* [19], and *Issatchenkia orientalis* [21]. To achieve a sustainable business case, the engineered host combined with an optimized process configuration must deliver the highest performance on KPIs (yield, productivity and titer) with a reasonable R&D effort.

3.2.1 Host and Process Considerations

The choice of the microbial host often dictates choices of the process configuration. For example, bacterial hosts usually require a near-neutral pH fermentation process, where SA is completely dissociated (Fig. 3.2). Eukaryotes such as yeast and filamentous fungi are in general tolerant to low pH and high concentrations of organic acids [21]. This allows a significantly lower pH (e.g., pH 3) in fermentation, where most SA is undissociated. The process pH leads to different downstream process (DSP) options for the recovery of SA from the fermentation broth, as reviewed in [22]. A neutral pH process requires more unit operations in DSP, giving rise to higher investment costs and increased yield loss. Furthermore, titrant use and salt formation—inherent to a neutral pH process—will increase the variable costs. In contrast, a low pH process allows for a more simple DSP, wherein undissociated SA in the broth is recovered by direct crystallization. The latter offers a fundamental cost-advantage, as well as significantly lower GHG (greenhouse gas) emissions and NREU (nonrenewable energy use), as compared to neutral pH process variants or a typical petrochemical process [23].

Besides a lower fermentation pH, eukaryotic hosts are in general insensitive to bacteriophage infections, thereby increasing the robustness in long-term operation of the fermentation process. Currently, Reverdia and BioAmber have both adopted a low pH yeast-based fermentation processes, while Myriant and Succinity employ a bacterial process at neutral pH.

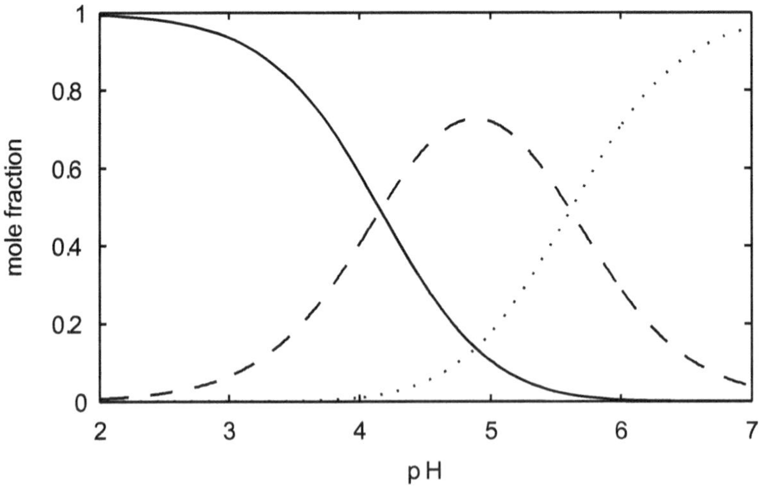

Fig. 3.2 Effect of pH on the distribution of different SA species. *Solid* undissociated acid (H_2SA), *dashed* monovalent anion of SA (HSA^-), *dotted* divalent anion of SA (SA^{2-}). pKa1 = 4.16, pKa2 = 5.6

3.2.2 Metabolic Engineering of Yeast for SA Production

Following the rationale of an attractive low pH process, we will further provide an overview of approaches and challenges to genetically engineer yeast toward SA production.

3.2.2.1 Pathway Design

Among the KPI's, product yield is usually the most critical determinant for the overall cost price of a bio-based building block. The maximum theoretical yield of SA ($C_4H_6O_4$) on glucose is 1.7 mol/mol, or 1.12 g/g (Eq. 1). At this theoretical maximum, no oxygen consumption is required, implying the possibility to develop an anaerobic fermentation process. This clearly provides advantages at large-scale, where energy requirements for mixing, aeration, and heat removal are significantly reduced. In addition, in the optimal case, a net fixation of CO_2 takes place.

$$1\ C_6H_{12}O_6 + \frac{6}{7}CO_2 \rightarrow \frac{12}{7}C_4H_6O_4 + \frac{6}{7}H_2O + 171\ \text{kJ/mol} \quad (1)$$

In a typical cell, major pathways toward SA are the reductive TCA cycle (rTCA cycle), the oxidative TCA cycle (oTCA), and the glyoxylate shunt (Fig. 3.3). Table 3.1 summarizes the differences between these pathways in terms of yield, CO_2 fixation/production, and redox requirement. By combining the rTCA cycle with either the oTCA cycle or the glyoxylate shunt, the maximum theoretical yield of 1.7 mol/mol can be reached while maintaining the redox balance.

Engineering part of the rTCA cycle by overexpression of pyruvate carboxylase (PYC), malate dehydrogenase (MDH), and a dicarboxylic acid transporter leads to high-level production of malate, two-steps upstream of SA in the rTCA cycle [24]. A strategy combining the rTCA cycle and the glyoxylate shunt has been successfully applied in *E. coli* [25, 26] and has been described for yeast [27]. Strategies employing one of the three pathways alone will likely not lead to a high yield, for example, producing succinate via the oTCA [20, 28, 29], or coupling SA production to growth via the glyoxylate shunt [30].

3.2.2.2 Compartmentation

In eukaryotic cells, proteins can be localized in the cytosol or targeted to various compartments, such as mitochondria, peroxisomes, and vacuoles. When the heterologous SA pathways are localized in the cytosol, the product can be directly transported out of the cell by a suitable transporter protein. However, if succinate would be produced by mitochondrial pathways, transport across the inner and outer mitochondrial membrane, as well as the plasma membrane, will be necessary to

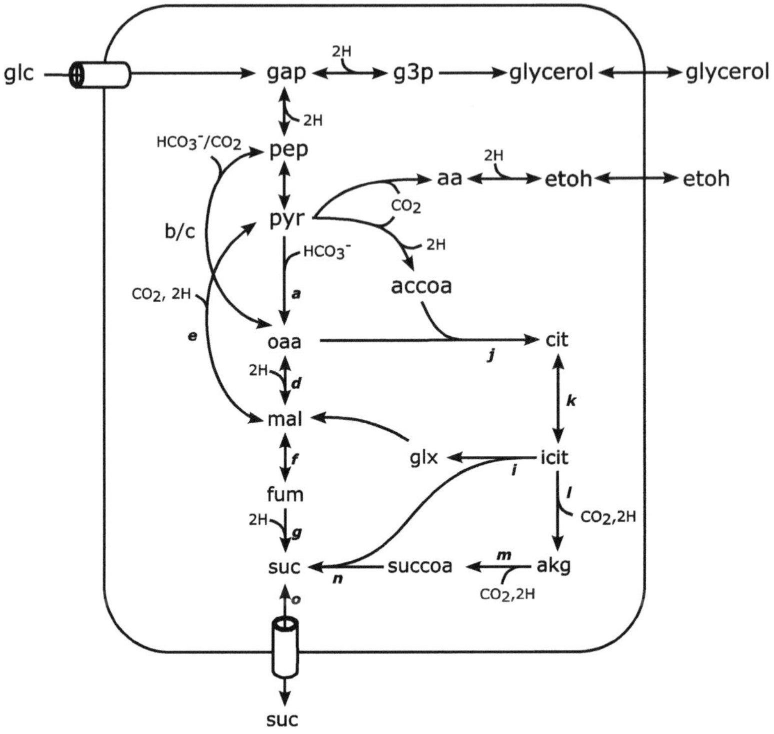

Fig. 3.3 Pathways leading to SA. *glc* glucose, *gap* glyceraldehyde 3-phosphate, *g3p* glycerol 3-phosphate, *pep* phosphoenolpyruvate, *pyr* pyruvate, *aa* acetaldehyde, *etoh* ethanol, *accoa* acetyl-CoA, *oaa* oxaloacetate, *mal* malate, *fum* fumarate, *suc* succinate, *cit* citrate, *icit* isocitrate, *akg* α-ketoglutarate, *succoa* succinyl-CoA, *H* reducing equivalent. *a* pyruvate carboxylase, *b* PEP carboxylase, *c* PEP carboxykinase, *d* malate dehydrogenase, *e* malic enzyme, *f* fumarase, *g* fumarate reductae, *h* malate synthase, *i* isocitrate lyase, *j* citrate synthase, *k* aconitase, *l* isocitrate dehydrogenase, *m* α-ketoglutarate dehydrogenase complex, *n* succinyl-CoA ligase, *o* dicarboxylate transporter

Table 3.1 Comparison between different pathways leading to SA

Pathway	Key steps[a]	SA/glucose, mol/mol	CO_2/glucose, mol/mol	H[b]/glucose, mol/mol
rTCA cycle	d, f, g	2	−2	−4
oTCA cycle	j, k, l, m, n	1	2	10
Glyoxylate shunt	h, i	1	2	10

[a]Key steps are depicted in Fig. 3.3
[b]Reducing equivalent

secrete the product. This might lead to secretion bottleneck and also adds complexity to strain engineering. In addition, a cytosolic production pathway can directly utilize reducing equivalents generated in the cytosol by the primary metabolism. Pathway gene candidates can be modified to ensure cytosolic localization in yeast, for example, putative mitochondrial targeting sequence was removed from a *Rhizopus* fumarase [31]; peroxisomal targeting sequence was removed from the malate dehydrogenase encoded by the yeast MDH3 gene [24]; putative glycosomal targeting sequence was removed from a *Trypanosoma* fumarate reductase [19, 32].

3.2.2.3 Redox Balance

Reducing equivalents generated by the primary metabolism are needed for the reduction steps catalyzed by MDH and fumarate reductase (FRD) in the rTCA cycle. NADH- and NADPH-dependent MDHs are found widely in nature [33], while this is not the case for FRD. FRDs are typically found as membrane-bound protein complexes with quinones as cofactor, accepting electrons from NADH via the membrane-bound NADH:quinone oxidoreductase [34]. It would be extremely challenging to engineer both membrane-bound protein complexes and the quinone biosynthesis pathway into a heterologous host like yeast. FRDs native to *S. cerevisiae* are soluble proteins but strictly $FADH_2$-dependent. Yeast FRDs are not involved in reoxidation of excess cytosolic NADH under anaerobic conditions [35]. Rather, they have been associated with reoxidation of $FADH_2$ during protein folding in ER [36]. Either deletion or overexpression of yeast FRDs has not shown any impact on SA levels under anaerobic, glucose excess conditions [37, 38]. Soluble NADH-dependent FRDs have only been reported in parasites such as *Trypanosoma brucei* [39, 40], where they are responsible for succinate production in glycosome and mitochondria. Indeed, overexpression of the truncated FRD that originates from *T. brucei* in yeast and *Aspergillus saccharolyticus* leads to increased accumulation of SA [19, 32].

3.2.2.4 CO_2 Fixation

CO_2 fixation, concomitant with the conversion of a C3-precursor (phosphoenolpyruvate [PEP] or pyruvate) to a C4-precursor (oxaloacetate or malate), is a key step in SA production. A variety of enzymes can catalyze this conversion, differing in reactants involved, metabolic cost for CO_2 fixation, as well as reversibility (Table 3.2). PYC and PPC (PEP carboxylase) have been successfully applied for SA production in bacteria, e.g., *E. coli* [17] and *C. glutamicum* [18]. In yeast, overexpression of PYC, in combination with the rTCA cycle, leads to increased titer and yield in a SA-producing *S. cerevisiae* (Fig. 3.4). The energy conserving PEP carboxykinase (PCK) is responsible for CO_2 fixation in natural SA-producing bacteria, such as *A. succinogenes* and *M. succiniciproducens*. However, in

Table 3.2 CO_2 fixation enzymes for SA production

Enzyme	Reaction	Reversibility	Metabolic cost[a]
PEP carboxylase (PPC)	HCO_3^- + PEP → OAA + Pi	Irreversible	1 ATP
pyruvate carboxylase (PYC)	HCO_3^- + ATP + Pyr → OAA + ADP + Pi + H^+	Irreversible	1 ATP
PEP carboxykinase (PCK)	CO_2 + PEP + ADP ↔ OAA + ATP	Reversible	None
Malic enzyme (ME)	Pyr + CO_2 + NAD(P)H ↔ MAL + NAD^+	Reversible	None

[a]mol/mol CO_2 fixed. Benchmark against the most energy efficient pathway from PEP to malate via PCK and MDH, which produces 1 ATP and consumes 1 NAD(P)H

S. cerevisiae it can only replace PYC as the sole anaplerotic enzyme after adaptive evolution and under elevated CO_2 partial pressure [41]. The same holds for NADPH-dependent malic enzyme (ME) localized to the cytosol in yeast [42]. The flux through PCK or ME after evolution is still much lower than what could be achieved by overexpression of PYC, which indicates potential thermodynamic and kinetic limitations.

To improve the thermodynamic and/or kinetic driving force for the carboxylation reactions, a higher CO_2 partial pressure (pCO_2) can be applied to the bioreactor. This leads to an increase in intracellular concentrations of dissolved CO_2 as well as HCO_3^-, the two major species of liquid phase CO_2 at a cytosolic pH between 6.5 and 7 [43]. Note that PCK and ME use dissolved CO_2, while PYC and PPC use HCO_3^- as substrate (Table 3.2). The high pCO_2 can be realized in the lab by, e.g., sparging CO_2 or adding carbonate [44]. For industrial scale fermenters, the techno-economic implications of supplying large amount of CO_2 gas has been discussed in detail by Jansen and van Gulik [22].

Dissolved CO_2 is spontaneously hydrated to carbonic acid, which subsequently dissociates into H^+ and HCO_3^-. The spontaneous hydration can be dramatically accelerated by the enzyme carbonic anhydrase, thereby increasing the availability of HCO_3^- for PYC and PPC [45–47].

3.2.2.5 Transport

SA, preferably produced in the cytosol, needs to be exported out of the cells. At a cytosolic pH of 6.5-7, SA is almost completely dissociated and present as HSA^- and SA^{2-} (Fig. 3.2). For these anions to cross cell membrane, a carrier protein is needed. A dicarboxylate:proton symporter has been reported in fungi [48–51]. Overexpression of the *S. pombe* malate transporter encoded by SpMAE1 in *S. cerevisiae* enabled reversible transport of the mono-anion of malate [52], which implies that such proton-symporters can be used for dicarboxylic acid export.

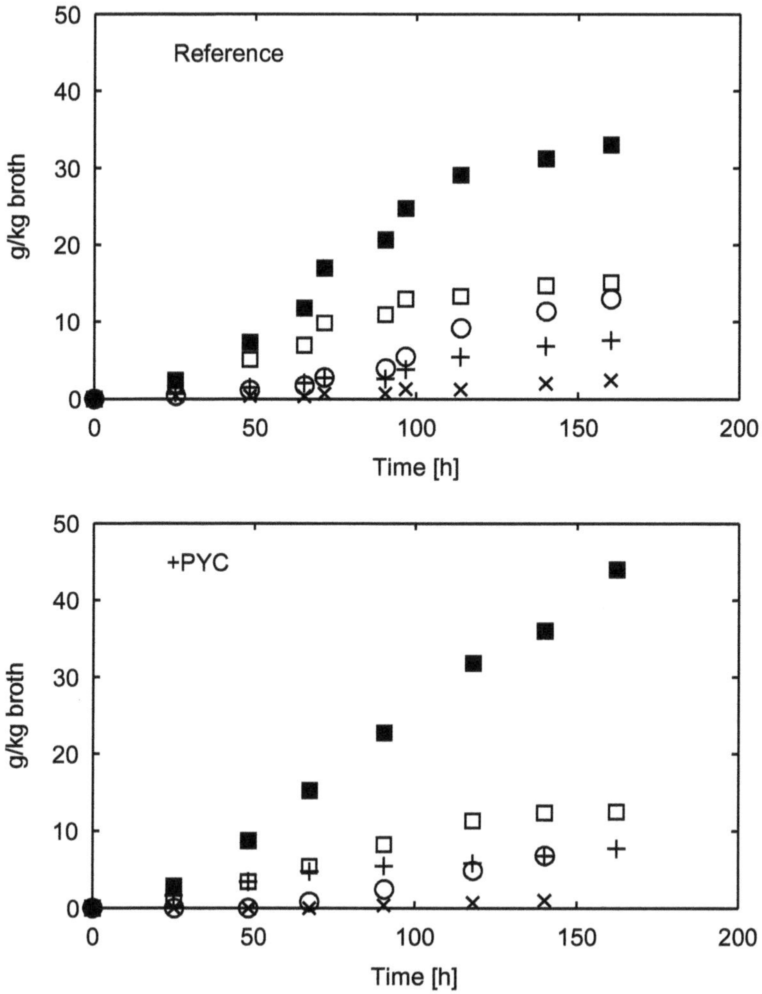

Fig. 3.4 Effect of PYC over expression in an engineered yeast strain. *Filled squares* SA, *empty squares* ethanol, *circles* glycerol, *pluses* malate, *crosses* 2,3-butanediol.

Indeed, overexpression of SpMAE1 along with PYC and MDH in *S. cerevisiae* leads to malate (~60 g/L) and succinate (8 g/L) accumulation [24]. Overexpression of an *A. niger* homologue of SpMAE1 (coded DCT02) in *S. cerevisiae*, along with PYC and the complete rTCA cycle, leads to significantly more SA, as compared to similar strains without a functional transporter (Fig. 3.5). In addition, the *A. niger* DCT02 transporter fused with green fluorescent protein (GFP) is shown to be localized to the cell membrane (Fig. 3.5).

Besides heterologous transporters, a sodium-dicarboxylate transporter has been characterized for succinate uptake in a wild-type *S. cerevisiae* [53]. However, the encoding gene has not been identified so far. This putative transporter seems to

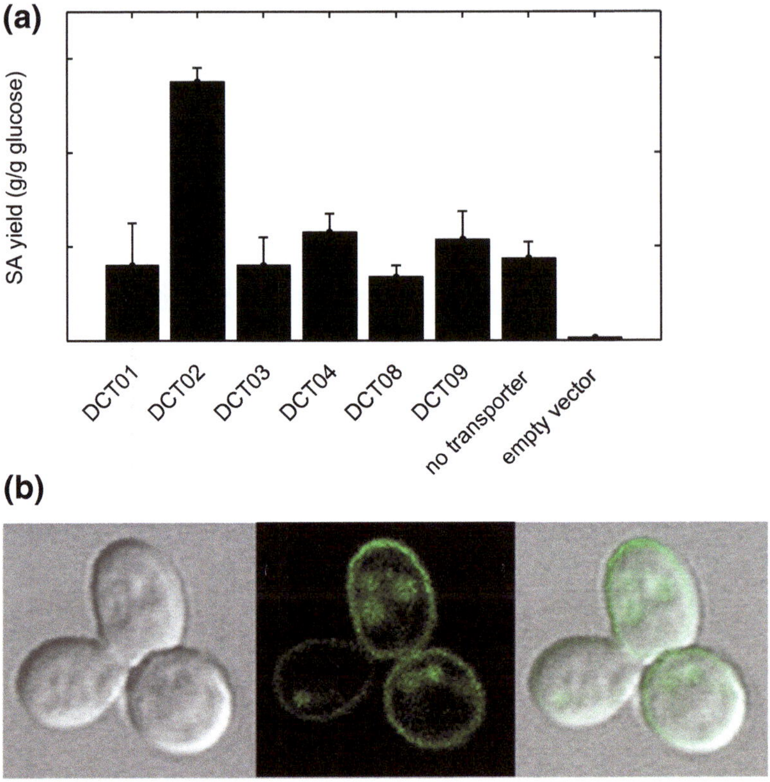

Fig. 3.5 **a** Test of putative dicarboxylate transporters in combination with the rTCA cycle. **b** Localization of DCT02-GFP fusion protein

Table 3.3 Possible biological transport mechanisms for HSA^-

Mechanism	Stoichiometry	Driving force	ATP/SA
A	HSA^- [in] + H^+ [in] $\leftrightarrow$ HSA^- [ex] + H^+ [ex]	$-\Delta S + Z\,\Delta pH$	0
B[a]	HSA^- [in] $\leftrightarrow$ HSA^- [ex]	$-\Delta S + \Delta\Psi$	1
C[a]	HSA^- [in] + ATP [in] $\leftrightarrow$ HSA^- [ex] + ADP [in] + Pi [in]	$-\Delta S + \Delta\Psi + \Delta G_{ATP}/F$	2
D	HSA^- [in] + H^+ [ex] $\leftrightarrow$ HSA^- [ex] + H^+[in]	$-\Delta S + 2\Delta\Psi - Z\,\Delta pH$	2

ΔS is the chemical potential difference of the transported solute across the membrane, calculated as $Z \cdot \log_{10}([HSA^-_{in}]/[HSA^-_{out}])$, where Z is ln10·R·T/F and F is the Faraday constant. $\Delta\Psi$ is electrical potential difference across the membrane potential (Ψin − Ψout). ΔpH is the pH gradient across the membrane ($pH_{in} - pH_{out}$). ΔG_{ATP} is the free-energy of ATP hydrolysis (J/mol)

[a]Additional ATP needed to export cations, e.g., H^+, to maintain electroneutrality. Assumed ATP/H^+ stoichiometry is 1

allow low level malate and succinate export in strains where (part of) the rTCA cycle is overexpressed [24].

To allow succinate export under process conditions (low pH, high extracellular concentration of undissociated SA), the carrier protein needs to have a transport mechanism with a downhill thermodynamic driving force (characterized by a negative ΔG). Different transport mechanisms exist for weak acids, as reviewed in [54]. Table 3.3 summarizes for example the possible biological transport mechanisms of HSA^-. Jamalzadeh et al. [55] concluded after a theoretical evaluation that the thermodynamically feasible succinate export mechanism is proton:succinic antiport (mechanism D in Table 3.3). However, such a transport mechanism has never been reported for dicarboxylates. A proton symport (mechanism A), as reported for SpMAE1, is thermodynamically infeasible according to Jamalzadeh et al. [55] under process conditions. This indicates a profound knowledge gap of the actual transport mechanism involved, as well as the thermodynamic determinants of the dicarboxylate transport process.

At low pH, the majority of the produced SA outside the cell is undissociated and can in principle diffuse back into the cell. Once inside the cell at a cytosolic pH between 6 and 7, SA will dissociate, leading to intracellular acidification. To maintain intracellular pH homeostasis, it is necessary to pump the protons out of the cells again, either by the introduced proton:dicarboxylate symporter if this is thermodynamically feasible, or at the expense of ATP by the plasma membrane H^+-ATPase. The rate of diffusion depends on the extracellular concentration of undissociated acid, as well as the membrane permeability of SA. The latter has shown to be highly pH dependent [50]. This so-called weak acid uncoupling effect has two consequences for the low pH process. First, the product yield on substrate will decrease as more substrate needs to be diverted for ATP generation; second, a complete anaerobic process will likely be infeasible due to limited ATP generation, hence supply of O_2 might be required [56].

3.2.2.6 Gene Selection and Overexpression

It is essential to ensure sufficient expression of pathway enzymes in the host. This can be achieved by proper design of expression constructs, e.g., choice of promoters and terminators, as reviewed in [57], and optimized codon usage [58], combined with careful selection of the genomic integration locus [59]. In addition, protein variants subject to post-translational inactivation in yeast need to be avoided or modified such that they remain active. For example, the first 12 N-terminal amino acids in the *S. cerevisiae* MDH2 protein needs to be deleted to make the protein less-susceptible to glucose-induced degradation [60]. The yeast endogenous isocitratelyase is degraded upon glucose addition to derepressed cells [61]. Instead, the *K. lactis* isocitrate lyase can be used as it is not subject to catabolite inactivation in *S. cerevisiae* [62]. Figure 3.6 shows that with proper design of the expression constructs and protein variants, all proteins in the rTCA cycle are stably expressed in *S. cerevisiae* during a fed-batch fermentation.

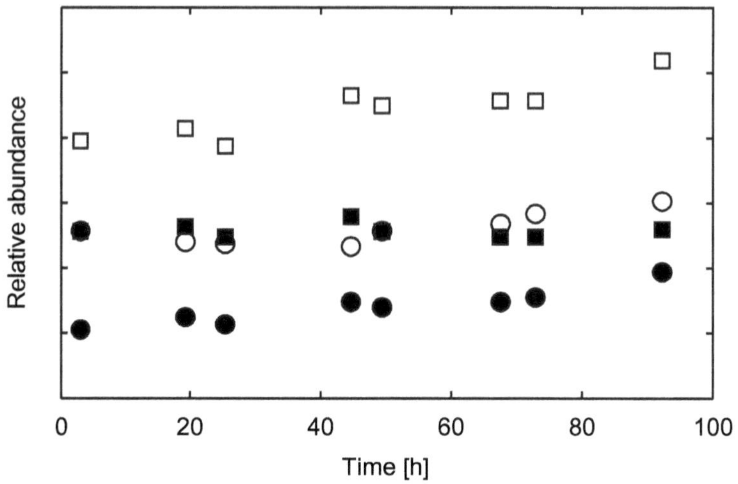

Fig. 3.6 Abundance of rTCA cycle proteins during a lab-scale fed-batch fermentation. *Filled squares* PYC, *empty squares* MDH, *filled circles* fumarase, *empty circles* FRD

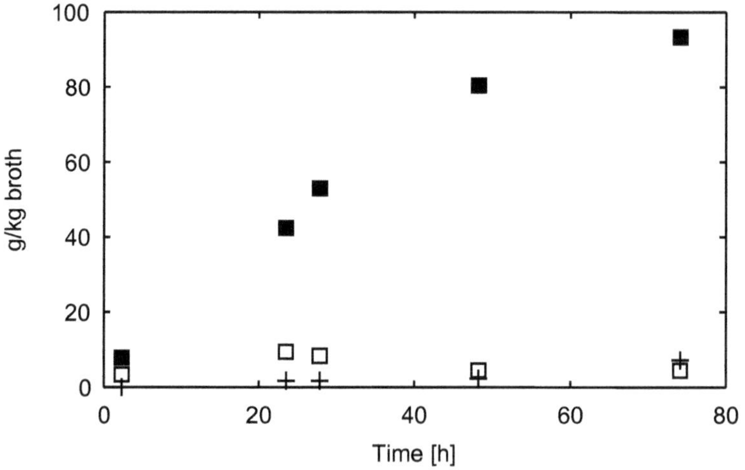

Fig. 3.7 Lab-scale fed-batch production of succinic acid at low pH by an engineered yeast strain at DSM. *Filled squares* SA, *empty squares* ethanol, *pluses* malate

3.2.2.7 Byproduct Removal

Removal of byproducts, e.g., acetate and lactate, has been a successful strategy in engineering prokaryotes for SA production [17, 18]. In yeast, ethanol and glycerol formation have been reduced by targeted knockouts [19]. Ethanol formation can be completely prevented in a PDC-negative, C2-independent, and glucose tolerant strain [63, 64]. However, a pyruvate decarboxylase (PDC) negative strain equipped with the rTCA cycle still produced large amount of pyruvate [44, 65].

High-level SA production in yeast can be achieved by combining one or more of the above-mentioned metabolic engineering strategies. Figure 3.7 shows that a further optimized yeast strain is able to produce up to 96 g/kg of SA in a fed-batch low pH fermentation running for 3 days.

3.3 Adipic Acid and Caprolactam

Both adipic acid (AA) and caprolactam (CL) are non-natural compounds, i.e., these do not occur in known natural biosynthetic pathways. The challenge for a fermentation process starts therefore with identifying enzymes that can be assembled into a non-natural metabolic pathway toward the target compounds. This can be done by the so-called 'retrobiosynthesis' approach, i.e., 'walk' backwards from the target molecules through the known chemical transformation rules to identify potential precursors in the cellular metabolism [66]. Pioneered by the BNICE framework [67] and the UM-BBD database [68], several computational tools have been developed to enumerate all possible pathways according to a predefined set of biochemical transformation rules, and to reduce the vast amount of putative pathways by incorporating thermodynamic and yield constraints [66].

3.3.1 Synthetic Pathways Toward Adipic Acid and Caprolactam

Figure 3.8 summarizes all reported biosynthetic routes toward AA, which have been experimentally demonstrated, or comprise of individual steps with sufficient biochemical evidence. The conversion sequence of each route is detailed in Table 3.4 and will be discussed in detail below.

3.3.1.1 Reverse Beta-Oxidation Pathway

We took a retrobiosynthesis approach to identify non-natural pathways for AA biosynthesis and started with the proposed AA degradation pathway. In bacteria and fungi, AA can be degraded to acetyl-CoA (AcCoA) and succinyl-CoA (SucCoA) (Fig. 3.8) via beta-oxidation [69–71]. Reverse beta-oxidation pathway should therefore start with CoA-esters which are then condensed and transformed to the corresponding fatty acid. Examples of such pathways in nature include, for example, butyrate and butanol biosynthesis in *Clostridium acetobutylicum*,

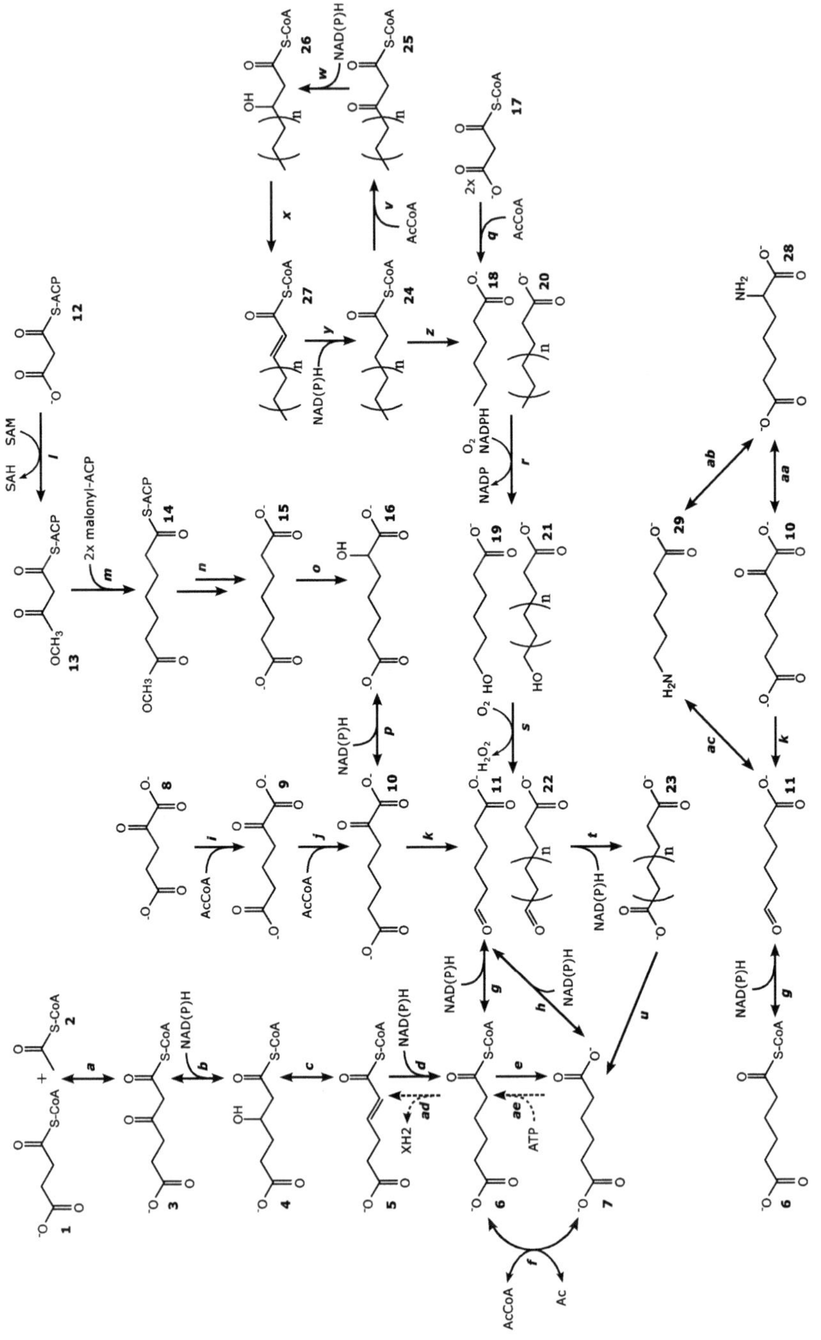

◀ **Fig. 3.8** Synthetic pathways toward adipic acid. Metabolites: 1. succinyl-CoA, 2. acetyl-CoA, 3. 3-oxoadipyl-CoA, 4. 3-hydroxyadipyl-CoA, 5. 2,3-dehydroadipyl-CoA, 6. adipyl-CoA, 7. adipate, 8. α-ketoglutarate, 9. α-ketoadipate, 10. α-ketopimelate (AKP), 11. adipate semialdehdye, 12. malonyl-ACP, 13. malonyl-ACP methyl-ester, 14. Me-pimeloyl-ACP, 15. pimelate (or pimeloyl-CoA/ACP), 16. 2-hydroxypimelate (or 2-hydroxypimeloyl-CoA/ACP), 17. malonyl-CoA, 18. hexanoate, 19. 6-hydroxyhexanoate, 20. long-chain fatty acid, 21. ω-hydroxy fatty acid, 22. ω-oxo fatty acid, 23. α,ω-dicarboxylate, 24. acyl-CoA (hexanyl-CoA), 25. 3-oxoacyl-CoA, 26. 3-hydroxyacyl-CoA, 27. *trans*-2,3-enoyl-CoA, 28. 2-aminopimelate, 29. 6-aminocaproate (6-ACA). Enzymes: a. 3-oxoadipyl-CoA thiolase, b. 3-hydroxyadipyl-CoA dehydrogenase, c. 3-hydroxyadipyl-CoA dehydratase, d. 2,3-dehydroadipyl-CoA reductase, e. adipyl-CoA hydrolase, f. acetate/succinate-adipate CoA-transferase, g. adipate semialdehyde dehydrogenase (CoA-acetylating), h. adipate semialdehyde dehydrogenase, i. (*R*)-homocitrate synthase, (*R*)-homocitrate dehydratase, *cis*-homoaconitate hydratase, and *threo*-isohomocitrate dehydrogenase, j. (homo)$_2$citrate synthase, dihomocitrate dehydratase, *cis*-(homo)$_2$aconitate hydratase and *threo*-iso(homo)$_2$citrate dehydrogenase, k. α-ketopimelate decarboxylase, l. malonyl-ACP methyltransferase, m. 2 successive cycles of fatty acid synthesis, n. pimeloyl-ACP methyl-ester esterase, o. pimelate/pimeloyl-CoA 2-hydroxylase, p. 2-hydroxypimelate dehydrogenase, q. hexanoate synthase, r. fatty acid ω-hydroxylase, s. ω-hydroxy fatty acid oxidase, t. ω-oxo fatty acid dehydrogenase, u. consecutive rounds of beta-oxidation, v. thiolase, w. 3-hydroxyacyl-CoA dehydrogenase, x. 3-hydroxyacyl-CoA dehydratase, y. *trans*-enoyl-CoA reductase, z. thioesterase, aa. α-ketopimelate aminotransferase, ab. 2-aminopimelate decarboxylase, ac. adipate semialdehyde aminotransferase, ad. acyl-CoA dehydrogenase (with FAD as electron acceptor), or acyl-CoA oxidase (with O_2 as electron acceptor, forming H_2O_2), ae. acyl-CoA synthetase. Dashed arrows: reactions specific to the adipate degradation pathway.

synthesis of wax-esters by *Euglena gracilis* [72], as well as synthesis of short chain carboxylic acids (valerate, caproate) by *Clostridium kluyveri* [73, 74].

Reverse beta-oxidation pathway for AA biosynthesis has first been disclosed in two patent applications filed independently by DSM and Genomatica [75, 76]. The

Table 3.4 Summary of reaction sequence of the identified adipic acid pathway variants

Product	Pathway variant	Reaction sequence[1]
–	Adipate degradation	ae-ad-c-b-a f-ad-c-b-a
Adipic acid	Reverse beta-oxidation	a-b-c-d-e a-b-c-d-f a-b-c-d-g-h
	α-Ketopimelate pathway via C1-elongation	i-j-k-h
	α-Ketopimelate pathway via pimelate	l-m-n-o-p-k-h
	Hexanoate pathway	q-r-s-h v-w-x-y-z-r-s-h
	From renewable oils	r-s-t-u
	From adipyl-CoA	g-ac
Caprolactam (6-ACA)	From α-ketopimelate	k-ac aa-ab
	From adipate semialdehyde	ac

[1]Biochemical conversions are depicted in Fig. 3.8

Table 3.5 Thermodynamic properties of reaction in the adipic acid degradation pathway [106]

Enzyme	Reaction	Δ*G* (kcal/mol)
3-Oxoadipyl-CoA thiolase	SucCoA + AcCoA ↔ 3-oxoadipyl-CoA + CoA	7.1
3-Hydroxyadipyl-CoA dehydrogenase	3-Oxoadipyl-CoA + NADH + H^+ ↔ 3-hydroxyadipyl-CoA + NAD^+	−1.1
3-Hydroxyadipyl-CoA dehydratase	3-Hydroxyadipyl-CoA ↔ 2,3-didehydroadipyl-CoA + H_2O	2.2

biosynthetic pathway shares several reactions (step a, b and c) with the degradation pathway, which are readily reversible due to their close to zero standard ΔG values (Table 3.5). However, two key steps in beta-oxidation are irreversible under physiological conditions.

The first irreversible step converts adipate to adipyl-CoA (AdCoA) at the expense of ATP (step ae) [77]. For the reverse reaction, three different routes can be employed (e, f, g-h in Fig. 3.8). The first route involves a reversible CoA-transferase implicated in adipate degradation [69], with succinate or acetate as the CoA-group acceptor. The second route makes use of promiscuous CoA-hydrolases (thioesterases) with reported activity on AdCoA [78]. The third route comprises of two enzymatic reactions with adipate semialdehyde as an intermediate. For the first reaction (step g), Turk et al. [79] identified a known succinate semialdehyde dehydrogenase (CoA-acylating) with low activity on AdCoA. A suitable candidate for the second reaction (step h) is NAD(P) H-dependent adipate semialdehyde dehydrogenase, found, e.g., in cyclohexanol degradation pathway [80].

The second irreversible step in adipate degradation concerns the oxidation of AdCoA to *trans*-2,3-dehydroadipyl-CoA with O_2 as terminal electron acceptor (step ad). The reverse reaction, i.e., reduction of the *trans*-enoyl-CoA to AdCoA, has not been reported. Several known biochemical conversions are similar to the reverse reaction, e.g., reduction of crotonyl-CoA to butyryl-CoA in butanol biosynthesis by butyryl-CoA dehydrogenase (bcd), or NADPH-dependent enoyl-CoA reductases (ecd) acting on short/medium chain *trans*-2-enoyl-CoA's [81–84]. Wu et al. [75] screened 16 enzyme candidates and identified 4 ecd's from various sources able to convert *trans*-2,3-dehydroadipyl-CoA to AdCoA.

By overexpressing a selected set of genes in the proposed reverse beta-oxidation pathway, a breakthrough of direct AA biosynthesis from glucose was demonstrated by DSM in both *E. coli* and *S. cerevisiae* [85–87], with a titer of up to a few hundred mg/L in shake flask cultures. Recently, Deng and Mao reported a native occurring reverse beta-oxidation pathway toward AA in the soil bacterium *Thermobifida fusca* B6, likely with a bcd-homologue for *trans*-enoyl-CoA reduction [88]. This indicates opportunities to further optimize the reverse beta-oxidation pathway by exploring natural diversity.

The reaction sequence of the reverse beta-oxidation pathway resembles the KS (ketosynthase), KR (ketoreductase), DH (dehydratase), ER (enoyl reductase) and

TE (thioesterase) modules in a polyketide synthase. Recently, proof-of-principle of in vitro AA production by an engineered PKS enzyme has been delivered [89]. A succinyl- and a malonyl-moiety are condensed in the initial step of the synthetic PKS to form 3-oxoadipyl-ACP, releasing CO_2. This is similar to the initial step in fatty acid biosynthesis and provides a larger thermodynamic driving force as compared to the non-decarboxylative condensation in the reverse beta-oxidation, albeit at the expense of carbon and energy efficiency.

3.3.1.2 α-Ketopimelate Pathway via C1-Elongation

A second route toward AA starts by decarboxylation of α-ketopimelate (AKP) to form adipate semialdehyde, which can be further converted to AA by aldehyde dehydrogenases [90]. The *L. lactis* branched-chain 2-ketoacid decarboxylase (encoded by the KdcA gene) has been identified to be highly active on AKP [79]. AKP can be derived from the biosynthetic pathway of coenzyme B in methanogenic Archaea [91], which consists of three rounds of chain elongation from α-ketoglutarate (C5, AKG) through α-ketoadipate (C6, AKA), AKP (C7) to α-ketosuberate (C8, AKS). To avoid further conversion of AKP to AKS, the NifV gene product from *Azotobacter vinelandii* can be used, which catalyzes selectively the condensation of AcCoA with AKG and AKA, but not AKP [92]. Overexpression of a selected set of genes in the proposed AKP pathway in *E. coli* led to direct production of AA from glucose, up to ~700 mg/L in non-optimized shake flask cultures [79]. Protein engineering of Kdc led to increased specificity toward AKP as compared to other keto-acids in the pathway (e.g., AKA), which allows for further decrease in byproducts and increase in yield [93].

3.3.1.3 α-Ketopimelate pathway via Pimelate/Pimeloyl-CoA

A variation of the AKP pathway has been disclosed in [94], where AKP is derived from 2-hydroxypimelate by a 2-hydroxyacid dehydrogenase. The 2-hydroxypimelate precursor can be derived from pimelate via a putative pimelate 2-hydroxylase (pimelate 2-monooxygenase). Although such enzyme activity has not been identified yet, similar reactions can be catalyzed by, e.g., ER-membrane-bound fatty acid 2-hydroxylases. In addition, phytanoyl-CoA 2-monooxygenase is found to have side activity on a number of straight chain acyl-CoA's with both even and odd chain length [95]. Pimelate or its CoA- or ACP-thioester is an intermediate in the biotin biosynthesis pathway and can be synthesized from malonyl-CoA [96], although other pathway variants may exist [97].

3.3.1.4 Hexanoate Pathway

Verdezyne disclosed in [98] a pathway toward AA via hexanoate, which is derived from two molecules of malonlyl-CoA and one molecule of AcCoA. This conversion is catalyzed by hexanoate synthase, a type I fatty acid synthase identified in *Aspergillus flavus* aflatoxin B1 biosynthesis [99]. Alternatively, hexanoate can be synthesized by reverse beta-oxidation from AcCoA [100]. Subsequent ω-oxidation would convert hexanoate to AA. Although proof-of-principle of the complete pathway in a single microbe has never been shown, all individual steps are biochemically feasible.

3.3.1.5 Adipic Acid from Renewable Oils

Verdezyne has demonstrated AA production from fatty acids by engineered *Candida* yeast [12, 98]. The fatty acids can be derived from vegetable oils, e.g., coconut oil, palm oil, or soapstock from edible oil processing. Fatty acids undergo ω-oxidation to form the corresponding long-chain dicarboxylic acids, which is converted to AA by consecutive rounds of beta-oxidation. To prevent further beta-oxidation of the product, an acyl-CoA oxidase with broad chain-length specificity (4-20 carbons) is deleted, leaving in the host only acyl-oxidase(s) with negligible activity on substrates below 6 carbons. The engineered microbe is able to produce $\sim$50 g/L AA in a fed-batch fermentation, the highest AA titer reported so far [12].

3.3.1.6 Synthetic Pathways Toward 6-ACA

Caprolactam (CL) is the cyclic form of 6-aminocaproic acid (6-ACA), the actual monomeric units in Nylon 6. In the CL degradation pathway, 6-ACA is converted to adipate semialdehyde by an aminotransferase. Thus, all pathways toward AA, where adipate semialdehyde is involved, can be used to produce 6-ACA. Suitable aminotransferases active on adipate semialdehyde have been identified in [79, 90]. Connecting the aminotransferase with the AKP pathway (via C1-elongation) in *E. coli* leads to, for the first time, direct fermentative production of 6-ACA from glucose, up to 160 mg/L in lab-scale batch fermentations [79]. In [90], a slight variation of the above-mentioned pathway has been disclosed, in which AKP is first converted to 2-aminopimelate, followed by decarboxylation to 6-ACA. This irreversible final step might be favorable as compared to the reversible aminotransferase to maintain high flux through the entire pathway against increasing production concentrations.

Table 3.6 Carbon and oxygen yield of different pathways toward AA simulated using an *E. coli* metabolic network

Pathway	Yield (Cmol/Cmol-substrate)	O_2 demand (mol/Cmol-substrate)
Reverse beta-oxidation	0.92	0
α-Ketopimelate via C1-elongation	0.50	0.46
Hexanoate via hexanoate synthase	0.67	0.38
Renewable oils	0.50[a]	0.88
Muconic acid[b]	0.74	0.32

[a]Assume starting from C12:0 fatty acid
[b]Exclude further chemical hydrogenation of muconic acid to produce AA

3.3.2 *Pathway Comparison and Process Considerations*

There are more synthetic pathways toward AA disclosed in patent literature, however, the pathways described above have either been demonstrated in the lab, or have substantial biochemical evidence. Therefore, we will evaluate only these pathways in terms of yield/carbon efficiency. By elemental and degree of reduction balances, a maximal AA ($C_6H_{10}O_4$) yield of 0.92 mol/mol can be achieved on glucose ($C_6H_{12}O_6$) anaerobically:

$$C_6H_{12}O_6 \rightarrow \frac{24}{26}C_6H_{10}O_4 + \frac{12}{26}CO_2 + \frac{36}{26}H_2O$$

Table 3.6 shows that this theoretical maximum can only be achieved by the reverse beta-oxidation pathway.

All feasible pathways investigated so far have led to yield levels still far below what would be required for commercialization. From the calculations above, the reverse beta-oxidation seems most attractive in terms of carbon efficiency. For commercial development, significant metabolic engineering efforts will be required to optimize precursor supply, fine tune expression of each pathway step, as well as identification and overexpression transport proteins to boost product secretion. The precursors AcCoA and SucCoA for the reverse beta-oxidation pathway are mainly generated in the mitochondria in eukaryotes. Thus, if the reverse beta-oxidation pathway is localized in the cytosol in a eukaryote host, supply of AcCoA and SucCoA needs to be optimized. To this end major efforts have been made to improve the cytosolic AcCoA supply in, e.g., *S. cerevisiae* [101–103], as well as to increase the SucCoA supply by overexpressing a succinyl-CoA ligase targeted to the cytosol [85].

A eukaryote host could allow for a low pH AA process, which would be advantageous as compared to a neutral pH process, for the same reason as SA production. The low solubility of AA in water (24 g/L at 25 °C) is favorable for downstream processing of such a low pH process. Strain tolerance to low pH and undissociated AA need to be addressed, both due to the weak acid uncoupling effect, as well as AA toxicity.

Unlike AA and SA, 6-ACA is a non-natural amino acid, which can be potentially produced in a neutral pH process, similar to industrial amino acid fermentations by bacteria (e.g., glutamate, lysine, etc.). *E. coli* showed the best tolerance to 6-ACA as compared to *B. subtilis* and *S. cerevisiae* [79] and might be a suitable host. 6-ACA produced by fermentation needs first to be cyclized to CL before polymerization. This cyclization can be carried out directly using (pretreated) fermentation broth, without the need of prior purification of 6-ACA, as disclosed in [104]. Further purification of CL can be achieved by, e.g., distillation.

3.4 Perspective

Large-scale fermentation process is one of the promising alternatives to produce chemical building blocks in a sustainable way. We have described that careful considerations up-front of process configuration, host choice, metabolic pathways, as well as strain designs, are critical to the success of such an endeavor. Significant progress has been made in developing strain and process technology to produce SA by fermentation. The challenge remains to further establish end user applications toward creating a substantial market, justifying more large-scale operations that can fully benefit from the economies of scale. Fermentative production of AA and CL is still in an early phase, and might require a substantial R&D effort.

Novel engineering biology practices hold the promise of significantly reducing the R&D effort and time needed between proof-of-principle and commercially attractive KPI's. Integrated design-build-test-learn cycles allow the much needed efficient exploration of the multidimensional strain and process optimization space. For example, a combined strain and process optimization driven by design of experiments (DOE) and high-throughput experimentation has been shown for 6-ACA production in *E. coli* [105].

Next to engineering biology practices, fundamental biological knowledge will always be required. For example, (synthetic) pathway enzymes should be characterized by reliable and high-throughput enzyme assays to understand the biochemistry, to diagnose potential bottlenecks, and to support protein engineering efforts leading to improved biocatalysts. Moreover, biological transport mechanisms and the associated thermodynamic driving force need to be better understood, which might give rise to engineered transporters with higher energy efficiency and specificity. If a eukaryote host is used, detailed understanding of metabolite transport and flux distribution across different compartments should be obtained by, e.g., high-resolution ^{13}C-flux analysis and metabolomics.

In the future, these large-scale bioprocesses will likely be integrated into biorefineries, which make use of second generation, non-food/non-feed raw materials. Biochemicals would become part of the integrated biomass value chain, next to other biorefinery products such as biofuels, feed, power, and/or heat. An integral optimization of the value chain calls for careful techno-economic feasibility analysis to benchmark against alternative processes and products. Robust and cost-effective technologies need to be developed for biomass pretreatment as well as downstream product purification from these complex streams. In addition, stress tolerant microorganisms that are able to feed on multiple available substrates need to be engineered.

References

1. Aeschelmann F, Carus M and nova team (2015) Bio-based building blocks and polymers in the world: capacities, production and applications—status quo and trends towards 2020. nova-Institut GmbH, Germany
2. Nakamura CE, Whited GM (2003) Metabolic engineering for the microbial production of 1, 3-propanediol. Curr Opin Biotechnol 14:454–459
3. Leavell MD, McPhee DJ, Paddon CJ (2016) Developing fermentative terpenoid production for commercial usage. Curr Opin Biotechnol 37:114–119
4. Ryan C (2015) Success and challenges in the commercial production of isobutanol and its hydrocarbon derivatives. Cph Biosci Conf Cell Fact Biosustain 7:41–42
5. Whited GM, Feher FJ, Benko DA, Cervin MA, Chotani GK, McAuliffe JC, LaDuca RJ, Ben-Shoshan EA, Sanford KJ (2010) Technology update: development of a gas-phase bioprocess for isoprene-monomer production using metabolic pathway engineering. Ind Biotechnol 6:152–163
6. Yim H, Haselbeck R, Niu W, Pujol-Baxley C, Burgard A, Boldt J, Khandurina J, Trawick JD, Osterhout RE, Stephen R, Estadilla J, Teisan S, Schreyer HB, Andrae S, Yang TH, Lee SY, Burk MJ, van Dien S (2011) Metabolic engineering of *Escherichia coli* for direct production of 1,4-butanediol. Nat Chem Biol 7:445–452
7. Werpy T, Petersen G (2004) Top value added chemicals from biomass. US Department of Energy, Office of Scientific and Technical Information, no. DOE/GO-102004-1992
8. Bozell JJ, Petersen GR (2010) Technology development for the production of biobased products from biorefinery carbohydrates—the US Department of Energy's "top 10" revisited. Green Chem 12:539–554
9. Beerthuis R, Rothenberg G, Shiju NR (2015) Catalytic routes towards acrylic acid, adipic acid and ε-caprolactam starting from biorenewables. Green Chem 17:1341–1361
10. Vink ET, Davies S (2015) Life cycle inventory and impact assessment data for 2014 Ingeo™ polylactide production. Ind Biotechnol 11:167–180
11. Draths KM, Frost JW (1994) Environmentally compatible synthesis of adipic acid from D-glucose. J Am Chem Soc 116:399–400
12. Beardslee T, Picataggio S (2012) Bio-based adipic acid from renewable oils. Lipid Technol 24:223–225
13. van Dien S (2013) From the first drop to the first truckload: commercialization of microbial processes for renewable chemicals. Curr Opin Biotechnol 24:1061–1068
14. Burk MJ, van Dien S (2016) Biotechnology for chemical production: challenges and opportunities. Trends Biotechnol 34:187–190
15. Beauprez JJ, De Mey M, Soetaert WK (2010) Microbial succinic acid production: natural versus metabolic engineered producers. Process Biochem 45:1103–1114
16. Scholten E, Renz T, Thomas J (2009) Continuous cultivation approach for fermentative succinic acid production from crude glycerol by *Basfia succiniciproducens* DD1. Biotechnol Lett 31:1947–1951

17. Thakker C, Martínez I, San K, Bennett GN (2012) Succinate production in *Escherichia coli*. Biotechnol J 7:213–224
18. Tsuge Y, Hasunuma T, Kondo A (2015) Recent advances in the metabolic engineering of *Corynebacterium glutamicum* for the production of lactate and succinate from renewable resources. J Ind Microbiol Biotechnol 42:375–389
19. Verwaal R, Wu L, Damveld RA, Sagt CMJ (2009) Succinic acid production in a eukaryotic cell. WO2009065778
20. Yuzbashev TV, Yuzbasheva EY, Sobolevskaya TI, Laptev IA, Vybornaya TV, Larina AS, Matsui K, Fukui K, Sineoky SP (2010) Production of succinic acid at low pH by a recombinant strain of the aerobic yeast *Yarrowia lipolytica*. Biotechnol Bioeng 107:673–682
21. Xiao H, Shao Z, Jiang Y, Dole S, Zhao H (2014) Exploiting *Issatchenkia orientalis* SD108 for succinic acid production. Microb Cell Fact 13:121
22. Jansen ML, van Gulik WM (2014) Towards large scale fermentative production of succinic acid. Curr Opin Biotechnol 30:190–197
23. Cok B, Tsiropoulos I, Roes AL, Patel MK (2014) Succinic acid production derived from carbohydrates: an energy and greenhouse gas assessment of a platform chemical toward a bio-based economy. Biofuels Bioprod Biorefin 8:16–29
24. Zelle RM, de Hulster E, van Winden WA, de Waard P, Dijkema C, Winkler AA, Geertman JM, van Dijken JP, Pronk JT, van Maris AJ (2008) Malic acid production by *Saccharomyces cerevisiae*: engineering of pyruvate carboxylation, oxaloacetate reduction, and malate export. Appl Environ Microbiol 74:2766–2777
25. Sánchez AM, Bennett GN, San K (2005) Novel pathway engineering design of the anaerobic central metabolic pathway in *Escherichia coli* to increase succinate yield and productivity. Metab Eng 7:229–239
26. Skorokhodova AY, Morzhakova AA, Gulevich AY, Debabov VG (2015) Manipulating pyruvate to acetyl-CoA conversion in *Escherichia coli* for anaerobic succinate biosynthesis from glucose with the yield close to the stoichiometric maximum. J Biotechnol 214:33–42
27. Jansen MLA, Heijnen JJ, Verwaal R (2012) Process for preparing dicarboxylic acids employing fungal cells. WO2013004670
28. Raab AM, Gebhardt G, Bolotina N, Weuster-Botz D, Lang C (2010) Metabolic engineering of *Saccharomyces cerevisiae* for the biotechnological production of succinic acid. Metab Eng 12:518–525
29. Ito Y, Hirasawa T, Shimizu H (2014) Metabolic engineering of *Saccharomyces cerevisiae* to improve succinic acid production based on metabolic profiling. Biosci Biotechnol Biochem 78:151–159
30. Otero JM, Cimini D, Patil KR, Poulsen SG, Olsson L, Nielsen J (2013) Industrial systems biology of *Saccharomyces cerevisiae* enables novel succinic acid cell factory. PLoS ONE 8:54144
31. Verwaal R, Wu L, Damveld RA, Sagt CMJ (2009) Dicarboxylic acid production in a recombinant yeast. WO2009065779
32. Yang L, Lübeck M, Ahring BK, Lübeck PS (2016) Enhanced succinic acid production in *Aspergillus saccharolyticus* by heterologous expression of fumarate reductase from *Trypanosoma brucei*. Appl Microbiol Biotechnol 100:1799–1809
33. Goward CR, Nicholls DJ (1994) Malate dehydrogenase: a model for structure, evolution, and catalysis. Protein Sci 3:1883–1888
34. Tielens AG, Van Hellemond JJ (1998) The electron transport chain in anaerobically functioning eukaryotes. Biochim Biophys Acta 1365:71–78
35. Camarasa C, Faucet V, Dequin S (2007) Role in anaerobiosis of the isoenzymes for *Saccharomyces cerevisiae* fumarate reductase encoded by OSM1 and FRDS1. Yeast 24:391–401
36. Liu Z, Osterlund T, Hou J, Petranovic D, Nielsen J (2013) Anaerobic alpha-amylase production and secretion with fumarate as the final electron acceptor in *Saccharomyces cerevisiae*. Appl Environ Microbiol 79:2962–2967

37. Bergdahl B, Gorwa-Grauslund MF, van Niel EW (2014) Physiological effects of over-expressing compartment-specific components of the protein folding machinery in xylose-fermenting *Saccharomyces cerevisiae*. BMC Biotechnol 14: 28-6750-14-28
38. Rezaei MN, Aslankoohi E, Verstrepen KJ, Courtin CM (2015) Contribution of the tricarboxylic acid (TCA) cycle and the glyoxylate shunt in *Saccharomyces cerevisiae* to succinic acid production during dough fermentation. Int J Food Microbiol 204:24–32
39. Besteiro S, Biran M, Biteau N, Coustou V, Baltz T, Canioni P, Bringaud F (2002) Succinate secreted by *Trypanosoma brucei* is produced by a novel and unique glycosomal enzyme, NADH-dependent fumarate reductase. J Biol Chem 277:38001–38012
40. Coustou V, Besteiro S, Riviere L, Biran M, Biteau N, Franconi JM, Boshart M, Baltz T, Bringaud F (2005) A mitochondrial NADH-dependent fumarate reductase involved in the production of succinate excreted by procyclic *Trypanosoma brucei*. J Biol Chem 280:16559–16570
41. Zelle RM, Trueheart J, Harrison JC, Pronk JT, van Maris AJ (2010) Phosphoenolpyruvate carboxykinase as the sole anaplerotic enzyme in *Saccharomyces cerevisiae*. Appl Environ Microbiol 76:5383–5389
42. Zelle RM, Harrison JC, Pronk JT, van Maris AJ (2011) Anaplerotic role for cytosolic malic enzyme in engineered *Saccharomyces cerevisiae* strains. Appl Environ Microbiol 77:732–738
43. Canelas AB, van Gulik WM, Heijnen JJ (2008) Determination of the cytosolic free NAD/NADH ratio in *Saccharomyces cerevisiae* under steady-state and highly dynamic conditions. Biotechnol Bioeng 100:734–743
44. Zelle RM, de Hulster E, Kloezen W, Pronk JT, van Maris AJ (2010) Key process conditions for production of C(4) dicarboxylic acids in bioreactor batch cultures of an engineered *Saccharomyces cerevisiae* strain. Appl Environ Microbiol 76:744–750
45. Wang D, Li Q, Li W, Xing J, Su Z (2009) Improvement of succinate production by overexpression of a cyanobacterial carbonic anhydrase in *Escherichia coli*. Enzyme Microb Technol 45:491–497
46. Winkler AA, de Hulster AF, van Dijken JP, Pronk JT, Trueheart J, Madden KT, Rodriguez CG, Mauriz CF, van Maris AJA, Harrison JC (2009) Organic acid production by fungal cells. WO2009011974
47. McFarland S, Brown S, Luttringer S (2013) Recombinant microorganisms for production C4-dicarboxylic acids. WO2013028512
48. Sousa MJ, Mota M, Leão C (1992) Transport of malic acid in the yeast *Schizosaccharomyces pombe*: evidence for proton-dicarboxylate symport. Yeast 8:1025–1031
49. Grobler J, Bauer F, Subden RE, van Vuuren HJ (1995) The *mae1* gene of *Schizosaccharomyces pombe* encodes a permease for malate and other C4 dicarboxylic acids. Yeast 11:1485–1491
50. Corte-Real M, Leao C (1990) Transport of malic acid and other dicarboxylic acids in the yeast *Hansenula anomala*. Appl Environ Microbiol 56:1109–1113
51. Queiros O, Casal M, Althoff S, Moradas-Ferreira P, Leao C (1998) Isolation and characterization of *Kluyveromyces marxianus* mutants deficient in malate transport. Yeast 14:401–407
52. Camarasa C, Bidard F, Bony M, Barre P, Dequin S (2001) Characterization of *Schizosaccharomyces pombe* malate permease by expression in *Saccharomyces cerevisiae*. Appl Environ Microbiol 67:4144–4151
53. Aliverdieva D, Mamaev D, Bondarenko D, Sholtz K (2006) Properties of yeast *Saccharomyces cerevisiae* plasma membrane dicarboxylate transporter. Biochemistry (Moscow) 71:1161–1169
54. van Maris AJ, Konings WN, van Dijken JP, Pronk JT (2004) Microbial export of lactic and 3-hydroxypropanoic acid: implications for industrial fermentation processes. Metab Eng 6:245–255

55. Jamalzadeh E, Taymaz-Nikerel H, Heijnen J, van Gulik W, Verheijen P (2012) A thermodynamic analysis of dicarboxylic acid production in microorganisms. In: von Stockar U, van der Wielen LAM, Prausnitz JM (eds) Thermodynamics in biochemical engineering. EPFL Press, Lausanne
56. Jansen MLA, Verwaal R (2010) Dicarboxylic acid production by fermentation at low pH. WO2010003728
57. Redden H, Morse N, Alper HS (2015) The synthetic biology toolbox for tuning gene expression in yeast. FEMS Yeast Res 15:1–10
58. Lanza AM, Curran KA, Rey LG, Alper HS (2014) A condition-specific codon optimization approach for improved heterologous gene expression in *Saccharomyces cerevisiae*. BMC Syst Biol 8:33
59. Bai Flagfeldt D, Siewers V, Huang L, Nielsen J (2009) Characterization of chromosomal integration sites for heterologous gene expression in *Saccharomyces cerevisiae*. Yeast 26:545–551
60. Minard KI, McAlister-Henn L (1992) Glucose-induced degradation of the MDH2 isozyme of malate dehydrogenase in yeast. J Biol Chem 267:17458–17464
61. Ordiz I, Herrero P, Rodicio R, Moreno F (1996) Glucose-induced inactivation of isocitrate lyase in *Saccharomyces cerevisiae* is mediated by the cAMP-dependent protein kinase catalytic subunits Tpk 1 and Tpk2. FEBS Lett 385:43–46
62. Lopez ML, Redruello B, Valdes E, Moreno F, Heinisch JJ, Rodicio R (2004) Isocitrate lyase of the yeast *Kluyveromyces lactis* is subject to glucose repression but not to catabolite inactivation. Curr Genet 44:305–316
63. van Maris AJ, Geertman JM, Vermeulen A, Groothuizen MK, Winkler AA, Piper MD, van Dijken JP, Pronk JT (2004) Directed evolution of pyruvate decarboxylase-negative *Saccharomyces cerevisiae*, yielding a C2-independent, glucose-tolerant, and pyruvate-hyperproducing yeast. Appl Environ Microbiol 70:159–166
64. Oud B, Flores C, Gancedo C, Zhang X, Trueheart J, Daran J, Pronk JT, van Maris A (2012) An internal deletion in MTH1 enables growth on glucose of pyruvate-decarboxylase negative, non-fermentative *Saccharomyces cerevisiae*. Microb Cell Fact 11:131
65. Yan D, Wang C, Zhou J, Liu Y, Yang M, Xing J (2014) Construction of reductive pathway in *Saccharomyces cerevisiae* for effective succinic acid fermentation at low pH value. Bioresour Technol 156:232–239
66. Hadadi N, Hatzimanikatis V (2015) Design of computational retrobiosynthesis tools for the design of *de novo* synthetic pathways. Curr Opin Chem Biol 28:99–104
67. Hatzimanikatis V, Li C, Ionita JA, Henry CS, Jankowski MD, Broadbelt LJ (2005) Exploring the diversity of complex metabolic networks. Bioinformatics 21:1603–1609
68. Ellis LB, Roe D, Wackett LP (2006) The University of Minnesota Biocatalysis/Biodegradation Database: the first decade. Nucleic Acids Res 34:D517–D521
69. Parke D, Garcia MA, Ornston LN (2001) Cloning and genetic characterization of *dca* genes required for beta-oxidation of straight-chain dicarboxylic acids in *Acinetobacter* sp. strain ADP1. Appl Environ Microbiol 67:4817–4827
70. Thykaer J, Christensen B, Nielsen J (2002) Metabolic network analysis of an adipoyl-7-ADCA-producing strain of *Penicillium chrysogenum*: elucidation of adipate degradation. Metab Eng 4:151–158
71. Harris DM, Westerlaken I, Schipper D, van der Krogt ZA, Gombert AK, Sutherland J, Raamsdonk LM, van den Berg MA, Bovenberg RA, Pronk JT (2009) Engineering of *Penicillium chrysogenum* for fermentative production of a novel carbamoylated cephem antibiotic precursor. Metab Eng 11:125–137
72. Inui H, Miyatake K, Nakano Y, Kitaoka S (1985) The physiological role of oxygen-sensitive pyruvate dehydrogenase in mitochondrial fatty acid synthesis in *Euglena gracilis*. Arch Biochem Biophys 237:423–429
73. Seedorf H, Fricke WF, Veith B, Bruggemann H, Liesegang H, Strittmatter A, Miethke M, Buckel W, Hinderberger J, Li F, Hagemeier C, Thauer RK, Gottschalk G (2008) The

genome of *Clostridium kluyveri*, a strict anaerobe with unique metabolic features. Proc Natl Acad Sci USA 105:2128–2133
74. Smith GM, Kim BW, Franke AA, Roberts JD (1985) ^{13}C NMR studies of butyric fermentation in *Clostridium kluyveri*. J Biol Chem 260:13509–13512
75. Wu L, Trefzer AC, de Wildeman SM, van den Berg MA (2009) Adipate (ester or thioester) synthesis. WO2009113853
76. Burgard AP, Pharkya P, Osterhout RE (2009) Microorganisms for the production of adipic acid and other compounds. WO2009151728
77. Koetsier MJ, Gombert AK, Fekken S, Bovenberg RA, van den Berg MA, Kiel JA, Jekel PA, Janssen DB, Pronk JT, van der Klei IJ (2010) The *Penicillium chrysogenum aclA* gene encodes a broad-substrate-specificity acyl-coenzyme A ligase involved in activation of adipic acid, a side-chain precursor for cephem antibiotics. Fungal Genet Biol 47:33–42
78. Westin MA, Hunt MC, Alexson SE (2005) The identification of a succinyl-CoA thioesterase suggests a novel pathway for succinate production in peroxisomes. J Biol Chem 280:38125–38132
79. Turk SC, Kloosterman WP, Ninaber DK, Kolen KP, Knutova J, Suir E, Schürmann M, Raemakers-Franken PC, Müller M, de Wildeman SM, Raamsdonk LM, van der Pol R, Wu L, Temudo MF, van der Hoeven RA, Akeroyd M, van der Stoel RE, Noorman HJ, Bovenberg RA, Trefzer AC (2016) Metabolic engineering toward sustainable production of Nylon-6. ACS Synth Biol 5:65–73
80. Iwaki H, Hasegawa Y, Teraoka M, Tokuyama T, Bergeron H, Lau PCK (1999) Identification of a transcriptional activator (ChnR) and a 6-oxohexanoate dehydrogenase (ChnE) in the cyclohexanol catabolic pathway in *Acinetobacter* sp. strain NCIMB 9871 and localization of the genes that encode them. Appl Environ Microbiol 65:5158–5162
81. Torkko JM, Koivuranta KT, Miinalainen IJ, Yagi AI, Schmitz W, Kastaniotis AJ, Airenne TT, Gurvitz A, Hiltunen KJ (2001) *Candida tropicalis* Etr1p and *Saccharomyces cerevisiae* Ybr026p (Mrf1'p), 2-enoyl thioester reductases essential for mitochondrial respiratory competence. Mol Cell Biol 21:6243–6253
82. Torkko JM, Koivuranta KT, Kastaniotis AJ, Airenne TT, Glumoff T, Ilves M, Hartig A, Gurvitz A, Hiltunen JK (2003) *Candida tropicalis* expresses two mitochondrial 2-enoyl thioester reductases that are able to form both homodimers and heterodimers. J Biol Chem 278:41213–41220
83. Miinalainen IJ, Chen ZJ, Torkko JM, Pirila PL, Sormunen RT, Bergmann U, Qin YM, Hiltunen JK (2003) Characterization of 2-enoyl thioester reductase from mammals. An ortholog of YBR026p/MRF1'p of the yeast mitochondrial fatty acid synthesis type II. J Biol Chem 278:20154–20161
84. Hoffmeister M, Piotrowski M, Nowitzki U, Martin W (2005) Mitochondrial *trans*-2-enoyl-CoA reductase of wax ester fermentation from *Euglena gracilis* defines a new family of enzymes involved in lipid synthesis. J Biol Chem 280:4329–4338
85. den Dulk B, Hans M, Wu L (2012) Process to increase the production of a succinyl-CoA derived compound. WO2012089613
86. Wu L (2012) Biobased production of polyamide 6 and polyamide 6,6 monomers. In: Abstracts of Metabolic Engineering IX, Biarritz, France, 3–7 June, 2012
87. Wu L (2013) Sustainable production of dicarboxylic acids by fermentation. In: Abstracts of Biochemical and Molecular Engineering XVIII, Beijing, China, 16–20 June, 2013
88. Deng Y, Mao Y (2015) Production of adipic acid by the native-occurring pathway in *Thermobifida fusca* B6. J Appl Microbiol 119:1057–1063
89. Hagen A, Poust S, Rond Td, Fortman JL, Katz L, Petzold CJ, Keasling JD (2016) Engineering a polyketide synthase for in vitro production of adipic acid. ACS Synth Biol 5:21–27
90. Raemakers-Franken PC, Schürmann M, Trefzer AC, de Wildeman SMA (2009) Preparation of 6-aminocaproic acid from 5-formyl valeric acid. WO2009113855

91. Howell DM, Harich K, Xu H, White RH (1998) α-Keto acid chain elongation reactions involved in the biosynthesis of coenzyme B (7-mercaptoheptanoyl threonine phosphate) in methanogenic Archaea. Biochemistry 37:10108–10117
92. Zheng L, White RH, Dean DR (1997) Purification of the *Azotobacter vinelandii nifV*-encoded homocitrate synthase. J Bacteriol 179:5963–5966
93. Jacobs DI, van der Laan JM, de Lange I, Trefzer AC, Turk SCHJT (2012) Preparation of 5-formyl valeric acid from alpha-ketopimelic acid. WO2012131059
94. Raemakers-Franken PC, Schürmann M, Trefzer AC, de Wildeman SMA (2011) Preparation of alpha-ketopimelic acid. WO2011031146
95. Mukherji M, Kershaw NJ, Schofield CJ, Wierzbicki AS, Lloyd MD (2002) Utilization of sterol carrier protein-2 by phytanoyl-CoA 2-hydroxylase in the peroxisomal α-oxidation of phytanic acid. Chem Biol 9:597–605
96. Agarwal V, Lin S, Lukk T, Nair SK, Cronan JE (2012) Structure of the enzyme-acyl carrier protein (ACP) substrate gatekeeper complex required for biotin synthesis. Proc Natl Acad Sci USA 109:17406–17411
97. Lin S, Cronan JE (2011) Closing in on complete pathways of biotin biosynthesis. Mol Bio Syst 7:1811–1821
98. Picataggio S, Beardslee T (2011) Biological methods for preparing adipic acid. WO2011003034
99. Hitchman TS, Schmidt EW, Trail F, Rarick MD, Linz JE, Townsend CA (2001) Hexanoate synthase, a specialized type I fatty acid synthase in aflatoxin B1 biosynthesis. Bioorg Chem 29:293–307
100. Clomburg JM, Blankschien MD, Vick JE, Chou A, Kim S, Gonzalez R (2015) Integrated engineering of β-oxidation reversal and ω-oxidation pathways for the synthesis of medium chain ω-functionalized carboxylic acids. Metab Eng 28:202–212
101. Kozak BU, van Rossum HM, Luttik MA, Akeroyd M, Benjamin KR, Wu L, de Vries S, Daran JM, Pronk JT, van Maris AJ (2014) Engineering acetyl coenzyme A supply: functional expression of a bacterial pyruvate dehydrogenase complex in the cytosol of *Saccharomyces cerevisiae*. MBio 5:e01696–14
102. Kozak BU, van Rossum HM, Benjamin KR, Wu L, Daran JG, Pronk JT, van Maris AJ (2014) Replacement of the *Saccharomyces cerevisiae* acetyl-CoA synthetases by alternative pathways for cytosolic acetyl-CoA synthesis. Metab Eng 21:46–59
103. Krivoruchko A, Zhang Y, Siewers V, Chen Y, Nielsen J (2015) Microbial acetyl-CoA metabolism and metabolic engineering. Metab Eng 28:28–42
104. Guit RPM, van der Does T, Raamsdonk LM (2011) Preparation of caprolactam from 6-amino caproic acid obtained in a fermentation process. WO2011078668
105. Zhou H, Vonk B, Roubos JA, Bovenberg RA, Voigt CA (2015) Algorithmic co-optimization of genetic constructs and growth conditions: application to 6-ACA, a potential nylon-6 precursor. Nucleic Acids Res 43:10560–10570
106. Caspi R, Altman T, Billington R, Dreher K, Foerster H, Fulcher CA, Holland TA, Keseler IM, Kothari A, Kubo A, Krummenacker M, Latendresse M, Mueller LA, Ong Q, Paley S, Subhraveti P, Weaver DS, Weerasinghe D, Zhang P, Karp PD (2014) The MetaCyc database of metabolic pathways and enzymes and the BioCyc collection of Pathway/Genome Databases. Nucleic Acids Res 42:D459–D471

Chapter 4
Specialty Enzymes for Chemical Needs

Dunming Zhu and Ling Hua

Abstract Chemical processes are vital for the manufacturing of goods that meet the human's growing needs; on the other hand, they have resulted in increasing air pollution and environmental contamination. It is desirable to develop green chemical processes for the sustainable development of chemical industry. In this context, industrial biotechnology, which deciphers the secrets of nature's engineering and redesigns the biological systems for exploitation in the industrial manufacturing, is becoming an exciting frontier in modern science and technology that ensures sustainable economic development in a world facing increasing environmental challenges and resource scarcity. The core of industrial biotechnology is enzyme catalysis, which possesses several advantages over traditional chemical reactions, such as high chemo-, regio- and stereoselectivity, and mild reaction conditions. As such, enzymes catalyze some reactions which are difficult to be achieved by traditional chemical reactions. Enzyme catalysis can reduce reaction steps by eliminating the protection and de-protection steps or redesign the synthetic route. In this chapter, we first discuss the unique features of enzyme catalysis compared to traditional chemical reactions. This is followed by several examples of enzyme application in the production of important chemicals to show their positive impacts in reducing chemical waste, energy consumption and production cost, thus contributing to cleaner environment, industrial sustainability, and quality living.

Keywords Enzyme catalysis · Enzyme application in chemical industry · Multienzymatic reaction · Chemoenzymatic cascade · Green chemical process · Sustainable synthetic technology

D. Zhu (✉)
National Engineering Laboratory for Industrial Enzymes
and Tianjin Engineering Research Center of Biocatalytic Technology,
Tianjin Institute of Industrial Biotechnology, Chinese Academy of Sciences,
Tianjin 300308, China
e-mail: zhu_dm@tib.cas.cn

L. Hua
DuPont Industrial Biosciences, Experimental Station, Wilmington, DE 19880, USA

P.C.K. Lau (ed.), *Quality Living Through Chemurgy and Green Chemistry*,
Green Chemistry and Sustainable Technology,
DOI 10.1007/978-3-662-53704-6_4

4.1 Advantages of Enzyme Catalysis

4.1.1 Chemoselectivity

Enzyme catalysis is usually highly chemospecific, thus the specific transformation of some functional groups can be realized. For example, in the traditional chemical hydrolysis of nitrile group to carboxylic acid that must be performed under strong basic or acidic conditions at elevated temperature, it is almost impossible to realize chemoselective hydrolysis of nitrile group in the presence of reactive functional groups such as ester in the same molecule without hydrolyzing it. On the other hand, nitrilases catalyze the hydrolysis of the nitriles to give the corresponding carboxylic acids under neutral conditions, and this is particularly useful for the transformation of compounds containing other acid- or base-sensitive functional groups [1–3]. The hydrolysis of ethyl (*R*)-4-cyano-3-hydroxybutyate (Scheme 4.1) proceeded smoothly with the recombinant *E. coli* cells harboring a nitrilase gene from *Arabidopsis thaliana* (AtNIT2) at the optimized conditions of 1.5 mol/L (235.5 g/L) substrate concentration, 6.0 wt% wet cells loading, pH 8.0 and 25 °C, and 100 % conversion was achieved within 4.5 h with ester group remaining intact [4]. Immobilization of the whole cells enhanced the substrate tolerance, stability, and reusability of the biocatalyst up to 16 batches under the optimized conditions (25 °C, 100 mmol/L Tris-HCl buffer, pH 8.0). Ethyl (*R*)-3-hydroxyglutarate, a key intermediate for synthesis of rosuvastatin, was synthesized with the biocatalyst, productivity and the space-time productivity were 55.6 g g^{-1} wet cells weight, and 625.5 g L^{-1} d^{-1}, respectively.

Chemoselective biotransformation of nitriles can also be achieved by using whole cells of naturally occurring microorganisms. The resting cells of *Rhodococcus equi* A4 catalyzed the chemoselective hydrolysis of methyl 3-cyanobenzoate and methyl 4-cyanobenzoate to produce mono-methyl isophtalate and mono-methyl terephtalate, respectively. However, the presence of other enzymes in the whole cells may deteriorate the chemoselectivity of the biotransformation. For example, the use of dried immobilized whole cells *Rhodococcus equi* A4 resulted in an increased formation of the unwanted 3-cyanobenzoic and 4-cyanobenzoic acids, due to the activated lipase activity in these dried whole cells [5, 6].

Reductions of aldehydes and ketones are common and fundamental chemical transformation in organic chemistry. However, the chemoselective reduction of either aldehyde or ketone by traditional chemical means is usually challenging and needs careful selection of the reducing agent and control of the reaction conditions. In contrast, many aldehyde reductases or ketone reductases have been found to reduce

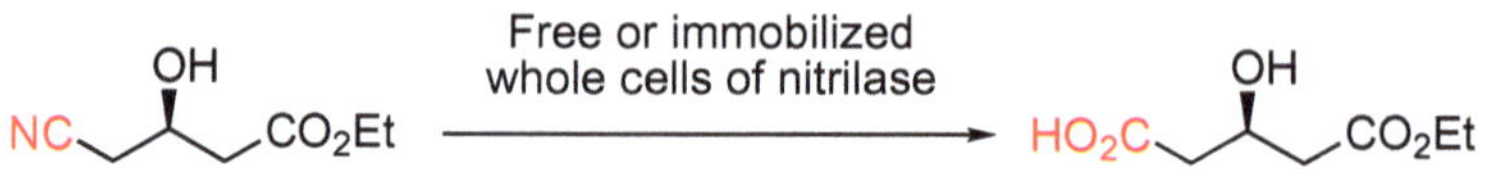

Scheme 4.1 Nitrilase-catalyzed chemospecific hydrolysis of ethyl (*R*)-4-cyano-3-hydroxybutyate

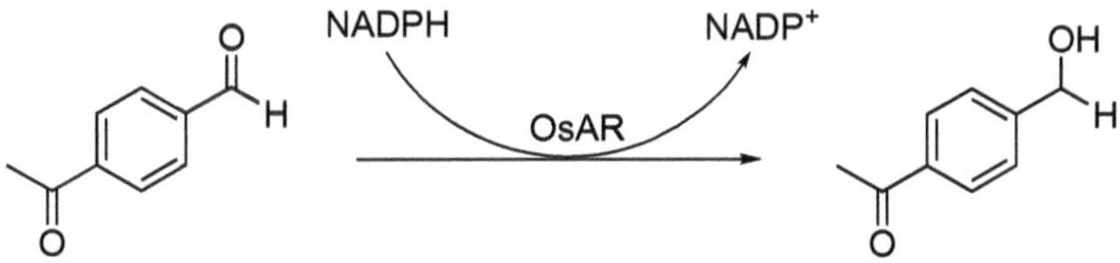

Scheme 4.2 The reduction of 4-acetylbenzaldehyde catalyzed by OsAR

aldehydes [7–9] or ketones chemospecifically to give the desirable alcohols. A NADPH-dependent aldehyde reductase from *Oceanospirillum* sp. MED92(OsAR) exhibited high activity toward a variety of aromatic and aliphatic aldehydes. This enzyme catalyzed the chemoselective reduction of aldehydes in the presence of ketones. For example, OsAR catalyzed the reduction of 4-acetylbenzaldehyde to exclusively give 4-acetylbenzyl alcohol (Scheme 4.2), and for the reaction of a mixture of hexanal and 2-nonanone, hexanal was reduced to hexanol but 2-nonanone remained intact [7].

Reduction of carboxylic acids usually requires strong reducing agents such as $LiAlH_4$, $NaBH_4$, or their derivatives [10, 11]. These reducing agent can also reduce C=O, C=N and other functional groups. As such, it is difficult to achieve the chemoselective reduction of carboxylic acid group without affecting other reducible functional groups in the molecule. Furthermore, the reduction of carboxylic acid usually does not stop at the first formation of aldehyde as the product, but further reduction to the corresponding alcohol [12, 13]. On the contrary, enzymatic reduction of carboxylic acids is a highly selective reaction under mild conditions [14]. A few carboxylic acid reductases (CAR, E.C.1.2.1.30) have been biochemically characterized and found to convert a wide range of aromatic and aliphatic acids into the corresponding aldehydes [15–18]. Other functional groups such as keto groups and C=C double bonds remain unaffected [16].

4.1.2 Regioselectivity

Regioselectivity of chemical reactions is very important for organic synthesis and enzymatic reactions are usually highly regioselective. A few examples of how enzymes differentiate the same functional group at different positions in a molecule are provided below. This often simplifies the synthetic route of a target compound. Among the tested commercially available ketoreductases (KREDs), six KREDs have been found to catalyze the reduction of trifluoromethyl keto group in the methyl/trifluoromethyl diketones to give either the *R* or *S* enantiomer with >99 % enantiomeric excess (ee), while one enzyme reduces the methyl keto group to afford *S*-configurated alcohol with 98 % ee (Scheme 4.3) [19]. A mutant 11β-hydroxysteroid dehydrogenase (11β-HSDH) from guinea pig effectively catalyzes highly regio- and stereo-specific reduction of the keto group at 11 position

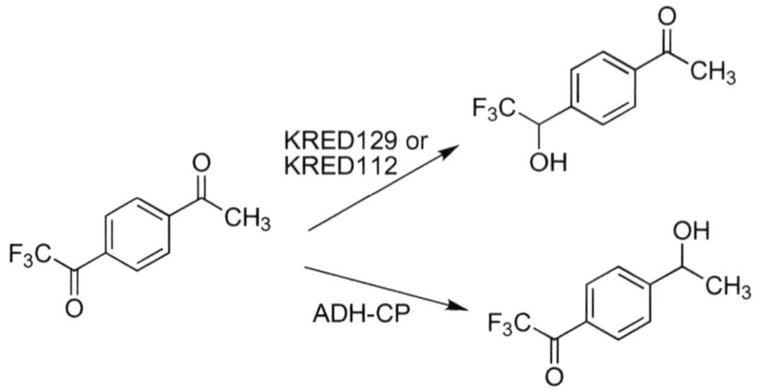

Scheme 4.3 Enzymatic regiospecific reduction of the methyl/trifluoromethyl diketone

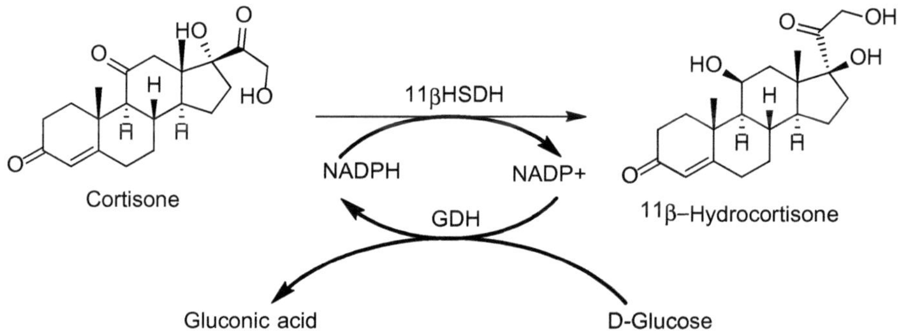

Scheme 4.4 Enzymatic regio- and stereospecific reduction of cortisone

among three keto groups in the molecule of cortisone, generating 11β-hydrocortisone (Scheme 4.4) [20].

In some cases, enzyme can act exclusively on one of the two identical functional groups. Many nitrilases catalyze the hydrolysis of one CN group in dinitriles to produce the corresponding cyanocarboxylic acids [21–24]. Scheme 4.5 shows the regio-specific hydrolysis of dinitriles by a nitrilase(bll6402) from *B. japonicum* [24, 25]. This would be impossible using traditional chemical hydrolysis. Furthermore, a few nitrilases have been found to catalyze dissymmetric hydrolysis of prochiral 3-substituted glutaronitriles to produce optically active 3-substituted-4-cyanobutanoic acids that can serve as important and versatile chiral intermediates for the synthesis of pharmaceutically important GABA derivatives. 3-Isobutylglutaronitriles and 3-(4′-chlorophenyl)glutaronitriles are hydrolyzed to form optically active 3-(cyanomethyl)-5-methylhexanoic acid and 3-(4′-chlorophenyl)-4-cyanobutanoic acid with high ee, respectively. These intermediates can be converted into two currently marketed drugs, (*S*)-Pregabalin and (*R*)-Baclofen by Curtius rearrangement followed by acidic hydrolysis (Scheme 4.6) [26].

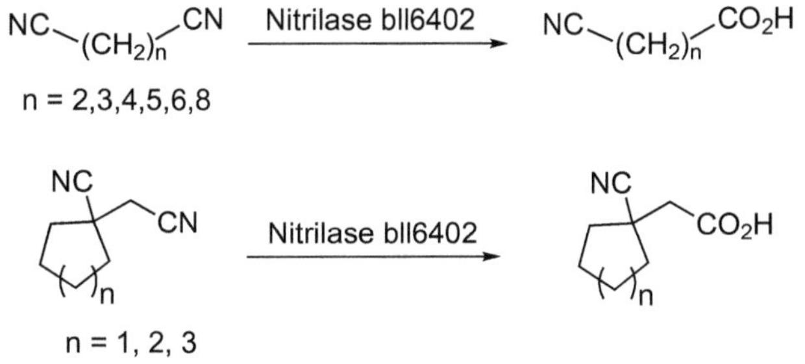

Scheme 4.5 Nitrilase bll6402 catalyzing the regiospecific hydrolysis of dinitriles

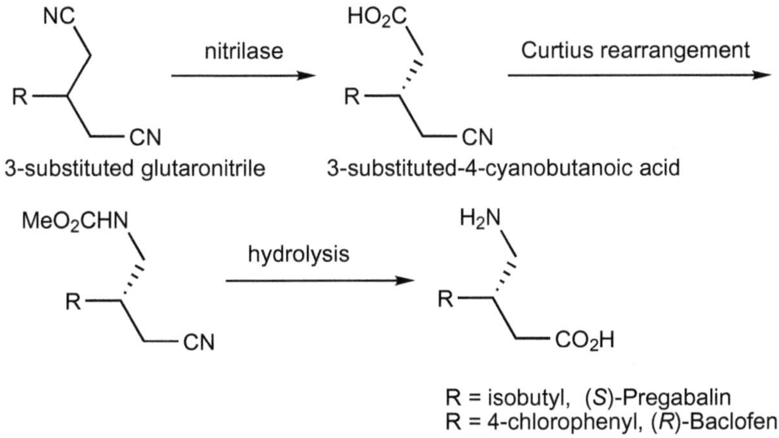

Scheme 4.6 Nitrilases catalyze desymmetric hydrolysis of prochiral 3-substituted glutaronitriles

4.1.3 *Stereoselectivity*

In 2006, a team from three leading pharmaceutical companies reported that of the 128 drug molecules analyzed, 69 (54 %) molecules contain at least one stereogenic center [27]. As such, asymmetric synthesis is of great importance in the production of chiral compounds as single enantiomers/diastereoisomers for pharmaceutical applications. Although asymmetric reactions using metal catalysts and organocatalysts have secured extensive applications in the synthesis of single enantiomeric/diastereoisomeric chiral chemicals, the exquisite stereoselectivities of some enzymes outperform the traditional chemical catalysis, making enzyme catalysis an alternative or complimentary toolbox for these transformations [28, 29]. High enantioselectivity of ketone reduction is achieved only when the two substituents flanking the carbonyl function are sterically different. The transition metal-catalyzed hydrogenation of diaryl ketones usually requires an *ortho*-substituent on one of the

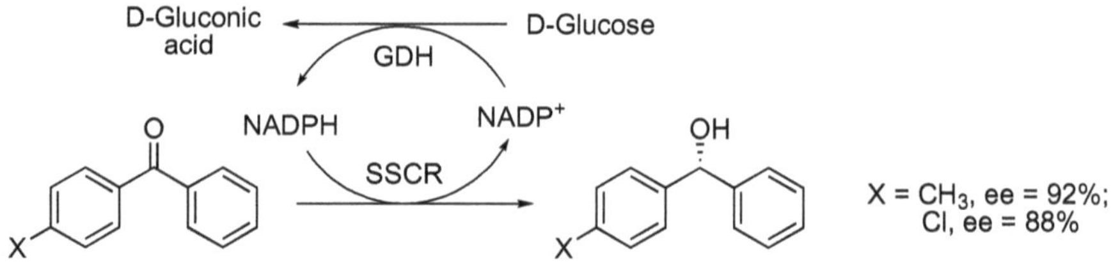

Scheme 4.7 SSCR-catalyzed reduction of diarylketones

aryl groups to achieve high enantiocontrol in the products. The reduction of diaryl ketones with only a *para*- or *meta*-substituent on one of the aryl groups affords the diarylmethanol in low enantiomeric purity (ee often being less than 50 %) [30, 31]. On the contrary, diaryl ketones with a *para*-substituent on one of the phenyl groups were reduced with high enantioselectivity (up to 92 % ee) by using commercially available carbonyl reductases, and one from red yeast *Sporobolomyces salmonicolor* AKU4429 (SSCR) and its mutant enzymes (Scheme 4.7) [32, 33].

Simple ketone tetrahydrothiophene-3-one is nearly spatially symmetrical and its asymmetric reductions usually resulted in 23–82 % ee by chemical means. It was originally prepared by multistep hazardous processes starting from the chiral pool compound L-aspartic acid. Although the native ketoreductase from *Lactobacillus kefir* catalyzed the reduction of tetrahydrothiophene-3-one to give (*R*)-tetrahydrothiophene-3-ol in 63 % ee, the chiral alcohol was obtained in >99 % ee by using a mutant enzyme [34]. (*R*)-Tetrahydrothiophene-3-ol is a key component of insulopenem, a potent antibacterial prodrug developed by Pfizer.

Chiral amines are very important intermediates for the synthesis of pharmaceuticals, agrichemicals and other fine chemicals, and about 40 % of marketed drugs contain a chiral amine moiety [35]. A straightforward method for the synthesis of chiral amines is the asymmetric reductive amination of carbonyl compounds. However, in traditional organic chemistry this reaction using ammonia as the amino donor is still challenging. In contrast, Nature has developed amino acid dehydrogenases for the reductive amination of α-keto acids using ammonia as the amino donor to synthesize nature amino acids, which have been used in industry for synthesis of natural and some unnatural L-amino acids [36, 37]. Recently D-amino acid dehydrogenases (D-AADH) [38–40], β-amino acid dehydrogenases (β-AADH) [41] and amine dehydrogenases (ADH) [42–44] have been discovered or evolved in the laboratory, and their applications in the synthesis of optically pure D-amino acids, β-amino acids and chiral amines have been explored by coupling with D-glucose dehydrogenase (GDH) for co-factor regeneration (Scheme 4.8) [41, 45, 46].

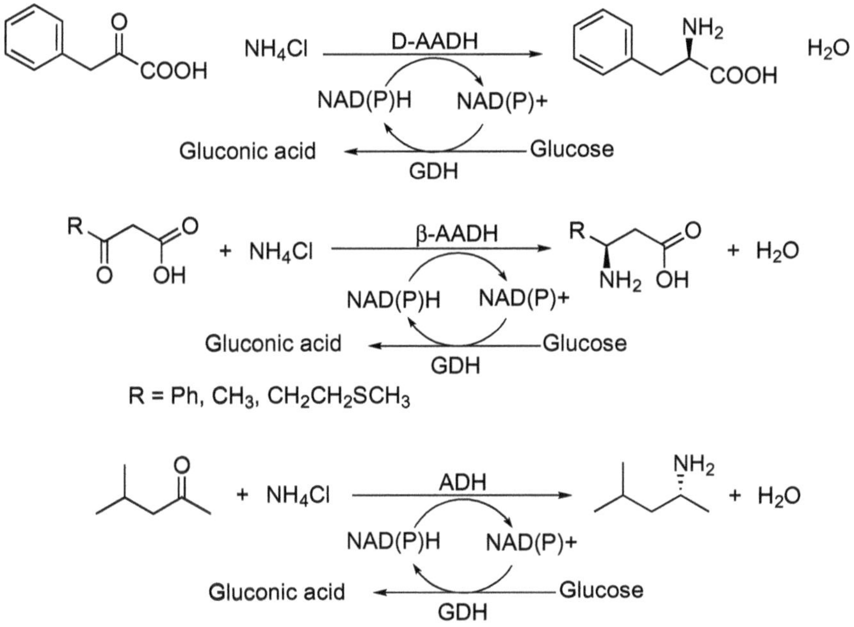

Scheme 4.8 Enzymatic reductive aminations of α-, β-keto acids and ketones

4.1.4 Mild Reaction Conditions

Enzymatic reactions not only are highly chemo-, regio- and stereoselective, but also operate at neutral or close to neutral pH and mild conditions, under which the labile functional groups in the molecule remain unaffected. Unlike chemical synthetic methods, the protection/deprotection steps are avoided in the synthesis of the target compounds, or the reaction can proceed much cleanly and efficiently without formation of unwanted by-products. In the reduction of α-bromoacetophenones using metal hydrides as the reducing agent, loss of the bromo group often occurs concomitantly, leading to low yield of α-bromohydrins and difficulty in product isolation. Although whole cell biocatalysis also resulted in the loss of bromo group [47], isolated carbonyl reductase from *Candida magnolia* have been used together with a D-glucose dehydrogenase/D-glucose cofactor regeneration system in a two-phase reaction medium to effectively prevent the side reactions in the reduction of α-bromoacetophenones, and the optically pure α-bromohydrins have been obtained in high yields [48].

Chemical hydrolysis of β-hydroxy nitriles often results in the undesirable elimination of OH group, yielding unsaturated by-products, because strong basic or acidic conditions and elevated reaction temperature are usually required [49]. Biocatalytic hydrolysis of nitriles can be performed at neutral pH condition and room temperature, offering possibility of hydrolyzing nitriles bearing functionalities

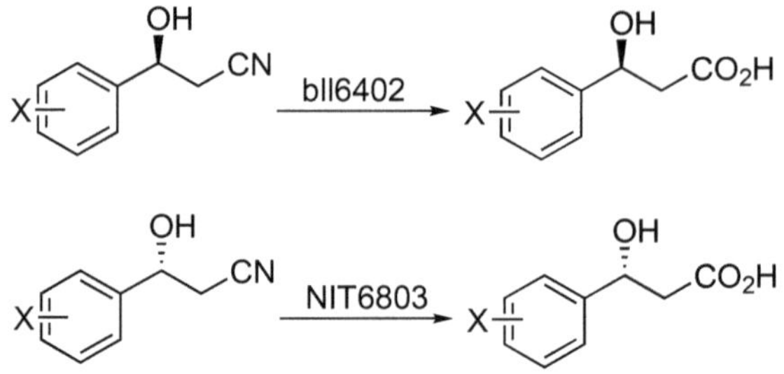

Scheme 4.9 Nitrilase-catalyzed hydrolysis of β-hydroxy nitriles

Table 4.1 Nitrilase-catalyzed hydrolysis of β-hydroxynitriles

X	Yield (%)[a]	ee (%)[a]	Yield (%)[b]	ee (%)[b]
4-H	90	99 (*S*)	85	98 (*R*)
4-F	93	99 (*S*)	86	99 (*R*)
2,4-F_2	91	99 (*S*)	89	98 (*R*)
4-Cl	92	99 (*S*)	86	99 (*R*)
4-Br	89	99 (*S*)	85	99 (*R*)
4-CH_3	91	99 (*S*)	85	98 (*R*)
4-CN	92	95 (*S*)	80	99 (*R*)[c]
4-NO_2	87	97 (*S*)	88	99 (*R*)
3-NO_2	88	99 (*S*)	87	99 (*R*)
3-CH_3O	88	95 (*S*)	90	97 (*R*)
4-CH_3O	91	99 (*S*)		

[a]Catalyzed by bll6402
[b]Catalyzed by NIT6803
[c]The product was 3-(4′-carboxyphenyl)-3-hydroxypropanoic acid

that cannot tolerate harsh conditions. For example, nitrilases from *B. japonicum* USDA110 (bll6402), cyanobacterium *Synechocystis* sp. strain PCC 6803 (NIT6803), and other sources effectively catalyzed the hydrolysis of β-hydroxy nitriles to afford the corresponding β-hydroxy carboxylic acids in excellent yield (Scheme 4.9; Table 4.1) [50–53].

4.2 Enzyme Applications in Chemical Production

Chemical products such as pharmaceuticals, agrichemicals, and cosmetics are ubiquitously involved in our daily life. As the world population increases constantly, there are ever-increasing demands for these products, which require development of highly efficient and green synthetic technologies for the production of these chemicals to address the resource and environmental issues that we are facing. As illustrated in the previous section, enzyme catalysis usually has unparalleled chemo-, regio-, and stereoselectivity and the reactions are often carried out under mild conditions. Enzymes can catalyze some transformations which are either difficult or cannot be achieved by traditional chemical synthetic methods. Therefore, introduction of enzyme catalysis to the production of target compounds can shorten the synthetic route by redesigning it, and facilitate product purification by preventing by-product formation, thus making the process more efficient and sustainable. Although enzymes have been used for human benefits since ancient time, it was only in the last decade or two that organic chemists started to embrace biocatalysis as an environmentally friendly alternative to conventional chemical catalysis in organic synthesis. Modern advances in molecular biology and fermentation technology are major contributing factors. All these developments resulted in myriad industrial applications in the production of pharmaceuticals, agrichemicals, and other specialty chemicals including some bulky chemicals when high reaction selectivity on complex molecules is required. Some of these examples are presented below and organized according to the enzyme type.

4.2.1 Lipases and Esterases

Lipase (triacyl glycerol acyl hydrolase, EC 3.1.1.3), which catalyzes the hydrolysis of esters, esterification and transesterification of carboxylic acids and alcohols, is one of the dominant enzymes extensively explored for their use in the synthesis of important chemicals for pharmaceutical, agrichemical, flavor, fragrance, and other industries. Currently, many lipases are commercially available and find applications

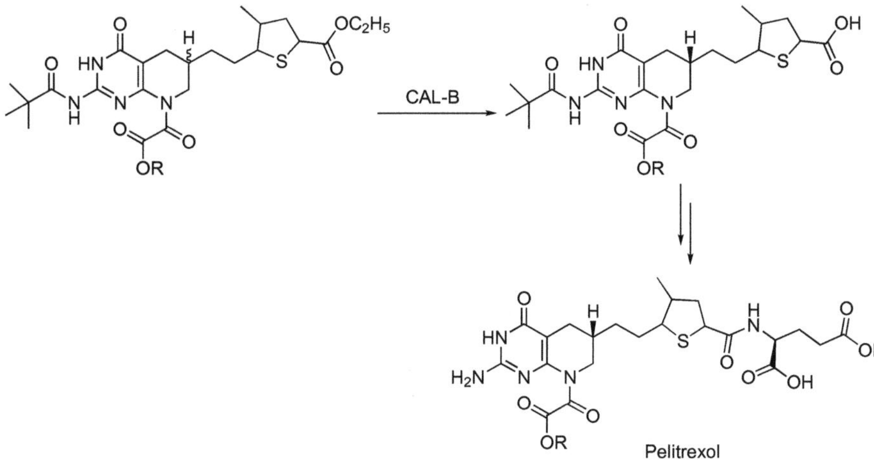

Scheme 4.10 Lipase CAL-B catalyzed chiral resolution of the intermediate for the synthesis of pelitrexol

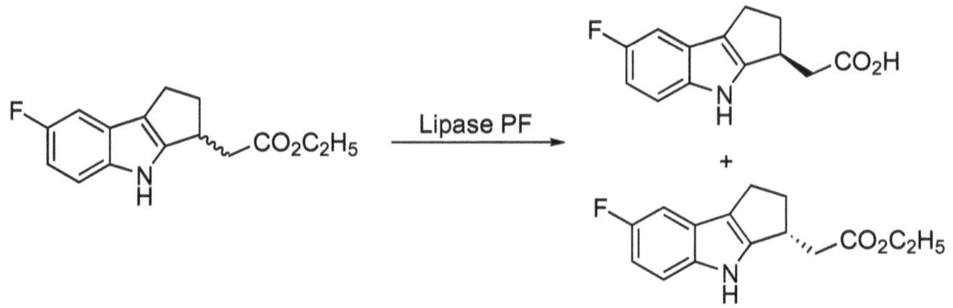

Scheme 4.11 Lipase catalyzed resolution of an indole-ethyl ester

in numerous industrial processes including areas of detergents, food, leather and paper processing. In the chemical industry, lipase-catalyzed reactions offer the proven environmentally benign access to chiral alcohols and amines, which are very important building blocks for synthesis of pharmaceuticals and agrichemicals.

In the synthesis of pelitrexol, a novel GARFT (glycinamide ribonucleotide formyltransferase) inhibitor, lipase CAL-B was applied to the chiral resolution of the molecule with a stereocenter in the tetrahydropterin moiety by hydrolyzing the terminal ester group. The introduction of an oxalamic ester adjacent to the stereocenter dramatically enhanced the enzyme's enantioselectivity, and the desired *S*-acid was obtained with high optical purity and yield (Scheme 4.10). The other enantiomer was recycled via a dehydrogenation/hydrogenation strategy [54]. The (*R*)-enantiomer of an indole-ethyl ester is the key intermediate of a prostaglandin D2 receptor (DP) antagonist targeted against allergic rhinitis. A research team at Merck developed a resolution process using *Pseudomonas fluorescens* lipase to produce this chiral compound with 99.75 % ee at 50 % conversion (Scheme 4.11) [55].

Using a commercially available lipase, *rac*-2-carboxyethyl-3-cyano-5-methylhexanoic acid ethyl ester was resolved to give 2-carboxyethyl-3-cyano-5-methylhexanoic acid, which was decarboxylated to generate (*S*)-3-cyano-5-methylhexanoic acid ethyl ester, followed by hydrogenation to afford Pregabalin, the active ingredient in antiepileptic drug Lyrica. Compared to the first-generation manufacturing process, this chemo-enzymatic route dramatically improved process efficiency (overall yield increased from 25–29 to 40–45 %), and resulted in substantial reductions of waste streams corresponding to a fivefold decrease in the Environmental (*E*) factor from 86 to 17 and energy use by 83 % [56]. *E* factor is defined as total waste (kg) divided by-product (kg). The enzymatic process was scaled up to 10 ton scale, and in 2006 Pfizer received the Astrazeneca Award for Excellence in Green Chemistry and Engineering for developing this process.

Lipases are also used in the production of herbicides, where only one enantiomer is biologically active. For example, only the (*S*)-enantiomer of indanofan exhibits herbicidal activity against grass weeds in paddy fields. (*S*)-Indanofancan be efficiently synthesized via lipase-catalyzed resolution and chiral inversion of the unwanted isomer (Scheme 4.12), in 99 % yield and >99 % chemical and optical purity [57].

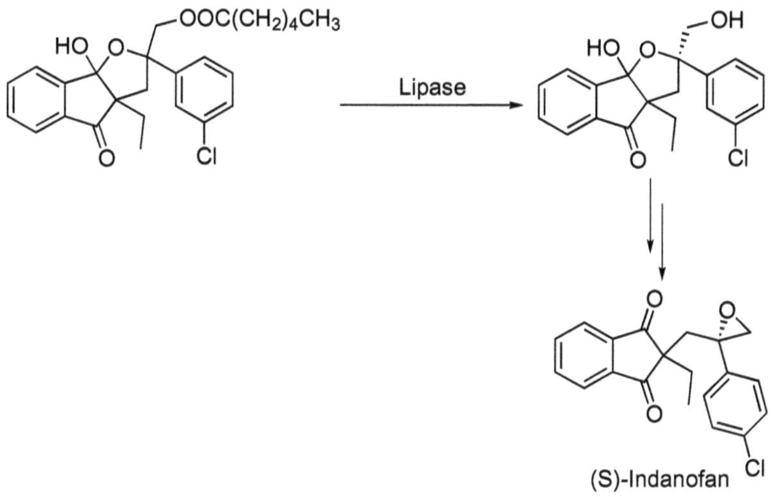

Scheme 4.12 Lipase catalyzed chiral resolution of the intermediate for the synthesis of (*S*)-Indanofan

Similarly, lipase-catalyzed enantioselective hydrolysis of the esters of 2-halopropionic acids afforded optically pure 2-halopropionic acids, which are widely used as precursors for the production of optically pure(*R*)-2-phenoxypropionic acids and their esters such as very selective weed killers, dichloprop, mecoprop, and silvex [58]. Use of *R*-isomer (the biological active component) of these herbicides instead of the racemic mixture greatly reduces the dosage and side effects, thus exerting less impact onto the environment.

Other hydrolases/esterases were also effective biocatalysts for the chiral resolution of pharmaceutically important building blocks. After screening various hydrolases, Hu et al. [59] found that pig liver esterase (PLE) efficiently catalyzed the enantioselective hydrolysis of 4,4-difluoro-3,3-dimethyl-*N*-benzyl-proline methyl ester (Scheme 4.13), leading to a process for the production of (2*S*)-4,4-difluoro-3,3-dimethyl-*N*-Boc-proline, a key intermediate for the synthesis of HIV protease inhibitors. Compared to their previously used protease resolution process, the space time yield was enhanced 57 fold and the production cost was reduced to 5 % [59].

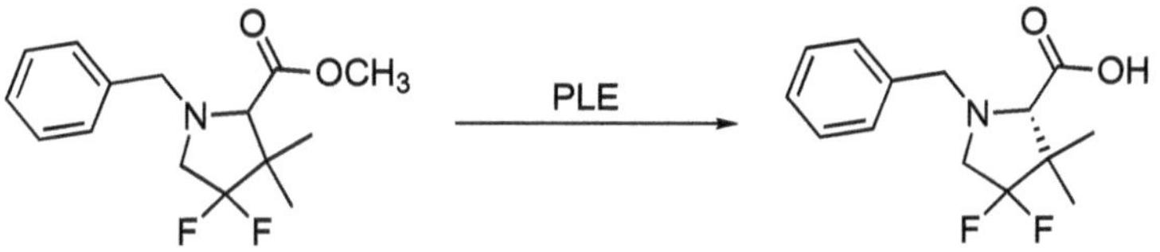

Scheme 4.13 PLE-catalyzed enantioselective hydrolysis of the substituted proline methyl ester

4.2.2 Nitrilases, Nitrile Hydratases, and Amidases

Nitrilases catalyze the hydrolysis of organic nitriles under mild conditions and often in a chemoselective, regioselective, or stereoselective manner. These enzymes have been widely used as valuable alternatives to chemical catalysts for the synthesis of carboxylic acids from nitriles. (*R*)-Mandelic acid is an important chiral building block for the production of various pharmaceutical and agricultural products, and also used as optical resolving agents. An industrial process based on the enzymatic stereoselective hydrolysis of racemic mandelonitrile had been adopted by BASF to produce (*R*)-mandelic acid in several hundred metric tons per year. Under proper reaction conditions, racemic mandelonitrile was dynamically resolved to give (*R*)-mandelic acid up to 100 % yield.

An efficient nitrilase-catalyzed desymmetrisation of 3-hydroxyglutaronitrile had been achieved to give ethyl (*R*)-4-cyano-3-hydroxybutyrate (Scheme 4.14), a potential intermediate in the synthesis of Atorvastatin (Lipitor). The nitrilase reaction could be carried out at 3 M (330 g/L) substrate concentration with an enzyme loading of 6 wt%, and 100 % conversion and 99 % ee product were obtained in 16 h [60–62]. Nitrilases also catalyze the desymmetrisation of other 3-substituted glutaronitriles, resulting in optically active 3-substituted-4-cyanobutanoic acids, which can be converted to β-substituted γ-aminoacids such as (*S*)-Pregabalin and (*R*)-Baclofen of pharmaceutical and therapeutic importance [26].

Immobilized *Acidovorax facilis* 72W cells with nitrilase activity was successfully employed in a chemo-enzymatic process for the preparation of 1,5-dimethyl-2-piperidone, in which the whole cells catalyzed the hydrolysis of 2-methylglutaronitrile to give 4-cyanopentanoic acid ammonium salt with >98 % regioselectivity at 100 % conversion. Hydrogenation of 4-cyanopentanoic acid ammonium salt produced 1,5-dimethyl-2-piperidone, which may be useful in electronics, coatings and solvent applications [63].

Nitrile hydratase and amidase are two hydrating and hydrolytic enzymes responsible for the sequential metabolism of nitriles. Nitrile hydratase catalyzes the hydration of nitriles to their corresponding amides, while amidase catalyzes the hydrolysis of amides to the carboxylic acids. In addition to the well-known industrial production of acrylamide from acrylonitrile, and removal of nitriles from wastewater, nitrile hydratases have also been employed in the synthesis of other fine chemicals. Lonza has developed a biocatalytic process with *Rhodococcus rhodochrous* J1 cells immobilized in polyacrylamide for the conversion of nicotinonitrile to nicotinamide, an essential nutrient in animal and human nutrition. The process has been used in the production of several thousand tons of nicotinamide per year

Scheme 4.14 Nitrilase catalyzed desymmetrisation of 3-hydroxyglutaronitrile

[64]. Nicotinamide can be hydrolyzed to nicotinic acid by using amidase as the biocatalyst [65].

A biocatalytic process for the regioselective hydration of adiponitrile to 5-cyanovaleramide has been developed with *Pseudomonas chlororaphis* B23 cells immobilized in calcium alginate beads. The biotransformation was conducted for 58 consecutive batches to convert a total of 12.7 metric tons of adiponitrile to 5-cyanovaleramide (93 % yield, 96 % selectivity) with biocatalyst productivity of 3150 kg/kg (dry cell weight). 5-Cyanovaleramide is the starting material for the synthesis of azafenidin, a broad spectrum herbicide. The chemical process for the production of 5-cyanovaleramide is based on the hydration of adiponitrile using manganese dioxide as the catalyst. The conversion of adiponitrile is only 25 % with about 20 % of adipamide as by-product. The bioprocess generates significantly less waste products than the chemical one, and avoids the difficulty in the purification of the product [66].

4.2.3 Epoxide Hydrolases

Epoxide hydrolase converts epoxides to *trans*-dihydrodiols. Epoxides are cytochrome P450 oxidase metabolites of unsaturated carbon–carbon bonds, and the enzyme thus functions in detoxification during drug metabolism. Epoxide hydrolase are found in bacteria, yeast, fungi, plants, and insects. The microbial enzymes usually exhibit high enantioselectivity and activity, enabling the preparation of enantiopure epoxides in a very simple way from easily available racemic epoxides. The transformation of epoxides to chiral diols is also an important reaction in organic synthesis. As such, epoxide hydrolases serve as highly appealing tools for the preparation of fine chemicals [67]. Synthesis of the *R*- and *S*-enantiomers of biologically active propranolol, a typical β-blocker, has been achieved in high optical purity via an epoxide hydrolase-catalyzed resolution of racemic α-naphthylglycidyl ether. Using a variant of *Bacillus megaterium* epoxide hydrolase ($Bm\mathrm{EH}_{\mathrm{F128T}}$) in a biphasic system (isopropyl ether/isooctane/aqueous), the racemic epoxide was resolved to give enantiopure (*S*)-epoxide (>99 % *ee*) and (*R*)-3-(1′-naphthyloxy)-propane-1,2-diol (>99 % *ee*) in 45.3 and 42.4 % yields, respectively, which were subsequently converted to (*S*)- and (*R*)-propranolol (>99 % *ee*). The enzymatic resolution of the racemic epoxide proceeded with a total turnover number of 70,000 and high space-time yields [136 g L^{-1} d^{-1} for (*S*)-epoxide and 139 g L^{-1} d^{-1} for (*R*)-diol] [68].

Coumarin derivatives possess various biological properties such as antifungal, anticoagulant and anti-HIV properties, and are widely used in the field of medicine. (*R*)-(+)-Marmin is a simple coumarin derivative isolated from the trunk bark of a highly reputed Ayurvedic medicinal plant, *Aeglemarmelos* Correa, a sacred tree dedicated to Lord Shiva. This compound was prepared in 95 % ee via the stereoselective hydrolysis of racemic trisubstituted terpenoidoxiranes by employing the epoxide hydrolase activity of *Rhodococcus* or *Streptomyces* spp [69].

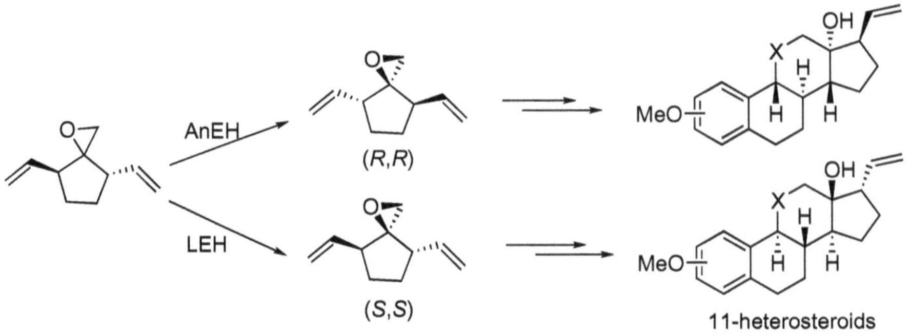

Scheme 4.15 Kinetic resolution of *rac*-spiroepoxide

Spiroepoxide is a strategic key building block allowing the synthesis of 11-heterosteroids, a family of compounds which are of substantial current interest due to their high biological and medicinal potential. Two microbial epoxide hydrolases from *Aspergillus niger* (AnEH) and *Rhodococcus erythropolis* (the so-called “Limonene EH”: LEH) exhibited opposite enantioselectivity for the hydrolytic kinetic resolution of *rac*-spiroepoxide, allowing isolation of the (*R*,*R*)- (from AnEH) and the (*S*,*S*)-isomer (from LEH) of spiroepoxide in nearly enantiopure forms (>98 %) (Scheme 4.15). In the case of AnEH, the biocatalytic reaction could be carried out at the substrate concentration up to 200 g/L by using a biphasic (water/isooctane) reaction medium [70]. This two liquid–liquid phase methodology also allowed the AnEH-catalyzed resolution of trifluoromethyl-substituted styrene oxide derivatives to proceed at a very high substrate concentration of 360 g/L (1.8 M) [71]. These results clearly demonstrated that, the epoxide hydrolase-catalyzed hydrolytic kinetic resolution of epoxides is a very mild, cheap and easy-to-use “green chemistry” methodology applicable for cost effective industrial preparation of optically pure epoxides and their corresponding vicinal diols, two classes of important chiral building blocks in the synthesis of a wide range of fine chemicals and pharmaceutical molecules.

4.2.4 Carbonyl Reductases

Chiral alcohols are important and valuable building blocks in the synthesis of pharmaceuticals, agrichemicals, and other fine chemicals. A straightforward approach to access the enantiometrically pure alcohols is the asymmetric reduction of prochiral ketones, which can be achieved by using a variety of chiral metal catalysts, organocatalysts, and biocatalysts. As in other enzyme catalyzed synthesis, the biocatalytic reduction offers advantages such as mild and environmentally benign reaction conditions, high chemo-, regio- and stereoselectivity, and void of residual metal in the products. Therefore, great efforts and significant advances have

Table 4.2 Enzymatic preparation of both enantiomers of chiral chlorohydrins

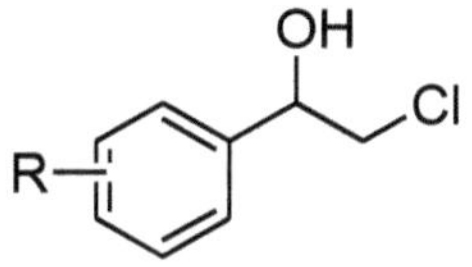

R	Ketoreductase	Yield (%)	ee (%) (configuration)
H	KRED112	72	>99(*S*)
H	KRED130	94	>99(*R*)
3′-Cl	KRED112	69	>99(*S*)
3′-Cl	KRED130	99	>99(*R*)
4′-Cl	KRED112	97	>99(*S*)
4′-Cl	KRED130	81	>99(*R*)
3′,4′-Cl_2	KRED112	79	>99(*S*)
3′,4′-Cl_2	KRED130	88	>99(*R*)
4′-NO_2	KRED107	87	>99(*S*)
4′-NO_2	KRED130	76	>99(*R*)
4′-CH_3SO_2NH	KRED113	96	>99(*S*)
4′-CH_3SO_2NH	KRED130	89	>99(*R*)

been made to develop biocatalytic ketone reduction processes in recent years. Chiral halohydrins such as 2-chloro-1-phenylethanol and its analogs are valuable synthetic intermediates for the preparation of a large group of antidepressants and α- and β-adrenergic drugs. For example, 2-chloro-1-phenylethanol is the precursor for the synthesis of fluoxetine, tomoxetine and nisoxetine, while 2-chloro-1-(3′-chlorophenyl)ethanol has been used to synthesize AJ-9677, a potent and selective β3 adrenergic receptor agonist. Various commercially available ketoreductases were examined toward the asymmetric reduction of α-chloroketone and its analogues, and both enantiomers of the product halohydrins could be obtained in greater than 99 % ee with high to excellent yields (Table 4.2) [72]. These chiral halohydrins can also be synthesized via hydrogen transfer reduction of α-chloroacetophenones by using an alcohol dehydrogenase from the hyperthermophilic archaeon *Pyrococcus furiosus* in a coupled substrate approach using *iso*-propanol as the hydrogen donor [73]. These results demonstrated yet another green chemistry method to access enantiomerically pure α-chloro alcohols of pharmaceutical importance.

Another substituted phenylethanol, (*S*)-3,5-bistrifluoromethylphenyl ethanol, which is a pharmaceutically important intermediate for the synthesis of NK-1 receptor antagonists, can also be synthesized via asymmetric enzymatic reduction with an alcohol dehydrogenase from *R. erythropolis*. The poorly water soluble ketone substrate was reduced at the substrate concentration of 580 mM with excellent ee (>99.9 %) and a space time yield of 260 g L^{-1} d^{-1}, and the product alcohol was isolated with >90 % yield [74].

Chiral α- and β-hydroxyester moieties are widely found in natural products and have been frequently used as building blocks for the synthesis of biologically active chemicals. Ethyl (*S*)-4-chloro-3-hydroxybutanoate [(S)-CHBE)] is an important precursor for the preparation of statin drugs such as atorvastatin. Asymmetric reduction of ethyl 4-chloro-3-oxobutanoate offers a straightforward approach to access this compound. As such, the bioreduction of ethyl 4-chloro-3-oxobutanoate has been extensively studied, and many ketoreductases of diverse origins have been found to be efficient catalysts for this reduction. The recombinant *E. coli* cells co-expressing the carbonyl reductase (S1) gene from *Candida magnolia* and the glucose dehydrogenase (GDH) gene from *B. megaterium* effectively catalyzed the asymmetric reduction of ethyl 4-chloro-3-oxobutanoate to (*S*)-CHBE [75]. In an organic solvent–water two-phase system, (*S*)-CHBE partitioned in the organic layer at concentration of 2.58 M (430 g/L) with a molar yield of 85 %. In an aqueous mono-phase system with continuous feeding of substrate, (*S*)-CHBE accumulated only up to 1.25 M (208 g/L). However, the ee value of the product in both cases was 100 %, thus leading to a simple and practical synthesis of the compound. Ethyl 4-chloro-3-oxobutanoate (COBE) could also be reduced by *E. coli* cells expressing a reductase (ScCR) from *Streptomyces coelicolor* to afford enantiopure (*S*)-CHBE. In this case, isopropanol was used as a co-substrate for regenerating NADH with no added enzyme. The reaction was performed in a toluene–aqueous biphasic system (1:1, v/v) on a pilot scale, producing (*S*)-CHBE in 85.4 % yield and 99.9 % ee, with specific production of 36.8 $g_{product}/g_{dcw}$ [76]. The ethyl (*R*)-4-chloro-3-hydroxybutanoate [(R)-ECHB] enantiomer was produced via reduction of ethyl 4-chloroacetoacetate using recombinant whole cells of *E. coli* expressing a secondary alcohol dehydrogenase of *Candida parasilosis* as the biocatalyst [77]. Cofactor NADH was regenerated by using 2-propanol as the hydrogen donor. The space productivity of (*R*)-ECHB reached 36.6 g/L (>99 % ee, 95.2 % conversion) without addition of NADH to the reaction mixture.

Optically active ethyl 2-hydroxy-4-phenylbutyrate (HPBE) is an important intermediate for the synthesis of angiotensin-converting enzyme (ACE) inhibitors, such as Enalapril, Lisinopril, Cilapril and Benazepril, which are widely used as antihypertensive drugs and in the therapy for congestive heart failure. Ciba company developed an industrial process for the production of (*R*)-2-hydroxy-4-phenylbutyric acid [(*R*)-HPBE] in 91 % yield and 99.9 % ee from the corresponding α-ketoester with an alcohol dehydrogenase from *Staphylococcus epidermidis* [14]. A new reductase, CgKR2 was reported to reduce ethyl 2-oxo-4-phenylbutyrate to [(*R*)-HPBE] with >99 % ee [78]. The bioreduction proceeded at 1 M substrate concentration without external addition of cofactors with the use of recombinant *E. coli* cells containing CgKR2 and glucose dehydrogenase (GDH) as the biocatalyst, reaching a space-time yield of 700 g L^{-1} d^{-1}. Similarly, the *S*-enantiomer of this α-hydroxy ester was prepared from ethyl 2-oxo-4-phenylbutyrate using *E. coli* cells that coexpressed a β-ketoacyl-ACP reductase (FabG) encoding gene from *Bacillus* sp. and that of GDH. As much as

620 g L^{-1} of substrate were reduced to ethyl (*S*)-2-hydroxy-4-phenylbutyrate with >99 % ee without external addition of a cofactor [79]. These results demonstrated that both enantiomers of this important α-hydroxy acid/ester could be easily produced at large scale via enzymatic reduction.

Clopidogrel is marketed by Bristol-Myers Squibb and Sanofi under the trade name Plavix, which is an oral, thienopyridine class antiplatelet agent. The drug has been widely administered to atherosclerotic patients with the risk of a heart attack or stroke caused by blood clots. Before the expiry of its patent, clopidogrel was the second best-selling drug in the world and grossed over US$9 billion in global sales in 2010. Methyl (*R*)-o-chloromandelate is the key chiral intermediate for the synthesis of (*S*)-clopidogrel. Biocatalytic reduction of methyl *o*-chlorobenzoyl formate offers an efficient method for the preparation of methyl (*R*)-*o*-chloromandelate. The recombinant *E coli* cells co-expressing a carbonyl reductase (SCR) gene from *Saccharomyces cerevisiae* (bakers' yeast) and GDH gene served as an effective biocatalyst for this reduction to afford methyl (*R*)-*o*-chloromandelate with >99 % ee and space productivity of 178 g L^{-1} [80]. Recently, the bioreduction of methyl *o*-chlorobenzoylformate was achieved by using *E. coli* whole cells co-expressing both aldo-keto reductase from *Bacillus* sp. and GDH [81].

Chiral γ-amino alcohols are valuable precursors for the synthesis of selective serotonin reuptake inhibitors, which serve as the key antidepressant drugs, such as fluoxetine, tomoxetine, nisoxetine, and duloxetine. This has inspired considerable efforts to search for methods of producing these γ-amino alcohols in optically pure form. Among the various approaches, the enantioselective bioreduction of β-amino ketones provides a straightforward method to access this class of chiral compounds. However, there was no report addressing the biocatalytic reduction of β-amino ketones, compounds such as 3-(dimethylamino)-1-phenylpropan-1-one and 3-(dimethylamino)-1-(2-thienyl)-propan-1-one, until recently. In 2010, Codexis reported in a patent that some engineered keto reductases catalyzed the conversion of 3-(dimethylamino)-1-(2-thienyl)-propan-1-one to the corresponding (*S*)-alcohol for the synthesis of duloxetine. In a recent study, we found that a double mutant M242F/Q245T of a carbonyl reductase (SSCR)from *Sporobolomyces salmonicolor* AKU4429 catalyzed the reduction of 3-(dimethylamino)-1-phenylpropan-1-one to the (*S*)-alcohol although with only 28 % ee. However, after combinatorial active-site saturation mutagenesis of the enzyme two resulting mutants, P170R/L174Y and P170H/L174Y, were found able to catalyze the reduction of 3-(dimethylamino)-1-phenylpropan-1-oneand 3-(dimethylamino)-1-(2-thienyl)-propan-1-one, thus generating the (*R*)-γ-amino alcohols with up to 95 % ee. The individual site saturation mutagenesis of Pro170 and Leu174 revealed that Pro170 did not significantly affect the enzyme enantioselectivity. Mutant L174 W led to the reduction of 3-(dimethylamino)-1-phenylpropan-1-one and 3-(dimethylamino)-1-(2-thienyl)-propan-1-one to the corresponding (*S*)-γ-amino alcohol with 96 and 65 % ee, respectively. Mutant L174Y, however, exhibited an (*R*)-preference of the two products with ee values of 88 and 95 %, respectively [82].

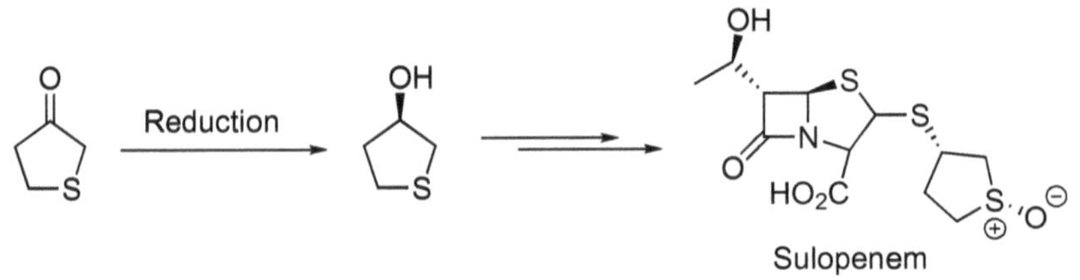

Scheme 4.16 Reduction of tetrahydrothiophene-3-one

Some aliphatic chiral alcohols are also important components in various agricultural or antimicrobial products. (*S*)-Pentane-1,2-diol is the key intermediate of triazole fungicide propiconazole with high biological activity. This diol is accessible via asymmetric reduction of 1-hydroxy-2-pentanone catalyzed by a self-sufficient whole cell biocatalyst containing a carbonyl reductase (CMCR) from *C. magnolia* and GDH from *B. subtilis* for co-factor regeneration. This biocatalyst also catalyzed the reduction of a series of α-hydroxy aromatic ketones to afford (*S*)-1-aryl-1,2-ethanediols in up to 99 % ee [83]. Another interesting example is (*R*)-tetrahydrothiophene-3-ol, which is a key building block for the synthesis of sulopenem, a potent antibacterial prodrug being developed by Pfizer. In Pfizer's original synthesis, this chiral alcohol was obtained from L-aspartic acid in five steps that involved hazardous reaction conditions and reagents. The asymmetric reduction of tetrahydrothiophene-3-one appeared to be the most straightforward approach to prepare this compound (Scheme 4.16). However, reduction of the nearly spatially symmetrical ketone to optically pure alcohol presented great challenge for both chemical and biocatalytic methods. After several rounds of directed evolution of an (*R*)-selective ketoreductase from *Lactobacillus kefir*, a mutant enzyme was found capable of carrying out the reduction of tetrahydrothiophene-3-one to the corresponding (*R*)-alcohol in a highly enantioselective manner (>99 % ee). This bioprocess allowed the production of this key component to a family of potent antibiotics starting from a commodity chemical at 100 kg scale [34].

4.2.5 Ene-Reductases

Asymmetric C=C hydrogenation generates up to two stereogenic centers, thus is one of the most widely used industrial reactions. Biocatalytic alkene reduction catalyzed by NAD(P)H-dependent enzymes ('ene'-reductases or ERs) has been extensively explored in recent years. The FMN-containing Old Yellow Enzyme (OYE) family of oxidoreductases and oxygen-sensitive FAD and [4Fe-4S]-containing NADH oxidases have been used to catalyze these biocatalytic hydrogenations to afford optically pure chiral compounds that are valuable in pharmaceutical, agricultural and other industries. For example, enantiomerically pure (*S*)-2-bromobutanoic acid and the corresponding methyl ester are useful key intermediates in the synthesis of

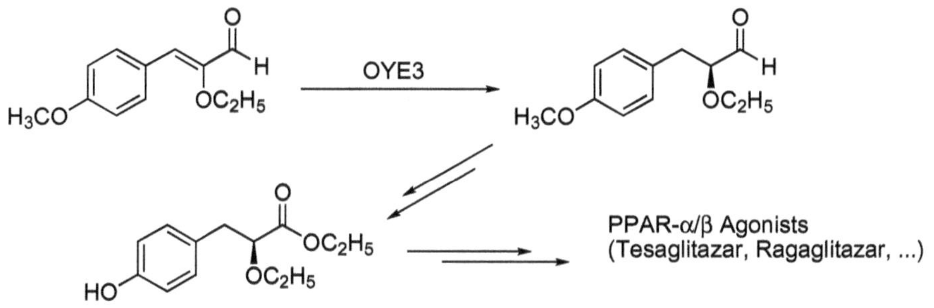

Scheme 4.17 Biocatalytic synthesis of the precursor to PPAR-α/β agonists

therapeutic agents for the treatment of non-insulin-dependent type 2 diabetes mellitus (T2DM). The methyl ester can be prepared in 97 % ee via the hydrogenation of methyl (*Z*)-2-bromocrotonate catalyzed by ER of the OYE family. The bioreductions of both (*Z*)- and (*E*)-diastereoisomers of α-bromounsaturated esters afford the same enantiomer of the product [84]. The OYE3 from baker's yeast effectively catalyzed the hydrogenation of C=C bond in α-enol ether aldehydes to give the corresponding *S*-configurated saturated aldehyde with 98 % ee and a productivity of 59.4 $g\,L^{-1}\,d^{-1}$, which can be converted to ethyl (*S*)-2-ethoxy-3-(p-methoxyphenyl)propanoate, an important precursor of several PPAR-α/β agonists such as Tesaglitazar and Ragaglitazar (Scheme 4.17) [85].

Dihydrocarvones are potential inhibitors of bacterial and fungal growth, as well as prospective insect repellents. For example, they have shown insecticidal activity against the rice weevil *Sitophilus oryzae* (L.), one of the most widespread insect pests of stored cereals. Recently, dihydrocarvones are also used as important renewable building blocks in the synthesis of functional materials such as shape memory polyesters. ERs of various sources have shown high activity and stereoselectivity toward the C=C reduction of carvone, resulting in dihydrocarvone with up to 98 % diastreomeric excess (de) [86–88]. A cascade reaction employing an ER from *Lactobacillus casei* (LacER) and a carbonyl reductase from *S. salmonicolor* (SSCR) or *C. magnolia* (CMCR) converted (*R*)-carvone or (*S*)-carvone to dihydrocarveols. For (*R*)-carvone, (1*S*,2*R*,5*R*)-dihydrocarveol was produced as the sole product with >99 % conversion, while (*S*)-carvone was double reduced to (1*S*,2*R*,5*S*)-dihydrocarveol as the major product with a lower de [89]. Dihydrocarveol is a fragrance ingredient widely used in cosmetics, fine fragrances, and household products. Carvone is currently extracted from caraway, dill, and spearmint seeds, providing a natural source for these valuable chemicals. While transformations of carvone into dihydrocarvones and dihydrocarveol have been extensively studied, the chemo-, regio-, and stereospecificity of these transformations still present a challenging task. (*R*)- or (*S*)-Carvone possesses three unsaturated bonds and a C-5 stereogenic center, and 17 different hydrogenation products are possible with no control over selectivity. This exemplifies the capability of biocatalytic transformations to address such demanding selectivity problems.

4.2.6 Transaminases

Transaminases (also known as amino acid aminotransferases) are pyridoxal-5′-phosphate-dependent enzymes that catalyze the transfer of an amino group between an amino acid and an α-keto acid. Transaminases usually show a broad substrate spectrum, and amines and ketones can also be the amino donor or acceptor, respectively. More profoundly, transaminases make optically pure primary amines and amino acids became accessible from the corresponding ketones or keto acids in one single reaction step. This is difficult by traditional chemical methods. Therefore, these enzymes are very important biocatalysts in the synthesis of amino acids and amines, which are key structural components widely found in numerous natural products as well as pharmaceuticals, agrochemicals, and other fine chemicals. The equilibrium constant of the aminotransfer reaction between an amino acid and an α-keto acid is about 1, so that shifting the equilibrium to the product direction is required to assure the efficiency of transaminase used in asymmetric synthesis of α-amino acids. The amino transfer from α-amino acid to ketone is more thermodynamically unfavorable with the equilibrium constant for the amino transfer reaction between acetophenone and alanine being 8.8×10^{-4}. As such, transaminases have often been used in the kinetic resolution of racemic amines, while the asymmetric synthesis of optically pure amines from the corresponding ketones requires shifting the equilibrium to product formation. The equilibrium shift can be achieved by coupling with a biocatalytic or chemical reaction to remove the product α-keto acid [37].

(*R*)-2-Amino-3-(7-methyl-1*H*-indazol-5-yl)propanoic acid is a key precursor for the antagonists of calcitonin gene-related peptide receptors, which are potentially useful for the treatment of migraine and other maladies. This D-amino acid was prepared in 79 % yield and > 99 % ee via amino transfer reaction to 2-oxo-3-(7-methyl-1*H*-indazol-5-yl)propanoic acid by using a commercially available transaminase as biocatalyst and racemic alanine as the amino donor. However, a large excess of alanine (4.9 equiv) was needed to achieve high conversion. Although

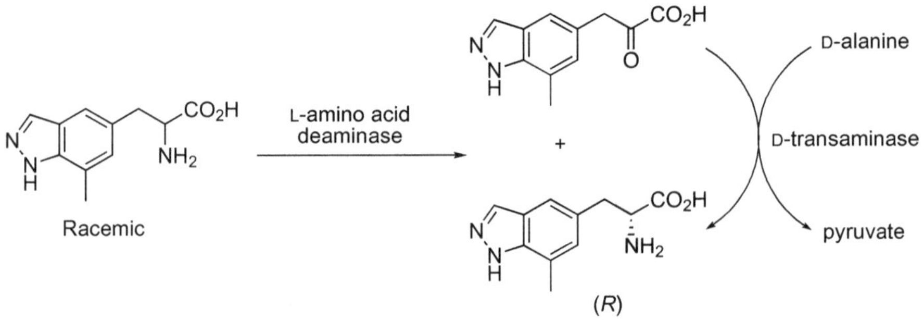

Scheme 4.18 Cascade reaction for the preparation of (*R*)-2-amino-3-(7-methyl-1*H*-indazol-5-yl) propanoic acid

the conversion of the amino transfer reaction could be retained with a small excess of alanine by addition of lactate dehydrogenase and a NADH regeneration system to remove pyruvate produced during the reaction, the lactate dehydrogenase resulted in a by-product hydroxyacid, causing an impurity problem. In this context, a D-transaminase from *Bacillus thuringiensis* SC16569 was found to be less sensitive to inhibition by pyruvate than a commercially available transaminase. The former enzyme effectively catalyzed the conversion of 2-oxo-3-(7-methyl-1*H*-indazol-5-yl) propanoic acid to (*R*)-2-amino-3-(7-methyl-1*H*-indazol-5-yl) propanoic acid at twice the keto acid concentration, half the equivalents of racemic alanine and almost ten times less of pyridoxal phosphate used for the commercial enzyme [90]. Furthermore, this amino transfer reaction could be cascaded with L-amino acid deaminase-catalyzed reaction to convert racemic 2-amino-3-(7-methyl-1*H*-indazol-5-yl)propanoic acid to the optically pure (*R*)-enantiomer (Scheme 4.18).

Enantiopure amines have broad spectrum of biological activities and also serve as chiral building blocks for the synthesis of more complex structures. For instance, *ortho*- and *meta*-halogenated 1-phenylethylamines are intermediates of potent potassium channel openers, modulators of hypertension, calcimimetic agents to treat hyperparathyroidism, or antiarthritic drugs. These substituted 1-phenylethylamines can be readily prepared in enantiopure form via amination of the corresponding substituted acetophenones by employing commercial transaminases as biocatalyst and isopropylamine as amino donor [91]. Interestingly, transaminases efficiently catalyze regio- and stereoselective mono-amination of diketones without protecting group. A series of 2,6-diketones were regioselectively aminated at 2-keto group to chiral amine with high ee values. The generated amino group reacted with the other keto group to form cyclic imines, which were reduced to give optically pure *syn*-2-methyl-6-substituted piperidine derivatives (Scheme 4.19) [92]. Furthermore, the triketone could be regioselectively aminated to afford either enantiomer of the cyclic imine by employing different transaminases (Scheme 4.20) [93]. The cyclic imines were further transformed to synthesize (+)- and (−)-enantiomer of the pyrrolizidine alkaloid xenovenine, a natural product with various bioactivities.

A prominent transaminase-based industrial process is the recently implemented biocatalytic synthesis of the antidiabetic drug sitagliptin. An *R*-selective transaminase from *Arthrobacter* sp., which lacked any activity toward the pro-sitagliptin

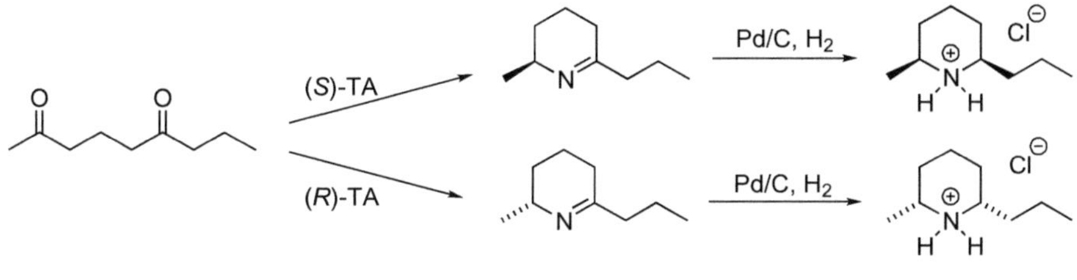

Scheme 4.19 Transaminase catalyzed regio- and stereoselective mono-amination of diketone

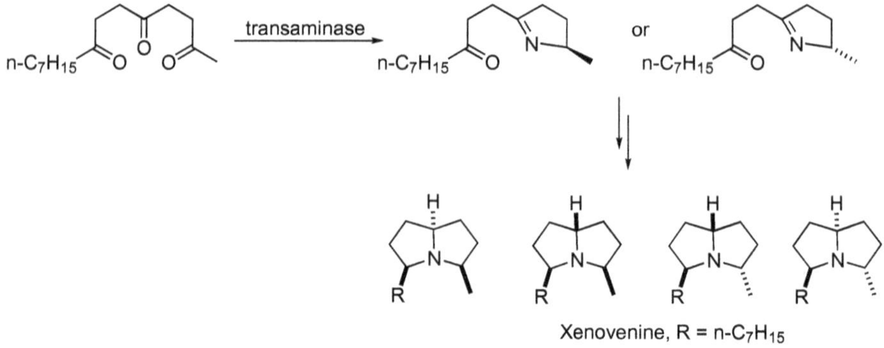

Scheme 4.20 Transaminase catalyzed regio- and stereoselective mono-amination of triketone

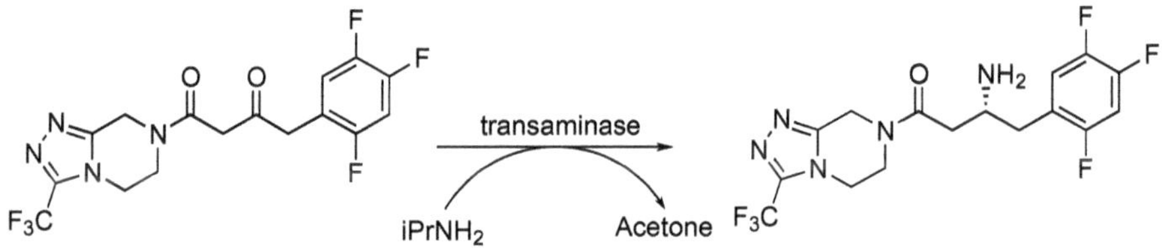

Scheme 4.21 Transaminase catalyzed synthesis of sitagliptin

ketone, was subjected to directed evolution and a variant was obtained to have a broad substrate range and increased tolerance to high concentrations of *iso*-$PrNH_2$ and organic solvents. Under optimal conditions, pro-sitagliptin ketone was aminated to sitagliptin with >99.95 % ee (Scheme 4.21). The biocatalytic process had been applied in a manufacturing setting to replace the previously used rhodium-catalyzed process, resulting in the following benefits: 10–13 % increase in overall yield of sitagliptin; 53 % increase in productivity (kg/L per day); 19 % reduction in total waste; and reduction in total manufacturing cost. The enzymatic reaction eliminated the use of heavy metals and high-pressure hydrogenation reactor [94]. This provided a paradigm for transaminases enabling efficient, economical, and environmentally benign processes for the manufacture of pharmaceuticals.

4.2.7 Amino Acid Dehydrogenases

L-Amino acid dehydrogenases catalyze the reversible aminations of α-ketoacids to their L-amino acid products in the presence of NADH or NADPH. The equilibrium for the amino acid dehydrogenase reactions lies far to the amination side with the K_{eq} values in the range 10^{14}–10^{18}. Thus the reactions are favorable for asymmetrical synthesis of amino acids from the prochiral oxo analogs and ammonia. Indeed, many of these dehydrogenases have been used in the production of non-natural L-α-amino acids. The most prominent representative of amino acids prepared by this method is L-*tert*-leucine (L-2-amino-3,3-dimethyl-butanoic acid), which is an important building block for pharmaceutical industry and produced in tons scale [95, 96]. Another similar L-amino acid of pharmaceutical interest, L-neopentylglycine (L-2-amino-4,4-dimethyl-pentanoic acid), has also been synthesized via a reductive amination of the corresponding α-keto acid catalyzed by recombinant whole cells bearing a leucine dehydrogenase and formate dehydrogenase. The reaction proceeded smoothly with >95 % conversion and >99 % ee at substrate concentrations of 88 g/L [36].

D-Amino acids also serve as important components or building blocks in the production of pharmaceuticals and other fine chemicals. The use of D-amino acid dehydrogenase (D-AADH) should offer a straightforward access to D-amino acids from the reductive amination of 2-keto acids. However, NAD(P)H-dependent D-amino acid dehydrogenase is much less abundant in nature than its L-amino acid counterpart and largely unexplored. The most-known D-AADH is *meso*-diaminopimelate dehydrogenase (*meso*-DAPDH), a key enzyme in the lysine biosynthetic

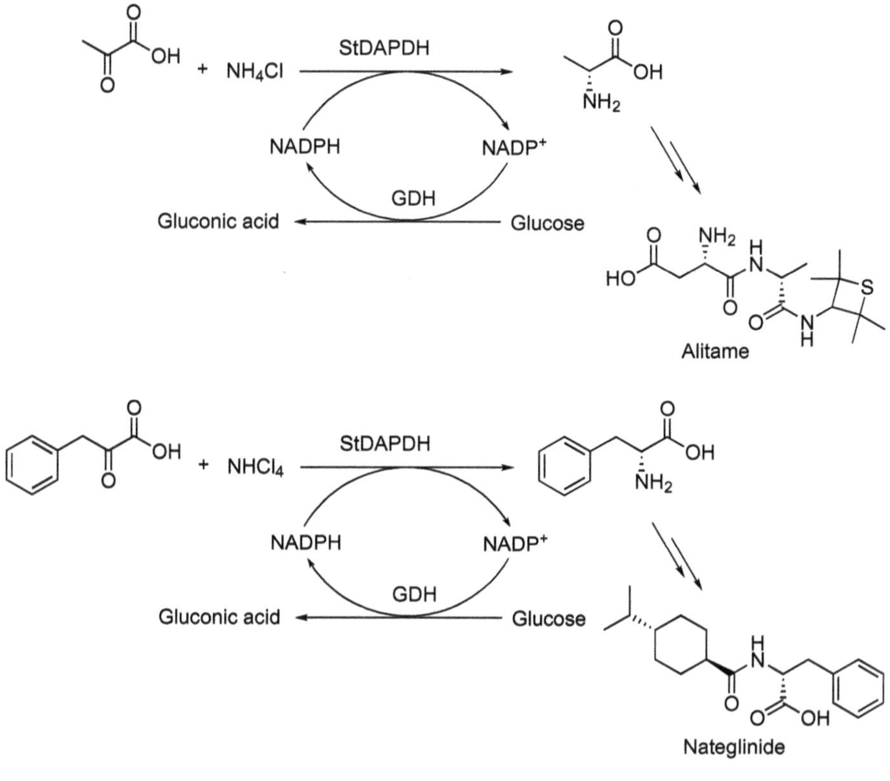

Scheme 4.22 D-Amino acid dehydrogenase catalyzed synthesis of intermediates for alitame and nateglinide

pathway. *meso*-DAPDH is NADP-dependent and catalyzes the reversible oxidative deamination on the D-configuration center of *meso*-2,6-diaminopimelate (*meso*-DAP) to yield L-2-amino-6-oxopimelate. However, most wild-type *meso*-DAPDHs are generally specific toward *meso*-DAP and not applicable for the synthesis of D-amino acids. The *meso*-diaminopimelate dehydrogenase (StDAPDH) from *Symbiobacterium thermophilum* was recently reported to be the only wild-type *meso*-DAPDH with a more relaxed substrate specificity and potential for D-amino acid synthesis [38, 97]. This enzyme and its mutant were applied to the synthesis of D-alanine and D-phenylalanine from the corresponding α-keto acids (Scheme 4.22) [39], respectively. The dipeptide artificial sweeter alitame contains a D-alanine moiety, and D-phenylalanine is the important chiral component of nateglinide, a drug for the treatment of type 2 diabetes. Given the importance of D-amino acids in the fine chemical industry and the advantages of D-AADHs in the synthesis of these compounds, the *meso*-DAPDH gene from *Corynebacterium glutamicum* was subjected to one round of mutagenesis targeted at the substrate binding site and two rounds of random mutagenesis over the entire gene. As a result, a broad substrate range and highly stereospecific D-amino acid dehydrogenase was obtained and applied to the synthesis of D-cyclohexylalanine [40]. Similarly, a mutant D-AADH was created by modification of the *meso*-DAPDH gene from *Bacillus sphaericus* and employed as the biocatalyst for the production of (*R*)-5,5,5-trifluoronorvaline, an intermediate for a γ-secretase inhibitor (BMS-708163) [98].

4.2.8 Aldolases

Stereoselective C–C bond formation is of utmost importance in organic synthesis. Aldolase catalyzes the reversible aldol addition of a nucleophilic donor onto an electrophilic aldehyde accept to generate a new β-hydroxyl carbonyl compound with up to two new stereogenic centers. As such, aldolase is an important C–C bond formation tool in building up complexity of organic molecules such as carbohydrates, amino acids, and their analogs. In reaction mechanism, aldolase first activates the nucleophilic donor substrate to form an enamine or enolate, which then attacks the carbonyl carbon of the acceptor to generate the new C–C bond. According to the different activation mechanisms of the nucleophilic component, aldolases are classified as Class I and II. For Class I aldolases, a conserved lysine residue in the active site forms a Schiff base intermediate with the donor component to generate an enamine nucleophile. In Class II enzymes, a divalent metal ion promotes the enolization of the donor substrate via Lewis acid complexation. While the acceptor of aldolases can be a broad range of aldehydes, the nucleophilic donor is often structurally limited to a few compounds. As such, aldolases are also grouped according to their donor specificity into pyruvate/2-oxobutyrate aldolases, dihydroxyacetone phosphate (DHAP) aldolases, dihydroxyacetone (DHA) aldolases, glycine/alanine aldolases, and acetaldehyde aldolases. These various aldolases allow assembly of complex chiral molecules from smaller starting materials and diversity-oriented synthesis of various structures and stereochemistries.

N-Acetylneuraminic acid and its analogs are involved in a number of physiological functions and pathological processes, such as cellular recognition and communication, bacterial and viral infection, and tumor metastasis. N-Acetylneuraminic acid lyase (NeuA), a Class I pyruvate-dependent aldolase, reversibly catalyzes the aldol addition of pyruvate to a variety of polyhydroxylated aldehydes yielding α-oxo acids. This enzyme and its mutants have been extensively exploited for the synthesis of *N*-acetylneuraminic acid and its analogs [99, 100].

D-Fructose-6-phosphate aldolase catalyzes the stereoselective reversible aldol addition of DHA on D-glyceraldehyde 3-phosphate (D-G3P) to afford D-fructose 6-phosphate. This enzyme and mutants were found to be active toward a variety of aldehyde acceptors, showing remarkable catalytic potential in the asymmetric synthesis of polyhydroxylated compounds such as iminocyclitols and carbohydrates. Compared to DHAP-dependent aldolases, this DHA aldolase catalyzes aldol additions of unphosphorylated DHA onto the acceptors, avoiding the use of expensive DHA phosphate [101].

Threonine aldolase (ThrA) catalyzes the aldol addition of glycine to aldehyde to afford chiral β-hydroxy-α-amino acid, a structural motif widely found in drug molecules and their intermediates. For instance, (2*R*,3*S*)-2-amino-3-hydroxy-3-(pyridin-4-yl)-propanoic acid is a key intermediate in the synthesis of a developmental drug candidate, (2R,3S)-2-amino-3-hydroxy3-(pyridin-4-yl)-1-(pyrrolidin-1-yl) propan-1-one. An efficient D-threonine aldolase-catalyzed synthesis of the β-hydroxy-α-amino acid was developed, and the resulting product was used in the production of an active pharmaceutical ingredient with a very high ee (>99 %) and de (99.8 %, Scheme 4.23) [102].

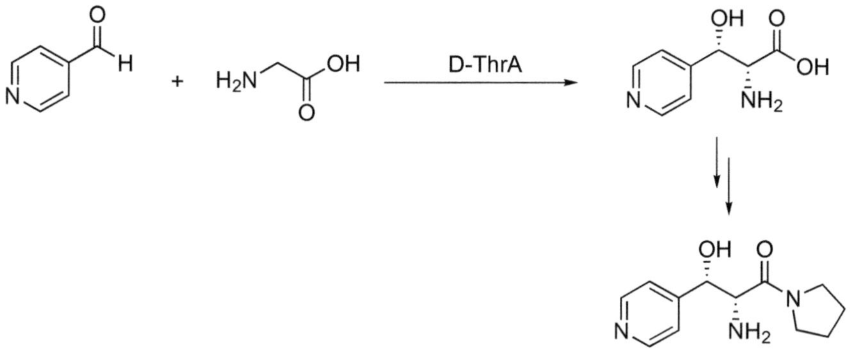

Scheme 4.23 Threonine aldolase catalyzed synthesis of precursor for a developmental drug candidate

4.2.9 Miscellaneous Enzymes

In addition to the above enzymes, there are others that have attracted considerable attention due to their remarkable application potential in the synthesis of valuable chemicals used in various industries.

Hydroxy nitrilelyases catalyze the reversible release of a cyanide group in the form of HCN from cyanohydrin to give the corresponding aldehyde or ketone. The reverse reaction has been used for the synthesis of enantiomerically pure cyanohydrins of great importance in chemical industry, such as (*R*)- or (*S*)-mandelonitrile [103].

An enantioselective amine oxidase can be used in combination with a nonselective reducing agent to effect the stereo inversion of the *S* to *R* enantiomer via the prochiral imine intermediate. The applications of a monoamine oxidase (MAO) from *Aspergillus niger,* a bacterial cyclohexylamine oxidase (CHAO) from *Brevibacterium oxydans* IH-35A, and their mutant enzymes have led to a series of novel biocatalytic methods for the preparation of several optically active chiral amines of pharmaceutical importance by deracemization of the corresponding racemic mixture with borane-ammonia hydride complex, for example, as a reducing agent [35, 104, 105, 106]. Similarly, deracemization and stereoinversion of DL-α-amino acids was achieved by using D-amino acid oxidase (DAAO) and a hydride reducing agent such as $NaBH_3CN$–$NaBH_4$ [107].

Recently, imine reductase (IRED) has become a hot topic in the field of biocatalysis [108]. The asymmetric reduction of imines provides an alternative method for synthesizing chiral amines. Imine reductases are NADPH-dependent oxidoreductases that catalyze the asymmetric reduction of cyclic imines to amines with excellent stereoselectivity. Both (*S*)- and (*R*)-imine reductases are available and have been applied to the production of both (*S*) and (*R*) cyclic secondary amines. For example, (*R*)-IRED from *Streptomyces* sp. GF3587 and (*S*)-IRED from *Streptomyces* sp. GF3546 were shown to catalyze the enantioselective reduction of 2-substituted cyclic imines to give enantio complementary 2-substituted cyclic amines [109–111].

4.3 Enzymatic and Chemoenzymatic Cascades to Green Synthesis

4.3.1 Enzymatic Cascade

In the preceding section, some prime examples of individually enzyme-catalyzed reactions for the synthesis of chemicals were presented. In order to develop more environmentally benign and sustainable chemical processes, the biocatalysis community has paid increasing attention to "systems biocatalysis", where more than one enzyme from various organisms are put together to perform a cascade of reactions that can produce structurally complex molecules from simple starting materials. The multienzyme reactions can proceed in two major approaches: the use of isolated enzymes (in vitro) or the use of whole cells with the enzymes (in vivo).

A large number of important and versatile enzymes have been employed in multienzymatic cascade routes to synthesize high-value chemicals that have great potential of reducing production cost and wastes by avoiding time-consuming and yield-reducing isolation and purification of intermediates. For example, an ER can be used with ADHs in combination with a GDH in a one-pot consecutive two-enzyme sequential cascade to concurrently reduce both C=C and C=O bonds. By employing this process, chiral γ-butyrolactones were obtained in up to 90 % yields and with 98–99 % ee from ethyl 4-oxo-pent-2-enoates, which are readily accessible via Wittig-type reactions [112]. Similarly, (*R*)- or (*S*)-carvone had been converted to (1*S*,2*R*,5*R*)-dihydrocarveol or (1*S*,2*R*,5*S*)-dihydrocarveol with high efficiency and stereoselectivity [89], and the most odorous stereoisomers of the chiral commercial fragrance Muguesia® [(2*R*,3*S*)-3-methyl-4-phenylbutan-2-ol)] had also been prepared [113].

Ethyl (*R*)-3-hydroxyglutarate is a key intermediate for the synthesis of rosuvastatin. An effective one-pot bienzymatic synthesis of ethyl (R)-3-hydroxyglutarate from ethyl (S)-4-chloro-3-hydroxybutyrate was achieved using recombinant *E. coli* cells expressing separately or co-expressing a mutant halohydrin dehalogenase gene from *Agrobacterium radiobacter* AD1 and a nitrilase gene from *A. thalian.* In the fed-batch process, a high accumulative product concentration of up to 0.9 mol $L^{-1}$of ethyl (*R*)-3-hydroxyglutarate was achieved in this one-pot process by fed-batch of ethyl (*S*)-4-chloro-3-hydroxybutyrate and NaCN to get rid of their inhibition on the nitrilase activity [114]. Nitrilase had also been combined with hydroxynitrile lyase or carbonyl reductase to form biocatalytic cascades for the asymmetric synthesis of chiral α- or β-hydroxycarboxylic acids [50, 53, 115].

A cascade biotransformation involving multiple reactions has been developed as a synthetic biology tool for one-pot conversion of cycloalkanes to their corresponding lactones (Scheme 4.24). The synthetic pathway consists of a cytochrome P450 monooxygenase (CYP450) for initial oxyfunctionalization of the cycloalkane, an ADH for ketone production and a Baeyer–Villiger monooxygenase (BVMO) for lactone formation [116]. Poly-ε-caprolactone (PCL) is a biodegradable polyester with various applications such as an additive for resins to improve their processing characteristics and their end use properties. For example, it is widely used in the manufacture of specialty polyurethanes to impart good water, oil, solvent, and chlorine resistance to the product. This polymer also finds many applications in the hobbyist market, and attracts increasing attention in biomedical applications because of its compatibility with physiological conditions. ε-Caprolactone is currently chemically produced by using hazardous peracetic acid as an oxidation reagent. BVMO would offer the enzymatic synthesis of ε-caprolactone (ε-CL) directly from cyclohexanone with molecular oxygen, but the substrate and product inhibitions to the enzyme activity are part of the contributing factors to their industrial application. ADH enabled an efficient in situ conversion of cyclohexanol to cyclohexanone, removing the substrate inhibition. The apparent product inhibition was prevented by a subsequent direct ring-opening oligomerization of the formed ε-CL by using lipase A from *C. antarctica*. Using this three-enzyme cascade oligo-ε-CL at more than 20 g $L^{-1}$was produced starting at 200 mM of cyclohexanol [117]. ε-Caprolactone could also be hydrolyzed by lipase to liberate the alcohol moiety to give 6-hydroxyhexanoic acid, which was then oxidized to 6-oxohexanoic acid. Following amination using an ω-transaminase as biocatalyst and L-alanine as amino donor, 6-aminohexanoic acid was produced. This terminal amino acid is the open-chain analog of ε-caprolactam, the building block of polymer nylon-6 [118]. In principle, this study showed a tremendous application potential of a multienzymatic process for the synthesis of chemicals like nylon for our daily use.

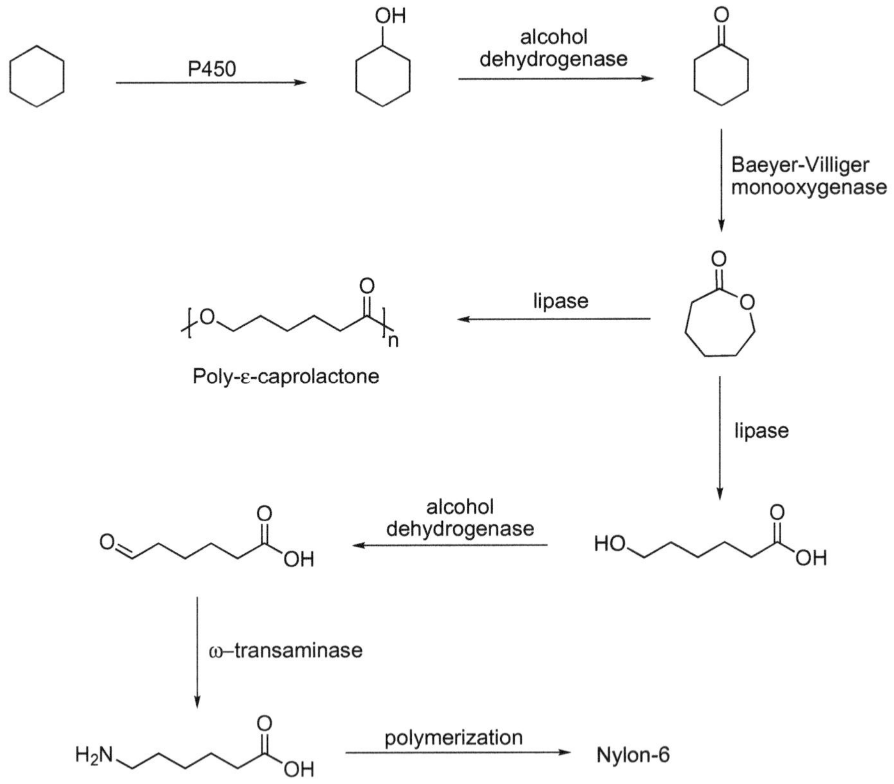

Scheme 4.24 Enzymatic cascades for the synthesis of poly-ε-caprolactone and Nylon-6 from cyclohexane

4.3.2 *Chemo-Enzymatic Cascade*

The chemo-enzymatic transformations are usually carried out in one of the three modes: separate-pot-two-step, one-pot-two-step, and one-pot-one-step. By separate-pot two-step, it means that the chemical reaction and biotransformation are carried out in two steps and separate pots, and the intermediate needs to be isolated. For example, enzymatic reduction of α-azidoacetophenone derivatives catalyzed by a recombinant carbonyl reductase from *C. magnolia* (CMCR) or an ADH from *S. cerevisiae* (Ymr226c) generated 2-azido-1-arylethanols with excellent optical purity. The isolated (*S*)-2-azido-1-(*p*-chlorophenyl)ethanols were then subjected to react with alkynes employing "click" chemistry to afford optically pure triazole-containing β-adrenergic receptor blocker analogs [119].

In the case of one-pot-two-step mode, the chemical reaction and biotransformation are carried out sequentially in one pot without the isolation of intermediates. The bioreduction of ketones and Cu-catalyzed "click" reaction can also be performed in one-pot-two-step mode. Some chiral 1,2,3-triazole-derived diols were synthesized in high yields and excellent enantio- and diastereoselectivities under very mild conditions in aqueous medium by this one-pot tandem reaction [120].

The most efficient mode of chemoenzymatic cascade is one-pot-one-step, in which the chemical reaction and biotransformation proceed concertedly without high accumulation of the intermediate. An example is our recent report that deracemization of 2-methyl-1,2,3,4-tetrahydroquinoline was achieved by using a mutant cyclohexylamine oxidase (CHAO) with reducing agent $BH_3{\cdot}NH_3$. The bio-oxidation of (*S*)-2-methyl-1,2,3,4-tetrahydroquinoline led to an imine intermediate, which was immediately reduced by the chemical reducing agent, resulting in the production of (*R*)-2-methyl-1,2,3,4-tetrahydroquinoline with 76 % isolated yield and 98 % ee (Scheme 4.25) [106]. Tetrahydroquinoline is a "privileged" scaffold or substructure in many biologically active natural products and therapeutic agents used in cancer drug development.

Both chemical synthesis and biotransformation have their own features, including advantages and disadvantages. Chemoenzymatic cascade reactions can give full play to their advantages, and thus offer tremendous opportunity for developing efficient and sustainable processes by designing novel synthetic routes

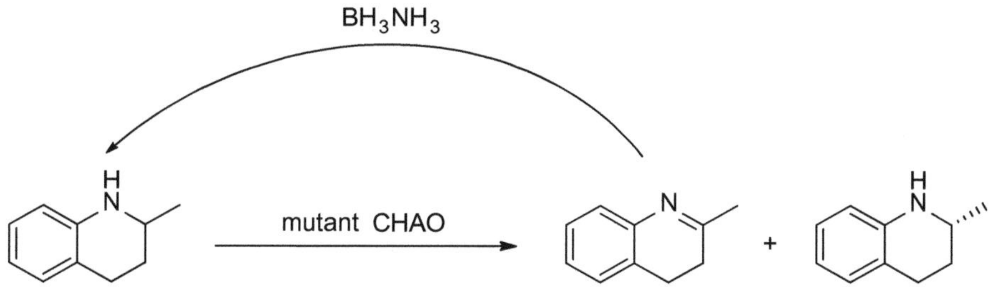

Scheme 4.25 Deracemization of 2-methyl-1,2,3,4-tetrahydroquinoline

from the starting materials to the products with retrosynthesis strategy, a disconnection approach that involves breaking down of the target molecule into simpler structures which then become the new target structure by some backward (antithetical) reactions. At same time, there are many challenges, such as the reaction condition compatibility of chemical reaction with biotransformation, the inhibition or inactivation of (bio)catalysts, and so on, which must be addressed before industrial application. In this context, scientists and engineers are designing chemical catalysts for use in aqueous reaction medium and evolving enzyme catalysts with tolerance of organic solvents, so that these catalysts can work under the same reaction conditions. Reaction engineering has also been carried out for developing efficient and economical chemoenzymatic cascade processes [121].

4.4 Concluding Remarks

Nature has created and evolved a diversity of enzymes, which catalyze different reactions in live organisms and show various advantages over traditional chemical reactions. As our understanding of the enzymes' properties and reaction mechanisms have deepened and the technologies in molecular biology and fermentation engineering have developed further, more and more useful enzymes have been produced and utilized in various industries. In chemical industry, because of the unparalleled selectivity and mild reaction conditions, enzymes can be expected to continue to play an exciting role in the advancement of "green chemistry". Enzymatic and chemoenzymatic cascade strategies will continue to offer unprecedented opportunity for developing efficient and sustainable synthetic technology to address the issues of health, environment, energy, and security that we face today.

References

1. de Raadt A, Klempier N, Faber K, Griengl H (1992) Chemoselective enzymatic hydrolysis of aliphatic and alicyclic nitriles. J Chem Soc Perkin Trans 1(1):137–140
2. Debabov VG, Yanenko AS (2011) Biocatalytic hydrolysis of nitriles. Ref J Chem 1(4): 385–402
3. Martínková L, Křen V (2002) Nitrile- and amide-converting microbial enzymes: Stereo-, regio- and chemoselectivity. Biocatal Biotransfor 20(2):73–93
4. Yao P, Li J, Yuan J, Han C, Liu X, Feng J, Wu Q, Zhu D (2015) Enzymatic synthesis of a key intermediate for rosuvastatin by nitrilase-catalyzed hydrolysis of ethyl (*R*)-4-cyano-3-hydroxybutyate at high substrate concentration. ChemCatChem 7(2):271–275
5. Martínková L, Klempier N, Prepechalov I, Prikrylová V, Ovesná M, Griengl H, Kren V (1998) Chemoselective biotransformation of nitriles by *Rhodococcus equi* A4. Biotechnol Lett 20(10):909–912
6. Martínková, L, Klempier, N, Bardakji, J, Kandelbauer, A, Ovesná, M, Podailová, T, Kuzma, M, Pepechalová, I, Griengl, H, Kren, VR (2001) Biotransformation of 3-substituted methyl

(*R,S*)-4-cyanobutanoates with nitrile- and amide-converting biocatalysts. J Mol Catal B Enzymatic 14 (4–6):95–99
7. Li G, Ren J, Wu Q, Feng J, Zhu D, Ma Y (2013) Identification of a marine NADPH-dependent aldehyde reductase for chemoselective reduction of aldehydes. J Mol Catal B Enzymatic 90:17–22
8. Salvano MS, Cantero JJ, Vázquez AM, Formica SM, Aimar ML (2011) Searching for local biocatalysts: Bioreduction of aldehydes using plant roots of the province of Córdoba (Argentina). J Mol Catal B Enzymatic 71(1–2):16–21
9. Sello G, Orsini F, Bernasconi S, Di Gennaro P (2006) Selective enzymatic reduction of aldehydes. Molecules 11(5):365–369
10. Dub PA, Ikariya T (2012) Catalytic reductive transformations of carboxylic and carbonic acid derivatives using molecular hydrogen. ACS Catal 2(8):1718–1741
11. Seyden-Penne J (1997) Reductions by the alumino- and borohydrides in organic synthesis. Wiley-VCH Inc, New York
12. Brewster TP, Miller AJM, Heinekey DM, Goldberg KI (2013) Hydrogenation of carboxylic acids catalyzed by half-sandwich complexes of iridium and rhodium. J Am Chem Soc 135(43):16022–16025
13. Otsuka T, Ishii A, Dub PA, Ikariya T (2013) Practical selective hydrogenation of α-fluorinated esters with bifunctional pincer-type ruthenium(II) catalysts leading to fluorinated alcohols or fluoral hemiacetals. J Am Chem Soc 135(26):9600–9603
14. Hollmann F, Arends IWCE, Holtmann D (2011) Enzymatic reductions for the chemist. Green Chem 13(9):2285–2314
15. Akhtar MK, Turner NJ, Jones PR (2013) Carboxylic acid reductase is a versatile enzyme for the conversion of fatty acids into fuels and chemical commodities. Proc Natl Acad Sci 110(1):87–92
16. Duan Y, Yao P, Chen X, Liu X, Zhang R, Feng J, Wu Q, Zhu D (2015) Exploring the synthetic applicability of a new carboxylic acid reductase from *Segniliparus rotundus* DSM 44985. J Mol Catal B: Enzymatic 115:1–7
17. Duan Y, Yao P, Du Y, Feng J, Wu Q, Zhu D (2015) Synthesis of α, β-unsaturated esters via a chemo-enzymatic chain elongation approach by combining carboxylic acid reduction and wittig reaction. Beil J Org Chem 11:2245–2251
18. He A, Li T, Daniels L, Fotheringham I, Rosazza JPN (2004) *Nocardia* sp. Carboxylic acid reductase: Cloning, expression, and characterization of a new aldehyde oxidoreductase family. Appl Environ Microbiol 70(3):1874–1881
19. Grau BT, Devine PN, DiMichele LN, Kosjek B (2007) Chemo- and enantioselective routes to chiral fluorinated hydroxyketones using ketoreductases. Org Lett 9(24):4951–4954
20. Zhang D, Zhang R, Zhang J, Chen L, Zhao C, Dong W, Zhao Q, Wu Q, Zhu D (2014) Engineering a hydroxysteroid dehydrogenase to improve its soluble expression for the asymmetric reduction of cortisone to 11β-hydrocortisone. Appl Microbiol Biotechnol 98 (21):8879–8886
21. Bayer S, Birkemeyer C, Ballschmiter M (2011) A nitrilase from a metagenomic library acts regioselectively on aliphatic dinitriles. Appl Microbiol Biotechnol 89(1):91–98
22. Mukherjee C, Zhu D, Biehl ER, Parmar RR, Hua L (2006) Enzymatic nitrile hydrolysis catalyzed by nitrilase ZmNIT2 from maize. An unprecedented β-hydroxy functionality enhanced amide formation. Tetrahedron 62(26):6150–6154
23. Veselá A, Rucká L, Kaplan O, Pelantová H, Nešvera J, Pátek M, Martínková L (2016) Bringing nitrilase sequences from databases to life: The search for novel substrate specificities with a focus on dinitriles. Appl Microbiol Biotechnol 100(5):2193–2202
24. Zhu D, Mukherjee C, Biehl ER, Hua L (2007) Nitrilase-catalyzed selective hydrolysis of dinitriles and green access to the cyanocarboxylic acids of pharmaceutical importance. Adv Synth Catal 349(10):1667–1670
25. Zhu D, Mukherjee C, Biehl ER, Hua L (2007) Discovery of a mandelonitrile hydrolase from *Bradyrhizobium japonicum*USDA110 by rational genome mining. J Biotechnol 129(4):645–650

26. Duan Y, Yao P, Ren J, Han C, Li Q, Yuan J, Feng J, Wu Q, Zhu D (2014) Biocatalytic desymmetrization of 3-substituted glutaronitriles by nitrilases. A convenient chemoenzymatic access to optically active (*S*)-pregabalin and (*R*)-baclofen. Sci China Chem 57 (8):1164–1171
27. Carey JS, Laffan D, Thomson C, Williams MT (2006) Analysis of the reactions used for the preparation of drug candidate molecules. Org Biomol Chem 4(12):2337–2347
28. Moore JC, Pollard DJ, Kosjek B, Devine PN (2007) Advances in the enzymatic reduction of ketones. Acc Chem Res 40(12):1412–1419
29. Wildeman SMAD, Sonke T, Schoemaker HE, May O (2007) Biocatalytic reductions: From lab curiosity to "first choice". Acc Chem Res 40(12):1260–1266
30. Ohkuma T, Koizumi M, Ikehira H, Yokozawa T, Noyori R (2000) Selective hydrogenation of benzophenones to benzhydrols. Asymmetric synthesis of unsymmetrical diarylmethanols. Org Lett 2(5):659–662
31. Wu J, Ji J-X, Guo R, Yeung C-H, Chan ASC (2003) Chiral [$RuCl_2$(dipyridylphosphane)(1,2-diamine)] catalysts: applications in asymmetric hydrogenation of a wide range of simple ketones. Chem Eur J 9 (13):2963–2968
32. Li H, Zhu D, Hua L, Biehl ER (2009) Enantioselective reduction of diaryl ketones catalyzed by a carbonyl reductase from *Sporobolomyces salmonicolor* and its mutant enzymes. Adv Synth Catal 351(4):583–588
33. Truppo MD, Pollard D, Devine P (2007) Enzyme-catalyzed enantioselective diaryl ketone reductions. Org Lett 9(2):335–338
34. Liang J, Mundorff E, Voladri R, Jenne S, Gilson L, Conway A, Krebber A, Wong J, Huisman G, Truesdell S, Lalonde J (2010) Highly enantioselective reduction of a small heterocyclic ketone: Biocatalytic reduction of tetrahydrothiophene-3-one to the corresponding (*R*)-alcohol. Org Proc Res Dev 14(1):188–192
35. Ghislieri D, Turner NJ (2014) Biocatalytic approaches to the synthesis of enantiomerically pure chiral amines. Top Catal 57(5):284–300
36. Groeger H, May O, Werner H, Menzel A, Altenbuchner J (2006) A "second-generation process" for the synthesis of L-neopentylglycine: asymmetric reductive amination using a recombinant whole cell catalyst. Org Proc Res Dev 10(3):666–669
37. Zhu D, Hua L (2009) Biocatalytic asymmetric amination of carbonyl functional groups—a synthetic biology approach to organic chemistry. Biotechnol J 4(10):1420–1431
38. Gao X, Chen X, Liu W, Feng J, Wu Q, Hua L, Zhu D (2012) A novel *meso*-diaminopimelate dehydrogenase from *Symbiobacterium thermophilum*: overexpression, characterization, and potential for D-amino acid synthesis. Appl Environ Microbiol 78(24):8595–8600
39. Gao X, Huang F, Feng J, Chen X, Zhang H, Wang Z, Wu Q, Zhu D (2013) Engineering the *meso*-diaminopimelate dehydrogenase from *Symbiobacterium thermophilum* by site saturation mutagenesis for D-phenylalanine synthesis. Appl Environ Microbiol 79(16):5078–5081
40. Vedha-Peters K, Gunawardana M, Rozzell JD, Novick SJ (2006) Creation of a broad-range and highly stereoselective D-amino acid dehydrogenase for the one-step synthesis of D-amino acids. J Am Chem Soc 128(33):10923–10929
41. Zhang D, Chen X, Zhang R, Yao P, Wu Q, Zhu D (2015) Development of β-amino acid dehydrogenase for the synthesis of β-amino acids via reductive amination of β-keto acids. ACS Catal 5(4):2220–2224
42. Abrahamson MJ, Vázquez-Figueroa E, Woodall NB, Moore JC, Bommarius AS (2012) Development of an amine dehydrogenase for synthesis of chiral amines. Angew Chem Int Ed 51(16):3969–3972
43. Abrahamson MJ, Wong JW, Bommarius AS (2013) The evolution of an amine dehydrogenase biocatalyst for the asymmetric production of chiral amines. Adv Synth Catal 355(9):1780–1786
44. Ye LJ, Toh HH, Yang Y, Adams JP, Snajdrova R, Li Z (2015) Engineering of amine dehydrogenase for asymmetric reductive amination of ketone by evolving *Rhodococcus* phenylalanine dehydrogenase. ACS Catal 5(2):1119–1122

45. Chen F-F, Liu Y-Y, Zheng G-W, Xu J-H (2015) Asymmetric amination of secondary alcohols by using a redox-neutral two-enzyme cascade. ChemCatChem 7(23):3838–3841
46. Mutti FG, Knaus T, Scrutton NS, Breuer M, Turner NJ (2015) Conversion of alcohols to enantiopure amines through dual-enzyme hydrogen-borrowing cascades. Science 349 (6255):1525–1529
47. Rocha L, Ferreira H, Pimenta E, Berlinck R, Rezende M, Landgraf M, Seleghim M, Sette L, Porto A (2010) Biotransformation of α-bromoacetophenones by the marine fungus *Aspergillus sydowii*. Mar Biotechnol 12(5):552–557
48. Ren J, Dong W, Yu B, Wu Q, Zhu D (2012) Synthesis of optically active α-bromohydrins via reduction of α-bromoacetophenone analogues catalyzed by an isolated carbonyl reductase. Tetrahedron Asymmetry 23(6–7):497–500
49. Hann EC, Sigmund AE, Fager SK, Cooling FB, Gavagan JE, Ben-Bassat A, Chauhan S, Payne MS, Hennessey SM, DiCosimo R (2003) Biocatalytic hydrolysis of 3-hydroxyalkanenitriles to 3-hydroxyalkanoic acids. Adv Synth Catal 345 (6+7):775–782
50. Ankati H, Zhu D, Yang Y, Biehl ER, Hua L (2009) Asymmetric synthesis of both antipodes of β-hydroxy nitriles and β-hydroxy carboxylic acids via enzymatic reduction or sequential reduction/hydrolysis. J Org Chem 74(4):1658–1662
51. Coady TM, Coffey LV, O'Reilly C, Owens EB, Lennon CM (2013) A high throughput screening strategy for the assessment of nitrile-hydrolyzing activity towards the production of enantiopure β-hydroxy acids. J Mol Catal B Enzymatic 97:150–155
52. Kamila S, Zhu D, Biehl ER, Hua L (2006) Unexpected stereorecognition in nitrilase-catalyzed hydrolysis of β-hydroxy nitriles. Org Lett 8(20):4429–4431
53. Zhu D, Ankati H, Mukherjee C, Yang Y, Biehl ER, Hua L (2007) Asymmetric reduction of β-ketonitriles with a recombinant carbonyl reductase and enzymatic transformation to optically pure β-hydroxy carboxylic acids. Org Lett 9(13):2561–2563
54. Hu S, Kelly S, Lee S, Tao J, Flahive E (2006) Efficient chemoenzymatic synthesis of pelitrexol via enzymic differentiation of a remote stereocenter. Org Lett 8(8):1653–1655
55. Truppo MD, Journet M, Shafiee A, Moore JC (2006) Optimization and scale-up of a lipase-catalyzed enzymatic resolution of an indole ester intermediate for a prostaglandin D2 (DP) receptor antagonist targeting allergic rhinitis. Org Proc Res Dev 10(3):592–598
56. Martinez CA, Hu S, Dumond Y, Tao J, Kelleher P, Tully L (2008) Development of a chemoenzymatic manufacturing process for pregabalin. Org Proc Res Dev 12(3):392–398
57. Tanaka K, Yoshida K, Sasaki C, Osano YT (2002) Practical asymmetric synthesis of the herbicide (s)-indanofan via lipase-catalyzed kinetic resolution of a diol and stereoselective acid-catalyzed hydrolysis of a chiral epoxide. J Org Chem 67(9):3131–3133
58. Aleu J, Bustillo AJ, Hernandez-Galan R, Collado IG (2006) Biocatalysis applied to the synthesis of agrochemicals. Curr Org Chem 10(16):2037–2054
59. Hu S, Martinez CA, Kline B, Yazbeck D, Tao J, Kucera DJ (2006) Efficient enzymatic process for the production of (2*S*)-4,4-difluoro-3,3-dimethyl-N-Boc-proline, a key intermediate in the synthesis of HIV protease inhibitors. Org Proc Res Dev 10(3):650–654
60. Bergeron S, Chaplin DA, Edwards JH, Ellis BSW, Hill CL, Holt-Tiffin K, Knight JR, Mahoney T, Osborne AP, Ruecroft G (2006) Nitrilase-catalysed desymmetrisation of 3-hydroxyglutaronitrile: preparation of a statin side-chain intermediate. Org Proc Res Dev 10 (3):661–665
61. DeSantis G, Wong K, Farwell B, Chatman K, Zhu Z, Tomlinson G, Huang H, Tan X, Bibbs L, Chen P, Kretz K, Burk MJ (2003) Creation of a productive, highly enantioselective nitrilase through gene site saturation mutagenesis (gssm). J Am Chem Soc 125(38):11476–11477
62. DeSantis G, Zhu Z, Greenberg WA, Wong K, Chaplin J, Hanson SR, Farwell B, Nicholson LW, Rand CL, Weiner DP, Robertson DE, Burk MJ (2002) An enzyme library approach to biocatalysis: Development of nitrilases for enantioselective production of carboxylic acid derivatives. J Am Chem Soc 124(31):9024–9025

63. Cooling FB, Fager SK, Fallon RD, Folsom PW, Gallagher FG, Gavagan JE, Hann EC, Herkes FE, Phillips RL, Sigmund A, Wagner LW, Wu W, DiCosimo R (2001) Chemoenzymatic production of 1,5-dimethyl-2-piperidone. J Mol Catal B Enzymatic 11 (4–6):295–306
64. Shaw NM, Robins KT, Kiener A (2003) Lonza: 20 years of biotransformations. Adv Synth Catal 345(4):425–435
65. Cantarella L, Gallifuoco A, Malandra A, Martínková L, Spera A, Cantarella M (2011) High-yield continuous production of nicotinic acid via nitrile hydratase–amidase cascade reactions using cascade csmrs. Enzyme Microb Technol 48(4–5):345–350
66. Hann EC, Eisenberg A, Fager SK, Perkins NE, Gallagher FG, Cooper SM, Gavagan JE, Stieglitz B, Hennessey SM, DiCosimo R (1999) 5-cyanovaleramide production using immobilized *Pseudomonas chlororaphis* b23. Bioorg Med Chem 7(10):2239–2245
67. Steinreiber A, Faber K (2001) Microbial epoxide hydrolases for preparative biotransformations. Curr Opin Biotechnol 12(6):552–558
68. Kong X-D, Yu H-L, Yang S, Zhou J, Zeng B-B, Xu J-H (2015) Chemoenzymatic synthesis of (*R*)- and (*S*)-propranolol using an engineered epoxide hydrolase with a high turnover number. J Mol Catal B Enzymatic 122:275–281
69. Edegger K, Mayer SF, Steinreiber A, Faber K (2004) Chemo-enzymatic enantio-convergent asymmetric synthesis of (*R*)-(+)-marmin. Tetrahedron 60(3):583–588
70. Bottalla A-L, Ibrahim-Ouali M, Santelli M, Furstoss R, Archelas A (2007) Epoxide hydrolase-catalysed kinetic resolution of a spiroepoxide, a key building block of various 11-heterosteroids. Adv Synth Catal 349(7):1102–1110
71. Deregnaucourt J, Archelas A, Barbirato F, Paris J-M, Furstoss R (2007) Enzymatic transformations 63. High-concentration two liquid-liquid phase *Aspergillus niger* epoxide hydrolase-catalysed resolution: Application to trifluoromethyl-substituted aromatic epoxides. Adv Synth Catal 349 (8–9):1405–1417
72. Zhu D, Mukherjee C, Hua L (2005) 'Green' synthesis of important pharmaceutical building blocks: Enzymatic access to enantiomerically pure α-chloroalcohols. Tetrahedron Asymmetry 16(19):3275–3278
73. Zhu DM, Hyatt BA, Hua L (2009) Enzymatic hydrogen transfer reduction of α-chloro aromatic ketones catalyzed by a hyperthermophilic alcohol dehydrogenase. J Mol Catal B Enzym 56(4):272–276
74. Pollard D, Truppo M, Pollard J, Chen C-y, Moore J (2006) Effective synthesis of (s)-3,5-bistrifluoromethylphenyl ethanol by asymmetric enzymatic reduction. Tetrahedron Asymmetry 17 (4):554–559
75. Kizaki N, Yasohara Y, Hasegawa J, Wada M, Kataoka M, Shimizu S (2001) Synthesis of optically pure ethyl (*S*)-4-chloro-3-hydroxybutanoate by *Escherichia coli* transformant cells coexpressing the carbonyl reductase and glucose dehydrogenase genes. Appl Microbiol Biotechnol 55(5):590–595
76. Pan J, Zheng G-W, Ye Q, Xu J-H (2014) Optimization and scale-up of a bioreduction process for preparation of ethyl (*S*)-4-chloro-3-hydroxybutanoate. Org Proc Res Dev 18 (6):739–743
77. Yamamoto H, Matsuyama A, Kobayashi Y (2002) Synthesis of ethyl (*R*)-4-chloro-3-hydroxybutanoate with recombinant *Escherichia coli* cells expressing (*S*)-specific secondary alcohol dehydrogenase. Biosci Biotechnol Biochem 66(2):481–483
78. Shen N-D, Ni Y, Ma H-M, Wang L-J, Li C-X, Zheng G-W, Zhang J, Xu J-H (2012) Efficient synthesis of a chiral precursor for angiotensin-converting enzyme (ACE) inhibitors in high space–time yield by a new reductase without external cofactors. Org Lett 14(8):1982–1985
79. Ni Y, Li C-X, Zhang J, Shen N-D, Bornscheuer UT, Xu J-H (2011) Efficient reduction of ethyl 2-oxo-4-phenylbutyrate at 620 g L^{-1} by a bacterial reductase with broad substrate spectrum. Adv Synth Catal 353(8):1213–1217
80. Ema T, Okita N, Ide S, Sakai T (2007) Highly enantioselective and efficient synthesis of methyl (*R*)-*o*-chloromandelate with recombinant *E. coli*: toward practical and green access to clopidogrel. Org Biomol Chem 5(8):1175–1176

81. Xu Y-P, Guan YH, Yu H-L, Ni Y, Ma B-D, Xu J-H (2014) Improved o-chlorobenzoylformate bioreduction by stabilizing aldo-keto reductase YtBe with additives. J Mol Catal B Enzymatic 104:108–114
82. Zhang D, Chen X, Chi J, Feng J, Wu Q, Zhu D (2015) Semi-rational engineering a carbonyl reductase for the enantioselective reduction of β-amino ketones. ACS Catal 5(4):2452–2457
83. Chen X, Mei T, Cui Y, Chen Q, Liu X, Feng J, Wu Q, Zhu D (2015) Highly efficient synthesis of optically pure (*S*)-1-phenyl-1,2-ethanediol by a self-sufficient whole cell biocatalyst. ChemistryOpen 4(4):483–488
84. Brenna E, Gatti FG, Manfredi A, Monti D, Parmeggiani F (2012) Enoate reductase-mediated preparation of methyl (*S*)-2-bromobutanoate, a useful key intermediate for the synthesis of chiral active pharmaceutical ingredients. Org Proc Res Dev 16(2):262–268
85. Bechtold M, Brenna E, Femmer C, Gatti FG, Panke S, Parmeggiani F, Sacchetti A (2012) Biotechnological development of a practical synthesis of ethyl (*S*)-2-ethoxy-3-(p-methoxyphenyl)propanoate (EEHP): Over 100-fold productivity increase from yeast whole cells to recombinant isolated enzymes. Org Proc Res Dev 16(2):269–276
86. Fryszkowska A, Toogood H, Sakuma M, Gardiner JM, Stephens GM, Scrutton NS (2009) Asymmetric reduction of activated alkenes by pentaerythritol petranitrate peductase: specificity and control of stereochemical outcome by reaction optimisation. Adv Synth Catal 351:2976–2990
87. Gao X, Ren J, Wu Q, Zhu D (2012) Biochemical characterization and substrate profiling of a new NADH-dependent enoate reductase from *Lactobacillus casei*. Enzyme Microb Technol 51(1):26–34
88. Padhi SK, Bougioukou DJ, Stewart JD (2009) Site-saturation mutagenesis of tryptophan 116 of *Saccharomyces pastorianus* old yellow enzyme uncovers stereocomplementary variants. J Am Chem Soc 131(9):3271–3280
89. Chen X, Gao X, Wu Q, Zhu D (2012) Synthesis of optically active dihydrocarveol via a stepwise or one-pot enzymatic reduction of (*R*)- and (*S*)-carvone. Tetrahedron Asymmetry 23(10):734–738
90. Hanson RL, Davis BL, Goldberg SL, Johnston RM, Parker WL, Tully TP, Montana MA, Patel RN (2008) Enzymatic preparation of a D-amino acid from a racemic amino acid or keto acid. Org Proc Res Dev 12(6):1119–1129
91. Paul CE, Rodríguez-Mata M, Busto E, Lavandera I, Gotor-Fernández V, Gotor V, García-Cerrada S, Mendiola J, de Frutos Ó, Collado I (2014) Transaminases applied to the synthesis of high added-value enantiopure amines. Org Proc Res Dev 18(6):788–792
92. Simon RC, Grischek B, Zepeck F, Steinreiber A, Belaj F, Kroutil W (2012) Regio- and stereoselective monoamination of diketones without protecting groups. Angew Chem 124 (27):6817–6820
93. Payer SE, Schrittwieser JH, Grischek B, Simon RC, Kroutil W (2016) Regio- and stereoselective biocatalytic monoamination of a triketone enables asymmetric synthesis of both enantiomers of the pyrrolizidine alkaloid xenovenine employing transaminases. Adv Synth Catal 358(3):444–451
94. Savile CK, Janey JM, Mundorff EC, Moore JC, Tam S, Jarvis WR, Colbeck JC, Krebber A, Fleitz FJ, Brands J, Devine PN, Huisman GW, Hughes GJ (2010) Biocatalytic asymmetric synthesis of chiral amines from ketones applied to sitagliptin manufacture. Science 329 (5989):305–309
95. Liese A, Seelbach K, Wandrey C (2000) Industrial biotransformations. Wiley-VCH Verlag GmbH&Co KGaA, Weinheim
96. Menzel A, Werner H, Altenbuchner J, Groeger H (2004) From enzymes to "designer bugs" in reductive amination: A new process for the synthesis of L-tert-leucine using a whole cell-catalyst. Eng Life Sci 4(6):573–576
97. Gao X, Ma Q, Zhu H (2015) Distribution, industrial applications, and enzymatic synthesis of D-amino acids. Appl Microbiol Biotechnol 99(8):3341–3349

98. Hanson RL, Johnston RM, Goldberg SL, Parker WL, Goswami A (2013) Enzymatic preparation of an *R*-amino acid intermediate for a γ-secretase inhibitor. Org Proc Res Dev 17 (4):693–700
99. Tao F, Zhang Y, Ma C, Xu P (2011) One-pot bio-synthesis: N-acetyl-D-neuraminic acid production by a powerful engineered whole-cell catalyst. Sci Reports 1:142
100. Xu P, Qiu JH, Zhang YN, Chen J, Wang PG, Yan B, Song J, Xi RM, Deng ZX, Ma CQ (2007) Efficient whole-cell biocatalytic synthesis of N-acetyl-D-neuraminic acid. Adv Synth Catal 349(10):1614–1618
101. Castillo JA, Guérard-Hélaine C, Gutiérrez M, Garrabou X, Sancelme M, Schürmann M, Inoue T, Hélaine V, Charmantray F, Gefflaut T, Hecquet L, Joglar J, Clapés P, Sprenger GA, Lemaire M (2010) A mutant D-fructose-6-phosphate aldolase (Ala129Ser) with improved affinity towards dihydroxyacetone for the synthesis of polyhydroxylated compounds. Adv Synth Catal 352(6):1039–1046
102. Goldberg SL, Goswami A, Guo Z, Chan Y, Lo ET, Lee A, Truc VC, Natalie KJ, Hang C, Rossano LT, Schmidt MA (2015) Preparation of β-hydroxy-α-amino acid using recombinant D-threonine aldolase. Org Proc Res Dev 19(9):1308–1316
103. Lanfranchi E, Steiner K, Glieder A, Hajnal I, Sheldon RA, van Pelt S, Winkler M (2013) Mini-review: recent developments in hydroxynitrile lyases for industrial biotechnology. Recent Pat Biotechnol 7(3):197–206
104. Ghislieri D, Green AP, Pontini M, Willies SC, Rowles I, Frank A, Grogan G, Turner NJ (2013) Engineering an enantioselective amine oxidase for the synthesis of pharmaceutical building blocks and alkaloid natural products. J Am Chem Soc 135(29):10863–10869
105. Leisch H, Grosse S, Iwaki H, Hasegawa Y, Lau PCK (2011) Cyclohexylamine oxidase as a useful biocatalyst for the kinetic resolution and dereacemization of amines. Can J Chem 90 (1):39–45
106. Li G, Ren J, Yao P, Duan Y, Zhang H, Wu Q, Feng J, Lau PCK, Zhu D (2014) Deracemization of 2-methyl-1,2,3,4-tetrahydroquinoline using mutant cyclohexylamine oxidase obtained by iterative saturation mutagenesis. ACS Catal 4(3):903–908
107. Beard TM, Turner NJ (2002) Deracemisation and stereoinversion of α-amino acids using D-amino acid oxidase and hydride reducing agents. Chem Commun 3:246–247
108. Scheller PN, Fademrecht S, Hofelzer S, Pleiss J, Leipold F, Turner NJ, Nestl BM, Hauer B (2014) Enzyme toolbox: novel enantiocomplementary imine reductases. ChemBioChem 15 (15):2201–2204
109. Grogan G, Turner NJ (2016) Inspired by nature: NADPH-dependent imine reductases (IREDs) as catalysts for the preparation of chiral amines. Chem A Eur J 22(6):1900–1907
110. Hussain S, Leipold F, Man H, Wells E, France SP, Mulholland KR, Grogan G, Turner NJ (2015) An (*R*)-imine reductase biocatalyst for the asymmetric reduction of cyclic imines. ChemCatChem 7(4):579–583
111. Leipold F, Hussain S, Ghislieri D, Turner NJ (2013) Asymmetric reduction of cyclic imines catalyzed by a whole-cell biocatalyst containing an (*S*)-imine reductase. ChemCatChem 5 (12):3505–3508
112. Classen T, Korpak M, Schölzel M, Pietruszka J (2014) Stereoselective enzyme cascades: an efficient synthesis of chiral γ-butyrolactones. ACS Catal 4(5):1321–1331
113. Brenna E, Crotti M, Gatti FG, Monti D, Parmeggiani F, Pugliese A, Santangelo S (2015) Multi-enzyme cascade synthesis of the most odorous stereoisomers of the commercial odorant muguesia®. J Mol Catal B: Enzymatic 114:37–41
114. Yao P, Wang L, Yuan J, Cheng L, Jia R, Xie M, Feng J, Wang M, Wu Q, Zhu D (2015) Efficient biosynthesis of ethyl (*R*)-3-hydroxyglutarate through a one-pot bienzymatic cascade of halohydrin dehalogenase and nitrilase. ChemCatChem 7(9):1438–1444
115. Sosedov O, Matzer K, Bürger S, Kiziak C, Baum S, Altenbuchner J, Chmura A, van Rantwijk F, Stolz A (2009) Construction of recombinant *Escherichia coli* catalysts which simultaneously express an (*S*)-oxynitrilase and different nitrilase variants for the synthesis of (*S*)-mandelic acid and (*S*)-mandelic amide from benzaldehyde and cyanide. Adv Synth Catal 351(10):1531–1538

116. Pennec A, Hollmann F, Smit MS, Opperman DJ (2015) One-pot conversion of cycloalkanes to lactones. ChemCatChem 7(2):236–239
117. Schmidt S, Scherkus C, Muschiol J, Menyes U, Winkler T, Hummel W, Gröger H, Liese A, Herz H-G, Bornscheuer UT (2015) An enzyme cascade synthesis of ε-caprolactone and its oligomers. Angew Chem Int Ed 54(9):2784–2787
118. Sattler JH, Fuchs M, Mutti FG, Grischek B, Engel P, Pfeffer J, Woodley JM, Kroutil W (2014) Introducing an in situ capping strategy in systems biocatalysis to access 6-aminohexanoic acid. Angew Chem Int Ed 53(51):14153–14157
119. Ankati H, Yang Y, Zhu D, Biehl ER, Hua L (2008) Synthesis of optically pure 2-azido-1-arylethanols with isolated enzymes and conversion to triazole-containing β-blocker analogues employing click chemistry. J Org Chem 73(16):6433–6436
120. Cuetos A, Bisogno FR, Lavandera I, Gotor V (2013) Coupling biocatalysis and click chemistry: one-pot two-step convergent synthesis of enantioenriched 1,2,3-triazole-derived diols. Chem Commun 49(26):2625–2627
121. Gröger H, Hummel W (2014) Combining the 'two worlds' of chemocatalysis and biocatalysis towards multi-step one-pot processes in aqueous media. Curr Opin Chem Biol 19:171–179

Chapter 5
Characterization and Engineering of Seaweed Degrading Enzymes for Biofuels and Biochemicals Production

Eva Garcia-Ruiz, Ahmet Badur, Christopher V. Rao and Huimin Zhao

Abstract Exploitation of natural sources and increasing concerns of environmental pollution are motivating a growing interest in renewable and sustainable feedstocks for biochemicals and biofuels. Marine macroalgae have many advantages over terrestrial plant biomass, including high carbohydrate content which converts seaweed in a cogent alternative feedstock. Algal carbohydrates show a diverse sugar composition, which implies that specialized enzymatic systems are required for their conversion into biofuels and chemicals. Discovery and characterization of degrading enzymes and assimilating the relevant pathways is a key step in the depolymerization of algal polysaccharides into fermentable sugars and their metabolism by fermenting microorganisms. Current advances in metabolic engineering have generated new microorganisms capable of efficiently metabolizing macroalgal carbohydrates while producing ethanol, the target product. However, more research is required to unlock the full potential of macroalgae biomass as a feedstock for biochemical and biofuels production. This book chapter provides an overview of seaweed polysaccharides properties, degrading enzymes, and their application in the bioconversion of macroalgae into biofuels and biochemicals.

Eva Garcia-Ruiz and Ahmet Badur have contributed equally.

E. Garcia-Ruiz · H. Zhao (✉)
Department of Chemical and Biomolecular Engineering, Carl R. Woese Institute for Genomic Biology, University of Illinois at Urbana-Champaign, Urbana, IL 61801, USA
e-mail: zhao5@illinois.edu

A. Badur · C.V. Rao
Department of Chemical and Biomolecular Engineering,
University of Illinois at Urbana-Champaign, Urbana, IL 61801, USA

H. Zhao
Departments of Chemistry, Biochemistry, and Bioengineering,
University of Illinois at Urbana-Champaign, Urbana, IL 61801, USA

Present Address:
A. Badur
Illumina, 5200 Illumina Way, San Diego, CA 92122, USA

P.C.K. Lau (ed.), *Quality Living Through Chemurgy and Green Chemistry*,
Green Chemistry and Sustainable Technology,
DOI 10.1007/978-3-662-53704-6_5

Keywords Seaweed degrading enzymes · Macroalgal polysaccharides · Macroalgal glucans · Alginate · Biofuels · Alginate lyases · Agarases · Laminarases · Glycoside hydrolase

5.1 Introduction

5.1.1 Next Generation Feedstocks

Current trends in the production of biofuels and biochemicals utilize feedstocks from recalcitrant sources. So-called first generation feedstocks, including sugar cane and corn, were easily metabolized by fermentative organisms. However, these sources came with drawbacks; these drawbacks are sought to be answered by second generation feedstocks, including lignocellulosic- and algal-based biomasses.

There is much work being done to engineer lignocellulosic feedstocks for the production of biofuels and biochemicals. The perennial grass *Miscanthus* × *giganteus* has been reported as having high yields in moderate climates [1], an attractive feature for large-scale cultivation. Other feedstocks, such as *Opuntia ficus-indica* and *Agave tequilana,* have also been demonstrated as having high yields for biomass with advantageous climate and water requirements [2]. Regardless of the particular lignocellulosic feedstock, these biomasses contain polymers that cannot be directly utilized by most fermentative organisms. These polymers, cellulose, hemicellulose, and lignin, form a complex network which cannot be easily solubilized nor degraded [3, 4]. Additionally, these polymers often contain constituent monomers which cannot be natively metabolized by common fermentation strains [5]. Thus, before these lignocellulosic feedstocks can be utilized at industrial scales, much work has to be done both in engineering the pretreatment process to degrade lignocellulosic polymers and in engineering fermentative strains to metabolize non-native carbon sources.

5.1.2 Algal-Based Feedstocks

Algae are an exciting target for the production of biofuels and biochemicals. These feedstocks have many advantages over current terrestrial biomass. As algae are cultivated from marine environments, their use in a fermentative process would avoid the conflict between food and fuel [6]. Additionally, algae are photosynthetic organism, and therefore they convert sunlight into chemical energy which is used to efficiently fix atmospheric CO_2 into biomass. Algal photosynthetic efficiency (6–8 %) is threefold higher than that of terrestrial biomass (1.8–2.2 %), which makes algae one of the fastest growing organisms on the planet [7, 8]. This ensures high productivity for their cultivation. As an example, brown macroalgae from

Saccharina sp. can grow as much as 50 cm/day and reach lengths of more than 80 m [9]. Most importantly, algae contain low amounts of crystalline cellulose and lignin [10, 11], which avoid the need of expensive pretreatment processes for their removal. Thus, algae-based feedstocks have advantages over those of terrestrial origin.

Carbon flux through an ecosystem is a useful measure to indicate how amenable is the cultivation of carbon feedstocks. If an ecosystem has a large carbon flux compared to potential feedstock cultivation, this ecosystem is expected to be less affected by large cultivation than one in which carbon flux is similar to feedstock cultivation. The carbon flux through algal bed, algal reef, and estuary ecosystems has been estimated as 10^9 tons of carbon per year [12]. These systems make up two thirds of all ocean carbon biomass, indicating that marine grasses and algae in these ecosystems are an important component of global carbon flux. Additionally, these ecosystems are dense stores of carbon, further exemplifying that algal-based feedstocks are attractive targets for industrial cultivation.

Algae can be broadly divided into two groups: microalgae and macroalgae. Microalgae are unicellular, photosynthetic microorganisms, whereas macroalgae are leafy multicellular, photosynthetic plants. Microalgae have been targeted for production of biofuels since they produce large quantities of lipids, oils, and other storage polymers. Various microalgae have been shown to contain up to 40 % of their dry cell weight in lipids and oils [13]. Additionally, microalgae can have low nutritional needs compared to terrestrial feedstocks; however, special bioreactors must be constructed to maximize growth rates and light penetration [14]. Microalgae can also be engineered to produce heterologous biofuels and biochemicals. Work has been done to engineer the cyanobacterium *Synechococcus elongatus* to produce isobutyraldehyde and isobutanol [15]. Atsumi and coworkers were able to engineer this microalgae to produce more than 6 mg l^{-1} h^{-1} of isobutyraldehyde and 3 mg l^{-1} h^{-1} of isobutanol by introducing a ketoacid decarboxylase gene, *kivd*, and overexpressing other metabolic genes. Further work engineering *Synechocystis* sp. has yielded ethanol production greater than 8 mg l^{-1} h^{-1} by introducing pyruvate decarboxylase, *pdc*, and overexpressing other metabolic genes [16]. These works indicate that microalgae are an achievable platform for the production of biofuels and biochemicals.

5.1.3 Macroalgae-Based Feedstocks

As microalgae necessitate bioreactor design to maximize light penetration [17], economies of scale found in typical fermentative processes are not expected in microalgae-based biofuel and biochemical production. Macroalgae-based bioprocesses would provide a compromise, utilizing marine feedstock cultivation and current industrial fermentation techniques. Macroalgae are primarily comprised of polysaccharides and, as such, must be degraded to a monomeric form to allow for metabolism by fermentative organisms. During this degradation, pathways and enzymes must be engineered to: (1) secrete polysaccharide degrading enzymes into the extracellular environment; (2) degrade the algal polysaccharides into oligomers; (3) transport the oligomers inside the cell; (4) degrade the oligomers into monomeric form; and (5) metabolize monomers into desired biofuel or biochemical product through metabolic engineering.

This chapter will discuss macroalgal polysaccharides properties and focus on biochemical and biophysical characterization and engineering of seaweed polysaccharides degrading enzymes and their applications for synthesis of biofuels and chemicals.

5.2 Properties of Seaweeds

They are especially abundant in coastal waters; however, due to their diverse evolutionary origin, they can be found in a wide variety of habitats, show different growth, and have varied chemical composition [18–22].

Marine macroalgae, referred as seaweeds, include a variety of multicellular photosynthetic organisms with a diverse evolutionary origin. Traditionally, they are classified in three different groups based on the color of their natural pigment content: green macroalgae (Chlorophyceae), red macroalgae (Rhodophyceae), and brown macroalgae (Phaeophyceae). Typical green, red and brown macroalgae-species are *Ulva* sp. and *Codium* sp.; *Gelidium* sp. and *Eucheuma* sp.; and *Laminaria* sp. and *Sargassum* sp., respectively.

Macroalgae as a photosynthetic organism convert sunlight into chemical energy, which is used to efficiently fix inorganic carbon (in form of CO_2 and/or H_2CO_3) into carbohydrates, which can be exploited for biorefinery. However, to consider macroalgae as a feedstock for biofuels and chemical production, it is essential to understand the chemical composition of macroalgal biomass, and their available content. Therefore, understanding the composition of macroalgal biomass helps to determine the appropriate process and technology to produce the biochemical of interest.

In general terms, macroalgae have low content of lignin, lipids, and proteins [19, 23–26]. However, they are rich in carbohydrates, the cell wall being the primarily source. The carbohydrate content can account for more than 60 % of dry weight [18]. Thus, the conversion of these carbohydrates is expected to be the main source for biofuel and chemical production.

Carbohydrates in macroalgae show a diverse chemical composition, and some of them are restricted to a certain group of macroalgae. For example, green seaweeds have a sulfated polysaccharide named ulvan as major non-glucan carbohydrate. Ulvan is a water-soluble sulfated polymer consisting of repeating units of L-rhamnose, D-xylose, D-glucose, D-glucuronic acid, and L-iduronic acid with diverse pattern depending on the species (Fig. 5.1). This type of carbohydrate can be found in the cell wall representing up to 29 % of dry weight. Nevertheless, green macroalgae also contain other common glucans as starch and cellulose. Starch is a

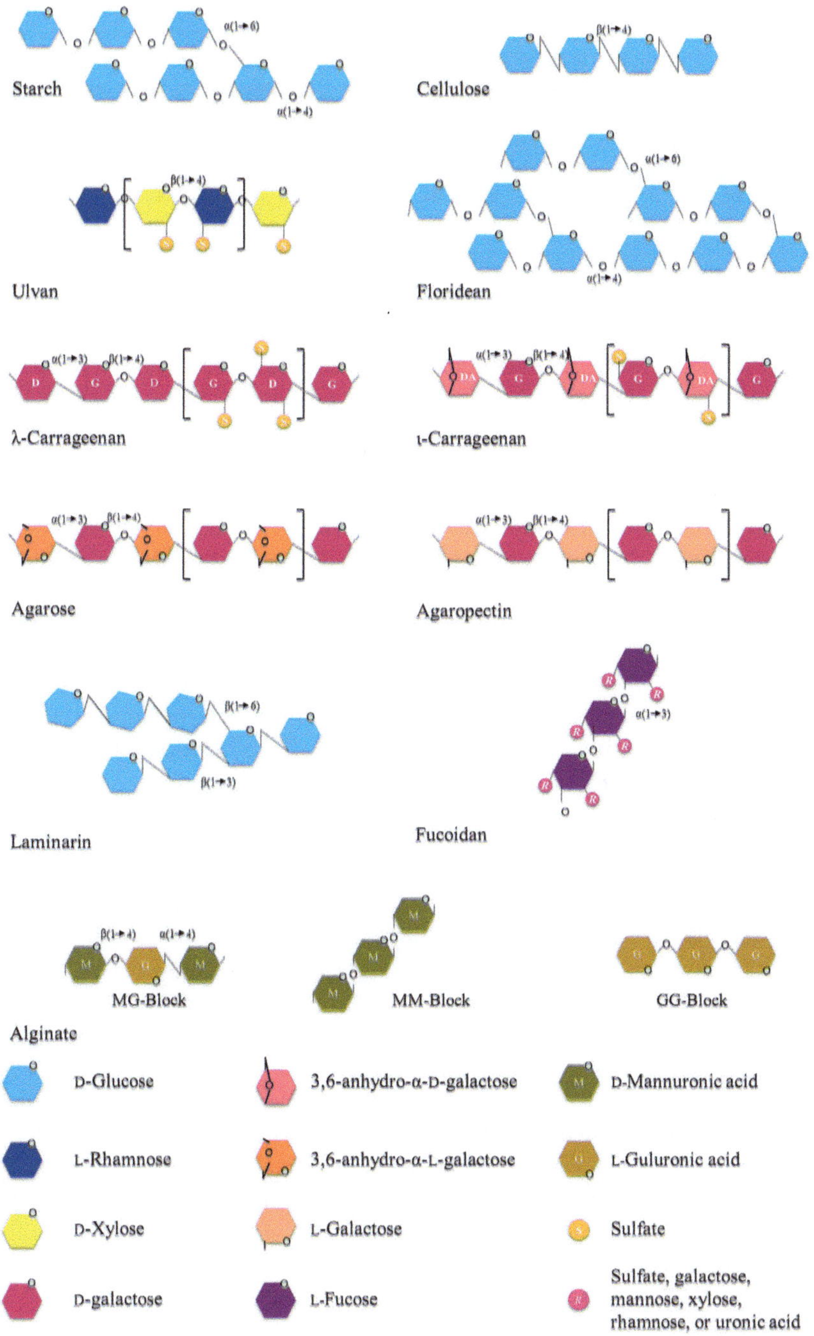
Starch
Cellulose
Ulvan
Floridean
λ-Carrageenan
ι-Carrageenan
Agarose
Agaropectin
Laminarin
Fucoidan
MG-Block
MM-Block
GG-Block
Alginate
D-Glucose
3,6-anhydro-α-D-galactose
D-Mannuronic acid
L-Rhamnose
3,6-anhydro-α-L-galactose
L-Guluronic acid
D-Xylose
L-Galactose
Sulfate
D-galactose
L-Fucose
Sulfate, galactose, mannose, xylose, rhamnose, or uronic acid

◀ **Fig. 5.1** Scheme of the structure of macroalgae polysaccharides. Sugars and linkages between them are represented. Main repeating disaccharide units are indicated between brackets. In λ-carrageenan: D, denotes D-units (α-D-galactose); G, denotes G-units (β-D-galactose). In ι-carrageenan: DA, denotes DA-units (3,6-anhydro-α-D-galactose). In alginate: M, denotes M-residues (β-D-mannuronicacid); G, denotes G-residues (α-L-guluronicacid). Fucoidan polymer is widely heterogeneous and composition of monosaccharides, sulphate positions, linkages, and branching is highly variable, thus it is only represented a homopolymer of α-1,3-L-fucose where R may be any of the following: sulfate, galactose, mannose, xylose, rhamnose, or uronic acid. For more information refer to the text

linear molecule of α-1,4-D-glucose with α-1,6 branches which is produced as an energy store. Cellulose is a water-insoluble linear chain of β-1,4-D-glucose that can be found in the cell wall as a structural polysaccharide (Fig. 5.1) [27–29].

Red macroalgae contain a unique type of glucan, floridean starch, a highly branched polymer of α-1,4-glucosidic linked D-glucose with α-1,6-branches (Fig. 5.1), which green and brown algae do not have. Floridean starch can be found as carbon reserve starch granules in the cytosol of red algae and represent up to 80 % of the cell volume [30]. However, red seaweeds contain carrageenan and agar as major non-glucan polysaccharides constituents in the cell wall, which can account for more than 50 % of dry weight [19, 31].

Carrageenans are sulfated linear polysaccharides of D-galactose and 3,6-anhydro-D-galactose. These polysaccharides are traditionally divided into six basic forms: Iota (ι)-, Kappa (κ)-, Lambda (λ)-, Mu (μ)-, Nu (ν)- and Theta (θ)-carrageenan. The μ-, ν-, and λ-carrageenans consist of a repeating disaccharide formed by 3-linked β-D-galactose (G-units) and 4-linked α-D-galactose (D-units), and have one, two or three sulfate ester groups, respectively. They are the natural biological precursor of κ-, ι-, and θ-carrageenans that consist of a repeating disaccharide formed by 3-linked β-D-galactose (G-units) and 4-linked 3,6-anhydro-α-D-galactose (DA-units) (Fig. 5.1) [26, 32].

Agar consists of the combination of agarose and agaropectin. Agarose is a linear polymer comprising of the repeating unit of agarobiose, which is a disaccharide of 3-linked β-D-galactose and 4-linked 3,6-anhydro-α-L-galactose, whereas agaropectin contains 3-linked β-D-galactose and 4-linked α-L-galactose (Fig. 5.1) [19, 26, 33–35].

Brown macroalgae contain alginate, mannitol and laminarin as major carbohydrates. However, other carbohydrates such as fucoidan and cellulose may also be present. The alginate content in brown algae cell wall can represent up to 40 % of dry weight. Alginate is a water-soluble, linear, unbranched polymer containing two types of uronic acid units: 4-linked β-D-mannuronic acid (M-residues), and 4-linked α-L-guluronic acid (G-residues) (Fig. 5.1). These units can be found forming G-blocks (repeating G units), M-blocks (repeating M units), and MG-blocks (alternating M and G units) [36, 37].

Laminarin is a water-soluble glucan carbohydrate produced as an energy store. It is composed of 3-linked β-D-glucose with β-1,6-branches (Fig. 5.1), and represents up to 35 % of dry weight of algae. There are two different types of laminarin chains depending on the terminal end: "M" laminarin contains a mannitol residue as a non-reducing end; "G" laminarin contains a glucose residue as a reducing end instead [19, 38]. Mannitol is the sugar alcohol of mannose, and together with laminarin, is the most readily accessible sugar in brown macroalgae.

Fucoidan is a sulfated polysaccharide composed chiefly of α-1,3-L-fucose alternating with other monosaccharides as galactose, mannose, xylose, rhamnose, and uronic acid (Fig. 5.1). Fucoidan structure is widely heterogeneous as the composition of monosaccharides, sulphate positions, linkages, and branching is highly variable [39–41].

The breakdown of macroalgae carbohydrate composition shows that the glucan content is low (around 20–25 % dry weight) compared to that of terrestrial biomass (30–45 % dry weight) [42], and hence use of these carbohydrates as a feedstock is not enough to achieve high productivities by their conversion into a biochemical of interest. Accordingly, the conversion of other macroalgae carbohydrates is needed to attain industrial productivities. Whereas glucans are easily fermentable carbohydrates, the uniqueness of non-glucan macroalgae carbohydrates makes them difficult to use.

A number of microorganisms have developed specialized enzymatic systems for the efficient utilization of these non-glucan carbohydrates as a carbon source. These carbohydrate-degrading enzymes, which will be discussed below, can be implemented in different biotechnological processes to make the macroalgae sugar content accessible to fermentative microorganism that eventually can be engineered to synthesize a biochemical of interest.

5.3 Enzymes Involved in Macroalgae Degradation

As the feedstocks found in macroalgae are polymers, degradation of these polymers is necessary for their utilization. The enzymes which accomplish this degradation can either be endolytic (acting within the fragment) or exolytic (acting from the outside of a fragment). The resulting degradation products must then be transported into the cell for metabolism. Here we present an overview of the proteins and enzymes necessary for an organism to metabolize a polysaccharide from macroalgae.

5.3.1 Alginate Lyases

5.3.1.1 Background and Families

Alginate lyases (EC 4.2.2.3 and 4.2.2.11) are enzymes that cleave the glycosidic bonds of alginate. These enzymes accomplish the cleavage of alginate through β-elimination. The cleavage of the glycosidic bonds occurs through a sequence of three reaction steps. First, the negative charge of the carboxyl anion is neutralized via the amino acid residues arginine, histidine, glutamine, glutamic acid, or tyrosine or a divalent cation. Second, the C-5 proton is abstracted from the sugar ring by either an aspartic acid, cysteine, glutamic acid, histidine, or lysine residue. Last, a tyrosine or arginine residue is used to accept the electrons from the C-4 and C-5 bond. The net result of these steps is the formation of a double bond between C-4 and C-5 in the cleaved guluronate or mannuronate [43, 44].

As alginate lyases are defined for their proclivity to degrade the bonds within alginate, these enzymes can have different 3-D structures, indicating that alginate lyases have evolved many times. Each of these structures can be classified into polysaccharide lyases (PL) families. These families show a wide variety in their structures, including β-helices and α/α barrels. There exist seven PL families typically associated with alginate lyases [44, 45], including PL5, PL6, PL7, PL14, PL15, PL17, and PL18. The families PL5 and PL16 contain an $(\alpha/\alpha)_6$ barrel structure, wherein the α barrels form a tube-like structure [46–48]. The families PL7, PL14, and PL18 contain a β-jelly roll with β-sheets in antiparallel, adjacent configuration [49, 50]. The β-sheets are arranged to form a cleft shape. The PL6 family contains a parallel β-helix structure with a long repeated series of β-strands forming a long tube [51]. The PL17 family has a structure with many α-helices forming a barrel structure [52].

5.3.1.2 Enzymatic Mechanism of Alginate Lyase Degradation

Alginate lyases cleave the glycosidic bonds within alginate using β-elimination. β-elimination is a reaction process wherein glycosidic bonds are cleaved by the formation of a reducing end on one end and an unsaturated ring on the other end of the broken bond [43]. This reaction mechanism differs from that of hydrolysis in that water is not used to break the glycosidic bond. Most broadly, an alginate lyase acts on the glycosidic bond at the C-4 position by first neutralizing the negative charge on the carboxylic group on C-5. This neutralization is accomplished by the action of a positively charged cation or amino acid residue. Once the neutralization occurs, the C-4 proton is abstracted and an electron transfer occurs which cleaves the glycosidic bond and forms a double bond between C-4 and C-5.

There are two general mechanisms for the neutralization of the positive charge by an alginate lyase. The first mechanism utilizes a divalent cation (such as Ca^{2+}) to neutralize the negative charge on the C-5 carboxylic group. A lysine group is then used to abstract the C-4 proton, while a lysine or arginine group accepts the required electrons. This metal-assisted mechanism is found in the PL6 family of alginate lyases [53].

The second catalytic mechanism employed by alginate lyases does not utilize a divalent cation for neutralization, but rather uses an asparagine, arginine, glutamic acid, or glutamine residue for neutralization. Independent of the amino acid residues used for carboxyl-group neutralization, the members of PL5 [54], PL7 [55], and PL15 [56] families of lyases utilize a histidine residue to abstract the proton from

C-4 and a tyrosine residue to accept the electrons which cleave the glycosidic bond. Interestingly, recent work has shown that the PL18 [57] family of alginate lyases utilizes a tyrosine residue to both abstract the proton from C-4 and then donate the same proton to the linked C-1 to cleave the glycosidic bond. Arginine and lysine are used to stabilize the carboxyl group. A similar mechanism has been proposed for a PL17 alginate lyase [52]. Two different tyrosine residues deprotonate the hydrogen from C-4 and accept the electrons to cleave the glycosidic bond. Asparagine and histidine residues were proposed to stabilize the carboxylic group.

The enzymatic mechanism for glycosidic cleavage of the PL14 family alginate lyases has not been fully elucidated. A representative of the PL14 family has been crystallized [50]. It was found that in this alginate lyase, mutagenesis of all the amino acids within the catalytic pocket reduced enzymatic activity. In addition, mutagenesis of K197A and S219A abolished enzymatic activity. These residues were conserved from other alginate lyases. This suggests that these amino acid residues might play a role in the cleavage of alginate.

(1) Oligoalginate Lyases

As alginate can contain up to 70 monomers within its structure, its degradation is expected to occur in several steps. First, the full alginate polymer is degraded into smaller fragments, called oligoalginate. These oligoalginate fragments are then degraded into the constitutive monomers by oligoalginate lyases which can act exolytically. The PL15 and PL17 families of alginate lyases are associated with the degradation of oligomers of alginate. A PL15 alginate lyase, Atu3025, from *Agrobacterium tumefaciens* has been crystallized [56] to understand how PL15 lyases act exolytically. Ochiai and coworkers found that this lyase contains a pocket-like structure which can recognize the terminal monomer within alginate [56]. Only when this terminal monomer is situated in the pocket, Atu3025 will adopt a "closed form" wherein β-elimination can occur. The enzyme then adopts an "open form" that releases the degraded monomer and a new alginate fragment can approach the enzyme.

Recent work has sought to characterize the oligoalginate degrading enzymes of *Vibrio splendidus* 12B01 [58]. This bacterium has three PL15 oligoalginate lyases (OalA, OalB, and OalC). OalA was found to have high catalytic efficiency (k_{cat}/K_m) on an alginate trisaccharide and less than 10 % catalytic efficiency on larger alginate fragments. It was confirmed that OalA acts on alginate exolytically since only monomers were released upon degradation of alginate. Other PL15 and PL17 oligoalginate lyases have also been expressed and characterized [52, 59].

(2) Specificity of Alginate Lyase Degradation

Since alginate contains different sequence combinations of L-guluronate and D-mannuronate, alginate lyases have evolved the ability to target specific regions of alginate. Alginate lyases targeting polyM [60], polyG [61, 62], and polyMG [63] regions within alginate have been identified. The specificity of an alginate lyase can be determined through multiple experimental techniques. Partial degradations of alginate can be chemically generated [64, 65], which results in the formation of polyG- and polyM-rich alginate substrates. The enzymatic activity towards these substrates can be then compared to that of the full alginate polymer. Another means by which substrate specificity can be determined is through NMR analysis of the enzymatically degraded alginate [66, 67]. In this procedure, the degraded products are analyzed for the formation and destruction of NMR peaks which correspond to specific linkages in alginate. Since alginate degradation is expected to target specific dyads within the structure, the dyad specificity can then be determined. This analysis has successfully identified the dyad specificity of four alginate lyases within a single bacterium [68].

5.3.1.3 Alginate Transport and Metabolism

Since alginate is a large molecule, a microorganism must utilize a means of transporting alginate or oligoalginate into the cell. The bacterium *Sphingomonas* sp. strain A1 contains a periplasmic alginate protein-dependent ATP binding cassette (ABC) transporter [69]. The overall transport mechanism contains three components: a pit on the cell surface which transports the full alginate polymer into the cell periplasm, proteins within the periplasm which bind alginate, and the aforementioned ABC transporter which is found in the inner-membrane. *Sphingomonas* sp. strain A1 then uses alginate lyases contained within the cytoplasm to degrade alginate into the constituent monomers.

Another mechanism by which degraded alginate is effectively transported into the cell can be found in the bacterium *V. splendidus* 12B01. This bacterium contains an outer-membrane porin KdgMN [70]. Long chain oligoalginate fragments diffuse into the periplasm of 12B01 through KdgMN. These long chains are then degraded into chains two to four units long within the periplasm by alginate lyases. The two- to four-mers are then transported into the cytoplasm by the symporter ToaABC. These short fragments are then degraded into monomeric form by oligoalginate lyases. The two transport mechanisms described illustrate how alginate must be carefully transported, either through assisted transport of the full polymer or initial degradation outside the cell followed by diffusion of the degraded oligoalginate.

Once alginate is degraded and transported into the cell, an organism must metabolize the alginate monomers. An alginate lyase will cleave alginate using β-elimination which results in an L-guluronate or a D-mannuronate and a non-reducing end. The generated non-reducing end is 4-deoxy-L-erythro-hex-4-enepyranosyluronate (DEH) [64]. L-Guluronate and D-mannuronate can also be converted to DEH non-enzymatically, so DEH serves as an important step in alginate metabolism. Once inside the cell, DEH is converted to 2-keto-3-deoxy-D-gluconic acid (KDG) by the NADPH-dependent enzyme DEH reductase [71]. KDG is then converted by the enzyme KDG kinase [70] to 2-keto-3-deoxy-6-phosphogluconate (KDPG). KDPG enters the Entner–Doudoroff pathway to be converted into pyruvate and glyceraldehyde-3-phosphate by the enzyme KDG-6-phosphate aldolase [70]. The final product of alginate metabolism is ATP and NADH.

5.3.1.4 Alginate Lyase Engineering

Many different alginate lyases from diverse organisms have been characterized, leading to a wealth of information on the enzyme kinetics and substrate binding of alginate lyases to alginate. However, attempts at protein engineering to improve alginate lyase enzymatic activity are less frequent. The alginate lyase, AlyVI, from *Vibrio* sp. QY101 was altered to improve enzymatic function [72]. Cho and coworkers built a homology model of this enzyme and modeled the binding of a repeating α-L-guluronic acid, trimer GGG, to the enzyme. Based on this model, they then mutated the amino acid residues within AlyVI predicted to be involved in enzymatic function; mutation of these residues abolished activity, which confirms their role in β-elimination. The authors also found that the double mutation L224V/D226G in AlyVI increased catalytic activity almost twofold without changing the binding affinity of GGG. Computational modeling of this double mutant indicated that the β-sheets of this PL7 lyase were rearranged allowing for increase in the enzymatic activity.

Another study unveiled the catalytic mechanism of the PL5 lyase Smlt1473 from *Stenotrophomonas maltophilia* k279a [73] to degrade polysaccharides. This lyase can degrade the polysaccharide hyaluronic acid along with polyM and polyG. Smlt1473 has optimal activity towards polyM at pH 9, while it has optimal activity towards polyG at pH 7. The authors sought to understand how this lyase degrades polyM and polyG preferentially by mutating non-catalytic residues within the catalytic cleft of the enzyme. They identified two single mutations, H221F and Y225F, which increased activity toward polyM twofold, and two other single mutations, Y115F and R312L, which increased activity toward polyG twofold. As polyM and polyG have different conformational structures, the authors concluded that these mutations resulted in subtle changes in Smlt1473 which increase activity towards either polyM or polyG.

These two mutational studies demonstrate that rational engineering of alginate lyases can lead to significant changes in enzymatic activity and specificity without sacrificing function. Other attempts at mutagenesis [74] have resulted in increased activity at the expense of weaker alginate binding. Specificity appears to be a parameter easily modified by rational design. However, random mutagenesis has been also used to modify specificity [75], albeit coming at the expense of enzymatic activity.

5.3.1.5 Multiplicity of Alginate Lyases

Alginate-degrading bacteria usually contain multiple putative lyases, thus some studies have sought to understand the purpose of these seemingly redundant enzymes. The marine bacterium *V. splendidus* 12B01 contains three oligoalginate lyases and four alginate lyases. Each of these enzymes was characterized. The three oligoalginate lyases, OalA, OalB, and OalC, were found to have optimal enzymatic activity between 16 and 35 °C and pH 6.5 and 7.5 [58]. Additionally, OalA was found to have both polyM and polyG specificity, while OalB and OalC were found to have either polyMG or polyM specificity, respectively. These enzymes act exolytically, cleaving the terminal moiety from alginate or oligoalginate. The four alginate lyases, AlyA, AlyB, AlyD, and AlyE were found to be optimally active between 20 and 25 °C and pH 7.5 and 8.5 [68]. Detailed NMR analysis identified that AlyA and AlyB act on the G-M dyad, corresponding to polyMG specificity, while AlyD and AlyE act on the G-G dyad, corresponding to polyG specificity. AlyA, AlyB, AlyD, and AlyE were found to act endolytically, cleaving alginate to an average degree of polymerization from 4 and 21. These alginate lyases also degraded alginate much slower than the oligoalginate lyases OalA, OalB, and OalC. Taking these two studies together, a picture of alginate degradation of *V. splendidus* 12B01 can be formed. AlyA, AlyB, AlyD, and AlyE degrade alginate to oligomers of average length between 4 and 21. These fragments are then transported into the cell and are degraded into monomeric form by OalA, OalB, and OalC with faster kinetics than the alginate lyases. Since the 12B01 enzymes have polyM, polyG, or polyMG specificity, these enzymes are expected to fully degrade alginate, since all portions of the polymer can be targeted.

Another study characterized two of the seven alginate lyases from the marine bacterium *Zobellia galactanivorans* [49]. One of these alginate lyases, $AlyA1_{PL7}$, was determined to act endolytically with a G-G dyad specificity. Conversely, AlyA5 was found to act exolytically on DEH, G, or M-terminal ends. These findings suggest that pairing of endolytic and exolytic alginate-degrading enzymes is advantageous for heterologous alginate degradation outside of its native host. Additionally, the pairing of enzymes with differing dyad specificities is expected to allow full alginate degradation.

In summary, enzyme multiplicity in alginate-degrading organisms aims to produce a set of enzymes with a wide spectrum of biochemical properties to fully degrade the alginate polymer.

5.3.2 *Laminarinases*

5.3.2.1 Background

Laminarinases (EC 3.2.1.6) are the enzymes capable of breaking the β-1,3- and β-1,6-glycosidic linkages of laminarin. These enzymes belong to the wider class of enzymes called glycoside hydrolyases (GHs) [76]. These enzymes are fundamentally different than the previously discussed alginate lyases, in that, they cleave the glycosidic bonds using water. More specifically, laminarinases cleave the bonds of laminarin in either a one-step or two-step process.

The one-step process is an inversion mechanism [76]. In this mechanism, the glycosidic linkage is simultaneously attacked by two amino acid residues, one acting as an acid while the other acts as a base. Through transfer of electrons from the basic amino acid residue to a water molecule, the water molecule attacks the anomeric center of the glucose residue. Simultaneously, the acidic amino acid residue accepts electrons from the glycosidic oxygen. The net result of this sequence is the cleavage of the glycosidic bond via hydrolysis. Since the newly reacted hydroxyl group from the water attack exists in the opposite configuration of the original β-glycosidic bond, the cleaved hemiacetal contains a 6C hydroxyl group in an α-configuration, hence the inversion mechanism.

The two-step reaction mechanism is a retaining mechanism. First, in a glycosylation step, one of the catalytic residues acts as a nucleophilic center by attacking the anomeric center of the linked glucose residue. Simultaneously, the other catalytic residue acts as an acid by protonating the glycosidic oxygen. The net result of these steps is the cleavage of the glycosidic bond [77]. Following the glycosylation step of laminarin cleavage, a complex is formed between one of the catalytic amino acid residues and the cleaved laminarin molecule. In order to complete the overall cleavage of the laminarin glycosidic bond, a second step must be performed: deglycosylation. In this step, a molecule of water is used to attack the complex, which frees the laminarinase catalytic residue from the cleaved laminarin [78]. Since the newly added hydroxyl group exists in the same β-configuration as the original β-glycosidic bond, this mechanism is a retaining mechanism. In laminarinases, the catalytic residues used by either the retaining or inverting mechanism are typically glutamic and aspartic acids [79, 80].

5.3.2.2 Laminarinase Characterization

Laminarinases from diverse organisms have been characterized. A laminarinase from the archaeon *Pyrococcus furiosus* has been overexpressed and purified [81]. This enzyme was shown to be optimally active between pH 6.0 and 6.5 and at 100 °C. As the glycosidic linkages found in laminarin are similar to other storage glucans, Gueguen and coworkers sought to determine if this laminarinase would degrade other glucans [81]. The authors found that the *P. furiosus* laminarinase had maximal activity on laminarin and approximately 90 % less activity on lichenan and barley β-glucan (both have β-1,3-1,4 glucose linkages). The laminarinase had no activity on glucans with only β-1,4 glucose linkage.

Another laminarinase from the actinobacterium *Streptomyces sioyaensis* was also characterized [82]. This laminarinase had optimal activity at pH 5.5 and 75 °C. The laminarinase was found to be most active on the β-1,3-linked glucans curdlan, laminarin, and pachyman, while the enzyme had fourfold less activity on lichenan (a β-1,3-1,4 glucan). A fungal laminarinase from *Phanerochaete chrysosporium* was purified from the secreted proteins and characterized [83]. This enzyme was tested to degrade various glucans and found to degrade laminarin and lichenan, β-1,3-1,6- and β-1,3-1,4-linkages, respectively. The laminarinase had no activity on cellulose (β-1,4 linkage).

5.3.3 Fucoidanases, Agarases, and Carrageenanases

Alginate and laminarin are the primary constituents of macroalgae; however, other polysaccharides are found within macroalgae. Fucoidanases (EC 3.2.1.44) are enzymes that degrade the glycosidic bonds within fucoidan. These enzymes have been purified from bacterial species and have shown optimal activity between pH 5 and 8 and between 30 and 45 °C [84, 85]. Fucoidan degrading enzymes are also found in marine invertebrates; an α-L-fucosidase from the marine mollusk *Pecten maximus* has been purified [86]. This enzyme was shown to have large enzymatic activity and high thermal stability. Additionally, the α-L-fucosidase was shown to have low fucoidan linkage specificity.

Enzymes have evolved to degrade the polysaccharides carrageen and agar. An agarase (EC 3.2.1.81) from *Pseudomonas* sp. was characterized and found to have optimal activity at pH 9 and 35 °C [87]. The agarase also had exolytic degradation. An agarase from the bacterium *Rhodococcus* sp. Q5 [88] was characterized and found to be optimally active at pH 6 and 40 °C.

Enzymes which degrade κ-, ι-, and λ-carrageen (EC 3.2.1.83, 3.2.1.157, and 3.2.1.162, respectively) have also been identified and purified. An ι-carrageenase was found to cleave ι-carrageen with a one-step inversion mechanism [89], in contrast to κ-carrageenases, which employ a retaining mechanism [90]. This indicates that enzymes have evolved to accommodate the specific degree of sulfation of carrageen. A λ-carrageenase has also been purified and characterized [91]. This enzyme was found to degrade only λ-carrageen and not κ- and ι-carrageen, again indicating that different enzymes have evolved to target the different forms of carrageen.

5.4 Gene Expression and Regulation

As polysaccharide degrading enzymes are expected to be expressed by their hosts in response to the presence of their corresponding substrates, a macroalgae degrading organism must be able to detect the presence of polysaccharides in the extracellular medium. A previous study has sought to understand how the hyperthermophilic bacterium *Thermotoga maritime* utilizes diverse carbon sources [92]. This bacterium is capable of metabolizing many simple sugars along with complex polysaccharides including β-1,3/1,4-glucan from barley, laminarin, and carboxymethylcellulose (CMC). The authors demonstrated that upon growth on different carbon sources, clusters of genes are overexpressed. Upon growth on laminarin, a laminarinase (Lam16), an exoglycosidase (Cel3 which cleaves exolytically), and a XlyR-related transcriptional regulator were overexpressed compared to the other growth conditions [92]. Similarly, upon growth on β-1,3/1,4-glucan from barley (β-1,3-1,4-linked glucose), β-1,4- and β-1,3-glucanases and Lam16 were overexpressed compared to other growth conditions. Conversely, when *T. maritime* was grown on CMC (β-1,4-linked glucose), intra- and extracellular endoglucanases were exclusively overexpressed. These findings indicate that *T. maritime* was able to detect the precise inter linkages of polysaccharides in the extracellular environment and then expressed the corresponding enzymes for degradation. Since a transcriptional regulator was identified upon growth on laminarin, a regulatory target exists for future genetic manipulation.

The alginate regulatory mechanism has also been investigated in the marine bacterium *V. splendidus* 12B01 [68]. This bacterium contains four alginate lyases and the gene expression of each of these enzymes was investigated. Each of the lyases was found to be conditionally expressed upon growth on alginate with higher gene expression at higher alginate loadings. Between 0.1 and 1.0 % alginate in the culture medium, the gene expression of three alginate lyases was found to increase by more than twofold. These findings indicate that polysaccharide degrading enzyme gene expression is often tied to detection of specific polysaccharides in the culture medium. Subsequently, these enzymes will degrade the carbon sources allowing the organism to metabolize the constituent monomers.

5.5 Carbohydrate Binding Modules

A common domain found within polysaccharide degrading enzymes is the carbohydrate binding modules (CBMs). These modules are implicated with binding to both monosaccharides and polysaccharides. A laminarinase, Lam16A, from *Thermotoga neapolitana* was found to contain two CBMs surrounding the catalytic domain [93]. By overexpressing and purifying the CBMs independently and with the catalytic domain, Zverlov and coworkers demonstrated that each CBM was bound to different glucans with different affinities [93]. One of the CBMs was bound to laminarin and barley glucan (β-1,3/1,4-linked glucose), while the other CBM was bound to curdlan (1,3-β-glucose), pustulan (1,6-β-glucose), and insoluble yeast cell wall glucan. When the catalytic domain was incubated with the tested glucans, no binding was found indicating that the CBM is essential for binding a catalytic domain to the substrate for eventual lytic action.

A further investigation of a CBM-containing β-1,3-glucanase found similar results [82]. This enzyme was cloned from *S. sioyaensis* and characterized. The authors found that the CBM was not essential for enzyme activity toward laminarin, but dramatically decreased activity toward curdlan and pachyman. They also found that the CBM was directly bound to these glucans. Curdlan and pachyman are insoluble glucans, so these findings indicate that the CBM of the *S. sioyaensis* glucanase is used to bind insoluble glucans, which facilitates their degradation.

The CtCBM11 CBM from the *Clostridium thermocellum* Lic26A-Cel5E glucanase was found to bind both β-1,4- and β-1,3-1,4-linked glucans [94]. The CBM was found to bind to the tested glucans with an association constant between 4.4×10^4 and $30.1 \times 10^4\ M^{-1}$. Additionally, through crystallization of the CBM, the authors were able to identify the putative amino acid residues responsible for binding. By mutating several of these residues, three amino acid residues were determined to independently abolish binding of the CBM to the tested glucans. Results of investigations into CBM binding indicated that CBMs are critical for the enzymatic activity of polysaccharide degrading enzymes, particularly toward insoluble polymers. Interestingly, CBMs are not found in all polysaccharide degrading enzymes.

5.6 Applications of Algal-Polysaccharide Degrading Enzymes

Macroalgae have high potential as a feedstock due to their high carbohydrate content among other advantages as mentioned above. As discussed in Sect. 5.2, because macroalgae contain various types of glucans they can be easily hydrolyzed to glucose for bioconversion into value added chemicals. However, due to the relatively low glucan content in macroalgae compared to that of terrestrial biomass, exploitation of glucan carbohydrates from macroalgae is not competitive enough for its implementation in biorefinery. On the other hand, the high content of non-glucan carbohydrates present in macroalgae makes the exploitation of seaweed feedstock feasible for biofuel and chemical production.

Despite the potential shown by macroalgal carbohydrates for the industrial sector, bioconversion of non-glucan polysaccharides into high value compounds is still challenging, and it hampers the development of this field. Along these lines, recent studies focused on harnessing specialized enzymatic pathways from those microorganisms that naturally metabolize algal-carbohydrates to transform them into energy metabolites and other biocompounds. For this purpose, different fermentation microorganisms have been engineered to produce macroalgal polysaccharides degrading enzymes. The action of these enzymes releases oligosaccharides and/or monosaccharides that could be directly used by the central carbon metabolism. However, due to the peculiarity of algal carbohydrates, the released oligosaccharides and/or monosaccharides were occasionally found not to be metabolized. So, further metabolic engineering of the appropriate microorganism is necessary.

In the following section, recent advances in implementation of polysaccharide-degrading enzymes and specialized metabolic pathways for biofuel and biochemical production from macroalgae are discussed.

5.6.1 *Biofuel Production*

Currently there are a number of technologies for conversion of macroalgal biomass into biofuels. Bioconversion, hydrothermal liquefaction (HTL), and pyrolysis have been developed for the production of biogas, bioethanol, biodiesel and bio-oil from macroalgal feedstock. HTL and pyrolysis are thermochemical conversion processes that require a high thermal treatment to decompose the biomass. These technologies do not require the use of seaweed polysaccharide-degrading enzymes, and thus they will not be discussed here. Readers are referred to various reviews on the subject [18–21, 95].

On the other hand, bioconversion harnesses the microbial enzymatic machinery to depolymerize macroalgal carbohydrates into sugars for production of liquid biofuels by fermentative microorganisms. Consequently, the choice of appropriate microorganisms for assimilation of these monomers is pivotal for successful fermentation. For example, green and red macroalgae contain relatively high levels of accessible fermenting sugars that can be fermented by common tractable microorganisms such as *Saccharomyces cerevisiae*. Conversely, biofuel production from brown macroalgae (with alginate as the main carbohydrate) is limited by the availability of microorganisms that can metabolize alginate.

Prior to fermentation by an appropriate ethanologenic microorganism, *S. cerevisiae*, for example, the carbohydrate fraction of macroalgae is usually released by physical and chemical treatments. The most widely adopted pretreatment is hydrothermal treatment with acid or alkali. These treatments release the polymers from the algal cell wall. However, these methods usually lead to the formation of fermentation inhibitors [96]. To reduce the toxicity of these treatments and also favor the availability of sugars to the fermentative microorganism, some studies have combined mild pre-treatments with enzymatic saccharification leading to improved ethanol productivities. Adams and coworkers combined enzymatic saccharification of extracts derived from the brown macroalgae, *Saccharina latissimi*, with fermentation by the ethanologenic *S. cerevisiae*. A laminarinase was also used to hydrolyze laminarin into glucose which was fermented to ethanol by the yeast. Ethanol yield obtained was 0.45 % (v/v) [97]. However, mannitol present in *S. latissimi* extracts was not consumed by *S. cerevisiae*. A microorganism with a more versatile sugar metabolism would allow more efficient conversion of the mixed carbohydrates in seaweed extracts. Thereby, simultaneous fermentation of mannitol and laminarin was performed by the yeast *Pichia angophorae* CBS 5830 in extracts of the brown algae *Laminaria hyperborean*. The *Pichia* yeast is able to convert mannitol and glucose into ethanol as well as hydrolyzing laminarin as it produces laminarinase. In this study, the maximum ethanol yield obtained was 0.43 g ethanol/g substrate [98]. In another study, extracts of *Laminaria digitata* were fermented by *P. angophorae* as in the above example. However, the addition of commercial laminarinase enzyme in the extract expedited the hydrolysis of laminarin and hence led to increase the ethanol yield to 0.89 % (v/v) [99].

Another microorganism capable of metabolizing mannitol and glucose is *Escherichia coli*. An engineered ethanologenic *E. coli* KO11 (ATCC55124) that contained the *Zymomonas mobilis* ethanol production genes [100] was used to produce ethanol from brown macroalgae. The acid lysates of *Saccharina japonica* (*Laminaria japonica*) treated with enzymes released high concentrations of glucose and mannitol, and thus simultaneous saccharification and fermentation (SSF) resulted in a final ethanol yield of up to 29 g/L [101].

In these studies, the main fermentation sugars were mannitol and glucose derived from laminarin, but in brown macroalgae the major carbohydrate content is alginate, which is underused due to the lack of an efficient alginate saccharification method. Jang and coworkers tried to increase the use of the macroalgae sugar content by employing *Bacillus* sp. JS-1 for saccharification, a marine bacterium isolated from seawater capable of degrading different seaweed polysaccharides such as alginate, fucoidan and laminaran. In this study, extracts from *S. japonica* treated with thermal acid hydrolysis were inoculated with *Bacillus* sp. JS-1 and *P. angophorae* KCTC 17574. The combination of thermal acid hydrolysis with saccharification by *Bacillus* sp. JS-1 increased the release of reducing sugars from *S. japonica* by threefold compared to treatment of thermal acid hydrolysis alone. Whereas, combination of thermal acid hydrolysis with enzymatic treatment using a commercial α-amylase (Termamyl 120 L) gave a twofold increase of sugar release. The maximum ethanol concentration obtained by SSF with *Bacillus* sp. JS-1 and *P. angophorae* KCTC 17574 was 9.8 mL/L [$\sim$1 %, (v/v)] [102].

These studies revealed that the customization of saccharification methods depending on the mixed carbohydrates from macroalgae together with the selection of an appropriate fermentative microorganism would increase bioethanol production. Pursuant to this goal, the most promising SSF approach is engineering the fermentative microorganism to combine saccharification enzyme production, carbohydrate degradation and monosaccharides metabolism, and ethanol formation in a single-process step.

Along these lines, *Sphingomonas* sp. A1 is an interesting candidate for metabolic engineering. *Sphingomonas* sp. A1 is a Gram-negative alginate-assimilating bacterium that directly incorporates alginate into the cytoplasm through a periplasmic alginate-binding protein-dependent ATP-binding cassette transporter. In the cytoplasm, alginate is degraded to monosaccharides by endo- and exotype alginate lyases. The released monosaccharides are then metabolized to pyruvate. By introduction of an ethanologenic pathway, pyruvate from alginate degradation can be redirected to produce ethanol. Takeda and coworkers engineered an ethanologenic *Sphingomonas* sp. A1 by introducing the pyruvate decarboxylase (*pdc*) and alcohol dehydrogenase B (*adhB*) genes from *Zymomonas mobilis* under the control of a strong constitutive promoter isolated from *Sphingomonas* sp. A1 [103]. Also, the lactate dehydrogenase gene (*ldh*) was deleted to redirect the carbon flux to production of ethanol. The combination of these modifications led to a maximum of 13 g/L of ethanol after 3 days of incubation [103].

One of the most remarkable breakthroughs in the field was the development of consolidated bioprocessing (CBP) for production of biofuels from macroalgal biomass by Wargacki and coworkers. In this work, an *E. coli* platform that can degrade, uptake, and metabolize alginate to produce ethanol was constructed [70].

Initially, an alginate lyase (Aly) secretion system was created. A truncated Aly from *Pseudo-alteromonas* sp. SM0524 (tSM0524 Aly) was fused to an engineered portion of the antigen 43 (Ag43) which allowed the extracellular release of the functional tSM0524 Aly. This step enabled the degradation of alginate into oligoalginate without the need of thermal and chemical pretreatment or enzymatic saccharification before fermentation. The next step was the identification and isolation of a putative alginate-metabolizing gene cluster from the marine bacterium *V. splendidus* 12B01 that enabled oligoalginate transport and conversion into pyruvate and glyceraldehyde-3-phosphate. The metabolic pathway was further optimized by expression of auxiliary genes. The putative alginate-metabolizing pathway allowed oligoalginates to pass through the outer membrane to the periplasmic space via porin (KdgM-KdgN). Then, oligoalginates were further degraded into smaller oligomers by alginate lyases (AlyABCD). The symporter (ToaABC) transported the oligomers to the cytoplasm where oligoalginate lyases (OalABC) degraded oligomers into monomers that were spontaneously transformed into DEH. DEH reductase (DehR) reduced DEH into KDG which is phosphorylated by KDG kinase (KdgK) into KDPG. Finally, KDG-6-phosphate aldolase (Eda) cleaved KDPG into pyruvate and glyceraldehyde-3-phosphate (Fig. 5.2) [70]. The alginolytic *E. coli* platform was further engineered to efficiently produce ethanol. To this end, the pyruvate decarboxylase (*pdc*) and alcohol dehydrogenase B (*adhB*) genes from *Zymomonas mobilis* were integrated into the genome. Additionally, *pflB-focA*, *frdABCD*, and *ldhA* genes were deleted to avoid the synthesis of fermentation by-products [70]. The *E. coli* strain and the copy number of the alginate-assimilating cluster and the ethanologenic pathway were optimized to enhance the microbial platform performance. *E. coli* ATCC8739 exhibited faster growth rates in pre-degraded alginate media than other strains evaluated. Also, single copy of each

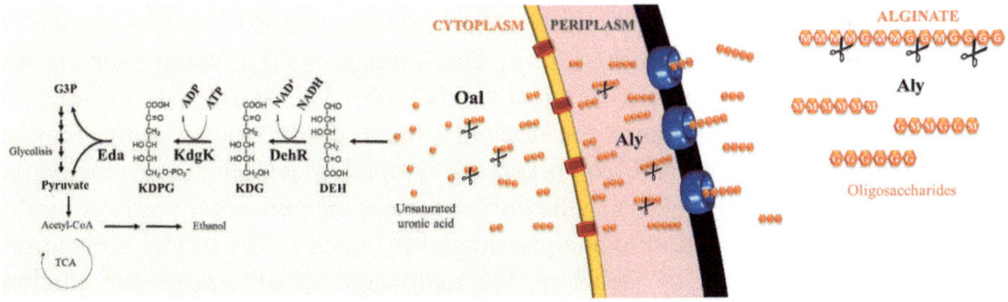

Fig. 5.2 Metabolic pathway for alginate degradation and ethanol production in an engineered *E. coli*. Alginate lyase (Aly, endo-type) degrades the alginate polymer into oligoalginate. These oligomers pass through the outer membrane (in *black*) to the periplasmic space via porin (KdgM-KdgN) (in *blue*). Then, oligoalginates are further degraded into smaller oligomers by alginate lyases. The symporter (ToaABC) (in *red*) transports the oligomers to the cytoplasm where oligoalginate lyases (Oal, exo-type) degrade oligomers into monomers of uronic acid that are spontaneously transformed into DEH (4-deoxy-L-erythro-5-hexoseulose uronic acid). DEH reductase (DehR) reduces DEH into KDG (2-keto-3-deoxy-D-gluconate) which is phosphorylated by KDG kinase (KdgK) into KDPG (2-keto-3-deoxy-6-P-gluconate). Finally, KDPG aldolase (Eda) cleaves KDPG into pyruvate and G3P (glyceraldehyde-3-phosphate). G3P is converted into pyruvate through the glycolysis metabolic pathway. Pyruvate can be redirected to be converted into ethanol by the introduction of an ethanologenic pathway. G, guluronic acid; M, mannuronic acid

pathway (alginate-metabolizing and ethanologenic) was found to produce higher ethanol titers [104]. The resulting strain (BAL1611) was evaluated for its ability to ferment *S. japonica* extracts. BAL1611 produced an average of 37 g/L of ethanol from dry milled brown macroalgae, which represented a final titer of ~4.7 % v/v of ethanol [70, 104].

Similarly, Enquist-Newman and coworkers developed a yeast platform that can simultaneously metabolize DEH and mannitol to produce ethanol [105]. Due to the lack of the metabolic machinery for alginate and/or DEH assimilation, the yeast *S. cerevisiae* cannot utilize either alginate or DEH. Therefore, several approaches were needed to achieve an ethanologenic DEH-assimilating strain. To this end, the DEH transporter AcDHT1 was identified from the alginolytic eukaryote *Asteromyces cruciatus*. This transporter was highly induced when *A. cruciatus* was grown on alginate. Also, yeast cells containing this transporter showed improved growth on DEH-containing media. Another important development was the identification of essential mannitol catabolism genes. The expression of an NAD^+-dependent mannitol-2-dehydrogenase (M2DH) (encoded by *yel070w/ynr073c*) and a mannitol transporter (encoded by *hxt13/hxt17*) were critical for mannitol assimilation in yeast. Finally, the encoding genes for DEH reductase, KDG kinase and KDPG aldolase, involved in DEH assimilation from different bacteria were studied. Those with maximum specific activity were overexpressed in the engineered yeast. Since the growth rate in DEH medium was low, the authors implemented an adaptive evolution strategy for several months to enhance the efficiency of DEH metabolism in aerobic and microaerobic conditions. BAL3215, the adapted strain containing the *Vibrio harveyi* DEH reductase (VhDehR), which uses both NADH and NADPH efficiently, stood out as the strain with improved ability to produce ethanol from mannitol and DEH in microaerobic conditions. Ethanol titers of 4.6 % (v/v) were obtained from fermentation of DEHU:mannitol 1:2 molar ratio, representing 83 % of the maximum theoretical yields [105]. This study also revealed the relevance of the cofactor balance for efficient ethanol production from macroalgae derived sugars. Contador and coworkers developed a genome-scale metabolic model that simulated the cellular behavior of the engineered yeast platform under macroalgal sugar fermentation to analyze the redox balance. The model pointed out the NADH-NADPH bias of DehR as the main determinant factor for efficient ethanol production. The model also detected other metabolic limitations that would be difficult to determine by other means. The authors suggested that this sort of metabolic model would help to improve the performance of the ethanol fermentation from alginate and mannitol [106].

Recently, the metabolic pathway for assimilation of 3,6-anhydro-L-galactose (AHG) has been elucidated. Red macroalgae contain high amounts of agar which is a combination of agarose and agaropectin. The enzymatic treatment of agarose with β-agarases and neoagarobiose hydrolase released β-D-galactose and AHG [107]. Galactose can be easily fermented by tractable microorganism, whereas AHG is an unusual sugar that is not fermented. To harness the full potential of red macroalgae for biofuel production, it was essential to discover the key enzymes that allow the catabolism of AHG. Yun and coworkers identified two essential enzymes for AHG assimilation from *Vibrio* sp. strain EJY3. A $NADP^+$-dependent AHG dehydrogenase (*Vej*AHGD) oxidized AHG to 3,6-anhydrogalactonate (AHGA) which is subsequently isomerized to 2-keto-3-deoxy-galactonate (KDGal) by a AHG Acycloisomerase (*Vej*ACI). Then, KDGal can be converted to 2-keto-3-deoxy-6-phosphogalactonate (KDPG), which is a key metabolite in the galactonate metabolic pathway. The authors reconstructed the AHG assimilation pathway in an ethanologenic *E. coli* KO11. This engineered strain was able to produce ethanol from a lysate of agarose (containing galactose and AHG), and showed a 24 % increased ethanol production than the control with an empty vector. Further optimization of the metabolic pathway will increase the ethanol production from red macroalgal carbohydrates [108].

Ethanol is not the only biofuel that has been produced by macroalgal carbohydrate fermentation. The so-called ABE fermentation produces acetone, butanol and ethanol by anaerobic fermentation of sugars. Bearing in mind the high content of sugars that macroalgae contain, it is presumed that the ABE fermentation of algal carbohydrates will produce acetone, butanol and ethanol. Thermal lysates of the green macroalgae *Ulva lactuca* treated with cellulases released glucose, rhamnose and xylose that were fermented by *Clostridium acetobutylicum* and *Clostridium beijerinckii* producing 0.35 g ABE/g sugars. The authors also found that 1,2-propanediol was produced. A study in more detail unveiled that *C. beijerinckii* produced 1,2-propanediol from rhamnose. The authors suggested that rhamnose-rich seaweeds can be used as feedstock for 1,2-propanediol production [109].

Anaerobic digestion (AD) is another approach that has been used to obtain biofuels from macroalgae. AD comprises a bacterial consortium that performs a number of sequential biological processes (hydrolysis, acidogenesis, acetogenesis and methanogenesis) that convert a feedstock into methane and carbon dioxide. In the first step, hydrolytic bacteria release sugars and amino acids. These are converted into volatile fatty acids (VFA) as carboxylic acids and alcohol by acidogenic bacteria. Then, acetic acid, hydrogen and CO_2 are produced from VFA, and subsequently converted in methane and CO_2 by methanogenic bacteria [95]. The biomethane yields reported from different macroalgae range between 0.12 to 0.48 m^3 CH_4/kg volatile solids (VL) [21]. However, little is known about the bacterial communities involved in the anaerobic degradation of macroalgal biomass. Recently, Kita and coworkers unveiled the bacterial consortium structure and alginate-degrading pathway in an alginate-assimilating microbial community obtained from sand from Hiroshima Gulf in Japan [110]. The metagenome study revealed that almost 98 % of the bacterial species were related to *Clostridium*, *Citrobacter* and *Dysgonomonas* genera, and two species related to Clostridiaceae bacterium SK082 and *Dysgonomonas capnocytophagoides* dominated the consortium followed by a third one related to *Citrobacter freundii*. The authors also determined the presence of all the genes involved in the metabolic pathway of alginate. They found 10 putative alginate lyase genes, 9 putative KDG reductase genes, 6 putative KDG kinase genes, and 6 putative KDPG aldolase genes which are the essential genes involved in the degradation of alginate to glyceraldehyde-3-phosphate and pyruvate. Further elucidation of the role of each strain in the degradation and metabolism of alginate was made. The *Clostridium* species seemed to contain alginate lyases, KDG reductases, KDG kinases, and KDPG aldolases. *Citrobacter* species contained KDG kinases and KDPG aldolases, so it was concluded that they were not involved directly in the degradation of alginate. The contribution of *D. capnocytophagoides* to alginate degradation was unclear as only KDG reductase genes were confirmed, which barely justify why it was a dominant strain in the consortium [110].

More studies of the metagenomic kind will uncover microorganisms and their algal-carbohydrate-metabolizing pathways generating a wealthy source of options for metabolic engineering of a tractable microorganism for efficient assimilation of macroalgal carbohydrates, and eventually their conversion not only into ethanol but also into value added chemicals.

5.6.2 Other Chemicals

Macroalgae biomass has great potential in the production of added value biochemical other than biofuels. In a recent study, 2,3 butanediol and acetoin (BA) were produced by an engineered *E. coli* from macroalgal biomass. *S. japonica* lysates treated with a cocktail of commercial enzymes were fermented with an *E. coli* strain containing a synthetic 2,3 butanediol pathway from *Enterobacter aerogenes* KCTC 2190. The final yield of the fermentation was 0.43 g/g of BA [111].

The production of L-lactate from macroalgal biomass has been also reported. In this case, the L-lactate dehydrogenase from *Streptococcus bovis/equinus* was introduced in an *E. coli* strain where synthesis of the competing by-product was blocked. *S. japonica* lysates treated with a cocktail of commercial enzymes were fermented with the engineered *E. coli* to produce 37.7 g/L of L-lactate with 80 % of the maximum theoretical yield [112]. Similarly, pyruvate was produced from alginate by an engineered *Sphingomonas* sp. strain A where the D-lactate dehydrogenase gene (*ldh*) was deleted. The concentration of pyruvate was 4.6 g/L, representing 18.6 % of theoretical yield [113].

Alginate lyases and oligoalginate lyases have been used to produce DEH that can be converted into 2,5-furandicarboxylic acid (FDCA). FDCA has generated great interest as it can be used as a precursor for the synthesis of polyethylene terephthalate (PET), nylons, jet fuels, and otherdiol-, diamine-, or dialdehyde-based chemicals [114].

On the other hand, algal carbohydrate degrading enzymes can be used to release oligosaccharides. Some of these oligosaccharides have shown bio-activity with applications in medicine and cosmetics. In this case, the use of enzymes is even more relevant due to their high specificity, and also the mild reaction conditions preserve the integrity of the native chemical structure (maintaining the bioactivity of the oligosaccharide). Further, physicochemical treatments used to release oligosaccharides lead to the formation of toxic by-products [96] that must be removed prior to use in medicine or cosmetics. Thus, by employing enzymatic treatments, expensive and tedious purification steps to eliminate these toxins can be avoided [40].

Fucoidanases or α-L-fucosidases can modify the algal polysaccharide fucoidan. These enzymes preserve the sulfation pattern, and thus the bioactivity of fucoidans is retained. It has been reported that fucoidan oligosaccharides obtained by enzymatic digestion of *Cladosiphon novae-caledoniaekylin* inhibited invasion and angiogenesis of tumor cells, showing a potential antitumor activity [115]. Also, fucoidan oligosaccharides can generate plant immunity against virus as described by Klarzynski and coworkers. The authors showed that fucoidan oligosaccharides prepared from enzymatic digestion of *Pelvetia canaliculata* stimulated defense against the tobacco mosaic virus (TMV) in *Nicotiana tabacum* plant [116]. Similarly, κ-carrageenases have been used to produce κ-carrageenans oligosaccharides that showed *in vivo* and *in vitro* anti-tumor and anti-angiogenic activity [117–119].

The application of algal-carbohydrate degrading enzymes to produce oligosaccharides is still in the early stages. Many enzymes are not presently commercially available thus making their implementation on an industrial scale not competitive enough, in part due to the high enzyme cost.

5.7 Conclusions and Perspective

Marine macroalgae are attracting great attention as a promising feedstock for production of both biofuels and chemicals, showing several advantages over terrestrial plant biomass feedstocks. Their potential in biorefinery lies in their high content of carbohydrates. These carbohydrates are generally complex and show diverse sugar composition, some of them exclusive to seaweed biomass. Thus, the depolymerization of these carbohydrates needs specific enzymes capable of releasing the uncommon sugars. Then, the transformation of these sugars into biofuels and chemicals by a fermenting microorganism requires specialized metabolic pathways that convert sugars into cellular metabolites. These specialized enzymes and metabolic pathways are not evenly distributed in nature. Actually, tractable microorganisms lack some of these assimilating mechanisms. Recent studies elucidated the alginate and agarose metabolism of a few microorganisms. However, more efforts are needed to discover the metabolism of agaropectin and carrageenan to exploit the full potential of seaweed polysaccharides.

Metabolic engineering approaches allowed the development of fermenting microorganisms that convert algal carbohydrates into biofuels. However, the limited number of known microbial mechanisms hinders further development. Therefore, the discovery, isolation and characterization of new degrading enzymes and pathways are key steps for the development of improved fermenting microorganisms capable of converting efficiently macroalgal carbohydrates into biofuels and biochemicals.

The exploitation of seaweeds, and hence their carbohydrates in biorefinery not only depends on improved enzymes and engineered microorganisms, but requires improved seaweed farming and harvesting technologies. The biochemical composition, carbohydrate content and growth rate of macroalgae vary depending on the species, the period of harvesting and growth temperature. Thus, the farming technology should consider these features to obtain maximal mass-cultivated productivities to sufficiently supply feedstocks for industry [19].

Despite the challenges to be overcome before industrial exploitation, current developments mitigate key bottlenecks that hindered the practical implementation of macroalgal carbohydrates-degrading enzymes and assimilating pathways. It is envisaged that future efforts will make feasible the cost-effective exploitation of these systems for biofuel and value added chemical production.

Acknowledgments We thank the U.S. Department of Energy, Office of Science, Office of Biological and Environmental Research under Award Number ER65474 for financial support.

References

1. Beale CV, Long SP (1995) Can perennial C4 grasses attain high efficiencies of radiant energy conversion in cool climates? Plant Cell Environ 18(6):641–650
2. Borland AM, Griffiths H, Hartwell J, Smith JAC (2009) Exploiting the potential of plants with crassulacean acid metabolism for bioenergy production on marginal lands. J Exp Bot 60 (10):2879–2896
3. Laureano-Perez L, Teymouri F, Alizadeh H, Dale BE (2005) Understanding factors that limit enzymatic hydrolysis of biomass: characterization of pretreated corn stover. Appl Biochem Biotechnol 124(1–3):1081–1100
4. Saha BC (2003) Hemicellulose bioconversion. J Ind Microbiol Biotechnol 30(5):279–291
5. Bobleter O (1994) Hydrothermal degradation of polymers derived from plants. Prog Polym Sci 19(5):797–841
6. Singh A, Nigam PS, Murphy JD (2011) Renewable fuels from algae: an answer to debatable land based fuels. Bioresour Technol 102(1):10–16
7. Stephens E, de Nys R, Ross IL, Hankamer B (2013) Algae fuels as an alternative to petroleum. J Pet Environ Biotechnol 4:1–7
8. Aresta M, Dibenedetto A, Barberio G (2005) Utilization of macro-algae for enhanced CO_2 fixation and biofuels production: development of a computing software for an LCA study. Fuel Process Technol 86(14–15):1679–1693
9. Wei N, Quarterman J, Jin Y-S (2013) Marine macroalgae: an untapped resource for producing fuels and chemicals. Trends Biotechnol 31(2):70–77
10. Martone PT, Estevez JM, Lu F, Ruel K, Denny MW, Somerville C, Ralph J (2009) Discovery of lignin in seaweed reveals convergent evolution of cell-wall architecture. Curr Biol 19(2):169–175
11. Ge L, Wang P, Mou H (2011) Study on saccharification techniques of seaweed wastes for the transformation of ethanol. Renew Energy 36(1):84–89
12. Smith S (1981) Marine macrophytes as a global carbon sink. Science 211(4484):838–840
13. Demirbas A, Demirbas MF (2011) Importance of algae oil as a source of biodiesel. Energy Convers Manag 52(1):163–170
14. Grima EM, Fernández FA, Camacho FG, Chisti Y (1999) Photobioreactors: light regime, mass transfer, and scaleup. J Biotechnol 70(1):231–247
15. Atsumi S, Higashide W, Liao JC (2009) Direct photosynthetic recycling of carbon dioxide to isobutyraldehyde. Nat Biotechnol 27(12):1177–1180

16. Gao Z, Zhao H, Li Z, Tan X, Lu X (2012) Photosynthetic production of ethanol from carbon dioxide in genetically engineered cyanobacteria. Energy Environ Sci 5(12):9857–9865
17. Ho S-H, Ye X, Hasunuma T, Chang J-S, Kondo A (2014) Perspectives on engineering strategies for improving biofuel production from microalgae—a critical review. Biotechnol Adv 32(8):1448–1459
18. Song M, Pham HD, Seon J, Woo HC (2015) Marine brown algae: a conundrum answer for sustainable biofuels production. Renew Sustain Energy Rev 50:782–792
19. Jung KA, Lim S-R, Kim Y, Park JM (2013) Potentials of macroalgae as feedstocks for biorefinery. Bioresour Technol 135:182–190
20. Bharathiraja B, Chakravarthy M, Kumar RR, Yogendran D, Yuvaraj D, Jayamuthunagai J, Kumar RP, Palani S (2015) Aquatic biomass (algae) as a future feed stock for bio-refineries: a review on cultivation, processing and products. Renew Sustain Energy Rev 47:634–653
21. Chen H, Zhou D, Luo G, Zhang S, Chen J (2015) Macroalgae for biofuels production: Progress and perspectives. Renew Sustain Energy Rev 47:427–437
22. Samaraweera AM, Vidanarachchi JK, Kurukulasuriya MS (2011) Industrial applications of macroalgae. Handb Mar Macroalgae Biotechnol Appl Phycol 500–521
23. Jensen A (1993) Present and future needs for algae and algal products. In: Fourteenth international seaweed symposium. Springer, pp 15–23
24. Cho D-M, Kim D-S, Lee D-S, Kim H-R, Pyeun J-H (1995) Trace components and functional saccharides in seaweed-1-changes in proximate composition and trace elements according to the harvest season and places. Korean J Fish Aquatic Sci 28(1):49–59
25. Meinita MDN, Hong Y-K, Jeong G-T (2012) Detoxification of acidic catalyzed hydrolysate of *Kappaphycus alvarezii* (cottonii). Bioprocess Biosyst Eng 35(1–2):93–98
26. Yanagisawa M, Kawai S, Murata K (2013) Strategies for the production of high concentrations of bioethanol from seaweeds: production of high concentrations of bioethanol from seaweeds. Bioengineered 4(4):224–235
27. Lahaye M, Robic A (2007) Structure and functional properties of ulvan, a polysaccharide from green seaweeds. Biomacromolecules 8(6):1765–1774
28. Tako M, Tamanaha M, Tamashiro Y, Uechi S (2015) Structure of ulvan isolated from the edible green seaweed, *Ulva pertusa*. Adv Biosci Biotechnol 6(10):645
29. Robic A, Bertrand D, Sassi J-F, Lerat Y, Lahaye M (2009) Determination of the chemical composition of ulvan, a cell wall polysaccharide from *Ulva* spp. (Ulvales, Chlorophyta) by FT-IR and chemometrics. J Appl Phycol 21(4):451–456
30. Bojko M, Madsen F, Olsen CE, Engelsen SB (2002) Physico-chemical characterization of floridean starch of red algae. Starch/Stärke 54:66–74
31. Lobban CS, Wynne MJ (1981) The biology of seaweeds, vol 17. University of California Press, Oakland
32. Campo VL, Kawano DF, Silva Jr DBd, Carvalho I (2009) Carrageenans: biological properties, chemical modifications and structural analysis—a review. Carbohydr Polym 77(2):167–180
33. Lahaye M, Rochas C (1991) Chemical structure and physico-chemical properties of agar. In: Juanes JA, Santelices B, McLachlan JL (eds) International workshop on *Gelidium*: proceedings of the international workshop on gelidium held in Santander, Spain, September 3–8, 1990. Springer, Dordrecht, pp 137–148
34. Fu XT, Kim SM (2010) Agarase: review of major sources, categories, purification method, enzyme characteristics and applications. Mar Drugs 8(1):200–218
35. Wang T-P, Chang L-L, Chang S-N, Wang E-C, Hwang L-C, Chen Y-H, Wang Y-M (2012) Successful preparation and characterization of biotechnological grade agarose from indigenous *Gelidium amansii* of Taiwan. Process Biochem 47(3):550–554
36. Larsen B, Salem DMSA, Sallam MAE, Mishrikey MM, Beltagy AI (2003) Characterization of the alginates from algae harvested at the Egyptian Red Sea coast. Carbohydr Res 338(22):2325–2336
37. Pawar SN, Edgar KJ (2012) Alginate derivatization: a review of chemistry, properties and applications. Biomaterials 33(11):3279–3305

38. Mautner HG (1954) The chemistry of brown algae. Econ Bot 8(2):174–192
39. Hahn T, Lang S, Ulber R, Muffler K (2012) Novel procedures for the extraction of fucoidan from brown algae. Process Biochem 47(12):1691–1698
40. Holtkamp AD, Kelly S, Ulber R, Lang S (2009) Fucoidans and fucoidanases—focus on techniques for molecular structure elucidation and modification of marine polysaccharides. Appl Microbiol Biotechnol 82(1):1–11
41. Morya V, Kim J, Kim E-K (2012) Algal fucoidan: structural and size-dependent bioactivities and their perspectives. Appl Microbiol Biotechnol 93(1):71–82
42. Kawai S, Murata K (2016) Biofuel production based on carbohydrates from both brown and red macroalgae: recent developments in key biotechnologies. Int J Mol Sci 17(2):145
43. Gacesa P (1987) Alginate-modifying enzymes. FEBS Lett 212(2):199–202
44. Garron M-L, Cygler M (2010) Structural and mechanistic classification of uronic acid-containing polysaccharide lyases. Glycobiology 20(12):1547–1573
45. Lombard V, Bernard T, Rancurel C, Brumer H, Coutinho Pedro M, Henrissat B (2010) A hierarchical classification of polysaccharide lyases for glycogenomics. Biochem J 432(3):437–444
46. Yoon H-J, Mikami B, Hashimoto W, Murata K (1999) Crystal structure of alginate lyase A1-III from *Sphingomonas* species A1 at 1.78 Å resolution. J Mol Biol 290(2):505–514
47. Martinez-Fleites C, Smith NL, Turkenburg JP, Black GW, Taylor EJ (2009) Structures of two truncated phage-tail hyaluronate lyases from *Streptococcus pyogenes* serotype M1. Acta Crystallogr Sect F Struct Biol Cryst Commun 65(10):963–966
48. Smith NL, Taylor EJ, Lindsay A-M, Charnock SJ, Turkenburg JP, Dodson EJ, Davies GJ, Black GW (2005) Structure of a group A streptococcal phage-encoded virulence factor reveals a catalytically active triple-stranded β-helix. Proc Natl Acad Sci USA 102(49): 17652–17657
49. Thomas F, Lundqvist LCE, Jam M, Jeudy A, Barbeyron T, Sandström C, Michel G, Czjzek M (2013) Comparative characterization of two marine alginate lyases from *Zobellia galactanivorans* reveals distinct modes of action and exquisite adaptation to their natural substrate. J Biol Chem 288(32):23021–23037
50. Ogura K, Yamasaki M, Yamada T, Mikami B, Hashimoto W, Murata K (2009) Crystal structure of family 14 polysaccharide lyase with pH-dependent modes of action. J Biol Chem 284(51):35572–35579
51. Huang W, Matte A, Li Y, Kim YS, Linhardt RJ, Su H, Cygler M (1999) Crystal structure of chondroitinase B from *Flavobacterium heparinum* and its complex with a disaccharide product at 1.7 Å resolution. J Mol Biol 294(5):1257–1269
52. Park D, Jagtap S, Nair SK (2014) Structure of a PL17 family alginate lyase demonstrates functional similarities among exotype depolymerases. J Biol Chem 289(12):8645–8655
53. Michel G, Pojasek K, Li Y, Sulea T, Linhardt RJ, Raman R, Prabhakar V, Sasisekharan R, Cygler M (2004) The structure of chondroitin B lyase complexed with glycosaminoglycan oligosaccharides unravels a calcium-dependent catalytic machinery. J Biol Chem 279(31): 32882–32896
54. Yoon H-J, Hashimoto W, Miyake O, Murata K, Mikami B (2001) Crystal structure of alginate lyase A1-III complexed with trisaccharide product at 2.0 Å resolution. J Mol Biol 307(1):9–16
55. Osawa T, Matsubara Y, Muramatsu T, Kimura M, Kakuta Y (2005) Crystal structure of the alginate (poly α-L-guluronate) lyase from *Corynebacterium* sp. at 1.2 Å resolution. J Mol Biol 345(5):1111–1118
56. Ochiai A, Yamasaki M, Mikami B, Hashimoto W, Murata K (2010) Crystal structure of exotype alginate lyase Atu3025 from *Agrobacterium tumefaciens*. J Biol Chem 285 (32):24519–24528
57. Dong S, Wei T-D, Chen X-L, Li C-Y, Wang P, Xie B-B, Qin Q-L, Zhang X-Y, Pang X-H, Zhou B-C, Zhang Y-Z (2014) Molecular insight into the role of the N-terminal extension in the maturation, substrate recognition, and catalysis of a bacterial alginate lyase from polysaccharide lyase family 18. J Biol Chem 289(43):29558–29569

58. Jagtap SS, Hehemann J-H, Polz MF, Lee J-K, Zhao H (2014) Comparative biochemical characterization of three exolytic oligoalginate lyases from *Vibrio splendidus* reveals complementary substrate scope, temperature, and pH adaptations. Appl Environ Microbiol 80(14):4207–4214
59. Ryu M, Lee EY (2011) Saccharification of alginate by using exolytic oligoalginate lyase from marine bacterium *Sphingomonas* sp. MJ-3. J Ind Eng Chem 17(5–6):853–858
60. Sawabe T, Ohtsuka M, Ezura Y (1997) Novel alginate lyases from marine bacterium *Alteromonas* sp. strain H-4. Carbohydr Res 304(1):69–76
61. Kim DE, Lee EY, Kim HS (2009) Cloning and characterization of alginate lyase from a marine bacterium *Streptomyces* sp. ALG-5. Mar Biotechnol 11(1):10–16
62. Hu X, Jiang X, Hwang H-m (2006) Purification and characterization of an alginate lyase from marine bacterium *Vibrio* sp. mutant strain 510–64. Curr Microbiol 53(2):135–140
63. Yamasaki M, Moriwaki S, Miyake O, Hashimoto W, Murata K, Mikami B (2004) Structure and function of a hypothetical *Pseudomonas aeruginosa* protein PA1167 classified into family PL-7: a novel alginate lyase with a β-sandwich fold. J Biol Chem 279(30):31863–31872
64. Haug A, Larsen BR, Smidsrod O (1967) Studies on the sequence of uronic acid residues in alginic acid. Acta Chem Scand 21:691–704
65. Chhatbar M, Meena R, Prasad K, Siddhanta A (2009) Microwave assisted rapid method for hydrolysis of sodium alginate for M/G ratio determination. Carbohydr Polym 76(4):650–656
66. Zhang Z, Yu G, Zhao X, Liu H, Guan H, Lawson AM, Chai W (2006) Sequence analysis of alginate-derived oligosaccharides by negative-ion electrospray tandem mass spectrometry. J Am Soc Mass Spectrom 17(4):621–630
67. Lundqvist LCE, Jam M, Barbeyron T, Czjzek M, Sandström C (2012) Substrate specificity of the recombinant alginate lyase from the marine bacteria *Pseudomonas alginovora*. Carbohydr Res 352:44–50
68. Badur AH, Jagtap SS, Yalamanchili G, Lee J-K, Zhao H, Rao CV (2015) Alginate lyases from alginate-degrading *Vibrio splendidus* 12B01 are endolytic. Appl Environ Microbiol 81(5):1865–1873
69. Hashimoto W, He J, Wada Y, Nankai H, Mikami B, Murata K (2005) Proteomics-based identification of outer-membrane proteins responsible for import of macromolecules in *Sphingomonas* sp. A1: alginate-binding flagellin on the cell surface. Biochemistry 44(42): 13783–13794
70. Wargacki AJ, Leonard E, Win MN, Regitsky DD, Santos CNS, Kim PB, Cooper SR, Raisner RM, Herman A, Sivitz AB (2012) An engineered microbial platform for direct biofuel production from brown macroalgae. Science 335(6066):308–313
71. Takase R, Ochiai A, Mikami B, Hashimoto W, Murata K (2010) Molecular identification of unsaturated uronate reductase prerequisite for alginate metabolism in *Sphingomonas* sp. A1. Biochim Biophys Acta (BBA) Proteins Proteomics 1804(9):1925–1936
72. Hamza A, Piao YL, Kim M-S, Choi CH, Zhan C-G, Cho H (2011) Insight into the binding of the wild type and mutated alginate lyase (AlyVI) with its substrate: a computational and experimental study. Biochim Biophys Acta (BBA) Proteins Proteomics 1814(12):1739–1747
73. MacDonald LC, Berger BW (2014) Insight into the role of substrate-binding residues in conferring substrate specificity for the multifunctional polysaccharide lyase Smlt1473. J Biol Chem 289(26):18022–18032
74. Yamasaki M, Ogura K, Hashimoto W, Mikami B, Murata K (2005) A structural basis for depolymerization of alginate by polysaccharide lyase family-7. J Mol Biol 352(1):11–21
75. Tøndervik A, Klinkenberg G, Aarstad OA, Drabløs F, Ertesvåg H, Ellingsen TE, Skjåk-Bræk G, Valla S, Sletta H (2010) Isolation of mutant alginate lyases with cleavage specificity for di-guluronic acid linkages. J Biol Chem 285(46):35284–35292
76. Davies G, Henrissat B (1995) Structures and mechanisms of glycosyl hydrolases. Structure 3(9):853–859
77. Viladot J-L, de Ramon E, Durany O, Planas A (1998) Probing the mechanism of *Bacillus* 1,3-1,4-β-D-glucan 4-glucanohydrolases by chemical rescue of inactive mutants at catalytically essential residues. Biochemistry 37(32):11332–11342

78. Varghese JN, Garrett TP, Colman PM, Chen L, Høj PB, Fincher GB (1994) Three-dimensional structures of two plant beta-glucan endohydrolases with distinct substrate specificities. Proc Natl Acad Sci USA 91(7):2785–2789
79. Henrissat B, Coutinho PM, Davies GJ (2001) A census of carbohydrate-active enzymes in the genome of *Arabidopsis thaliana*. In: Plant Cell Walls. Springer, pp 55–72
80. Høj PB, Condron R, Traeger JC, McAuliffe JC, Stone BA (1992) Identification of glutamic acid 105 at the active site of *Bacillus amyloliquefaciens* 1,3-1,4-β-D-glucan 4-glucanohydrolase using epoxide-based inhibitors. J Biol Chem 267(35):25059–25066
81. Gueguen Y, Voorhorst WG, van der Oost J, de Vos WM (1997) Molecular and biochemical characterization of an endo-β-1, 3-glucanase of the hyperthermophilic Archaeon *Pyrococcus furiosus*. J Biol Chem 272(50):31258–31264
82. Hong T-Y, Cheng C-W, Huang J-W, Meng M (2002) Isolation and biochemical characterization of an endo-1,3-β-glucanase from *Streptomyces sioyaensis* containing a C-terminal family 6 carbohydrate-binding module that binds to 1,3-β-glucan. Microbiology 148(4):1151–1159
83. Kawai R, Igarashi K, Yoshida M, Kitaoka M, Samejima M (2005) Hydrolysis of β-1,3/1,6-glucan by glycoside hydrolase family 16 endo-1,3(4)-β-glucanase from the basidiomycete *Phanerochaete chrysosporium*. Appl Microbiol Biotechnol 71(6):898–906
84. Furukawa S-i, Fujikawa T, Koga D, Ide A (1992) Purification and some properties of exo-type fucoidanases from *Vibrio* sp. N-5. Biosci Biotechnol Biochem 56(11):1829–1834
85. Bakunina IY, Shevchenko LS, Nedashkovskaya OI, Shevchenko NM, Alekseeva SA, Mikhailov VV, Zvyagintseva TN (2000) Screening of marine bacteria for fucoidanases. Microbiology 69(3):303–308
86. Berteau O, McCort I, Goasdoué N, Tissot B, Daniel R (2002) Characterization of a new α-L-fucosidase isolated from the marine mollusk *Pecten maximus* that catalyzes the hydrolysis of α-L-fucose from algal fucoidan (*Ascophyllum nodosum*). Glycobiology 12(4):273–282
87. Gupta V, Trivedi N, Kumar M, Reddy CRK, Jha B (2013) Purification and characterization of exo-β-agarase from an endophytic marine bacterium and its catalytic potential in bioconversion of red algal cell wall polysaccharides into galactans. Biomass Bioenergy 49:290–298
88. Feng Z, Li M (2013) Purification and characterization of agarase from *Rhodococcus* sp. Q5, a novel agarolytic bacterium isolated from printing and dyeing wastewater. Aquaculture 372:74–79
89. Barbeyron T, Michel G, Potin P, Henrissat B, Kloareg B (2000) ι-Carrageenases constitute a novel family of glycoside hydrolases, unrelated to that of κ-Carrageenases. J Biol Chem 275(45):35499–35505
90. Potin P, Richard C, Barbeyron T, Henrissat B, Gey C, Petillot Y, Forest E, Dideberg O, Rochas C, Kloareg B (1995) Processing and hydrolytic mechanism of the *cgkA*-encoded κ-Carrageenase of *Alteromonas carrageenovora*. Eur J Biochem 228(3):971–975
91. Ohta Y, Hatada Y (2006) A novel enzyme, λ-carrageenase, isolated from a deep-sea bacterium. J Biochem 140(4):475–481
92. Chhabra SR, Shockley KR, Conners SB, Scott K, Wolfinger RD, Kelly RM (2002) Carbohydrate-induced differential gene expression patterns in the hyperthermophilic bacterium *Thermotoga maritima*. J Biol Chem 278(9):7540–7552
93. Zverlov VV, Volkov IY, Velikodvorskaya GA, Schwarz WH (2001) The binding pattern of two carbohydrate-binding modules of laminarinase Lam16A from *Thermotoga neapolitana*: differences in β-glucan binding within family CBM4. Microbiology 147(3):621–629
94. Carvalho AL, Goyal A, Prates JAM, Bolam DN, Gilbert HJ, Pires VMR, Ferreira LMA, Planas A, Romão MJ, Fontes CMGA (2004) The family 11 carbohydrate-binding module of *Clostridium thermocellum* Lic26A-Cel5E accommodates β-1,4- and β-1,3–1,4-mixed linked glucans at a single binding site. J Biol Chem 279(33):34785–34793
95. McKennedy J, Sherlock O (2015) Anaerobic digestion of marine macroalgae: a review. Renew Sustain Energy Rev 52:1781–1790

96. Schultz-Jensen N, Thygesen A, Leipold F, Thomsen ST, Roslander C, Lilholt H, Bjerre AB (2013) Pretreatment of the macroalgae *Chaetomorpha linum* for the production of bioethanol—comparison of five pretreatment technologies. Bioresour Technol 140:36–42
97. Adams JM, Gallagher JA, Donnison IS (2008) Fermentation study on *Saccharina latissima* for bioethanol production considering variable pre-treatments. J Appl Phycol 21(5):569–574
98. Horn S, Aasen I, Østgaard K (2000) Ethanol production from seaweed extract. J Ind Microbiol Biotechnol 25(5):249–254
99. Adams JMM, Toop TA, Donnison IS, Gallagher JA (2011) Seasonal variation in *Laminaria digitata* and its impact on biochemical conversion routes to biofuels. Bioresour Technol 102(21):9976–9984
100. Ohta K, Beall DS, Mejia JP, Shanmugam KT, Ingram LO (1991) Genetic improvement of *Escherichia coli* for ethanol production: chromosomal integration of *Zymomonas mobilis* genes encoding pyruvate decarboxylase and alcohol dehydrogenase II. Appl Environ Microbiol 57(4):893–900
101. Kim N-J, Li H, Jung K, Chang HN, Lee PC (2011) Ethanol production from marine algal hydrolysates using *Escherichia coli* KO11. Bioresour Technol 102(16):7466–7469
102. Jang J-S, Cho Y, Jeong G-T, Kim S-K (2011) Optimization of saccharification and ethanol production by simultaneous saccharification and fermentation (SSF) from seaweed, *Saccharina japonica*. Bioprocess Biosyst Eng 35(1):11–18
103. Takeda H, Yoneyama F, Kawai S, Hashimoto W, Murata K (2011) Bioethanol production from marine biomass alginate by metabolically engineered bacteria. Energy Environ Sci 4(7):2575–2581
104. Santos CNS, Regitsky DD, Yoshikuni Y (2013) Implementation of stable and complex biological systems through recombinase-assisted genome engineering. Nat Comm 4:2503
105. Enquist-Newman M, Faust AME, Bravo DD, Santos CNS, Raisner RM, Hanel A, Sarvabhowman P, Le C, Regitsky DD, Cooper SR, Peereboom L, Clark A, Martinez Y, Goldsmith J, Cho MY, Donohoue PD, Luo L, Lamberson B, Tamrakar P, Kim EJ, Villari JL, Gill A, Tripathi SA, Karamchedu P, Paredes CJ, Rajgarhia V, Kotlar HK, Bailey RB, Miller DJ, Ohler NL, Swimmer C, Yoshikuni Y (2014) Efficient ethanol production from brown macroalgae sugars by a synthetic yeast platform. Nature 505 (7482):239–243
106. Contador CA, Shene C, Olivera A, Yoshikuni Y, Buschmann A, Andrews BA, Asenjo JA (2015) Analyzing redox balance in a synthetic yeast platform to improve utilization of brown macroalgae as feedstock. Metab Eng Commun 2:76–84
107. Kim HT, Lee S, Kim KH, Choi I-G (2012) The complete enzymatic saccharification of agarose and its application to simultaneous saccharification and fermentation of agarose for ethanol production. Bioresour Technol 107:301–306
108. Yun EJ, Lee S, Kim HT, Pelton JG, Kim S, Ko H-J, Choi I-G, Kim KH (2015) The novel catabolic pathway of 3,6-anhydro-L-galactose, the main component of red macroalgae, in a marine bacterium. Environ Microbiol 17(5):1677–1688
109. van der Wal H, Sperber BLHM, Houweling-Tan B, Bakker RRC, Brandenburg W, López-Contreras AM (2013) Production of acetone, butanol, and ethanol from biomass of the green seaweed *Ulva lactuca*. Bioresour Technol 128:431–437
110. Kita A, Miura T, Kawata S, Yamaguchi T, Okamura Y, Aki T, Matsumura Y, Tajima T, Kato J, Nishio N, Nakashimada Y (2016) Bacterial community structure and predicted alginate metabolic pathway in an alginate-degrading bacterial consortium. J Biosci Bioeng 121(3):286–292
111. Mazumdar S, Lee J, Oh M-K (2013) Microbial production of 2, 3 butanediol from seaweed hydrolysate using metabolically engineered *Escherichia coli*. Bioresour Technol 136:329–336
112. Mazumdar S, Bang J, Oh M-K (2014) L-Lactate production from seaweed hydrolysate of *Laminaria japonica* using metabolically engineered *Escherichia coli*. Appl Biochem Biotechnol 172(4):1938–1952

113. Kawai S, Ohashi K, Yoshida S, Fujii M, Mikami S, Sato N, Murata K (2014) Bacterial pyruvate production from alginate, a promising carbon source from marine brown macroalgae. J Biosci Bioeng 117(3):269–274
114. Yoshikuni Y, Cooper SR, Wargacki AJ, Manzer LE, White JF, Kapicak L, Miller DJ, Peereboom L (2014) Methods for preparing 2,5-furandicarboxylic acid. Patent number WO2013049711 A1
115. Ye J, Li Y, Teruya K, Katakura Y, Ichikawa A, Eto H, Hosoi M, Hosoi M, Nishimoto S, Shirahata S (2005) Enzyme-digested fucoidan extracts derived from seaweed Mozuku of *Cladosiphon novae-caledoniae* kylin inhibit invasion and angiogenesis of tumor cells. Cytotechnology 47(1):117–126
116. Klarzynski O, Descamps V, Plesse B, Yvin J-C, Kloareg B, Fritig B (2003) Sulfated fucan oligosaccharides elicit defense responses in tobacco and local and systemic resistance against tobacco mosaic virus. Mol Plant Microbe Interact 16(2):115–122
117. Hu X, Jiang X, Aubree E, Boulenguer P, Critchley AT (2006) Preparation and *in vivo*. antitumor activity of κ-carrageenan oligosaccharides. Pharm Biol 44(9):646–650
118. Haijin M, Xiaolu J, Huashi G (2003) A κ-carrageenan derived oligosaccharide prepared by enzymatic degradation containing anti-tumor activity. J Appl Phycol 15(4):297–303
119. Yao Z, Wu H, Zhang S, Du Y (2014) Enzymatic preparation of κ-carrageenan oligosaccharides and their anti-angiogenic activity. Carbohydr Polym 101:359–367

Chapter 6
Ex Vivo Enzymatic Conversion of Non-food Cellulose Biomass to Starch

Chun You and Y.H. Percival Zhang

Abstract To meet the world's rising future food/feed needs, outputs of modern agriculture must grow substantially while minimizing agriculture's environmental footprint and conserving biodiversity. In this chapter, we propose an ex vivo synthetic enzymatic pathway to enable the transformation of non-food cellulose to amylose, a high-value linear starch, meanwhile glucose released by enzymatic hydrolysis of cellulose is used to produce ethanol and/or single-cell protein by yeast fermentation in the same vessel. The strategy of simultaneous enzymatic bio-transformation and microbial fermentation is the basis of new biomass biorefineries that would address the food, fuels, and environment trilemma by coproducing food/feed, biomaterials, and biofuels from the most abundant renewable bioresource—non-food lignocellulosic biomass. Toward this development, new directions pertaining to pretreatment of lignocellulosic biomass and advanced enzyme engineering are discussed to increase the efficiency of saccharification.

Keywords Bioeconomy · New biorefinery · Food and feed · In vitro synthetic biology · Synthetic amylose · Food-energy-water nexus

6.1 Introduction

The continuing growth of population and food consumption per capita means that the global demand for food could increase by 70 % by 2050 [1, 2]. In the developed countries, starch provides at least 35 % of man's daily intake, whereas, this value

C. You (✉) · Y.H.P. Zhang (✉)
Tianjin Institute of Industrial Biotechnology, Chinese Academy of Sciences,
32 West 7th Avenue, Tianjin Airport Economic Area, Tianjin 300308, China
e-mail: you_c@tib.cas.cn

Y.H.P. Zhang
e-mail: ypzhang@vt.edu

C. You · Y.H.P. Zhang
Biological Systems Engineering Department, Virginia Tech,
304 Seitz Hall, Blacksburg, VA 24061, USA

P.C.K. Lau (ed.), *Quality Living Through Chemurgy and Green Chemistry*,
Green Chemistry and Sustainable Technology,
DOI 10.1007/978-3-662-53704-6_6

may go to 80 % in most developing countries, especially in Africa and the Far East. Increase in the supply of starch food to fight against hunger is becoming a global challenge. However, approximately 30 % of the world's arable land and 70 % of the world's fresh water withdrawals are being used for the production of food and feed to support seven billion people [3]. It is difficult to greatly increase arable land to agricultural use due to its other use and increase freshwater withdrawals. Some solutions were proposed to solve potential food crisis, like increasing agricultural resource efficiency, closing yield gaps on underperforming lands, reducing food waste, and developing genetically modified (GM) crops. However, long-term impacts of GM cereals as staple food on human health are not clear and their wide acceptance is in heated debates, especially in China and Europe [4, 5]. On the other hand, modern agriculture is also environmentally destructive, because some forests have been cut down and wetlands have been drained, more use of fertilizers and pesticides have caused serious environmental problems, such as changes in ecosystems, nonpoint water runoff pollution, climate changes, and decreased biodiversity [6].

Cellulose and starch are both anhydroglucose polymers, which are major products derived from plant photosynthesis. Cellulose is the supporting material of plant cell walls and the most abundant renewable bioresource on earth. It is a linear glucan linked by β-1,4-glucosidic bonds. Annual production of cellulose is estimated to be 100 billion tons [7]. It can be derived from a variety of sources, such as wood, agriculture residues, perennial plants, and so on. Humans, unlike herbivores, cannot utilize cellulose-containing biomass as a food source. Starch is the most important diet component for humans because it provides more than 50 % calories of healthy diets. It is a polysaccharide consisting of a large number of anhydroglucose units joined together primarily by α-1,4-glucosidic bonds and α-1,6-glucosidic bonds. Three major crops for producing starch-rich seeds are maize, rice, and wheat. However, dedicated cereal crops produce approximately 2.8 billion tons of starch-rich cereals annually, much less than annual cellulose production (Fig. 6.1). Due to the same building block (i.e., anhydroglucose) of cellulose and starch, the cost-effective transformation of non-food cellulose to edible starch could drastically enhance food security, while revolutionizing agriculture from cultivating annual seed-harvesting plants to perennial cellulose-rich plants, maintaining biodiversity, minimizing agriculture's environmental footprint, decrease inputs of fertilizers and herbicide, lowering energy consumption of farming machinery, and conserving fresh water [8]. This transformation cannot be done efficiently by microorganisms because of inherent side-reactions of microorganisms and cellular membrane that can prevent large molecular weight polymeric compounds such as cellulose and starch to go through. However, ex vivo synthetic enzymatic biosystems can implement this seemingly mission impossible task by assembling cascade enzymes outside cell membranes. Such ex vivo systems have many appealing advantages, like high product yield, faster reaction rate, great engineering flexibility, and high tolerance to toxic compounds [9–15].

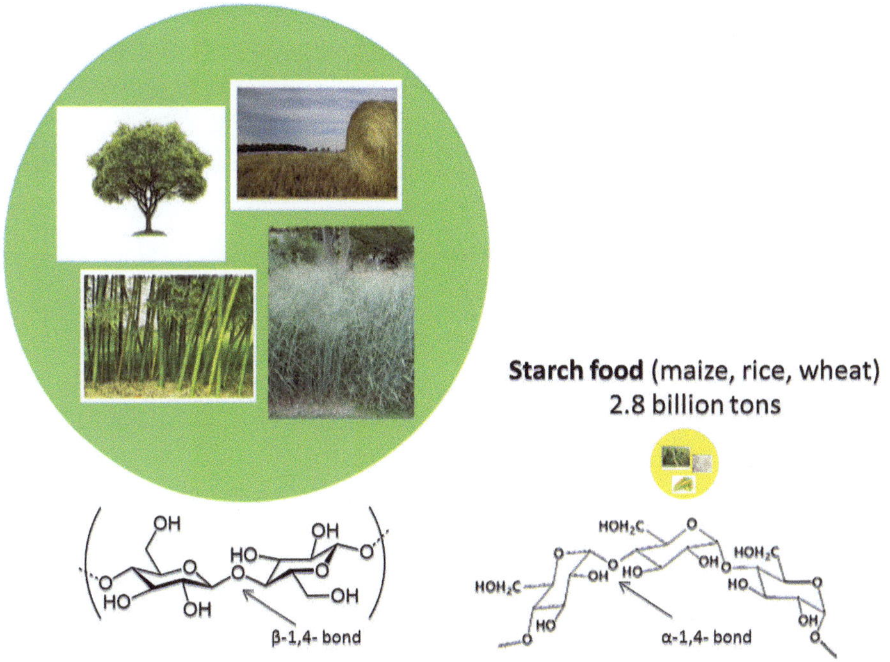

Fig. 6.1 The annual production of cellulose and starch on earth

Here we demonstrate simultaneous (ex vivo) enzymatic transformation and microbial fermentation that can transform cellulosic materials to starch, ethanol, and single-cell protein in one pot.

6.2 Transform Cellulose to Starch

An ex vivo enzymatic pathway has been designed to transform cellulose to synthetic amylose in an aqueous solution (Fig. 6.2a) [16]. This enzyme cocktail has two modules: (1) partial hydrolysis of cellulose to cellobiose by an optimized mixture of cellobiohydrolase (CBH) and endoglucanase (Endo), and (2) amylose synthesis by utilizing cellobiose phosphorylase (CBP) and potato alpha-glucan phosphorylase (PGP). In this system, CBP converts cellobiose to glucose 1-phosphate (G1P) and glucose in the presence of phosphate ions; PGP adds one glucose unit from G1P at the nonreducing end of amylose, and phosphate is recycled to maintain nearly constant pH and phosphate levels (Fig. 6.2a). To eliminate glucose inhibition, the ethanol-producing yeast *Saccharomyces cerevisiae* is added to the vessel because the yeast cannot utilize cellobiose and G1P [17, 18]. This bioprocess called simultaneous enzymatic biotransformation and microbial

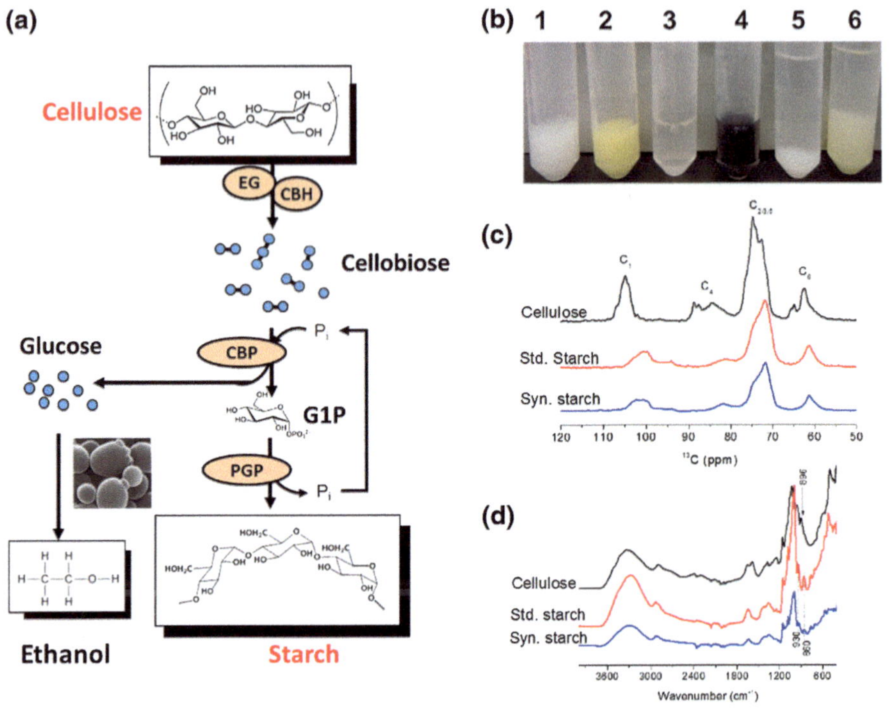

Fig. 6.2 The enzymatic transformation of converting cellulose to starch by endoglucanases (EGs), cellobiohydrolases (CBHs), cellobiose phosphorylase (CBP), and potato alpha-glucan phosphorylase (PGP), and the residual glucose was taken up by yeast to produce ethanol (**a**). Characterization of synthetic starch by iodine test (**b**), cross-polarization magic angle spinning (CP/M AS) ^{13}C-NMR (**c**) and FTIR (**d**). Tube 1: cellulose-suspended solution; tube 2: cellulose solution plus iodine/potassium iodide; tube 3: water-soluble synthetic starch solution made from cellulose mediated by the four-enzyme cocktail; tube 4: synthetic starch solution plus iodine/potassium iodide; tube 5: precipitated starch by ethanol addition; and tube 6: precipitated starch when the mixture was supplemented with glucose oxidase [16]

fermentation can transform pretreated biomass to amylose, ethanol, and yeast as a single-cell protein in one bioreactor.

It is important to identify the right enzymes suitable for this ex vivo biotransformation. Enzymatic cellulose hydrolysis usually requires synergistic action of EG, CBH, and beta-glucosidase (BG) to yield glucose. But in our process, beta-glucosidase is not allowed because this enzyme hydrolyzes cellobiose to glucose, which cannot be added to starch chains. We tested the combinations of two EGs [family 5 *Bacillus subtilis* EG (BsCel5) and *Trichoderma* spp. EG II (TrCel5A)] and three CBHs family 7 *Trichoderma* spp. CBH (TrCel7A), family 9

Clostridium phytofermentans CBH (CpCel9), and family 48 *C. phytofermentans* CBH (CpCel48). Based on cellobiose yield and the cellulose degradation, the best cellulase combination on regenerated amorphous cellulose [19] (RAC, high surface area, prepared from Avicel) to cellobiose was bacterial BsCel5 and fungal TrCel7A. For starch synthesis, three combinations of the *Clostridium thermocellum* CBP (CtCBP) and one of the three αGP (from potato, *Solanum tuberosum* and two thermophilic bacteria, *C. thermocellum* and *Thermotoga maritima*) were tested for the synthesis of amylose from cellobiose. Potato αGP (PGP) has the strongest ability to generate amylose from low concentration of cellobiose (5 mM), while αGP from *T. maritima* can only generate a little amylose from high concentration of cellobiose (50 mM); and αGP from *C. thermocellum* cannot generate any amylose from very high concentration (200 mM) of cellobiose. This study suggests the importance of identification of right enzymes used in such ex vivo pathways.

One-pot biotransformation of RAC to amylose was performed by four enzymes, Bscel5, TrCel5A, CtCBP, and PGP, in 0.5 ml of reaction volume (Fig. 6.2b). The RAC slurry (Fig. 6.2b, tube 1) was completely hydrolyzed and then converted into amylose (tube 3). The synthetic amylose exhibited a deep blue color in the presence of iodine (tube 4), whereas the negative control (cellulose/iodine) was yellow (tube 2). The soluble amylose was precipitated by the addition of ethanol (tube 5). The amylose yield was 14.4 % (wt/wt) (i.e., 0.144 g of amylose per gram

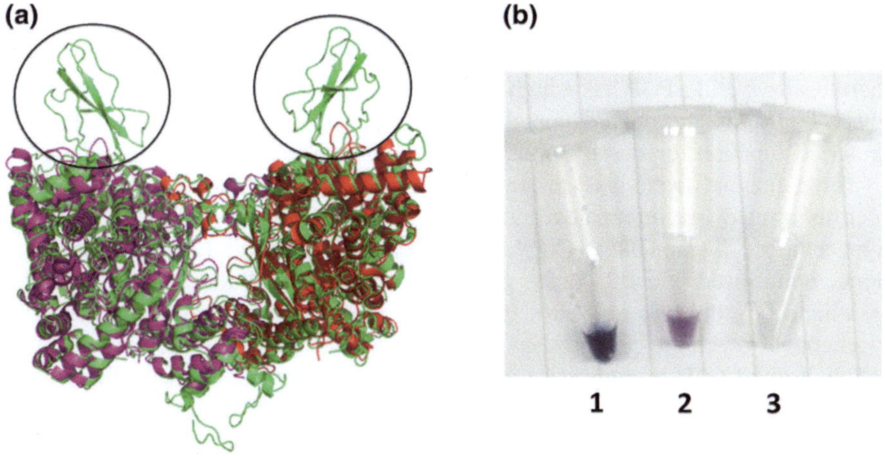

Fig. 6.3 Homology structure comparison between PGP (*green*) and *Clostridium thermocellum* alpha-glucan phosphorylase (*purple* and *red*), the circled regions represent cap structure of PGP (**a**) and photos of starch-synthesizing ability (**b**) based on cellobiose mediated by CBP and wild-type PGP (Tube 1), partially decapped PGP (Tube 2) or completely decapped PGP (Tube 3) [16]

of cellulose), and the number-average degree of polymerization (DP) was ~150. The addition of glucose oxidase to remove glucose, a strong inhibitor of CBP, resulted in a yield increase of 30.0 %. And the formation of amylose was also validated by ^{13}C-NMR and FTIR (Fig. 6.2c, d).

Alpha-glucan phosphorylase from *C. thermocellum* and *T. maritima* were tested for the negative production of amylose from cellulose. Comparison of the primary sequences of these three αGP indicates that the residues involved in substrate binding and catalysis are fairly conserved. However, homology modeling of the structures of the three αGPs reveals that potato αGP has a special cap consisting of residues from 478 to 561, which was absent in the other two bacterial enzymes (Fig. 6.3a). This cap structure can be observed in many plant αGPs, e.g., from sweet potato (*Ipomoea batatas*), spinach (*Spinacia oleracea*), rice (*Oryza sativa*), and wheat (*Triticum aestivum*) but as yet reported in bacteria. We hypothesized that the polypeptide cap on the catalytic site of PGP was responsible for driving low concentration G1P toward the synthesis of amylose. We designed two PGP deletion mutants—one had a part of the cap sequence removed and the other without the entire cap. In the buffer containing cellobiose and CBP, the partially "decapped" PGP (PGP-PDC) had decreased amylose synthesis ability compared to the wild-type enzyme (Fig. 6.3b, tube 2), whereas, the completely decapped PGP (PGP-CDC) lost all the activity (Fig. 6.3b, tube 3). The kinetic parameters, K_m and k_{cat} values of PGP, PGP-PDC, PGP-CDC, *C. thermocellum* αGP and *T. maritima* αGP are compared in Table 6.1. The complete removal of the cap of PGP decreased the k_{cat}/K_m values from 3.33 to 0.43 mM^{-1} s^{-1} in the starch synthesis direction and from 0.55 to 0.06 mM^{-1} s^{-1} in the starch degradation direction. Compared to the *C. thermocellum* αGP, PGP has a higher k_{cat}/K_m value in the synthesis direction and a lower k_{cat}/K_m in the degradation direction, suggesting that wild-type PGP has a preferred function for starch synthesis to degradation.

These enzymes were used to produce amylose from pretreated natural biomass, such as diluted acid-pretreated and cellulose solvent- and organic solvent-based lignocellulose fractionation (COSLIF)-pretreated corn stover [20]. In those reactions, nonutilized glucose units generated from the cellulase and CBP were assimilated by the baker's yeast; the baker's yeast can produce ethanol or other biochemical or single-cell proteins. Because typical baker's yeast cannot utilize cellobiose and G1P, adding yeast in this reaction system can increase the product yield by removing glucose which is the inhibitor of cellobiose phosphorylase [17, 18]. Under the tested conditions, the amylose yields were 23 and 2 % for COLISF-pretreated corn stover and dilute acid pretreated corn stover, respectively. This data indicated that starch can be produced from pretreated natural lignocellulose.

Table 6.1 Comparison of potato alpha-glucan phosphorylase and mutants, thermophilic *C. thermocellum* and *T. maritima* alpha-glucanphosphorylase

Enzymes	Synthesis			Degradation			Ratio (syn/deg)
	k_{cat} (s^{-1})	K_m (mM)	k_{cat}/K_m (mM^{-1} s^{-1})	k_{cat} (s^{-1})	K_m (mM)	k_{cat}/K_m (mM^{-1} s^{-1})	
PGP	5.83 ± 0.34	1.76 ± 0.18	3.33	0.90 ± 0.11	1.64 ± 0.13	0.55	6.5
PGP-PDC	4.88 ± 0.27	1.92 ± 0.21	2.54	0.83 ± 0.09	2.20 ± 0.21	0.38	5.9
PGP-CDC	0.95 ± 0.11	2.22 ± 0.41	0.43	0.19 ± 0.04	3.11 ± 0.23	0.06	5.0
CthαGP	6.6 ± 0.3	1.9 ± 0.2	3.50	8.1 ± 0.2	0.39 ± 0.01	21.0	0.8
TmαGP	12.14 ± 0.68	1.03 ± 0.13	11.77	14.44 ± 1.11	1.47 ± 0.11	9.85	0.8

6.3 Proposed Methods to Increase the Efficiency of Converting Cellulose to Starch

The cost-effective transformation of non-food cellulose to starch could revolutionize agriculture and the fledgling bioeconomy, while maintaining biodiversity, minimizing agriculture's environmental footprint, and conserving fresh water [8]. This transformation would not only promote the cultivation of plants chosen for rapid growth rather than those optimized for starch-rich seed production, but it would also efficiently utilize marginal land for the production of the biomass required to meet the increasing needs of biofuels and renewable materials [21–23]. Therefore, increasing the efficiency of converting cellulose to starch could lead to the sustainable agricultural revolution.

6.3.1 Increasing Cellulose Pretreatment Efficiency

The goal of biomass pretreatment is to improve the enzymatic digestibility of pretreated lignocellulosic biomass. Many factors, such as substrate accessibility to hydrolytic enzymes, lignin content, cellulose degree of polymerization, particle size, and so on, are related to its recalcitrance [24, 25]. Among these factors, substrate accessibility is the most important substrate property impacting the efficiency of enzymatic cellulose hydrolysis [20]. Solvent-based biomass pretreatments can greatly increase cellulose accessibility compared to conventional biomass pretreatments, e.g., dilute acid, steam explosion and hot water [26]. Solvent-based lignocellulose pretreatments by using concentrated phosphoric acid and ionic liquids have the apparent advantages of: high glucan digestibility at low enzyme loading; fast hydrolysis rate; and potential revenues from separated coproducts (e.g., hemicellulose, lignin). In such case, an ideal solvent should be able to dissolve cellulose at modest temperature (i.e., low energy input and less sugar degradation) and those of wet cellulose so that no biomass drying step is required. In addition, the solvent should be highly recyclable nonvolatile or highly volatile for easy recycling; thermostable and chemostable for an unlimited number recycling; and nontoxic to the sequential steps of enzymatic hydrolysis and microbial fermentation. High cellulose dissolution capacity (>10 wt% cellulose/vol) and fast diffusion rate in solid lignocellulose composite are additional desirable properties of such a solvent system [26].

6.3.2 Increasing PGP Thermostability

Because PGP loses its activity quickly above 45 °C, the reaction is better be performed at 37 °C despite the other enzymes can work at about 50 °C, resulting in low reaction rate of the whole process. High thermostable αGP which favors starch synthesis is required to increase the efficiency of this ex vivo enzymatic pathway. Plant αGPs should have the best ability for starch synthesis compared to microbial homologs. However, few plant αGPs can work at >50 °C. In lieu of using naturally occurring enzymes, an option is to engineer current PGP to a thermostable one by protein engineering via directed evolution or rational design. Accordingly, a mutant library of PGP was generated by error-prone PCR, and the mutated plasmid library was transformed in *Escherichia coli* BL21 (DE3). Screening for the more stable PGP mutants at 50 °C on agar plates containing 0.05 % soluble starch had been described by Yanase et al. [27]. Blue spots stained by iodine solution on the filter indicated the presence of active PGP after heat treatment. Three mutations (F39L, N135S, and T706I) combined together enhanced the thermostability of PGP significantly without compromising activity, retaining almost all the activity after heat treatment at 60 °C for 2 h while wild-type enzyme was completely inactivated.

6.3.3 Use of Commercial Cellulases

In our previous work, we used a bacterial cellulase, BsCel5 endoglucanase that was expressed and purified from recombinant *E. coli* strain, and a purified *Trichoderma* cellobiohydrolase (TrCel7A). Bacterial cellulase is not as good as the secreted *Trichoderma* cellulase, because of its high production cost based on *E. coli* cell culture. Novozymes and Genencor (Dupont) are selling less costly (e.g., $20/kg dry weight of cellulase) cellulase mixture products, like CTec2, HTec2, etc., that are highly efficient in converting amorphous cellulose to glucose. However, these commercial cellulases contain β-glucosidase activities that are designed to hydrolyze cellobiose to glucose. Hence, it is important to selectively remove β-glucosidase or produce β-glucosidase-free cellulase cocktail to meet our special need here.

6.3.4 Engineering of Yeast

The yeast *S. cerevisiae* used in simultaneous enzymatic biotransformation and microbial fermentation (SEBF) could be further improved by strain development. In general, yeast has been engineered to produce ethanol from cellulose directly by the expression of foreign cellulases including EG, CBH, and EG [28–33]. These three enzymes could be assembled as an enzyme complex by the interaction between cohesin and dockerin through synthetic mini-scaffolding that is displayed on the yeast cell surface. Instead, we just need to replace BG with CBP and PGP enzymes. So four enzymes (EG, CBH, CBP, and PGP) and synthetic scaffolding can be expressed in yeast, and displayed on the yeast cell surface. A cellulose-enzyme-microorganism (CEM) complex is expected to enable more efficient cellulose hydrolysis, starch synthesis, and ethanol production [34]. However, before this is realized, many factors should be taken into account to improve the efficiency of this ex vivo system, e.g., cellulase selection, the number of cohesions and their order in scaffolding, linker length, and enzyme orientation.

6.4 The Application of Synthetic Linear Amylose

The only way to synthesize pure amylose is by means of phosphorylase-catalyzed enzymatic polymerization [35, 36] because natural amylose isolated from seeds always contains some small α-1,6 branches. The length of synthetic linear amylose can be controlled by the concentration of primer oligosaccharides, G1P concentration, and reaction time. Synthetic amylose could have a variety of application, from low-value to high-value products (Fig. 6.4). (1) Top-quality synthetic amylose with a well-controlled degree of polymerization can be used as a chromatographic column matrix and drug capsule material in the pharmaceutical industry. (2) Amylose can be used to make biodegradable plastics, where starch-based plastics account for 50 % of the bioplastic market [37]. High-quality linear amylose is suitable for producing clear, transparent, and flexible low-oxygen diffusion plastic sheets and films [38]. (3) Amylose is an important thickener, water binder, emulsion stabilizer, and gelling agent in the food industry. (4) Food-grade amylose can be blended with regular cereals and processed to high-amylose tailored foods for meeting special dietary needs [36, 39]. These lower glycemic load foods can improve human health and lower the risk of serious noninfectious diseases (e.g., diabetes and obesity) [40]. Amylose can be processed to form a resistant starch, which resists digestion and passes through to the large intestine, where it acts as a dietary fiber [41]. Slowly digestible and resistant starch has some healthy benefits, including the prevention and alleviation of metabolic diseases and the prevention of colon cancer [42]. (5) Medium-quality amylose can be used as a high-density hydrogen carrier for the enzymatic production of hydrogen, that could solve the challenges associated with hydrogen production, hydrogen storage, infrastructure, and safety concerns [13, 43, 44]. (6) Low-quality amylose mixed with yeast cells can be used as animal feed for nonruminant animals, such as pigs and chickens, where yeast cells are a protein source. (7) Amylose could be converted into other food nutrients. To increase amylose digestion efficiency as food/feed, the addition of the starch-branching enzyme converts linear amylose to branched amylopectin [45] (Fig. 6.4) that is more water soluble and easily digested than amylose.

The production of the polysaccharides amylose and amylopectin rather than glucose is essential as key food components because over-consumption of simple sugars (e.g., glucose and fructose) is strongly associated with noninfectious diseases, such as diabetes and obesity. It is why humans have to eat starch as a major calories source instead of drinking simple sugar solutions. Starch may be an

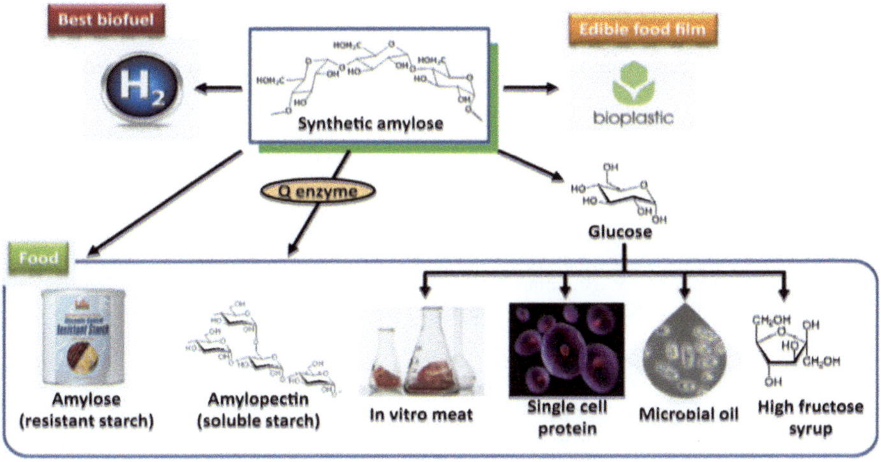

Fig. 6.4 The central role of synthetic amylose made from cellulose as new food sources (e.g., resistant starch, soluble amylopectin, in vitro meat, single-cell protein, microbialoil and high-fructose syrup), biodegradable plastic, and a hydrogen carrier. Reprinted from Ref. [6], Copyright 2015, with permission from Elsevier

important carbon source to produce other food nutrients, such as in vitro meat, single-cell proteins, and microbial oil. For example, slowly utilized starch is a better energy source for cell-free protein synthesis than glucose [46, 47].

6.5 Conclusion

To meet the growing needs of biofuels and renewable materials, as well as food and feed, we proposed an ex vivo enzymatic pathway along with microbial fermentation to produce edible starch from cellulose and other value-added products. This method provides a potential alternative to solve the potential food crisis because cellulose resource is approximately 40 times the starch food produced by cultivated crops, whose production requires large amounts of arable land, water, and energy. Whereas, cellulosic crops like dedicated perennial bioenergy crops (e.g., switchgrass, bamboo, poplar) can grow on low-quality arable land with higher productivity, higher water utilization efficiency and less energy-related inputs compared to cultivated annual crops. So the cost-effective transformation of non-food cellulose to starch could lead to sustainable agriculture and potentially shape the bioeconomy, while maintaining biodiversity, minimizing agriculture's environmental footprint, and conserving fresh water. New biorefineries based on this technology could help address the food, biofuels, and environment trilemma; decrease the negative impacts of growing food and feed consumption on the environment; provide healthier food; and promote the bioeconomy. Because ex vivo synthetic biosystems cannot duplicate themselves, the large-scale implementation of cellulose-to-starch in new biorefineries would most likely not raise the questions about ethics, biosecurity, and biosafety that often confront in vivo synthetic biology projects.

References

1. Foley JA, Ramankutty N, Brauman KA, Cassidy ES, Gerber JS, Johnston M, Mueller ND, O'Connell C, Ray DK, West PC et al (2011) Solutions for a cultivated planet. Nature 478:337–342
2. Godfray HCJ, Beddington JR, Crute IR, Haddad L, Lawrence D, Muir JF, Pretty J, Robinson S, Thomas SM, Toulmin C (2010) Food security: the challenge of feeding 9 billion people. Science 327:812–818
3. The World Economic Forum Water Initiative (2011) Water security: the water-food-energy-climate nexus. Island Press, Washington
4. Bruce TJ (2012) GM as a route for delivery of sustainable crop protection. J Exp Bot 63:537–541
5. Bagla P (2012) Negative report on GM crops shakes government's food agenda. Science 337:789
6. Chen H-G, Zhang YHP (2015) New biorefineries and sustainable agriculture: Increased food, biofuels, and ecosystem security. Renew Sustain Energy Rev 47:117–132
7. Varshney VK, Naithani S (2011) Chemical functionalization of cellulose derived from nonconventional sources. In: Kalia S, Kaith BS, Kaur I (eds) Cellulose fibers: bio- and nano-polymer composites. Springer, Berlin, pp 43–60
8. Somerville C, Youngs H, Taylor C, Davis SC, Long SP (2010) Feedstocks for lignocellulosic biofuels. Science 329:790–792
9. Wang Y, Huang W, Sathitsuksanoh N, Zhu Z, Zhang Y-HP (2011) Biohydrogenation from biomass sugar mediated by *in vitro* synthetic enzymatic pathways. Chem Biol 18:372–380
10. Xu Y, Masuko S, Takieddin M, Xu H, Liu R, Jing J, Mousa SA, Linhardt RJ, Liu J (2011) Chemoenzymatic synthesis of homogeneous ultralow molecular weight heparins. Science 334:498–501
11. Bujara M, Schümperli M, Pellaux R, Heinemann M, Panke S (2011) Optimization of a blueprint for *in vitro* glycolysis by metabolic real-time analysis. Nat Chem Biol 7:271–277
12. Swartz JR (2011) Transforming biochemical engineering with cell-free biology. AIChE J 58:5–13
13. Zhang Y-HP, Evans BR, Mielenz JR, Hopkins RC, Adams MWW (2007) High-yield hydrogen production from starch and water by a synthetic enzymatic pathway. PLoS ONE 2: e456
14. Guterl J-K, Garbe D, Carsten J, Steffler F, Sommer B, Reiße S, Philipp A, Haack M, Rühmann B, Kettling U et al (2012) Cell-free metabolic engineering—production of chemicals via minimized reaction cascades. ChemSusChem 5:2165–2172
15. Zhang Y-HP, Sun J-B, Zhong J-J (2010) Biofuel production by *in vitro* synthetic pathway transformation. Curr Opin Biotechnol 21:663–669
16. You C, Chen H, Myung S, Sathitsuksanoh N, Ma H, Zhang X-Z, Li J, Zhang Y-HP (2013) Enzymatic transformation of nonfood biomass to starch. Proc Natl Acad Sci USA 110:7182–7187
17. De Winter K, Cerdobbel A, Soetaert W, Desmet T (2011) Operational stability of immobilized sucrose phosphorylase: continuous production of α-glucose-1-phosphate at elevated temperatures. Proc Biochem 46:2074–2078

18. Galazka JM, Tian C, Beeson WT, Martinez B, Glass NL, Cate JHD (2010) Cellodextrin transport in yeast for improved biofuel production. Science 330:84–86
19. Zhang YHP, Cui J, Lynd LR, Kuang LR (2006) A transition from cellulose swelling to cellulose dissolution by o-phosphoric acid: evidence from enzymatic hydrolysis and supramolecular structure. Biomacromolecules 7:644–648
20. Rollin JA, Zhu Z, Sathisuksanoh N, Zhang Y-HP (2011) Increasing cellulose accessibility is more important than removing lignin: a comparison of cellulose solvent-based lignocellulose fractionation and soaking in aqueous ammonia. Biotechnol Bioeng 108:22–30
21. Sheppard AW, Gillespie I, Hirsch M, Begley C (2011) Biosecurity and sustainability within the growing global bioeconomy. Curr Opin Environ Sustain 3:4–10
22. French CE (2009) Synthetic biology and biomass conversion: a match made in heaven? J Roy Soc Interface 6:S547–S558
23. Casillas CE, Kammen DM (2010) The energy-poverty-climate nexus. Science 330:1181–1182
24. Himmel ME, Ding SY, Johnson DK, Adney WS, Nimlos MR, Brady JW, Foust TD (2007) Biomass recalcitrance: engineering plants and enzymes for biofuels production. Science 315:804–807
25. Chandra RP, Bura R, Mabee WE, Berlin A, Pan X, Saddler JN (2007) Substrate pretreatment: the key to effective enzymatic hydrolysis of lignocellulosics? Adv Biochem Eng Biotechnol 108:67–93
26. Sathitsuksanoh N, George A, Zhang YHP (2013) New lignocellulose pretreatments using cellulose solvents: a review. J Chem Technol Biotechnol 88:169–180
27. Yanase M, Takata H, Fujii K, Takaha T, Kuriki T (2005) Cumulative effect of amino acid replacements results in enhanced thermostability of potato type L alpha-glucan phosphorylase. Appl Environ Microbiol 71:5433–5439
28. Bae J, Kuroda K, Ueda M (2015) Proximity effect among cellulose-degrading enzymes displayed on the *Saccharomyces cerevisiae* cell surface. Appl Environ Microb 81:59–66
29. Tsai S, Oh J, Singh S, Chen R, Chen W (2009) Functional assembly of minicellulosomes on the *Saccharomyces cerevisiae* cell surface for cellulose hydrolysis and ethanol production. Appl Environ Microbiol 75:6087–6093
30. Wen F, Sun J, Zhao H (2010) Yeast surface display of trifunctional minicellulosomes for simultaneous saccharification and fermentation of cellulose to ethanol. Appl Environ Microbiol 76:1251–1260
31. Fujita Y, Ito J, Ueda M, Fukuda H, Kondo A (2004) Synergistic saccharification, and direct fermentation to ethanol, of amorphous cellulose by use of an engineered yeast strain codisplaying three types of cellulolytic enzyme. Appl Environ Microbiol 70:1207–1212
32. Nakatani Y, Yamada R, Ogino C, Kondo A (2013) Synergetic effect of yeast cell-surface expression of cellulase and expansin-like protein on direct ethanol production from cellulose. Microb Cell Fact 12:66
33. Yanase S, Yamada R, Kaneko S, Noda H, Hasunuma T, Tanaka T, Ogino C, Fukuda H, Kondo A (2010) Ethanol production from cellulosic materials using cellulase-expressing yeast. Biotechnol J 5:449–455
34. You C, Zhang XZ, Sathitsuksanoh N, Lynd LR, Zhang Y-HP (2012) Enhanced microbial utilization of recalcitrant cellulose by an *ex vivo* cellulosome-microbe complex. Appl Environ Microbiol 78:1437–1444
35. J-i Kadokawa (2011) Precision polysaccharide synthesis catalyzed by enzymes. Chem Rev 111:4308–4345
36. J-i Kadokawa (2012) Preparation and applications of amylose supramolecules by means of phosphorylase-catalyzed enzymatic polymerization. Polymers 4:116–133
37. van Soest JJG, Vliegenthart JFG (1997) Crystallinity in starch plastics: consequences for material properties. Trends Biotechnol 15:208–213
38. Frische R, Wollmann K, Gross-Lannert R, Schneider J, Best B (1994) Special amyloses and their use for producing biodegradable plastics. US Patent 5374304

39. Maki KC, Pelkman CL, Finocchiaro ET, Kelley KM, Lawless AL, Schild AL, Rains TM (2012) Resistant starch from high-amylose maize increases insulin sensitivity in overweight and obese men. J Nutr 142:717–723
40. Regina A, Bird A, Topping D, Bowden S, Freeman J, Barsby T, Kosar-Hashemi B, Li Z, Rahman S, Morell M (2006) High-amylose wheat generated by RNA interference improves indices of large-bowel health in rats. Proc Natl Acad Sci USA 103:3546–3551
41. Miao M, Jiang B, Cui SW, Zhang T, Jin Z (2015) Slowly digestible starch—a review. Crit Rev Food Sci Nutr 55:1642–1657
42. Gilbert R, Wu A, Sullivan M, Sumarriva G, Ersch N, Hasjim J (2013) Improving human health through understanding the complex structure of glucose polymers. Anal Bioanal Chem 405:8969–8980
43. Rollin JA, Martin del Campo J, Myung S, Sun F, You C, Bakovic A, Castro R, Chandrayan SK, Wu C-H, Adams MWW et al (2015) High-yield hydrogen production from biomass by in vitro metabolic engineering: mixed sugars coutilization and kinetic modeling. Proc Natl Acad Sci 112:4964–4969
44. Ye X, Wang Y, Hopkins RC, Adams MWW, Evans BR, Mielenz JR, Zhang Y-HP (2009) Spontaneous high-yield production of hydrogen from cellulosic materials and water catalyzed by enzyme cocktails. ChemSusChem 2:149–152
45. Peat S, Bourne EJ, Barker SA (1948) Enzymic conversion of amylose into amylopectin. Nature 161:127
46. Caschera F, Noireaux V (2014) Synthesis of 2.3 mg/ml of protein with an all *Escherichia coli* cell-free transcription-translation system. Biochimie 99:162–168
47. Wang Y, Zhang Y-HP (2009) Cell-free protein synthesis energized by slowly-metabolized maltodextrin. BMC Biotechnol 9:58

Chapter 7
Biorefinery of Lignocellulosics for Biofuels and Biochemicals

Mingyu Wang and Jin Hou

Abstract Sustainable development of the world's economy would require a paradigm shift from the current fossil fuel-based energy and chemical production model to that of a renewable supply-based model, where lignocellulosic biorefinery has a great potential. In this chapter, we review and discuss the current knowledge and progress on biorefinery of lignocellulosics for the production of biofuels and biochemicals, along with the socio-economic and technical problems this industry has to tackle. The chemical, microbiological, and technical aspects for the enzymatic conversion of lignocellulose to the platform sugars were examined in detail. In particular, we addressed the cost reduction issue of lignocellulose degradation, which is otherwise a major impediment of the biorefinery industry. We further discussed the biotechnological and bioengineering efforts to convert lignocellulosic sugars to biofuels and biochemicals. These include the predominant biofuel ethanol, advanced biofuels such as butanol with better compatibility to current infrastructure, and bulk and fine chemicals such as organic acids and isoprenoids. Various metabolic engineering strategies were also summarized to enable further development of microbial strains for biofuel and biochemical production. We believe that the renewable lignocellulosic biofuel and biochemical industry are critical contributors to a sustainable future that is independent of the fossil fuels and their derived products.

Keywords Lignocellulosic biomass · Lignocellulose degradation · Cellulase formation · Biofuel and biochemical · Lignocellulose ethanol production

---

M. Wang (✉) · J. Hou (✉)
State Key Laboratory of Microbial Technology, School of Life Sciences,
Shandong University, Jinan 250100, China
e-mail: wangmingyu@sdu.edu.cn

J. Hou
e-mail: houjin@sdu.edu.cn

P.C.K. Lau (ed.), *Quality Living Through Chemurgy and Green Chemistry*,
Green Chemistry and Sustainable Technology,
DOI 10.1007/978-3-662-53704-6_7

7.1 Biorefinery of Lignocellulosics, a Promising Solution to Potential Energy and Chemical Shortages

Two of the most significant events in human history are the industrial revolutions that fundamentally changed the way mankind lives. In approximately 300 years, the population of mankind increased by roughly ninefold [1], the life expectancy of people more than doubled [2], and we as a species have evolved from a grounded land animal to a species that frequently flies and even cruises outer space. These astonishing achievements were obtained in less than 1 % of our history as an intelligent species that uses tools and fire [3], which oftentimes amazes anyone who finds interest in anthropology.

Although we are rightful in being proud of all these miracles we created, we often mistakenly take these achievements for granted and assume they are irreversible while we continue neglecting one key fact: these achievements were almost entirely dependent on the use of fossil fuels, a resource that unfortunately is nonrenewable and will sooner or later deplete. To date, the majority of our energy sources are fossil fuels: analysis by experts from U.S. Energy Information Administration projected over 75 % of our energy supplies are based on fossil fuels from today to 2040 [4]. While this fossil fuel dependence of energy can be partially relieved by utilization of other renewable energy sources, a bigger problem of our society is that a major fraction of our chemicals, one of the pillars for modern life style and industry, are synthesized from fossil fuels such as petroleum. Imagine the day (projected between 2043 and 2116) when fossil fuels are depleted [5] (Table 7.1), without proper and adequate solutions, the human civilization as we know it will almost certainly collapse due to energy and chemical shortages.

To solve this sustainability problem for both energy and chemicals, we have to find reliable and sustainable sources for each necessity. All the energy sources that mankind can utilize now, including fossil fuels, solar energy, wind energy, bioenergy, tidal energy, nuclear energy, etc., are fundamentally from either nuclear reactions or the sun. Although nuclear energy is arguably clean, safe, and quite mature at this stage, it is unfortunately nonrenewable and sooner or later we will come across the same sustainability problem as the fossil fuels. On the other hand, solar energy will be almost eternally available to mankind, at least for a few more billion years. It would be reasonable to develop more efficient and reliable technologies for harvesting energy from the sun. Although there are many ways to trap

Table 7.1 Predicted fossil fuel depletion time. This table is redrawn using data from [5]

Model	Klass model			Shafiee model		
	Oil	Coal	Natural gas	Oil	Coal	Natural gas
Available time (in years, from 2009)	34	106	36	35	107	37
Depletion time	2043	2115	2045	2044	2116	2046

solar energy, these technologies do not address the need of chemical sustainability for the ready supply of organic compounds.

A viable option is to source biomass that is renewable and in abundance. Of all the available biomass resources on earth, lignocellulosics appear to be the ideal candidates for both sustainable energy and chemical sources. Lignocellulose is the primary component of plant cell wall and is therefore present in essentially every plant, making it the most abundant biomass on earth [6]. In China alone, annually 700 million metric tons of lignocellulosics are available in just the agricultural sector [7]. To make it even more appealing, unlike starch that is widely used in first-generation bioethanol production, lignocellulose is unedible. Hence, usage of this nonfood commodity would not impact food security.

The primary physiological function of lignocellulose is the protection of plant cells, making it relatively more difficult to decompose and has consequently fewer applications. On a positive side, this improves the availability of lignocellulosics, because the majority of lignocellulosics remain unutilized. For instance, approximately 40–50 % of the straws, one of the best utilized lignocellulosic biomass, produced in China (300 million metric tons each year) are either burned or wasted [8, 9].

Unlike solar energy and wind energy that can only be stored in the form of electricity and are therefore only suitable where a power grid is present, bioenergy from lignocellulose is stored in the form of chemical bonds and therefore has superior portability. Most importantly, lignocellulosics can be the source of chemicals as they are essentially organic matter. This is an unusual attribute that no other form of renewable resource possesses.

With these properties, lignocellulosics represent an excellent answer to the forseeable depletion of fossil fuels. It needs to be noted, however, that the full potential of lignocellulosic biomass can only satisfy approximately 30 % of the world's total energy need [10], and other forms of renewable energy need to be developed in parallel to fully solve the energy sustainability issue. People may argue that the refinery of lignocellulosics is not urgently needed as they are currently not economically competitive in comparison with fossil fuel-based energy and chemicals. However, with the industrialization and modernization of developing countries, the exponentially increasing demand for energy and chemicals may lead to a much quicker depletion of fossil fuels than the already short time frame projected. It will be too late to start developing technologies to replace fossil fuels when they are depleting, while it is never too early to invest on alternatives for a forseeable sustainability issue. The consequence of not having an alternative may be catastrophic. With the decrease of newly discovered oil reserves in recent years [11], the impact of fossil fuel shortage may strike earlier than expected. Hence, the whole world needs to take this issue with full seriousness.

7.2 Key to the Biorefinery Technologies

7.2.1 *Chemical, Biological, and Technological Basis for Lignocellulose Degradation*

Three polymeric substances, cellulose (40–50 %), hemicellulose (20–30 %), and lignin (10–30 %), along with other minor components, e.g., pectin, comprise lignocellulose [12, 13]. This complex substance is the primary structural component of plant cell wall that gives plants the rigid physical property and protects plants from external physical, chemical, and enzymatic damage [14]. Cellulose is a relatively simple polymer composed of glucose moieties connected by β-1,4 glycosidic linkages. It generally has a crystalline structure where fibers are organized in a highly ordered manner, although amorphous regions are also present [13, 15]. In contrast, hemicellulose has a more complicated composition. The backbone of hemicellulose is composed of a variety of pentoses and hexoses, such as glucose, xylose, mannose, arabinose, and galactose. Unlike cellulose, side chains are widespread along hemicellulose fibers. These side chains are composed of sugar moieties such as xylose, arabinose, and galactose, as well as other molecules including acetyl groups, glucuronic acids, and ferulic acids [16]. As depicted in Fig. 7.1, combinations of backbones and side chains lead to the formation of many types of hemicelluloses including glucuronoxylan, glucomannan, xyloglucan, galactoglucomannan, arabinoglucuronoxylan, arabinogalactan, arabinoxylan, and β-1,3/1,4-glucan [16]. While cellulose and hemicellulose are primarily composed of sugars, lignin is a polymer composed of aromatic building blocks guaiacyl, sinapyl, and *p*-hydroxylphenyl units, abbreviated as G, S, and H, respectively (Fig. 7.1) [17]. An additional significant difference between lignin and the aforementioned carbohydrates is the versatility and variability of lignin that makes it a diverse and nonlinear structure where a pattern of repeats is unavailable [18]. These three polymers are not isolated entities, but rather interconnected with noncovalent and covalent bonds [19]. These interconnections lead to the formation of a highly complex and rigid structure in lignocellulose.

The entanglement and interconnection between component polymers lead to a pervasive recalcitrance of lignocellulosic materials toward degradation by chemical, physical, or biological means. Biodegradation of lignocellulose is particularly difficult because of the limit of enzyme accessibility to their substrates as a result of the tight structure [20]. In nature, the cellulose component of lignocellulose is degraded by a group of enzymes termed cellulases. These enzymes are glycoside hydrolases that selectively and effectively degrade β-1,4 glycosyl bonds between glucose moieties. They include endoglucanases that primarily target the amorphous region of cellulose and break cellulose fibers into smaller fragments; cellobiohydrolases (also termed exoglucanases) that digest cellulose from either reducing or nonreducing ends of the cellulose fiber and liberate cellobiose; and β-glucosidases that degrade cellobiose to glucose [21].

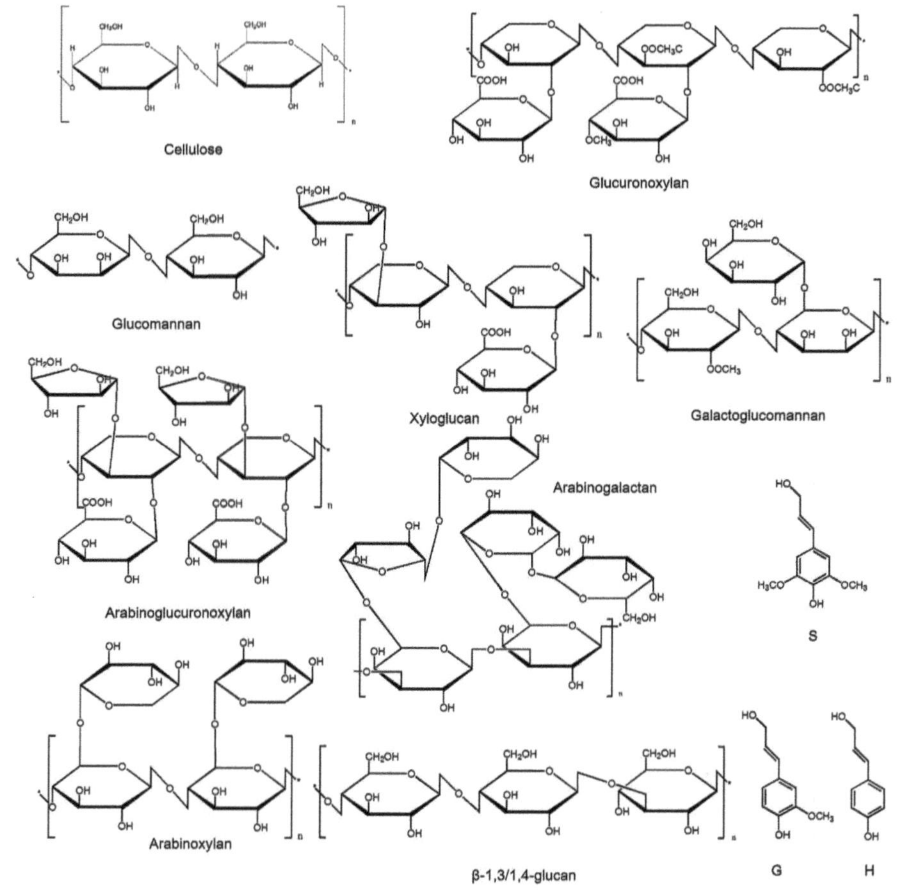

Fig. 7.1 Structures and building blocks of lignocellulose components

Hemicellulose is degraded in nature by hemicellulases that include a variety of enzymes belonging to either superfamilies of glucoside hydrolases or carbohydrate esterases targeting various chemical bonds. They are backbone-digesting enzymes, such as endo-xylanases, β-xylosidases, β-mannosidases, endo-mannases, β-glucosidases, and β-galactosidases, and sidechain digesting enzymes such as α-glucuronidases, α-galactosidases, acetyl xylanesterases, and ferulic acid esterases [22]. By and large, the degradation of hemicellulose is easier than cellulose because of a looser structure of hemicellulose, although it requires a more complex enzyme system.

On the other hand, lignin degradation is most difficult, even though the structure of lignin is less ordered and linear. The rich aromatic content and the mesh-like structure are major impediments for the biodegradation of lignin. Nature has provided oxidative enzymes such as peroxidases (lignin peroxidases, manganese peroxidases, and versatile peroxidases) along with laccases, many of which come

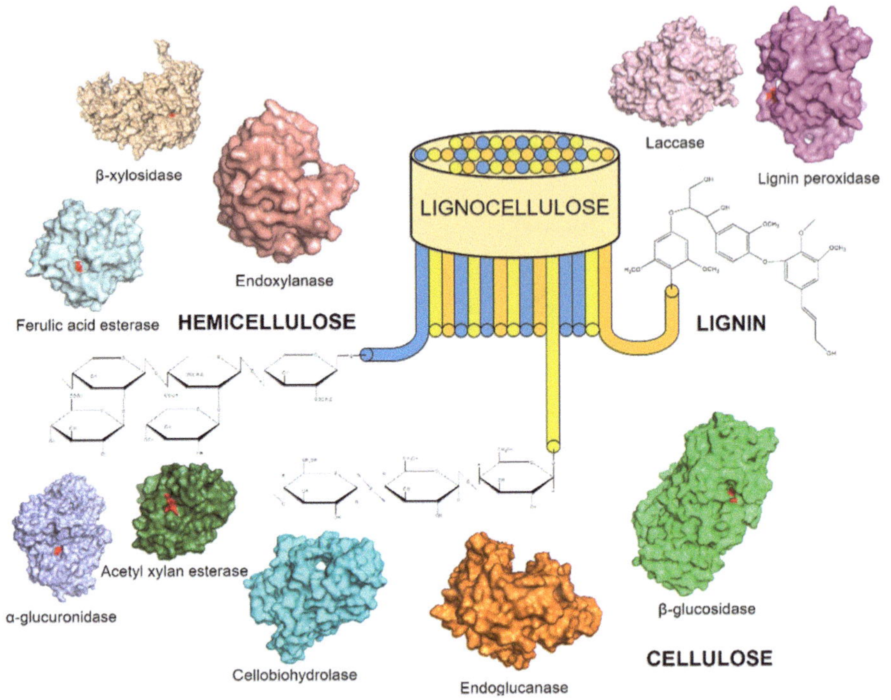

Fig. 7.2 Lignocellulolytic enzymes and their target polymers. Presented in this figure are the following enzymes: Cellobiohydrolase I from *T. reesei* (PDB accession: 1CEL) [25]; Endoglucanase from *Humicola insolens* (PDB accession: 1DYS) [26]; β-Glucosidase 1 from *T. reesei* (PDB accession: 3ZYZ) [27]; Xylanase I from *T. reesei* (PDB accession: 1XYN) [28]; β-Xylosidase from *Thermoanaerobacterium saccharolyticum* (PDB accession: 1PX8) [29]; α-Glucuronidase from *Geobacillus stearothermophilus* (PDB accession: 1K9D) [30]; Acetyl xylan esterase from *Clostridium thermocellum* (PDB accession: 2C71) [31]; Ferulic acid esterase from *Aspergillus niger* (PDB accession: 1USW) [32]; Lignin peroxidase from *Trametes cervina* (PDB accession: 3Q3U) [33]; Laccase from *Botrytis aclada* (PDB accession: 3SQR) [34]. The red patches indicate active sites

from brown-rot and white-rot fungi. However, none of these enzyme systems are particularly robust [23, 24]. A summary of lignocellulosic degrading enzymes and their target polymers is presented in Fig. 7.2.

Many other auxiliary enzymes also play key roles in lignocellulose degradation, among which the two most important groups are hydrolases (esterases) that break the interconnections between components of lignocellulose (e.g., Cip2 from *Trichoderma reesei*) and oxidoreductases (e.g., TaGH61 from *Thermoascus aurantiacus*) that catalyzes oxidoreductive degradation of lignocellulose [35–38]. The latter group is also termed lytic polysaccharide monooxygenases and classified as Auxiliary Activity family proteins (AA9, AA10, and AA11) [39].

The recalcitrance of lignocellulose indicates that it requires an unusually large dosage of enzymes for degradation. A variety of microbes are known to secrete enzymes for lignocellulose degradation, among which two groups are used in industry: filamentous fungi such as *T. reesei* and *Penicillium oxalicum* that secrete a very high level of enzymes—as much as 100 g/L for *T. reesei,* for example [40, 41]; and anaerobic bacteria such as *Clostridium thermocellum*, which assemble lignocellulases into highly efficient extracellular multimillion Dalton enzyme complexes, termed the cellulosome, to improve lignocellulose destruction efficiency [6].

Current lignocellulose decomposition technologies generally focus on the deconstruction of cellulose and hemicellulose into carbohydrates, while ignoring lignin due to the difficulties of its utilization [24]. Sugars such as glucose and xylose thus produced play the role of a central platform molecule that can be further utilized by various organisms for the production of a variety of products including energy molecules and other chemicals. Beside this, an alternative is the Consolidated BioProcessing (CBP) technology where a single microorganism degrades lignocellulose directly into target products like ethanol. This technology avoids the compatibility issue between cellulases and sugar-utilizing, product-yielding organisms, and theoretically improves the efficiency of lignocellulose conversion. For this purpose, there have been quite a few attempts in constructing engineered product-yielding organisms like *S. cerevisiae* that produces cellulases [42], or the application of natural lignocellulose-utilizing, ethanol-producing organisms such as *C. thermocellum* [43]. The former approach has yet to result in engineered strains with good enough cellulase yields for practical use. The latter approach is suffering from byproduct issue that hampers target product yields [44]. Therefore, there is still plenty of room for the improvement of CBP for industrial applications.

Another promising route is to avoid making sugar as the platform molecule but rather uses syngas (CO and H_2), the product of lignocellulose pyrolysis, for the production of fuels and chemicals by syngas-utilizing organisms such as *Clostridium ljungdahlii* [45]. Although this technical route successfully avoids the biodegradation recalcitrance issue for lignocellulose, it does not alleviate the byproduct problem encountered by a related organism, e.g., *C. thermocellum* [46]. In addition, the byproduct of pyrolysis such as tars can inhibit bacterial growth and product generation [47].

7.2.2 Technological–Economical Barriers for Lignocellulose Utilization

Several major technological–economical barriers prevent today's large-scale utilization of lignocellulose. The most important one results from the aforementioned lignocellulose recalcitrance toward degradation, which leads to a particularly high enzyme dosage and subsequently cost for the bioconversion of lignocellulosics to sugars. When compared with amylases that degrade starch to glucose for the first-generation starch-based bioethanol technology, the dosage of cellulases for the conversion of cellulose to glucose for the second-generation lignocellulose-based bioethanol technology is approximately 100-fold higher [21]. The cost for enzymes is estimated to be around 0.2—0.5 US dollars for each liter of lignocellulosic bioethanol [48–50]. In comparison, starch-based bioethanol technology only costs approximately 0.01 US dollars on enzymes for each liter of ethanol produced [51]. Considering a 3 USD/gallon retail gasoline price, even at 20 cents/L enzyme cost, lignocellulosic enzyme costs still take up over 25 % of the total acceptable production budget. Therefore, either a major technology breakthrough in enzyme technology takes place that reduces enzyme cost dramatically, or we will have to wait for oil price to at least double for the economical production of lignocellulosic bioethanol. While the latter scenario may be inevitable eventually, the former solution can actually be achieved in a shorter timeframe by developing efficient, inhibitor-free and cost-effective lignocellulose pretreatment technologies that are compatible with downstream enzymatic digestion to decrease the demand for enzymes, improving lignocellulase production so that the average enzyme price per FPU (Filter Paperase Unit) is decreased, and improving lignocellulase efficiency so that a lower dosage of enzymes may be used to efficiently degrade lignocellulose.

A second technological barrier for lignocellulose utilization is the simultaneous utilization of pentoses and hexoses. The primary product of hemicellulose degradation is xylose, a pentose. While there are numerous xylose-utilizing organisms in nature, nearly all of them prefer glucose, the product of cellulose degradation, over pentoses [52]. A common scheme is that organisms will first deplete any available glucose in the environment before switching to consuming xylose [53]. This phenomenon not only prolongs the production process, but also requires a dramatic shift of metabolism between the glucose-utilization phase and the xylose-utilization phase, thus resulting in energy waste and decreased efficiency. A glucose/xylose simultaneously consuming co-fermentation strategy is therefore preferred, which will be elaborated in Sect. 7.4.

A third technological–economical barrier for lignocellulose utilization relates to the efficiency of various pretreatment processes and the associated costs. The efficiency of cellulases digesting natural lignocellulose materials is quite low, therefore requiring the materials to be pretreated by chemical, physical, or biological methods to loosen the structure and improve enzyme accessibility [54]. These pretreatment technologies, while significantly improve enzymatic lignocellulose saccharification, elicit significant costs and lead to the production of inhibitory substances for downstream processes [55]. It is estimated that the cost for pretreatment is over 0.2 USD/gallon of bioethanol, over twice of the saccharification and fermentation costs [56]. This leaves plenty of room for optimization and improvement.

Finally, the availability and collection of raw materials is a major challenge in large-scale lignocellulosic industries. Most of the lignocellulosic materials used today are from agricultural, forestry, or urban wastes. Although a lot of them are underutilized today and the prices are reasonably low, the availability of these materials may be challenged in a large-scale scenario, thus driving the cost up. Additionally, these biomass materials are generally of lower densities, leading to the increase of collection and transportation costs. There have been attempts to grow crops that are specifically developed for lignocellulosic fuel and chemical purposes in places where regular crops are difficult to grow [57]. These attempts may alleviate or even solve the raw material problem.

The key to the improvement of lignocellulosic biorefinery technologies is to overcome the above-mentioned technological–economical barriers thus decreasing the average cost of lignocellulose- based biofuel and biochemical manufacturing. The following Sects. 7.3 and 7.4 discuss the various improvements that may be needed.

7.3 High-Efficiency, Low-Cost Production of Lignocellulose-Degrading Enzymes

Reducing the cost of lignocellulose-degrading enzymes, as discussed in Sect. 7.2.2, is undoubtedly the key for the reduction of production costs for lignocellulose-based fuels and chemicals. One way to mitigate the problem is using acid hydrolysis or pyrolysis for the degradation of lignocellulose, but neither of these two approaches has yet to see the potential of enzymatic saccharification technologies. Further research effort is therefore needed on the improvement of lignocellulose-degrading enzymes to increase efficiency therefore decreasing costs. To tackle this problem, interdisciplinary efforts in the areas of microbial physiology, genetics, protein biochemistry, metabolic/biochemical engineering, strain mutagenesis and development, as well as process design are warranted, be it of fundamental nature or applied. This makes the arena of lignocellulose-degrading enzymes an exemplary project of cross-field collaboration where ideas and innovations by scientists and engineers from all aspects of microbiology and biotechnology run into each other and shine when the job is done.

7.3.1 Cellulase-Producing Microorganisms and Their Enzymes

7.3.1.1 *Trichoderma reesei*

Originally known as "*Trichoderma viride*", *T. reesei* (syn. *Hypocrea jecorina*) has a long history of development since World War II [58]. Today, it is regarded as the most important cellulase-producing filamentous fungi, the best option for industrial cellulase production and the model organism for research in cellulase synthesis and production [59]. After many rounds of mutagenesis from various laboratories, the cellulase-producing ability of *T. reesei* has been improved by nearly an order of magnitude [21, 40, 58]. The genomic sequence of *T. reesei* was obtained in 2008. However, it was much surprising to find that fewer than expected lignocellulose-degradation-related genes are present than most lignocellulose-degrading filamentous fungi despite its effectiveness in lignocellulose degradation [60]. Key genes that are involved in lignocellulose degradation include the cellobiohydrolase-encoding genes *cbh1* and *cbh2*, the endoglucanase-encoding genes *eg1* and *eg2*, the β-glucosidase-encoding gene *bgl1*, along with other less important glycoside hydrolase coding genes [59, 60]. It came to an understanding that *T. reesei* achieves the goal of effective lignocellulose degradation by its copious secretion rather than the versatility of its enzyme cocktail composition. This conclusion was echoed by the finding that *T. reesei* enzymes actually lack β-glucosidase activities, which leads to accumulation of cellobiose and subsequently product inhibition for cellulases [59, 61]. Supplementation with additional β-glucosidases and/or genetic modification of *T. reesei* to improve β-glucosidase production had been carried out to address this issue [62, 63].

7.3.1.2 *Clostridium thermocellum*

Enzymes from *C. thermocellum* represent high-efficiency complexed cellulases that take a different lignocellulose degradation strategy from the quantity-based strategy adopted by filamentous fungi such as *T. reesei*. *C. thermocellum* is a strictly anaerobic thermophilic bacterium that produces cellulases and hemicellulases in the form of a highly ordered protein complex-cellulosome [6]. It is the best lignocellulose-degrading bacterium known to date [64]. Structural analysis has revealed that the cellulosome of *C. thermocellum* is composed of cellulases and hemicellulases attached to a scaffoldin protein CipA which is further attached on the cell wall via interactions between Cip2 and cell wall proteins such as SdbA, Orf2p, and OlpB [6]. These protein interactions are mediated by protein domains: Type I dockerin domains on enzymes interact with Type I cohesion domains on CipA, and Type II dockerin domains on CipA interact with Type II cohesion domains on cell wall proteins. This highly ordered complexed form leads to the high lignocellulose degradation efficiency—it is known that cellulosomes from *C. thermocellum* degrade cellulose faster than enzymes from *T. reesei* [15]. This high catalytic efficiency, along with its capability of directly converting cellulose to acids and alcohols without the need to first converting cellulose to sugars, makes *C. thermocellum* a promising candidate for industrial lignocellulose utilization.

7.3.1.3 *Penicillium oxalicum*

P. oxalicum (formerly known as *Penicillium decumbens*) is the only other filamentous fungus whose cellulases were applied in large-scale lignocellulosic bioethanol industry in practice. The wildtype *P. oxalicum* 114-2 strain was first isolated in China from decayed straw covered soil in 1979, and applied to industrial applications for almost 20 years [65–67]. It has since received repeated mutagenesis, leading to a significant improvement of cellulase production level to over 10 g/L [68]. This fungus, in comparison with the more widely acknowledged *T. reesei*, has a more diversified, balanced, and robust enzyme composition [66, 69]. In particular, it has a higher β-glucosidase component and is therefore free from the cellobiose accumulation and product inhibition issue for *T. reesei* [70].

7.3.1.4 Other Fungi and Bacteria

A series of other fungi and bacteria also possess the ability to produce cellulases, many of which have been investigated intensely. These fungi include members of the *Aspergillus* genus, e.g., *Aspergillus niger* [71] and *Aspergillus fumigatus* [72], members of the *Trichoderma* genus, e.g., *T. viride* [73] and *Trichoderma harzianum* [74], as well as members of the *Penicillium* genus like *Penicillium janthinellum* [75] and *Penicillium echinulatum* [76]. Other widely investigated fungi include *Humicola insolens* [77], *Acremonium cellulolyticus* [78], and *Neurospora crassa* [79]. Research on cellulolytic bacteria generally focused on the cellulosomes from *Clostridium* (such as *Clostridium cellulolyticum* and *Clostridium cellulovorans*) and *Ruminococcus* species [15]. An exception that received attention in recent years is the thermophilic *Thermoanaerobacterium* species that do not degrade cellulose but rather hemicellulose, partly due to their interactions with cellulolytic bacteria since the removal of hemicellulose can help exposing cellulose fibers for more efficient degradation by cellulolytic bacteria [80].

7.3.2 The Biological Basis for Cellulase Production

Understanding the biological basis for cellulase production is the starting point for making improvements to the enzyme technology and engineering more versatile cellulase-producing organisms. However, our knowledge on the biological process has been quite limited. While we do have some rudimentary knowledge on which factors are involved in lignocellulase production, the detailed mechanisms and pathways leading to their effects still need further work.

7.3.2.1 Induction and Inhibition Signals of Cellulase Production and Activities

Fungi

Cellulase production in fungi is strictly regulated by carbon sources. The substrate cellulose is a strong inducer of cellulase production and the product glucose is a strong inhibitor for cellulase production [81], neither of which is a surprise. The glucose inhibition, in particular, is a carbon catabolite repression that has been discovered in many organisms for many physiological processes [82]. What is unique in cellulase production regulation in fungi is that many disaccharides also act as inducers for cellulase production, including cellobiose that is a hydrolysis product of cellulose, as well as sophorose and lactose that are not direct products of cellulose degradation [81, 83–85]. On a first glance, these phenomena do not make evolutionary sense as these disaccharides should be absent in the natural environment where fungi grow. However, they may serve as hints of a more fundamental yet undiscovered mechanism that is involved in the induction of cellulase production.

Light has also been shown to be an inducing signal for cellulase production in fungi, particularly *T. reesei* [86]. Interestingly, this light-dependent induction has been shown to be cross-linked with carbon source regulation [87]. The mechanism of light regulation had been studied in detail, and this is discussed in Sect. 7.3.2.2.

Bacteria

Similarly with fungi, cellulases in bacteria are also regulated by carbon sources. Cellulose and laminaribiose have been shown to be inducers for cellulase formation in *C. thermocellum* [88–90]. However, in contrast, cellobiose is an inhibitor rather than an inducer for cellulase formation in bacteria [91].

Growth rate has also been shown to be a key factor regulating cellulase formation in *C. thermocellum.* With chemostats, scientists were able to conduct steady-state kinetic studies with *C. thermocellum*, which led to the finding that growth rate negatively regulates cellulase formation [88, 92, 93].

7.3.2.2 Transcription Regulation of Cellulase-Coding Genes

Fungi

A series of transcription factors are known to regulate the transcription of cellulase-coding genes. These transcription factors include CreA, Xyr1, XlnR, ACEI, ACEII, Clr2, and other less important transcription factors.

CreA (Cre1 in *T. reesei*) is a Cys_2His_2-type transcription factor that is involved in carbon catabolite repression (CCR) [94, 95]. It is homologous (albeit with low homology) to *S. cerevisiae* Mig1 whose role in CCR has been well established [96]. It has been shown that CreA is involved in the repression of cellulase-coding genes in presence of glucose, an easier and better substrate in comparison with cellulose [97]. Investigations on cellulase-hyperproducing *T. reesei* strains have shown the loss of Cre1 function [58, 98] demonstrating that Cre1 is an ideal target of genetic engineering for better cellulase production.

Xyrl is the most important and primary activator for cellulase-coding genes in cellulase-producing fungi. This protein is a Cys_2His_6-type transcription factor whose loss-of-function leads to elimination of nearly all cellulase activities [99]. Strains lacking *xyrl* had been shown to produce no cellulases while Xyrl-overproducing strains showed elevated levels of cellulases, again suggesting the critical role of Xyrl in the induction of cellulases [100, 101].

Clr2 is a recently discovered transcription factor that is essential in the induction of cellulase formation in *N. crassa*, *Aspergillus nidulans,* and *P. oxalicum* [41, 102, 103]. It is a fungal-specific zinc binuclear cluster containing transcription factor that is shown to be essential in the expression of cellulase-coding genes. The widespread presence of this protein in cellulase-producing filamentous fungi has been demonstrated.

Besides these more important transcription factors, a series of other transcription factors have been shown to regulate cellulase production. The Cys_2His_2 transcription factor ACEI and the fungal-type zinc binuclear transcription factor ACEII repressed and activated the transcription of cellulase-coding genes [104–106]. The pH-dependent Cys_2His_2 transcription factor PacC (TrPac1 in *T. reesei*) is involved in the activation of cellulase production [107]. The global nitrogen regulatory transcription factor AreA is an activator of cellulase production in *A. nidulans* [108]. The Hap2/3/5 complex binds to the promoter regions of *cbh2* and exerts a negative regulatory role in *xyn3* transcription [109, 110].

The transcription factors involved in cellulase production have quite a complex regulatory relationship among themselves. Cre1 is a phosphorylated protein of which phosphorylation is required for the binding of Cre1 to its target sequences for carbon catabolite repression [111]. In *A. nidulans,* CreA seems to be autoregulated (autorepressed) by binding to the promoter region of *creA* [112]. Cre1 binds to the promoter of *xyr1* and represses *xyr1* expression in *T. reesei* [97, 99, 113]. However, the expression of *xyr1* also requires the presence of Cre1 for full induction, which in fact is an indication that Cre1's function is not merely a carbon catabolite repressor, but rather more complicated than originally hypothesized [101]. It is the same case for *ace2* whose full induction requires the presence of Cre1 [101]. ACEI represses the expression of *xyr1* [113], while Cre1 in turn represses the expression of ACEI [101]. Both the regulation of Xyrl and Cre1 was also shown to be impacted by chromatin remodeling and nucleosome positioning that determines the binding efficiency of the transcription factors with their target sequences [96, 114–116]. A brief summary of the transcription factors and their mutual regulations is presented in Fig. 7.3.

The regulation of cellulase-coding genes is initiated by external signals, and executed by transcription factors and other DNA-interacting assemblies that are intrinsically nuclear while functioning. Signal transduction systems are therefore required for the passage of external signals to these proteins located in the nucleus. These systems, however, are less understood than the transcription factors. While cellulose is a strong inducer for cellulase production, it is unable to penetrate cell membrane. The functioning cellulase-inducing compound therefore has to be a product from cellulose hydrolysis and transformation. Our knowledge on this "true

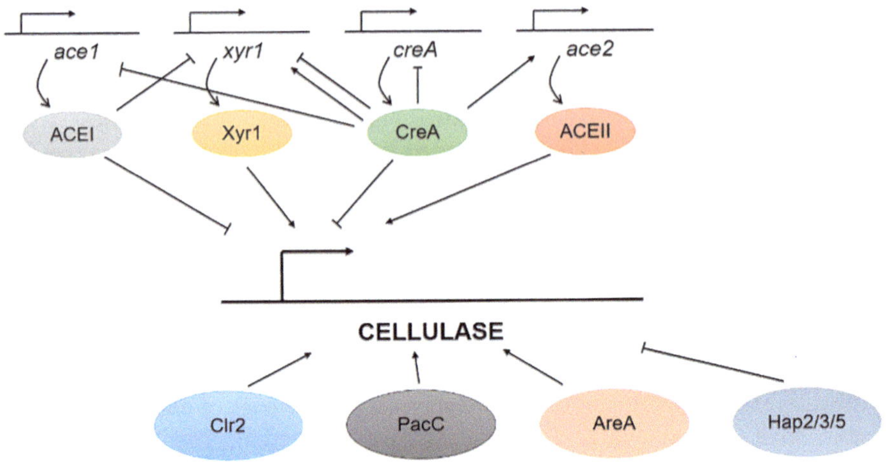

Fig. 7.3 Cellulase-regulating transcription factors and their mutual regulation in filamentous fungi

inducer" is still lacking. One possibility is that this "true inducer" is cellobiose that is a product of extracellular cellulose hydrolysis. This cellulose–cellobiose induction pathway is backed up by the finding that deletion of β-glucosidases, both extracellular and intracellular, stimulates cellulase induction from lactose and cellulose [117–119], and that cellodextrin transporters are important in cellulase induction [120]. However, cellobiose is merely a weak inducer and cannot lead to the same level of induction as cellulose. One hypothesis is that cellobiose is converted to sophorose, the strongest cellulose inducer, in vivo by transglycosylation reactions. There has been evidence on both the wet-lab experimental level and the omics level, however, that the induction patterns using cellulose, lactose, and sophorose are different [104, 121], and a simple unified cellobiose/lactose–sophorose scheme is unlikely. Additionally, the chemical property of the "true inducer" has yet to be identified experimentally despite in vivo metabolomic efforts [122], and downstream signal transduction pathways are yet to be connected to the cellulose–cellobiose induction scheme. All these evidence points toward a diversified induction mechanism possibly fine-tuned individually for each carbon source, although each of them likely start with the import of disaccharides.

The membrane heterotrimeric G proteins have been shown to modulate cellulase synthesis in filamentous fungi [123]. The G proteins are known to function in signal transduction cascades by picking up signals from the membrane and passing them downstream. The downstream target of G proteins has been shown to be the cAMP-PKA (cyclic AMP-protein kinase A) pathway, in which adenylatecyclase picks up the signal, produces the cellulase stimulating molecule cAMP, which further binds to PKA to pass signals downstream for cellulase stimulation [124]. This pathway could potentially be hypothesized to be a second mechanism of cellulase regulation by carbon sources other than the cellulose–cellobiose mechanism.

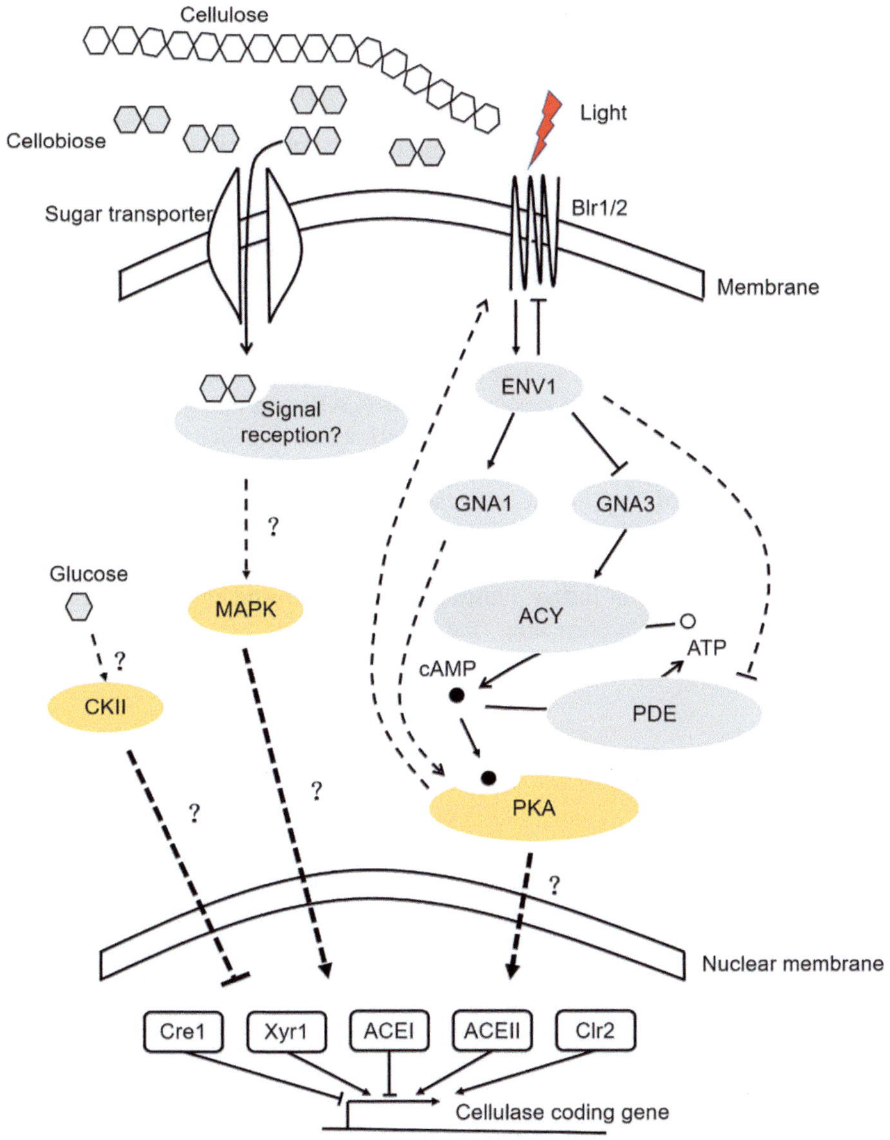

Fig. 7.4 Signal transduction for cellulase induction and repression in filamentous fungi. *Dotted lines* indicate unclear pathways. *Yellow ovals* indicate protein kinases

However, as summarized in Fig. 7.4, research on *T. reesei* by Monika Schmoll lab clearly suggested that this pathway is part of the light signal transduction cascade in which light is picked up by the photoreceptors Blr1 and Blr2, whose activation leads to the transcription of *env1*, the product of which is a PAS/LOV domain containing protein and the central light regulatory protein in *T. reesei* [87]. The Env1 protein then transcriptionally controls the expression of a G protein alpha subunit Gna3 which in turn leads to the activation of adenylatecyclases for the production of cAMP, as well as phosphodiesterases that degrades cAMP [125]. PKA then picks up the cAMP signal for further passages. Downstream signal transduction cascades remain obscure, although the involvement of a VelA/VeB/VeC/LaeA heterotetrameric complex in light-dependent cellulase synthesis may hint its involvement in this light signal transduction [126–128].

The understanding on carbon source signaling transduction pathways is still preliminary. A clear link between external signals/cellobiose with transcription regulation is still lacking. There are, however, several investigations that showed promise in understanding this connection. Experimental and bioinformatic studies have shown that transcription factors in *T. reesei* are phosphoproteins which contain phosphorylation motifs of casein kinase II (CKII) and mitogen-activated protein kinases (MAPKs) [111, 129]. Further experimental evidence showed that CKII and MAPKs are involved in transcription of cellulase-coding genes [129, 130]. In particular, the role of Tmk3 (but not other MAPKs, unpublished data from our laboratory) in cellulase induction and CKII in cellulase repression (unpublished data from our laboratory) in *T. reesei* are clearly established [129, 131, 132]. These investigations, pending on further illustration between upstream signals with these pathways, may lead to revealing the carbon source signaling pathway in cellulase-producing filamentous fungi.

Bacteria

Cellulolytic bacteria, most notably *C. thermocellum*, utilize a delicate σ/anti-σ system for the regulation of cellulase production. The anti-σ factors are membrane proteins that sense external polysaccharide signals, upon receipt of which structural transformation takes place, releasing σ factors from the membrane. The σ factors further binds to the promoters of cellulase and hemicellulase genes for regulation of their expression. This system has been well studied and documented in *C. thermocellum* [133]. Other transcription factors such as GlyR3 are also shown to regulate cellulase expression [90].

7.3.2.3 Posttranslational Modification

Cellulases are heavily glycosylated proteins [134, 135]. Not only is glycosylation critical in their activities, but they are also key to their affinities and production [136–140]. A detailed study in Yinbo Qu lab clearly suggested that different glycosylation forms with distinct activities synergistically coexist for the same protein in *P. oxalicum* [141]. Glycosylation is therefore one key consideration in producing fungal cellulases heterologously for the purpose of consolidated processing [142, 143], because glycosylation, unlike other aspects of the enzymes, are relatively more difficult to harness.

7.3.3 Improving Organisms for Better Cellulase Production

Two primary strategies are adopted for producing high-efficiency and low-cost cellulases. One strategy is to produce more enzymes so that the cost for each unit is brought down. The other strategy is to produce more active and efficient enzymes to reduce loading. The former strategy focuses on improving the organisms that produce cellulases, while the latter strategy focuses on improving the cellulases themselves. Both of these strategies are discussed below and in the following Sect. 7.3.4.

7.3.3.1 Mutagenesis

The conventional and the most effective method on improving organisms is the mutagenesis-screening approach. This method relies on carrying out random mutagenesis by chemical, physical, or genomic approaches and then developing screening method (preferably high-throughput methods) to identify the best performing organism. This approach has been successfully applied to both *T. reesei* and *P. oxalicum* over the past decades resulting in at least an order of magnitude of improvement in cellulase production ability [21].

Mutagenesis of *T. reesei* QM6a, the wildtype, was carried out by "bombing" the fungus with high-energy electron streams in a linear accelerator leading to the generation of *T. reesei* QM9123, the precursor of *T. reesei* QM9414, with double the FPA (Filter Paperase Activity) [144]. UV radiation was applied to *T. reesei* QM6a to generate *T. reesei* M7, which was further mutated chemically by nitrosoguanidine to generate *T. reesei* NG14 that had a sixfold increase in FPA than *T. reesei* QM9414 [145]. *T. reesei* NG14 was radiated by a second round of UV light to generate the "legendary" catabolite derepressed *T. reesei* Rut-C30 strain that produced approximately 20 g/L extracellular proteins, a strain widely used and studied [58]. With similar methods, the FPA in the wild-type *P. oxalicum* 114 was improved by 10-fold making *P. oxalicum* JU-A10-T an industrial strain [66]. Chemical treatment by ethyl methane sulfonate (EMS) on *Ashbya gossypii*, a cellulase-producing filamentous fungus, led to the generation of a strain with 1.4–2-fold increase in extracellular activities [146]. The same method was also applied to *T. reesei*, improving enzymatic activity by 1.5–2-fold [147]. Nitrous acid was used as a chemical reagent for mutagenesis of cellulase-producing *Bacillus* PC-BC9 to improve its cellulase production by 1.7-fold [148].

Besides these traditional chemical and physical mutagenic methods, more advanced methods of random mutagenesis have been used. Disparity mutagenesis is a genomic approach which artificially mutates DNA polymerase so that replication fidelity is damaged and therefore an increased mutagenesis rate is introduced. Using this method, Iwakuma et al. recently constructed a glucose-derepressed *T. reesei* strain [149]. Genome shuffling is another genomic method which basically "shuffles" the DNA fragments between the genomic DNA of different mutants, therefore improving the chances of combining the advantages of these mutants in one newly generated strain. With this new method, Cheng et al. were able to significantly improve the cellulase production of *P. oxalicum* [150]. ARTP (Atmospheric Room Temperature Plasma) is a recently developed physical mutagenesis approach, which creates plasma for mutagenesis under room temperature and atmospheric pressure. This method has been used on *P. oxalicum*, generating a strain with 28.2 % improved FPA [151].

In addition to carrying out mutagenesis on strains, extensive site-directed mutagenesis attempts have been carried out on cellulases, which aim to enhance efficiencies of these proteins. After a library of proteins is constructed this way, screening is performed to find the protein that best fits our anticipation. Nearly all the major cellulases in *T. reesei*, namely CBHI, EGI, EGII, and EGII were engineered with this approach, yielding more active enzymes [152–155].

Besides this mutagenesis-screening scheme, knowledge-based engineering has also been carried out on cellulases. These engineering attempts use previously known properties of cellulases and improve activity in a more rational manner. These knowledge on which engineering is based include their structures [156], posttranslational modification [157], and computational guidance [158]. Although this approach seems more accurate and reliable, protein function-to-structure relationship is far from being clearly and fully understood. Therefore, successful cases of this rational design technique are relatively rare in comparison with the mutagenesis-screening approach.

One bottleneck of the mutagenesis approach in generating cellulase hyperproducing strains is the limited efficiency for screening. Therefore, high-throughput screening methods have been used for the generation of cellulase hyperproducing strains. This can be traditionally carried out by plate-based colorimetric/morphological technologies, where cellulase-secreting colonies will show a different morphology on plates. For instance, cellulase-producing colonies can create a clear zone around them on cellulase-containing plates and the size of the halo may be used as an indication of cellulase-secreting capability [159]. Similarly, β-glucosidase-secreting colonies can digest esculin in the plate, forming esculetin that binds with ferric ions forming a black circle [160]. The size of this black circle may be used as an indication of β-glucosidase-secreting ability for the colonies. Modifications of these methods have taken place to include an additional microwell plate activity assay step, which increases accuracies on cellulase activity determination but does not improve the level of throughput [161]. These traditional plate-based techniques are generally very easy to execute, and the designs are easily understood, but they rarely go above a throughput of 10^2/experiment. The solution to this issue certainly needs further work that may include the combination of microwell plate and robotic techniques [162].

7.3.3.2 Rational Design, Genetic Engineering, and System Biology

Improved understanding on the physiology of cellulase-producing microbes together with improved genetic manipulation techniques have led to more precise and rational engineering approaches for the improvement of cellulase-producing strains.

Methodological improvement on genetic engineering technologies have led to breakthrough in both cellulolytic filamentous fungi and bacteria in the last few years. In filamentous fungi, disruption of the Non-Homologous End Joining (NHEJ) pathway had dramatically improved the efficiency of homologous recombination [163–165], which effectively leads to the possibility of large-scale genetic manipulations such as the construction of a full transcription factor knockout strain library in *P. oxalicum* [41]. The downside of this approach is that the construction of multiple deletion mutants requires more than one genetic marker, which is difficult in filamentous fungi. Techniques for reusing the same genetic marker such as the Cre/loxP system [166] have since been developed to address this issue. Other techniques aimed at improving transformation efficiency and to reduce expression levels for essential genes have also been developed [167–169]. More importantly, the development of the CRISPR-Cas9 systems, one of the most convenient and promising gene editing technologies because of its extraordinary efficiencies, for cellulolytic filamentous fungi has been under way [170, 171].

In cellulolytic *Clostridium* species, the bottleneck in genetic engineering has long been the transformation efficiency. This problem has been successfully solved approximately 5 years ago by adopting an electroporation protocol [172–174].

Other genetic modifications aim at making a better cellulase cocktail as well as promoter replacement or other regulatory element to increase gene expression. Because of the lack of β-glucosidases in *T. reesei* cellulase cocktails, a foreign β-glucosidase-encoding gene, *bgl1* was introduced in *T. reesei* to obtain a more balanced enzyme cocktail [175, 176]. In both *T. reesei* and *P. oxalicum*, promoter substitution for cellulase-coding genes for improved production of cellulases was carried out [177, 178]. Work was also carried out on optimizing the main cellulase regulators, Xyr1 and CreA, for better cellulase production. These efforts include: changing the promoter for *xyr1* to a copper responsive promoter P_{tcu1} for deregulation of *xyr1* and removal of catabolite repression [179]; carrying out mutagenesis on Xyr1 to deregulate cellulase gene expression regardless of the presence of inducers so that constitutive expression is achieved [180]; disrupting *creA* so that catabolite repression is lifted [98, 181, 182]; replacing the Cre1 binding motif in *cbh1* promoter with the binding motif of the cellulase activator ACEII for better cellulase production [183]; and, combining constitutive expression of *xyr1* with downregulation of *ace1* to enhance cellulase production by *T. reesei* Rut C-30 [184]. The most significant improvement on cellulase production by rewiring the cellulase regulation network took place in *P. oxalicum*, when Yao et al. disrupted *bgl2* and *creA* while overexpressing the *clr2* homologous *clrB* gene to achieve a 27-fold increase in secreted FPA [159]. Disruption of a variety of other genes directly or indirectly related to cellulase formation had also been carried out. These genes include the MAPK-coding *Trime2* and *tmk2* [131, 185], the ergosterol biosynthesis regulating SREBP pathway [186], and *cre2* which encodes a ubiquitin C-terminal hydrolase [187].

An alternative genetic engineering approach took place in glucose-consuming, ethanol-forming microbes, aiming at producing cellulases in these microbes so that they can perform cellulose hydrolysis and ethanol fermentation in one single cell. Microbes that have been modified for this purpose include *S. cerevisiae*, *Saccharomyces pastorianus,* and *Yarrowia lipolytica*, and the cellulases expressed are from both fungi and bacteria [188–191].

With the maturation of genetic manipulation techniques and the availability of high-throughput engineering options, systems biology has become a possibility to engineer microbes at a more sophisticated level. Initial steps have already been taken to develop the necessary tools [170, 171], and the modification of microbes at a larger scale [41].

7.3.4 Process Improvements for Better and More Abundant Cellulolytic Enzyme Preparations

Although improving cellulase-producing microbes is still the most prominent approach in improving cellulase titers and activities, other approaches have also been shown effective. These approaches include optimizing cellulase-producing or lignocellulose-degrading processes to improve lignocellulose saccharification performances, etc., in industrial applications. Both approaches aim at either improving cellulase production or improving cellulase performance.

Supplementation of a variety of chemicals and proteins with different functions has been shown to improve cellulase production and activities. Exogenous cAMP has been shown to effectively improve cellulase production, because of its function in cellulase induction signal transduction [192]. Addition of β-glucosidase improved the activity of enzymes because it effectively removes additional cellobiose in cellulose saccharification which may cause product inhibition for cellobiohydrolases [193, 194]. Bacterial expansins, their fungal equivalent swollenins, pectinases, α-L-arabinofuranosidases, and xylanases showed synergistic effects with cellulases and their supplementation leads to improved saccharification performance on lignocellulosic substrates [195–199]. This synergy is due to their function in the destruction of interconnecting bonds and minor components, which lead to the reduction of the recalcitrance of these substrates against enzymatic hydrolysis. Finally, addition of surfactants such as Triton X-100 and poly (ethylene glycol) effectively enhances cellulase production [200, 201], probably because they facilitated the cellulase secretion process.

Improvement of reaction systems also helped in cellulase production and activities. Immobilization of cellulases was carried out in several systems, which showed significantly better performance than free floating enzymes and potentials in significantly reducing lignocellulose saccharification costs because the enzymes could be recycled [202, 203]. The simultaneous saccharification and fermentation (SSF) process was applied, in which lignocellulose breakdown and end product formation from sugars took place simultaneously, to avoid the inhibition of sugars on saccharification, to reduce the amount of time needed for end product formation, and ultimately to bring down the cost for saccharification. This strategy has been shown to be effective in bioethanol production from waste fiber sludge [204]. Optimization of fermentation parameters such as agitation rate has been shown to be effective in enhancing cellulase production [68, 205]. Kinetic modeling has been performed on cellulase production processes to guide the improvement of cellulase production [206, 207]. Finally, on-site cellulase production, which means to produce cellulase on the site of lignocellulose saccharification rather than purchasing commercial enzymes, has been shown to be effective in bringing activities up and bringing costs down, primarily because enzyme storage, preservation, and delivery costs are drastically decreased [208].

7.4 Biofuel and Biochemical Production from Lignocellulosic Biomass

By enzymatic or chemical digestion, lignocellulosics can be degraded to platform molecules, the most important being sugars that can be utilized by a variety of microorganisms for the production of biofuels and biochemicals. The production of biofuels and biochemicals takes place in biorefineries, which integrates biomass conversion processes and fuel, power, heat, and value-added chemical production together. The key to biorefineries is the fermentation process whereby it is highly desirable to develop efficient microbial cell factories catalyzing biofuel and chemical production as discussed below.

7.4.1 Biofuel Production

7.4.1.1 Ethanol Production

Annually, 75 billion liters of bioethanol are produced worldwide, the majority is derived from cornstarch or cane juice in a fermentation process commonly known as the first-generation bioethanol. The utilization of food-based feedstock, however, triggered a food/feed versus fuel debate that led to the use of lignocellulose, a nonfood biomass as a preferred substrate. Biofuel thus produced has been referred to as the second-generation bioethanol. Lignocellulosic materials such as corn stover, wheat straw, grasses, sugarcane bagasse, wood chips, and other agricultural residues, are of great interest as feedstock because they are abundant, inexpensive, and renewable. However, the processes involved are much more complicated than those using cornstarch, and the costs of ethanol production from lignocellulosic materials are higher.

Ethanol production from lignocellulosic materials includes the following main steps: (1) lignocellulosic biomass pretreatment; (2) hydrolysis of cellulose and hemicellulose to fermentable sugars (including hexoses and pentoses); (3) microbial fermentation for ethanol production; (4) separation and concentration of ethanol by distillation and dehydration [14]. The inhibitors derived from the lignocellulosic material pretreatment often affect the growth of the fermentative strains. Therefore, inhibitor tolerance and hexose and pentose co-fermentation are two main issues for strain development.

Baker's yeast (*Saccharomyces cerevisiae*) has long been used in industry to produce ethanol from hexoses (six-carbon sugars). Due to the complexity of the carbohydrates present in lignocellulosic biomass, a significant amount of xylose and arabinose (five-carbon sugars derived from the hemicellulose portion of the lignocellulose) is also present in the lignocellulosic hydrolysate. For instance, in the hydrolysate of corn stover, approximately 30 % of the total fermentable sugar is xylose. Therefore, the ability of the fermenting microorganisms to use both C6 and C5 sugars available from the hydrolysate is very important to increase the economic competitiveness of lignocellulosic ethanol.

Recently, metabolic engineering has been applied to microorganisms such as *S. cerevisiae*, *Zymomonas mobilis*, and *E. coli* targeted for lignocellulosic ethanol production [209]. *S. cerevisiae*is especially attractive because it is a traditional ethanol producer that has very high tolerance to ethanol and inhibitors, and can grow at low pH values to reduce bacterial contamination. However, *S. cerevisiae* cannot naturally metabolize xylose. Consequently, in the past few decades, the introduction of xylose metabolic pathways into *S. cerevisiae* has been a focus of research [210, 211].

Two pathways have been widely for the utilization of D-xylose. In fungi and xylose metabolic yeasts, D-xylose is reduced to xylitol by NAD(P)H-dependent xylose reductase (XR) encoded by *XYL1*. Subsequently, xylitol is oxidized to D-xylulose by NAD^+-dependent xylitol dehydrogenase (XDH) encoded by *XYL2* [212–214]. The D-xylulose is converted to xylulose-5-phosphate by endogenous xylulose kinase (XK). Alternatively, D-xylose can also directly convert to xylulose via the xylose isomerase (XI) pathway which is cofactor independent in some bacteria and fungi [215–217]. Both pathways have been successfully introduced into *S. cerevisiae*, allowing the recombinant strains to produce ethanol from xylose (Fig. 7.5). However, the xylose consumption rate and ethanol production from xylose are still very low [212].

Introducing the *Scheffersomyces stipitis* XR-XDH pathway into *S. cerevisiae* has enabled the yeast to effectively utilize xylose [213, 214, 218]. However, the different cofactor dependence of XR and XDH leads to cofactor imbalance and accumulation of byproducts such as xylitol and glycerol. Therefore, much work has been done to balance intracellular cofactor levels or on modifying the cofactor specificities of XR or XDH to establish an oxidation–reduction cycle to decrease byproduct accumulation [212, 219–223]. Several approaches have been

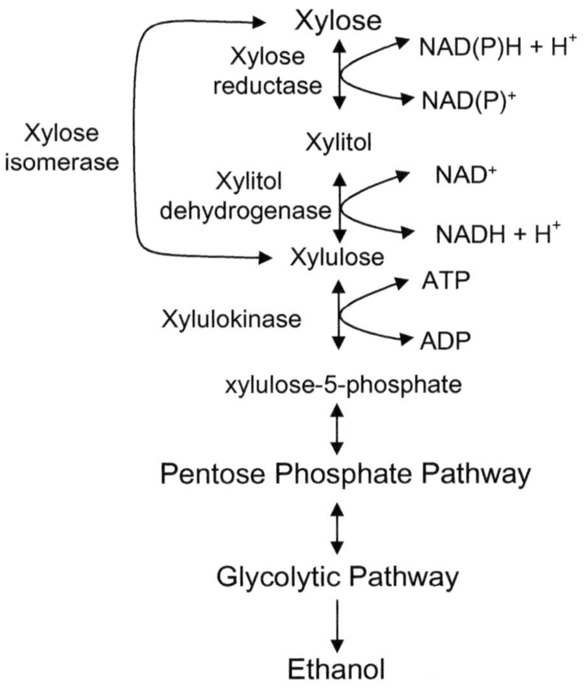

Fig. 7.5 Xylose metabolism pathways in recombinant *S. cerevisiae*

implemented for balancing intracellular cofactors in recombinant *S. cerevisiae*, including expressing the *Kluyveromyces lactis GPD1* gene which encodes a fungal $NADP^+$-dependent D-glyceraldehyde-3-phosphate dehydrogenase, expressing the *gapN* gene from *Streptococcus* mutants which encodes a non-phosphorylating $NADP^+$-dependent GAPDH, manipulating ammonia assimilation from being NADPH dependent to NADH dependent by replacing *GDH1* (encoding NADPH-dependent glutamate dehydrogenase) with *GDH2* (encoding NADH-dependent glutamate dehydrogenase), overexpressing the truncated *POS5* gene which encodes a cytosolic NADH kinase and expressing *noxE* gene which encodes a NADH oxidase [212, 222, 224–226].

An alternative pathway for xylose catabolism is the isomerase-based pathway. This pathway is cofactor independent, thus could potentially lead to higher ethanol yields. However, only recently, a few xylose isomerase genes have been successfully expressed in *S. cerevisiae*, mainly derived from bovine rumen metagenomic sequencing [227] as well as microorganisms such as *Piromyces* sp. [215, 228], *Orpinomyces* sp. [229], *Clostridium phytofermentans* [230], and *Prevotella ruminicola* [231].

Further engineering strategies, including overexpressing the endogenous xylulose kinase genes (*XKS1*), overexpressing genes involved in the non-oxidative pentose phosphate pathway and xylose transporters [232, 233], as well as adaptive evolutionary engineering [233], have been implemented to improve xylose consumption

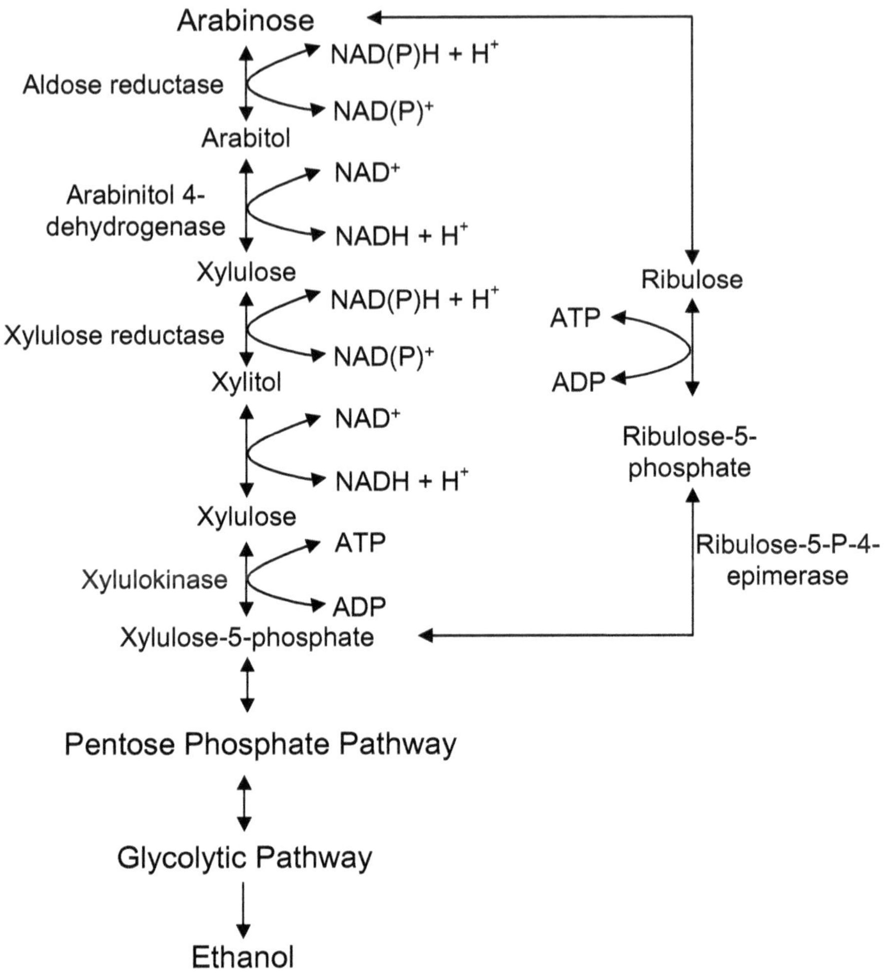

Fig. 7.6 The arabinose metabolism pathways in recombinant *S. cerevisiae*

and cell growth. The best reported industrial yeast strains can consume 80 g/L glucose and 40 g/L xylose within 24 h at an initial OD_{600} of 1.0 and produce 53 g/L ethanol [234].

Aside from glucose and xylose, lignocellulosic materials contain approximately 3–15 % L-arabinose. It is therefore necessary to construct L-arabinose fermenting microorganisms to facilitate its utilization.

There are two L-arabinose metabolic pathways in fungi and bacteria. The fungal L-arabinose metabolic pathway consists of the aldose reductase (AR), L-arabitol-4-dehydrogenase (LAD), L-xylulosereductase (LXR), and D-xylitol dehydrogenase (XDH) (Fig. 7.6). AR and LXR use NADPH as cofactor, while LAD and XDH use NAD^+ as cofactor. These reactions can also lead to cofactor imbalance. Similarly with the xylose metabolic pathway, xylulose produced in this pathway is also phosphorylated and enters the pentose phosphate pathway (PPP).

In contrast to the fungal L-arabinose metabolic pathway, the bacterial pathway is cofactor independent and comprises three enzymes, L-arabinose isomerase (AraA), L-ribulokinase (AraB), and L-ribulose-5-phosphate 4-epimerase (AraD). The D-xylulose-5-phosphate produced subsequently enters the PPP [235, 236].

Wild-type *S. cerevisiae* strains cannot metabolize arabinose. Therefore, the fungal and bacterial L-arabinose metabolic pathways have both been engineered into *S. cerevisiae*. Redox cofactor imbalance occurs in the recombinant *S. cerevisiae* strain containing the fungal L-arabinose metabolic pathway, accumulating high amounts of the byproduct L-arabitol.

The introduction of *araA*, *araB,* and *araD* genes from *E. coli* into *S. cerevisiae* was insufficient for yeast to utilize L-arabinose. However, after replacing the *E. coli* L-arabinose isomerase gene with *Bacillus subtilis araA* gene, the strain can grow on L-arabinose through "adaptive evolution" [237]. Recently, the *araA*, *araB*, and *araD* genes of *L. plantarum* were expressed in *S. cerevisiae* [238]. After further overexpression of the genes in PPP including *TAL1*, *TKL1*, *RPE1,* and *RKI1* and the transporter gene *GAL2*, as well as additional adaptive evolution, the strain could metabolize L-arabinose, resulting in an ethanol yield of 0.43 g/g consumed L-arabinose.

Although many pentose and hexose co-fermentation yeast strains have been constructed, the poor growth in hydrolysate of lignocellulosic materials poses a continuous challenge. Different pretreatment strategies such as acid pretreatment, alkali pretreatment, or stream explosion generate different inhibitors. For example, the acetic acid is the main inhibitor in acid pretreatment process. In general, the inhibitors can be divided into three categories: acids (e.g., acetic acid, formic acid, and levulinic acid), furan aldehyde (e.g., 5-hydroxymethylfurfural and furfural), and phenolics (e.g., vanillin). Attempts to improve the tolerance of inhibitors like acetic acid through evolutionary engineering have been carried out [239]. The evolutionary engineering has been done through sequential anaerobic, batch cultivation (pH 4) at increasing acetic acid concentrations or prolonged anaerobic continuous cultivation without pH control. After 400 generations, the strain can grow on xylose at pH $\leq$ 4 with 6 and 5 g L^{-1} acetic acid, respectively. However, improving the tolerance to just one kind of inhibitor may not contribute much to promote cell growth in the hydrolysate mixture. Therefore, strain evolutionary engineering in real lignocellulosic hydrolysate is a promising approach to develop robust strains.

7.4.1.2 Advanced Biofuels

Ethanol, as a traditional bulk chemical, is at present the main biofuel. However, it has a number of limitations because of its high hygroscopicity, low-energy density, and incompatibility with existing fuel infrastructure [240]. Therefore, advanced biofuels having similar properties with current petroleum-based transportation

Table 7.2 Representative advanced biofuels produced by engineered microorganisms

Biofuels	Engineered organisms	Titer	References
1-Butanol	*E. coli*	30 g/L	[250]
	C. saccharoperbutylacetonicum	32.8 g/L	[264]
Isobutanol	*E. coli*	20 g/L	[265]
3-Methyl-1-butanol	*E. coli*	9.5 g/L	[266]
1-Propanol	*E. coli*	3.5 g/L	[267]
3-Methyl-1-pentanol	*E. coli*	384.3 mg/L	[268]
Fatty acid ethyl esters	*E. coli*	1.5 g/L	[269]
Fatty alcohols	*E. coli*	1.725 g/L	[270]
Alkanes	*E. coli*	580.8 mg/L	[271]
Farnesene	*S. cerevisiae*	Not available	
Bisabolene	*S. cerevisiae*	994 mg/L	[272]

fuels, such as butanol, isopentanol, terpenes, and fatty acid ethyl esters, have been developed widely to overcome the problems of bioethanol [240]. Table 7.2 lists some of the advanced biofuels and their production in engineered microorganisms. As the first example of a commercialized product, we review the recent progress on metabolic engineering of microorganisms on butanol production.

7.4.1.3 Butanol Production

n-Butanol (or butyl alcohol) is a four-carbon, straight-chained alcohol. It is an important biofuel and chemical precursor for paints, polymers, and plastics. Butanol as a fuel is superior to ethanol in many regards such as higher energy density, lower volatility and hygroscopicity, and less corrosion to existing infrastructure. Therefore, it has been considered as an advanced biofuel [240]. The global market for *n*-butanol was estimated to be annually 3.8 million tons, with a worth of approximately 7 billion US dollars.

Butanol is fermentatively produced by solvent-producing Clostridia. The fermentative butanol production is usually called ABE fermentation because three products, acetone, butanol, and ethanol are produced at a ratio of 3:6:1. However, the fermentative butanol production pathway is not economically competitive compared to the petrochemical route. Fortunately, rapid progress in biological sciences and applied technologies provides us with more opportunities to improve the product titer, butanol ratio, and feedstock selection, thereby enabling the industrial fermentation of butanol (Fig. 7.7).

Similar to cellulosic ethanol production, cellulosic butanol conversion process can be also summarized as: (1) lignocellulosic biomass pretreatment; (2) enzymatic hydrolysis of the cellulose to monosaccharides (hexose and pentose); (3) sugar fermentation to butanol; and (4) product recovery by distillation. Hexose and pentose co-fermentation and inhibitor tolerance are also two main issues for efficient solvent production derived from the hydrolyzed sugars.

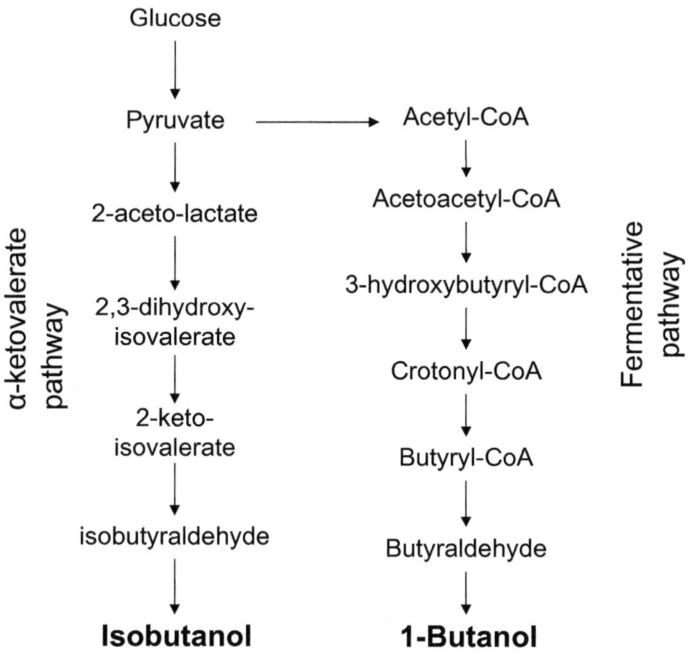

Fig. 7.7 The fermentative and nonfermentative butanol production pathways [245]

Many solvent-producing clostridia such as *Clostridium acetobutylicum* are able to utilize xylose, but its utilization rate is lower than that of glucose which represses the xylose pathway. Ren et al. [241] disrupted the *ccpA* gene encoding the pleiotropic regulator *ccpA* which enabled the bacterium to metabolize D-xylose and D-glucose simultaneously. In addition, co-overexpression of genes encoding D-xylose symporter, D-xylose isomerase, and xylulokinase improved D-xylose utilization by *C. acetobutylicum* [242]. Overexpression of the genes involved in PPP in *C. acetobutylicum* also significantly improved xylose-utilizing ability in the recombinant strain [243].

The inhibitors from pretreatment also impede the growth of strains and the solvent production significantly. Therefore, detoxification processes are required to remove toxic compounds. Otherwise, the processes using non-detoxified hydrolysates as the substrate will require higher levels of additional nutrients such as yeast extract, buffers, minerals, and vitamins [244]. Increasing the robustness of the strains to tolerate an array of inhibitors is therefore necessary to improve the industrial lignocellulosic butanol fermentation process.

ABE fermentation normally produces acetone, butanol, and ethanol at a ratio of 3:6:1 with a total production of less than 20 g/L. Thus, it is necessary to improve butanol yield, titer, and productivity. Strain selection is one strategy. A *C. acetobutylicum* strain EA2018 with a butanol- acetone- ethanol ratio of 7:2:1 instead of 6:3:1 was isolated in the 1980s from soil samples collected in a soybean field.

When this was applied in ABE fermentation plants in China, 18.5 g/L total solvent with a yield of 0.38 g/g sugar was recorded [246].

Metabolic engineering is a rational strategy to improve butanol production. For example, the disruption of the acetoacetate decarboxylase coding gene *adc* in *C. acetobutylicum* EA2018 blocked acetone formation and significantly increased the butanol ratio from 70 to 80 % [247]. Engineering *adhE1* and *ctfAB* genes to facilitate the formation of acetoacetate and butyrate also improved butanol ratio from 0.57 to 0.84 [228]. The disruption of *pta* and *buk* genes and overexpression of the *adhE1* gene can direct butanol production from acetyl-CoA by producing butyryl-CoA, leading to the production of 18.8 g/L butanol [248].

Developing butanol-producing strains from nonsolvent-producing species such as *E. coli* and *S. cerevisiae* can eliminate acetone and ethanol production [249]. These two organisms are therefore valid options besides clostridia. Two pathways have been developed for butanol synthesis, including a fermentation pathway that exists in solventogenic clostridia, and an α-ketovalerate pathway (Fig. 7.7). Commercial isobutanol production using *S. cerevisiae* had been demonstrated by Butalco, Butamax Advanced Biofuels (a joint venture of BP and Dupont), and Gevo. Among various studies, the most successful work on engineering *E. coli* led to the production of 30 g/L butanol [250]. In this study, a modified clostridial 1-butanol pathway was constructed in *E. coli* to provide an irreversible reaction catalyzed by trans-enoyl-coenzyme A (CoA) reductase (Ter) and created NADH and acetyl-CoA driving forces to direct the flux. The engineered strain produced 30 g/L butanol at a yield of 80 % of the theoretical maximum level [250].

Thus far the economics of fermentation processes is still not favorable compared to the cost of petrochemical-based *n*-butanol production. However, given the increasing attention from governments, bioproduction of *n*-butanol promises to have a competitive edge. The only commercialized cellulosic biobutanol factory in the world is the Lignicell Refining Biotechnologies Ltd. (formerly Laihe Rockley Bio-chemicals Co. Ltd.). The economic benefits of cellulosic butanol, however, remains to be proven in the long run.

Beside butanol, other advanced biofuels have also been developed recently. The company Amyris produces the sesquiterpene biofuel farnesene in *S. cerevisiae*. Farnesene can be used as diesel and jet fuel after saturation by hydrogenation. The company LS9 is actively pursuing production of alkanes and fatty acid ethyl esters (FAEEs) in *E. coli* [251, 252]. The production level of most of these biofuels is still too low to be commercially feasible. Therefore, further research is needed to develop efficient microorganisms for various advanced fuels production.

7.4.2 Production of Valuable Bio-Based Chemicals

Today, commodity chemicals are by far predominantly produced from fossil- based resources. However, the petrochemical- based production leads to concerns of the sustainability and negative environmental impact. Therefore, the bio-based

Table 7.3 Representative bio-based fine chemicals

Chemical category	Examples	Function
Organic acids	Succinic acid, lactic acid, 3-hydroxypropionic acid	Building block chemicals, food additive, feed additive
Amino acids	L-Tryptophan, L-valine, L-Phenylalanine	Nutritional supplement, food additive, feed additive
Sugar alcohols	Xylitol, mannitol	Nutritional supplement, flavoring agent, food additive
Polymers	Polyhydroxyalkanoates, polylactic acid	Biodegradable plastics

chemical production from renewable carbon sources is a desirable alternative to petrochemical-based production [253]. The chemicals produced through bio-based processes include bulk chemicals and fine chemicals. There have been several commercial examples, such as polylactic acid, polyhydroxyalkanoates, and 1,3-propanediol. With the development of biotechnology, more and more bio-based chemicals produced from renewable resources via conventional or lignocellulosic feedstocks will become cost competitive to replace petroleum-based chemicals. In this section, we review the production of several representative biochemicals and the metabolic engineering strategy used in their strain development.

7.4.2.1 Bulk Chemicals

With the development of microbial strains and downstream processes, the production of some of the bulk chemicals like organic acids and alcohols from renewable resources has already been commercialized (Table 7.3) The production of two organic acids, succinic acid and 3-hydroxypropionic acid is described briefly here.

Succinic Acid

Succinic acid was identified as one of the top 12 building block chemicals by the U. S. Department of Energy. It is a chemical intermediate in detergent/surfactant and additive in food and pharmaceutical. The current global market for succinic acid is about 30,000–50,000 tons per year with a market price of USD 2400–3000 per ton.

Three metabolic pathways have been implemented for bio-based succinic acid production, including the reductive branch of the tricarboxylic acid (TCA) cycle, the oxidative branch of the TCA cycle and the glyoxylate pathway (Fig. 7.8) [254]. Considerable efforts have been made to engineer microorganisms for succinic acid production. For example, Jantama et al. [255] developed a recombinant *E. coli* strain by rational design combined with adaptive evolution. The deletion of *ackA*, *focA*, *pflB*, *mgsA*, *poxB,* and metabolic evolution with over 2000 generations of

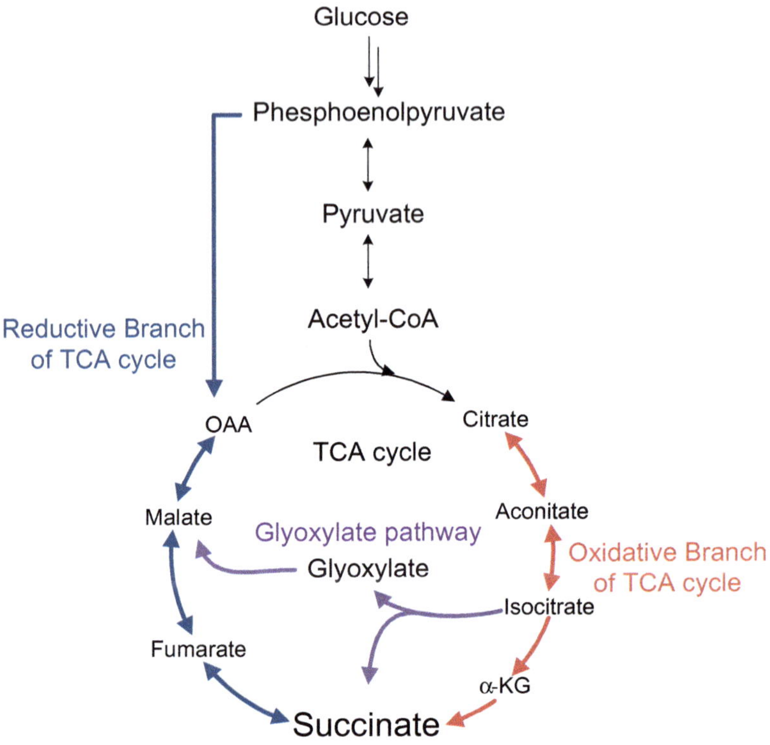

Fig. 7.8 The succinic acid production pathways [254]. *Blue*, the reductive branch of TCA cycle; red, the oxidative branch of TCA cycle; and *purple,* the glyoxylate pathway

growth-based selection resulted in a strain capable of producing 87 g/L succinic acid, with molar yields of 1.2–1.6 per mole of metabolized glucose. Engineered yeast and filamentous fungi strains have also been developed for succinate production, since they can grow at low pH which would benefit for downstream processing. BioAmber developed *Candida krusei* to produce succinate, while Myriant engineered *E. coli* to synthesis succinate. Reverdia, a company backed by DSM and Roquette, has developed a low-pH yeast-based fermentation process to produce succinic acid from glucose and carbon dioxide [254].

3-Hydroxypropionic Acid

3-Hydroxypropionic acid (3-HP) is a platform chemical that can be used to produce a variety of specialty chemicals, such as acrylic acid, 1,3-propanediol, methyl acrylate, propiolactone, malonic acid, acrylamide, and hydroxyamides. The biological production of 3-HP has gained considerable attention recently due to its wide applications.

No microorganism has been identified that naturally produces 3-HP as a major product, thus a number of biosynthetic pathways have been suggested for 3-HP production from renewable resource such as glucose and glycerol [256]. Among these pathways, significant efforts have been made to develop recombinant *E. coli* using the glycerol pathway. Heterologous glycerol dehydratase (DhaB), DhaB reactivase, and alpha-ketoglutaric semialdehyde dehydrogenase (AbSadH) were employed to construct the 3-HP biosynthetic pathway in *E. coli*, and 38.7 g/L 3-HP was obtained from glycerol under aerobic conditions [257]. However, this process requires the addition of vitamin B12 to the culture medium, making it unfeasible for industrial applications.

With the consideration of the utilization of lignocellulosic sugars, the pathways from glucose were also investigated widely in both *E. coli* and *S. cerevisiae*. *S. cerevisiae* is an attractive host as it can grow at low pH and can tolerant high titers of 3-HP. The biosynthetic pathways from intermediate metabolites malonyl-CoA and β-alanine were, respectively, constructed in *S. cerevisiae* recently, and the best strain with β-alanine pathway produced 13.7 g/L 3-HP in controlled fed-batch fermentation [258, 259].

7.4.2.2 Fine Chemicals

Fine chemicals are valuable bio-based chemicals such as pharmaceuticals, nutritional supplements, cosmetics, flavoring agents as well as additives for food, feed, and fertilizer. Although many of them are present naturally, most of them are not commercially available, because they are present in low abundance and very expensive to purify. The production of fine chemicals from microorganisms can be obtained through metabolic engineering. The advantage of fine chemical production from microbial fermentation is the low cost of feedstock and high productivity. In particular, coproduction of several fine chemicals from common carbon sources is more economical. The production of some representative bio-based fine chemicals is summarized in Table 7.4. Some recent advances in strain development for isoprenoids production are introduced here.

Table 7.4 Representative bio-based fine chemicals [273]

Chemical category	Examples	Function
Isoprenoids	Artemisinic acid, geraniol, miltiradine, taxadiene	Medicine, cosmetics, perfumes, flavoring agent, food additive
Alkaloids	Opioids, (S)-reticuline	Medicine
Polyphenols	Resveratrol, flavonoids	Medicine, cosmetics, nutritional supplement, flavoring agent, food additive
Aromatic compounds	Vanillin, cinnamic acid, *p*-hydroxycinnamic acid, caffeic acid	Medicine, cosmetics nutritional supplement, flavoring agent, food additive

Isoprenoids

Isoprenoids are a diverse group of natural compounds that have many different biological functions, and they have found various applications in fields like medicines, food additives, perfumes, and so on [260]. The production of isoprenoids has been limited by their low quantity in natural sources like plants. Moreover, the diversity of isoprenoids is enormous, which means that novel bioactive compounds can be produced by the use of new enzymes or chimeric pathways. Therefore, developing microbial-based production of isoprenoids by creating plant-like pathways in microorganism has become more and more attractive.

Isoprenoid molecules are typically synthesized from two common C5 intermediates isopentenyldiphosphate (IPP) and dimethylallyldiphosphate (DMAPP). The combination of them can generate bigger unit molecules such as geranyldiphosphate (GPP, ten carbons), farnesyldiphosphate (FPP, 15 carbons), and geranylgeranyldiphosphate (GGPP, 20 carbons), which are precursors for monoterpenes, sesquiterpenes, diterpenes, and carotenoids (Fig. 7.9) [261].

There are two biosynthetic pathways for IPP and DMAPP: the mevalonate-dependent (MVA) pathway that converts acetyl-CoA to IPP and the deoxyxylulose-5-phosphate (DXP) pathway that converts glyceraldehyde-3-phosphate and pyruvate to IPP and DMAPP. The MVA pathway is present in yeast, while the DXP pathway is mainly found in bacteria. Higher plants have both pathways, with the DXP pathway being present in plastid, and MVA pathway in the cytosol [260].

Considerable research efforts have been made by Jay Keasling's research group and Amyris to develop recombinant *S. cerevisiae* platforms for the production of plant-derived antimalarial precursor artemisinic acid, along with other products such as farnesene and squalene. To improve the flux of FPP, the promoters of all the genes in the MVA pathway have been changed. An additional copy of *tHMG1* expressing a truncated HMG-CoA synthase was expressed under the control of strong promoter (*pTDH3*). The *GAL* expression system has been modified to maintain constitutive strong expression in fermentation media containing glucose. The downstream pathway was repressed by replacing the native promoter of *ERG9* with the *CTR3* promoter which is controlled by copper [262, 263]. These modifications allowed the large increase of FPP pool. The introduction and optimization of artemisinic acid biosynthetic pathway resulted in 25 g/L artemisinic acid production [262, 263]. The platform developed here has also been applied to produce other plant sesquiterpenes from FPP such as farnesane, bisabolene, and santalene which can be used as biofuels, perfumes, and fine fragrances.

Aside from sesquiterpenes, the production of other larger terpenes have also been studied widely recently. Diterpenes such as paclitaxel precursors oxygenated taxanes, tanshinone precursor miltiradiene, and ginsenosides are produced from engineered *E. coli* or *S. cerevisiae*, and their production was improved significantly through the co-culture of engineered organisms. The microorganisms were also engineered to produce larger terpenes such as lycopene and carotenoids [249].

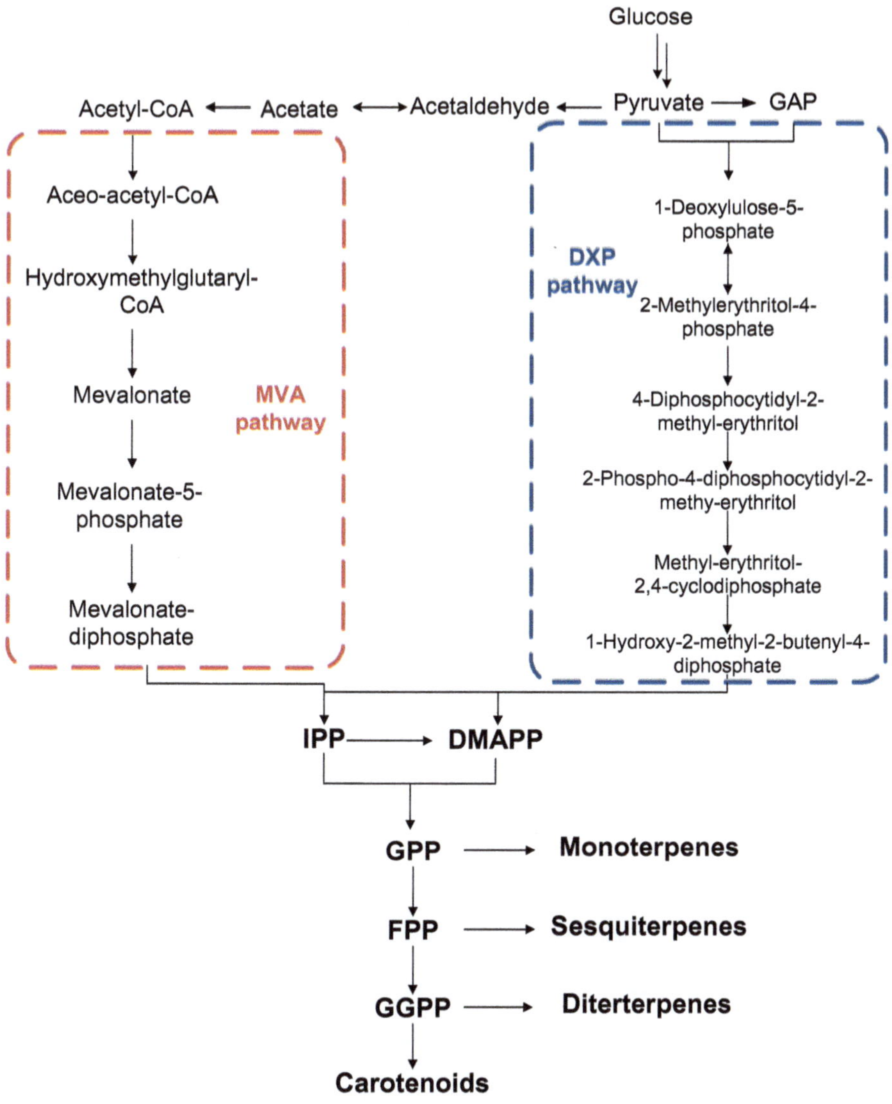

Fig. 7.9 The isoprenoid synthetic pathway in yeast (*red*, mevalonate-dependent (MVA) pathway) and in bacteria (*blue*, deoxyxylulose-5-phosphate (DXP) pathway) [249]. *IPP* isopentenyl diphosphate, *DMAPP* dimethylallyl diphosphate, *GPP* geranyl diphosphate, *FPP* farnesyl diphosphate, *GGPP* geranylgeranyl diphosphate

Amyris had developed engineered *S. cerevisiae* strains and processes to convert sugars to low cost and stable supplies of artemisnin antimalarial drug. In addition, they have converted sugarcane syrup into farnesene, and the industrial manufacturing production facility has begun operations in Brazil since 2012. There are over

30,000 known terpenoids in nature. With the development of synthetic biology, more and more compounds will be produced by engineered microorganisms with high yields.

7.5 Perspectives for Lignocellulosic Biofuel and Biochemical Production

The development of technologies for the production of biofuels and biochemical from lignocellulosics is driven by the ever increasing demand and the forseeable depletion of fossil fuels on which the current fuel and chemical industries are based. The shift of a fossil fuel economy to more sustainable and renewable resources is an answer to maintain our lifestyle, and to a more broad extent, our civilization. To do this, the renewable fuels and chemicals need to be produced in a bio-based route. Although bioethanol production from lignocellulosic materials, along with quite a few other bio-based chemicals, has been commercialized, the production of biofuels and biochemicals is still in its commercial infancy. A lot of work still needs to be done to increase the titer, yields, and productivities of the products. Advances in synthetic biology, systems biology, and improvement of molecular tools for genetic manipulation can further contribute to microbial strain engineering. With these promising improvements, we believe that a more sustainable and environmentally friendly fuel and chemical industry will be constructed to provide quality living for everyone.

References

1. Durand JD (1977) Historical estimates of world population: an evaluation. Popul Dev Rev 3 (3):253–296
2. Riley JC (2005) Estimates of regional and global life expectancy, 1800–2001. Popul Dev Rev 31(3):537–543
3. Bowman DM, Balch JK, Artaxo P, Bond WJ, Carlson JM, Cochrane MA, D'Antonio CM, Defries RS, Doyle JC, Harrison SP, Johnston FH, Keeley JE, Krawchuk MA, Kull CA, Marston JB, Moritz MA, Prentice IC, Roos CI, Scott AC, Swetnam TW, van der Werf GR, Pyne SJ (2009) Fire in the Earth system. Science 324(5926):481–484
4. Conti J, Holtberg P, Beamon JA, Napolitano S, Schaal AM, Turnure JT, Westfall L (2013) International energy outlook 2013. U.S. Energy Information Administration, Washington, DC
5. Shafiee S, Topal E (2009) When will fossil fuel reserves be diminished? Energy Policy 37:181–189
6. Demain AL, Newcomb M, Wu JH (2005) Cellulase, clostridia, and ethanol. Microbiol Mol Biol Rev 69(1):124–154
7. Xie G, Wang X, Ren L (2010) China's crop residues resources evaluation. Chin J Biotechnol 26(7):855–863
8. Zhu C, Zhang H, Xiao R, Chen Y, Liu D, Du F, Ying H, Ouyang P (2015) Research progress in catalytic valorization of lignocellulose. Sci Chin Chem 45(5):454–478

9. Plan for comprehensive utilization of agricultural straws in the 12th Five-Year Plan (2011)
10. Parikka M (2004) Global biomass fuel resources. Biomass Bioenergy 27:613–620
11. Sorrell S, Speirs J, Bentley R, Brandt A, Miller R (2010) Global oil depletion: a review of the evidence. Energy Policy 38:5290–5295
12. Bornscheuer U, Buchholz K, Seibel J (2014) Enzymatic degradation of (ligno)cellulose. Angew Chem Int Ed Engl 53(41):10876–10893
13. Wang M, Liu K, Dai L, Zhang J, Fang X (2013) The structural and biochemical basis for cellulose biodegradation. J Chem Technol Biotechnol 88:491–500
14. Sánchez C (2009) Lignocellulosic residues: biodegradation and bioconversion by fungi. Biotechnol Adv 27(2):185–194
15. Lynd LR, Weimer PJ, van Zyl WH, Pretorius IS (2002) Microbial cellulose utilization: fundamentals and biotechnology. Microbiol Mol Biol Rev 66(3):506–577
16. Peng F, Peng P, Xu F, Sun RC (2012) Fractional purification and bioconversion of hemicelluloses. Biotechnol Adv 30(4):879–903
17. Zeng Y, Zhao S, Yang S, Ding SY (2014) Lignin plays a negative role in the biochemical process for producing lignocellulosic biofuels. Curr Opin Biotechnol 27:38–45
18. Voxeur A, Wang Y, Sibout R (2015) Lignification: different mechanisms for a versatile polymer. Curr Opin Plant Biol 23:83–90
19. Iiyama K, Lam T, Stone BA (1994) Covalent cross-links in the cell wall. Plant Physiol 104 (2):315–320
20. Himmel ME, Ding SY, Johnson DK, Adney WS, Nimlos MR, Brady JW, Foust TD (2007) Biomass recalcitrance: engineering plants and enzymes for biofuels production. Science 315 (5813):804–807
21. Wang M, Li Z, Fang X, Wang L, Qu Y (2012) Cellulolytic enzyme production and enzymatic hydrolysis for second-generation bioethanol production. Adv Biochem Eng Biotechnol 128:1–24
22. Shallom D, Shoham Y (2003) Microbial hemicellulases. Curr Opin Microbiol 6(3):219–228
23. Brown ME, Chang MC (2014) Exploring bacterial lignin degradation. Curr Opin Chem Biol 19:1–7
24. Pollegioni L, Tonin F, Rosini E (2015) Lignin-degrading enzymes. FEBS J 282(7):1190–1213
25. Divne C, Stahlberg J, Reinikainen T, Ruohonen L, Pettersson G, Knowles JK, Teeri TT, Jones TA (1994) The three-dimensional crystal structure of the catalytic core of cellobiohydrolase I from *Trichoderma reesei*. Science 265(5171):524–528
26. Davies GJ, Brzozowski AM, Dauter M, Varrot A, Schülein M (2000) Structure and function of *Humicola insolens* family 6 cellulases: structure of the endoglucanase, Cel6B, at 1.6 Å resolution. Biochem J 348(Pt 1):201–207
27. Karkehabadi S, Helmich KE, Kaper T, Hansson H, Mikkelsen NE, Gudmundsson M, Piens K, Fujdala M, Banerjee G, Scott-Craig JS, Walton JD, Phillips GN Jr, Sandgren M (2014) Biochemical characterization and crystal structures of a fungal family 3 β-glucosidase, Cel3A from *Hypocrea jecorina*. J Biol Chem 289(45):31624–31637
28. Törrönen A, Rouvinen J (1995) Structural comparison of two major endo-1,4-xylanases from *Trichoderma reesei*. Biochemistry 34(3):847–856
29. Yang JK, Yoon HJ, Ahn HJ, Lee BI, Pedelacq JD, Liong EC, Berendzen J, Laivenieks M, Vieille C, Zeikus GJ, Vocadlo DJ, Withers SG, Suh SW (2004) Crystal structure of β-D-xylosidase from *Thermoanaerobacterium saccharolyticum*, a family 39 glycoside hydrolase. J Mol Biol 335(1):155–165
30. Golan G, Shallom D, Teplitsky A, Zaide G, Shulami S, Baasov T, Stojanoff V, Thompson A, Shoham Y, Shoham G (2004) Crystal structures of *Geobacillus stearothermophilus* α-glucuronidase complexed with its substrate and products: mechanistic implications. J Biol Chem 279(4):3014–3024
31. Taylor EJ, Gloster TM, Turkenburg JP, Vincent F, Brzozowski AM, Dupont C, Shareck F, Centeno MS, Prates JA, Puchart V, Ferreira LM, Fontes CM, Biely P, Davies GJ (2006) Structure and activity of two metal ion-dependent acetylxylan esterases involved in plant cell

wall degradation reveals a close similarity to peptidoglycan deacetylases. J Biol Chem 281 (16):10968–10975

32. Hermoso JA, Sanz-Aparicio J, Molina R, Juge N, Gonzalez R, Faulds CB (2004) The crystal structure of feruloyl esterase A from *Aspergillus niger* suggests evolutive functional convergence in feruloyl esterase family. J Mol Biol 338(3):495–506
33. Miki Y, Calviño FR, Pogni R, Giansanti S, Ruiz-Dueñas FJ, Martinez MJ, Basosi R, Romero A, Martinez AT (2011) Crystallographic, kinetic, and spectroscopic study of the first ligninolytic peroxidase presenting a catalytic tyrosine. J Biol Chem 286(17):15525–15534
34. Osipov E, Polyakov K, Kittl R, Shleev S, Dorovatovsky P, Tikhonova T, Hann S, Ludwig R, Popov V (2014) Effect of the L499 M mutation of the ascomycetous *Botrytis aclada* laccase on redox potential and catalytic properties. Acta Crystallogr D Biol Crystallogr 70(Pt 11):2913–2923
35. Li XL, Špániková S, de Vries RP, Biely P (2007) Identification of genes encoding microbial glucuronoyl esterases. FEBS Lett 581(21):4029–4035
36. Quinlan RJ, Sweeney MD, Lo Leggio L, Otten H, Poulsen JC, Johansen KS, Krogh KB, Jorgensen CI, Tovborg M, Anthonsen A, Tryfona T, Walter CP, Dupree P, Xu F, Davies GJ, Walton PH (2011) Insights into the oxidative degradation of cellulose by a copper metalloenzyme that exploits biomass components. Proc Natl Acad Sci USA 108(37):15079–15084
37. Harris PV, Welner D, McFarland KC, Re E, Navarro Poulsen JC, Brown K, Salbo R, Ding H, Vlasenko E, Merino S, Xu F, Cherry J, Larsen S, Lo Leggio L (2010) Stimulation of lignocellulosic biomass hydrolysis by proteins of glycoside hydrolase family 61: structure and function of a large, enigmatic family. Biochemistry 49(15):3305–3316
38. Langston JA, Shaghasi T, Abbate E, Xu F, Vlasenko E, Sweeney MD (2011) Oxidoreductive cellulose depolymerization by the enzymes cellobiose dehydrogenase and glycoside hydrolase 61. Appl Environ Microbiol 77(19):7007–7015
39. Hemsworth GR, Henrissat B, Davies GJ, Walton PH (2014) Discovery and characterization of a new family of lytic polysaccharide monooxygenases. Nat Chem Biol 10(2):122–126
40. Cherry JR, Fidantsef AL (2003) Directed evolution of industrial enzymes: an update. Curr Opin Biotechnol 14(4):438–443
41. Li Z, Yao G, Wu R, Gao L, Kan Q, Liu M, Yang P, Liu G, Qin Y, Song X, Zhong Y, Fang X, Qu Y (2015) Synergistic and dose-controlled regulation of cellulase gene expression in *Penicillium oxalicum*. PLoS Genet 11(9):e1005509
42. van Zyl WH, Lynd LR, den Hann R, McBride JE (2007) Consolidated bioprocessing for bioethanol production using *Saccharomyces cerevisiae*. Adv Biochem Eng Biot 108:205–235
43. Olson DG, McBride JE, Shaw AJ, Lynd LR (2012) Recent progress in consolidated bioprocessing. Curr Opin Biotechnol 23(3):396–405
44. Biswas R, Zheng T, Olson DG, Lynd LR, Guss AM (2015) Elimination of hydrogenase active site assembly blocks H_2 production and increases ethanol yield in *Clostridium thermocellum*. Biotechnol Biofuels 8:20
45. Shen Y, Jarboe L, Brown R, Wen Z (2015) A thermochemical-biochemical hybrid processing of lignocellulosic biomass for producing fuels and chemicals. Biotechnol Adv 33 (8):1799–1813
46. Martin ME, Richter H, Saha S, Angenent LT (2015) Traits of selected *Clostridium* strains for syngas fermentation to ethanol. Biotechnol Bioeng 113(3):531–539
47. Ahmed A, Cateni BG, Huhnke RL, Lewis RS (2006) Effects of biomass-generated producer gas constituents on cell growth, product distribution and hydrogenase activity of *Clostridium carboxidivorans* P7^T. Biomass Bioenerg 30(7):665–672
48. Macrelli S, Mogensen J, Zacchi G (2012) Techno-economic evaluation of 2nd generation bioethanol production from sugar cane bagasse and leaves integrated with the sugar-based ethanol process. Biotechnol Biofuels 5:22

49. Klein-Marcuschamer D, Oleskowicz-Popiel P, Simmons BA, Blanch HW (2012) The challenge of enzyme cost in the production of lignocellulosic biofuels. Biotechnol Bioeng 109(4):1083–1087
50. Littlewood J, Wang L, Turnbull C, Murphy RJ (2013) Techno-economic potential of bioethanol from bamboo in China. Biotechnol Biofuels 6(1):173
51. Galbe M, Sassner P, Wingren A, Zacchi G (2007) Process engineering economics of bioethanol production. Adv Biochem Eng Biotechnol 108:303–327
52. Kim SR, Park YC, Jin YS, Seo JH (2013) Strain engineering of *Saccharomyces cerevisiae* for enhanced xylose metabolism. Biotechnol Adv 31(6):851–861
53. Grimmler C, Held C, Liebl W, Ehrenreich A (2010) Transcriptional analysis of catabolite repression in *Clostridium acetobutylicum* growing on mixtures of D-glucose and D-xylose. J Biotechnol 150(3):315–323
54. Sun S, Sun S, Cao X, Sun R (2015) The role of pretreatment in improving the enzymatic hydrolysis of lignocellulosic materials. Bioresour Technol 199:49–58
55. Jonsson LJ, Martin C (2015) Pretreatment of lignocellulose: Formation of inhibitory by-products and strategies for minimizing their effects. Bioresour Technol 199:103–112
56. Gnansounou E, Dauriat A (2010) Techno-economic analysis of lignocellulosic ethanol: a review. Bioresour Technol 101:4980–4991
57. Liu W, Yan J, Li J, Sang T (2012) Yield potential of *Miscanthus* energy crops in the Loess Plateu of China. GCB Energy 4:545–554
58. Peterson R, Nevalainen H (2012) *Trichoderma reesei* RUT-C30-thirty years of strain improvement. Microbiology 158(Pt 1):58–68
59. Gusakov AV (2011) Alternatives to *Trichoderma reesei* in biofuel production. Trends Biotechnol 29(9):419–425
60. Martinez D, Berka RM, Henrissat B, Saloheimo M, Arvas M, Baker SE, Chapman J, Chertkov O, Coutinho PM, Cullen D, Danchin EG, Grigoriev IV, Harris P, Jackson M, Kubicek CP, Han CS, Ho I, Larrondo LF, de Leon AL, Magnuson JK, Merino S, Misra M, Nelson B, Putnam N, Robbertse B, Salamov AA, Schmoll M, Terry A, Thayer N, Westerholm-Parvinen A, Schoch CL, Yao J, Barabote R, Nelson MA, Detter C, Bruce D, Kuske CR, Xie G, Richardson P, Rokhsar DS, Lucas SM, Rubin EM, Dunn-Coleman N, Ward M, Brettin TS (2008) Genome sequencing and analysis of the biomass-degrading fungus *Trichoderma reesei* (syn. *Hypocrea jecorina*). Nat Biotechnol 26(5):553–560
61. Murphy L, Bohlin C, Baumann MJ, Olsen SN, Sorensen TH, Anderson L, Borch K, Westh P (2013) Product inhibition of five *Hypocrea jecorina* cellulases. Enzyme Microb Technol 52 (3):163–169
62. Ma L, Zhang J, Zou G, Wang C, Zhou Z (2011) Improvement of cellulase activity in *Trichoderma reesei* by heterologous expression of a β-glucosidase gene from *Penicillium decumbens*. Enzyme Microb Technol 49(4):366–371
63. Chen M, Qin Y, Liu Z, Liu K, Wang F, Qu Y (2010) Isolation and characterization of a β-glucosidase from *Penicillium decumbens* and improving hydrolysis of corncob residue by using it as cellulase supplementation. Enzyme Microb Technol 46(6):444–449
64. Gold ND, Martin VJ (2007) Global view of the *Clostridium thermocellum* cellulosome revealed by quantitative proteomic analysis. J Bacteriol 189(19):6787–6795
65. Qu Y, Gao P, Wang Z (1984) Screening of catabolite repression-resistant mutants of cellulase producing *Penicillium* spp. Acta Mycol Sinica 3:238–243
66. Liu G, Zhang L, Qin Y, Zou G, Li Z, Yan X, Wei X, Chen M, Chen L, Zheng K, Zhang J, Ma L, Li J, Liu R, Xu H, Bao X, Fang X, Wang L, Zhong Y, Liu W, Zheng H, Wang S, Wang C, Xun L, Zhao GP, Wang T, Zhou Z, Qu Y (2013) Long-term strain improvements accumulate mutations in regulatory elements responsible for hyper-production of cellulolytic enzymes. Sci Rep 3:1569
67. Fang X, Shen Y, Zhao J, Bao X, Qu Y (2010) Status and prospect of lignocellulosic bioethanol production in China. Bioresour Technol 101(13):4814–4819

68. Wang M, He D, Liang Y, Liu K, Jiang B, Wang F, Hou S, Fang X (2013) Factors involved in the response to change of agitation rate during cellulase production from *Penicillium decumbens* JUA10-1. J Ind Microbiol Biotechnol 40(9):1077–1082
69. Liu G, Zhang L, Wei X, Zou G, Qin Y, Ma L, Li J, Zheng H, Wang S, Wang C, Xun L, Zhao GP, Zhou Z, Qu Y (2013) Genomic and secretomic analyses reveal unique features of the lignocellulolytic enzyme system of *Penicillium decumbens*. PLoS ONE 8(2):e55185
70. Shen Y, Zhang Y, Ma T, Bao X, Du F, Zhuang G, Qu Y (2008) Simultaneous saccharification and fermentation of acid-pretreated corncobs with a recombinant *Saccharomyces cerevisiae* expressing β-glucosidase. Bioresour Technol 99(11):5099–5103
71. Jiang Y, Duarte AV, van den Brink J, Wiebenga A, Zou G, Wang C, de Vries RP, Zhou Z, Benoit I (2015) Enhancing saccharification of wheat straw by mixing enzymes from genetically-modified *Trichoderma reesei* and *Aspergillus niger*. Biotechnol Lett 38(1):65–70
72. Mehboob N, Asad MJ, Asgher M, Gulfraz M, Mukhtar T, Mahmood RT (2014) Exploring thermophilic cellulolytic enzyme production potential of *Aspergillus fumigatus* by the solid-state fermentation of wheat straw. Appl Biochem Biotechnol 172(7):3646–3655
73. Lan TQ, Wei D, Yang ST, Liu X (2013) Enhanced cellulase production by *Trichoderma viride* in a rotating fibrous bed bioreactor. Bioresour Technol 133:175–182
74. Rahnama N, Foo HL, Abdul Rahman NA, Ariff A, Md Shah UK (2014) Saccharification of rice straw by cellulase from a local *Trichoderma harzianum* SNRS3 for biobutanol production. BMC Biotechnol 14:103
75. Singhania RR, Saini JK, Saini R, Adsul M, Mathur A, Gupta R, Tuli DK (2014) Bioethanol production from wheat straw via enzymatic route employing *Penicillium janthinellum* cellulases. Bioresour Technol 169:490–495
76. dos Reis L, Fontana RC, Delabona Pda S, da Silva Lima DJ, Camassola M, Pradella JG, Dillon AJ (2013) Increased production of cellulases and xylanases by *Penicillium echinulatum* S1M29 in batch and fed-batch culture. Bioresour Technol 146:597–603
77. Karlsson J, Momcilovic D, Wittgren B, Schulein M, Tjerneld F, Brinkmalm G (2002) Enzymatic degradation of carboxymethyl cellulose hydrolyzed by the endoglucanases Cel5A, Cel7B, and Cel45A from *Humicola insolens* and Cel7B, Cel12A and Cel45A core from *Trichoderma reesei*. Biopolymers 63(1):32–40
78. Fang X, Yano S, Inoue H, Sawayama S (2009) Strain improvement of *Acremonium cellulolyticus* for cellulase production by mutation. J Biosci Bioeng 107(3):256–261
79. Roche CM, Glass NL, Blanch HW, Clark DS (2014) Engineering the filamentous fungus *Neurospora crassa* for lipid production from lignocellulosic biomass. Biotechnol Bioeng 111(6):1097–1107
80. Currie DH, Herring CD, Guss AM, Olson DG, Hogsett DA, Lynd LR (2013) Functional heterologous expression of an engineered full length CipA from *Clostridium thermocellum* in *Thermoanaerobacterium saccharolyticum*. Biotechnol Biofuels 6(1):32
81. Messner R, Kubicek CP (1991) Carbon source control of Cellobiohydrolase I and II formation by *Trichoderma reesei*. Appl Environ Microbiol 57(3):630–635
82. Conrad M, Schothorst J, Kankipati HN, Van Zeebroeck G, Rubio-Texeira M, Thevelein JM (2014) Nutrient sensing and signaling in the yeast *Saccharomyces cerevisiae*. FEMS Microbiol Rev 38(2):254–299
83. Leisola MS, Ulmer DC, Pitkanen K, Fiechter A (1985) Induction of cellulases in *Chaetomium cellulolyticum* by cellobiose. Biotechnol Bioeng 27(9):1389–1391
84. Ilmén M, Saloheimo A, Onnela ML, Penttilä ME (1997) Regulation of cellulase gene expression in the filamentous fungus *Trichoderma reesei*. Appl Environ Microbiol 63 (4):1298–1306
85. Sternberg D, Mandels GR (1979) Induction of cellulolytic enzymes in *Trichoderma reesei* by sophorose. J Bacteriol 139(3):761–769
86. Tisch D, Schmoll M (2013) Targets of light signalling in *Trichoderma reesei*. BMC Genom 14:657

87. Tisch D, Schuster A, Schmoll M (2014) Crossroads between light response and nutrient signalling: ENV1 and PhLP1 act as mutual regulatory pair in *Trichoderma reesei*. BMC Genom 15:425
88. Dror TW, Rolider A, Bayer EA, Lamed R, Shoham Y (2005) Regulation of major cellulosomal endoglucanases of *Clostridium thermocellum* differs from that of a prominent cellulosomal xylanase. J Bacteriol 187(7):2261–2266
89. Stevenson DM, Weimer PJ (2005) Expression of 17 genes in *Clostridium thermocellum* ATCC 27405 during fermentation of cellulose or cellobiose in continuous culture. Appl Environ Microbiol 71(8):4672–4678
90. Newcomb M, Chen CY, Wu JH (2007) Induction of the *celC* operon of *Clostridium thermocellum* by laminaribiose. Proc Natl Acad Sci USA 104(10):3747–3752
91. Johnson EA, Bouchot F, Demain AL (1985) Regulation of cellulase formation in *Clostridium thermocellum*. J Gen Microbiol 131:2303–2308
92. Dror TW, Morag E, Rolider A, Bayer EA, Lamed R, Shoham Y (2003) Regulation of the cellulosomal *CelS* (*cel48A*) gene of *Clostridium thermocellum* is growth rate dependent. J Bacteriol 185(10):3042–3048
93. Dror TW, Rolider A, Bayer EA, Lamed R, Shoham Y (2003) Regulation of expression of scaffoldin-related genes in *Clostridium thermocellum*. J Bacteriol 185(17):5109–5116
94. Felenbok B, Flipphi M, Nikolaev I (2001) Ethanol catabolism in *Aspergillus nidulans*: a model system for studying gene regulation. Prog Nucleic Acid Res Mol Biol 69:149–204
95. Strauss J, Mach RL, Zeilinger S, Hartler G, Stoffler G, Wolschek M, Kubicek CP (1995) Cre1, the carbon catabolite repressor protein from *Trichoderma reesei*. FEBS Lett 376(1–2):103–107
96. Portnoy T, Margeot A, Linke R, Atanasova L, Fekete E, Sandor E, Hartl L, Karaffa L, Druzhinina IS, Seiboth B, Le Crom S, Kubicek CP (2011) The CRE1 carbon catabolite repressor of the fungus *Trichoderma reesei*: a master regulator of carbon assimilation. BMC Genom 12:269
97. Antoniêto AC, dos Santos Castro L, Silva-Rocha R, Persinoti GF, Silva RN (2014) Defining the genome-wide role of CRE1 during carbon catabolite repression in *Trichoderma reesei* using RNA-Seq analysis. Fungal Genet Biol 73:93–103
98. Nakari-Setälä T, Paloheimo M, Kallio J, Vehmaanperä J, Penttilä M, Saloheimo M (2009) Genetic modification of carbon catabolite repression in *Trichoderma reesei* for improved protein production. Appl Environ Microbiol 75(14):4853–4860
99. Stricker AR, Grosstessner-Hain K, Wurleitner E, Mach RL (2006) Xyr1 (xylanase regulator 1) regulates both the hydrolytic enzyme system and D-xylose metabolism in *Hypocrea jecorina*. Eukaryot Cell 5(12):2128–2137
100. Lichius A, Bidard F, Buchholz F, Le Crom S, Martin J, Schackwitz W, Austerlitz T, Grigoriev IV, Baker SE, Margeot A, Seiboth B, Kubicek CP (2015) Genome sequencing of the *Trichoderma reesei* QM9136 mutant identifies a truncation of the transcriptional regulator XYR1 as the cause for its cellulase-negative phenotype. BMC Genom 16:326
101. Portnoy T, Margeot A, Seidl-Seiboth V, Le Crom S, Ben Chaabane F, Linke R, Seiboth B, Kubicek CP (2011) Differential regulation of the cellulase transcription factors XYR1, ACE2, and ACE1 in *Trichoderma reesei* strains producing high and low levels of cellulase. Eukaryot Cell 10(2):262–271
102. Coradetti ST, Xiong Y, Glass NL (2013) Analysis of a conserved cellulase transcriptional regulator reveals inducer-independent production of cellulolytic enzymes in *Neurospora crassa*. Microbiology Open 2(4):595–609
103. Coradetti ST, Craig JP, Xiong Y, Shock T, Tian C, Glass NL (2012) Conserved and essential transcription factors for cellulase gene expression in ascomycete fungi. Proc Natl Acad Sci USA 109(19):7397–7402
104. Aro N, Saloheimo A, Ilmén M, Penttilä M (2001) ACEII, a novel transcriptional activator involved in regulation of cellulase and xylanase genes of *Trichoderma reesei*. J Biol Chem 276(26):24309–24314

105. Aro N, Ilmén M, Saloheimo M, Penttilä M (2003) ACEI of *Trichoderma reesei* is a repressor of cellulase and xylanase expression. Appl Environ Microbiol 69(1):56–65
106. Saloheimo A, Aro N, Ilmén M, Penttilä M (2000) Isolation of the *ace1* gene encoding a Cys_2-His_2 transcription factor involved in regulation of activity of the cellulase promoter cbh1 of *Trichoderma reesei*. J Biol Chem 275(8):5817–5825
107. He R, Ma L, Li C, Jia W, Li D, Zhang D, Chen S (2014) *Trpac1*, a pH response transcription regulator, is involved in cellulase gene expression in *Trichoderma reesei*. Enzyme Microb Technol 67:17–26
108. Lockington RA, Rodbourn L, Barnett S, Carter CJ, Kelly JM (2002) Regulation by carbon and nitrogen sources of a family of cellulases in *Aspergillus nidulans*. Fungal Genet Biol 37 (2):190–196
109. Zeilinger S, Ebner A, Marosits T, Mach R, Kubicek CP (2001) The *Hypocrea jecorina* HAP 2/3/5 protein complex binds to the inverted CCAAT-box (ATTGG) within the *cbh2* (cellobiohydrolase II-gene) activating element. Mol Genet Genom 266(1):56–63
110. Würleitner E, Pera L, Wacenovsky C, Cziferszky A, Zeilinger S, Kubicek CP, Mach RL (2003) Transcriptional regulation of *xyn2* in *Hypocrea jecorina*. Eukaryot Cell 2(1):150–158
111. Cziferszky A, Mach RL, Kubicek CP (2002) Phosphorylation positively regulates DNA binding of the carbon catabolite repressor Cre1 of *Hypocrea jecorina* (*Trichoderma reesei*). J Biol Chem 277(17):14688–14694
112. Strauss J, Horvath HK, Abdallah BM, Kindermann J, Mach RL, Kubicek CP (1999) The function of CreA, the carbon catabolite repressor of *Aspergillus nidulans*, is regulated at the transcriptional and post-transcriptional level. Mol Microbiol 32(1):169–178
113. Mach-Aigner AR, Pucher ME, Steiger MG, Bauer GE, Preis SJ, Mach RL (2008) Transcriptional regulation of *xyr1*, encoding the main regulator of the xylanolytic and cellulolytic enzyme system in *Hypocrea jecorina*. Appl Environ Microbiol 74(21):6554–6562
114. Ries L, Belshaw NJ, Ilmén M, Penttilä ME, Alapuranen M, Archer DB (2014) The role of CRE1 in nucleosome positioning within the *cbh1* promoter and coding regions of *Trichoderma reesei*. Appl Microbiol Biotechnol 98(2):749–762
115. Mello-de-Sousa TM, Rassinger A, Pucher ME, dos Santos Castro L, Persinoti GF, Silva-Rocha R, Pocas-Fonseca MJ, Mach RL, Nascimento Silva R, Mach-Aigner AR (2015) The impact of chromatin remodelling on cellulase expression in *Trichoderma reesei*. BMC Genom 16:588
116. Xin Q, Gong Y, Lv X, Chen G, Liu W (2013) *Trichoderma reesei* histone acetyltransferase Gcn5 regulates fungal growth, conidiation, and cellulase gene expression. Curr Microbiol 67 (5):580–589
117. Zhou Q, Xu J, Kou Y, Lv X, Zhang X, Zhao G, Zhang W, Chen G, Liu W (2012) Differential involvement of β-glucosidases from *Hypocrea jecorina* in rapid induction of cellulase genes by cellulose and cellobiose. Eukaryot Cell 11(11):1371–1381
118. Xu J, Zhao G, Kou Y, Zhang W, Zhou Q, Chen G, Liu W (2014) Intracellular β-glucosidases CEL1a and CEL1b are essential for cellulase induction on lactose in *Trichoderma reesei*. Eukaryot Cell 13(8):1001–1013
119. Chen M, Qin Y, Cao Q, Liu G, Li J, Li Z, Zhao J, Qu Y (2013) Promotion of extracellular lignocellulolytic enzymes production by restraining the intracellular β-glucosidase in *Penicillium decumbens*. Bioresour Technol 137:33–40
120. Li J, Liu G, Chen M, Li Z, Qin Y, Qu Y (2013) Cellodextrin transporters play important roles in cellulase induction in the cellulolytic fungus *Penicillium oxalicum*. Appl Microbiol Biotechnol 97(24):10479–10488
121. Kubicek CP (2013) Systems biological approaches towards understanding cellulase production by *Trichoderma reesei*. J Biotechnol 163(2):133–142
122. Karaffa L, Coulier L, Fekete E, Overkamp KM, Druzhinina IS, Mikus M, Seiboth B, Novak L, Punt PJ, Kubicek CP (2013) The intracellular galactoglycome in *Trichoderma reesei* during growth on lactose. Appl Microbiol Biotechnol 97(12):5447–5456

123. Hu Y, Liu G, Li Z, Qin Y, Qu Y, Song X (2013) G protein-cAMP signaling pathway mediated by PGA3 plays different roles in regulating the expressions of amylases and cellulases in *Penicillium decumbens*. Fungal Genet Biol 58–59:62–70
124. Schuster A, Tisch D, Seidl-Seiboth V, Kubicek CP, Schmoll M (2012) Roles of protein kinase A and adenylate cyclase in light-modulated cellulase regulation in *Trichoderma reesei*. Appl Environ Microbiol 78(7):2168–2178
125. Tisch D, Kubicek CP, Schmoll M (2011) New insights into the mechanism of light modulated signaling by heterotrimeric G-proteins: ENVOY acts on *gna1* and *gna3* and adjusts cAMP levels in *Trichoderma reesei* (*Hypocrea jecorina*). Fungal Genet Biol 48 (6):631–640
126. Seiboth B, Karimi RA, Phatale PA, Linke R, Hartl L, Sauer DG, Smith KM, Baker SE, Freitag M, Kubicek CP (2012) The putative protein methyltransferase LAE1 controls cellulase gene expression in *Trichoderma reesei*. Mol Microbiol 84(6):1150–1164
127. Karimi-Aghcheh R, Németh Z, Atanasova L, Fekete E, Paholcsek M, Sandor E, Aquino B, Druzhinina IS, Karaffa L, Kubicek CP (2014) The VELVET A orthologue VEL1 of *Trichoderma reesei* regulates fungal development and is essential for cellulase gene expression. PLoS ONE 9(11):e112799
128. Liu K, Dong Y, Wang F, Jiang B, Wang M, Fang X (2015) Regulation of cellulase expression, sporulation and morphogenesis by velvet family proteins in *Trichoderma reesei*. Appl Microbiol Biotechnol 100(2):769–779
129. Wang M, Zhao Q, Yang J, Jiang B, Wang F, Liu K, Fang X (2013) A mitogen-activated protein kinase Tmk3 participates in high osmolarity resistance, cell wall integrity maintenance and cellulase production regulation in *Trichoderma reesei*. PLoS ONE 8(8): e72189
130. Lei Y, Liu G, Li Z, Gao L, Qin Y, Qu Y (2014) Functional characterization of protein kinase CK2 regulatory subunits regulating *Penicillium oxalicum* asexual development and hydrolytic enzyme production. Fungal Genet Biol 66:44–53
131. Wang M, Dong Y, Zhao Q, Wang F, Liu K, Jiang B, Fang X (2014) Identification of the role of a MAP kinase Tmk2 in *Hypocrea jecorina* (*Trichoderma reesei*). Sci Rep 4:6732
132. Wang M, Yang H, Zhang M, Liu K, Wang H, Luo Y, Fang X (2015) Functional analysis of *Trichoderma reesei* CKIIα2, a catalytic subunit of casein kinase II. Appl Microbiol Biotechnol 99(14):5929–5938
133. Nataf Y, Bahari L, Kahel-Raifer H, Borovok I, Lamed R, Bayer EA, Sonenshein AL, Shoham Y (2010) *Clostridium thermocellum* cellulosomal genes are regulated by extracytoplasmic polysaccharides via alternative sigma factors. Proc Natl Acad Sci USA 107(43):18646–18651
134. Harrison MJ, Nouwens AS, Jardine DR, Zachara NE, Gooley AA, Nevalainen H, Packer NH (1998) Modified glycosylation of cellobiohydrolase I from a high cellulase-producing mutant strain of *Trichoderma reesei*. Eur J Biochem 256(1):119–127
135. Zhou F, Olman V, Xu Y (2009) Large-scale analyses of glycosylation in cellulases. Genom Proteom Bioinform 7(4):194–199
136. Chen L, Drake MR, Resch MG, Greene ER, Himmel ME, Chaffey PK, Beckham GT, Tan Z (2014) Specificity of *O*-glycosylation in enhancing the stability and cellulose binding affinity of Family 1 carbohydrate-binding modules. Proc Natl Acad Sci USA 111(21):7612–7617
137. Wei W, Chen L, Zou G, Wang Q, Yan X, Zhang J, Wang C, Zhou Z (2013) *N*-glycosylation affects the proper folding, enzymatic characteristics and production of a fungal ß-glucosidase. Biotechnol Bioeng 110(12):3075–3084
138. Payne CM, Resch MG, Chen L, Crowley MF, Himmel ME, Taylor LE 2nd, Sandgren M, Stahlberg J, Stals I, Tan Z, Beckham GT (2013) Glycosylated linkers in multimodular lignocellulose-degrading enzymes dynamically bind to cellulose. Proc Natl Acad Sci USA 110(36):14646–14651
139. Beckham GT, Dai Z, Matthews JF, Momany M, Payne CM, Adney WS, Baker SE, Himmel ME (2012) Harnessing glycosylation to improve cellulase activity. Curr Opin Biotechnol 23(3):338–345

140. Taylor CB, Talib MF, McCabe C, Bu L, Adney WS, Himmel ME, Crowley MF, Beckham GT (2012) Computational investigation of glycosylation effects on a family 1 carbohydrate-binding module. J Biol Chem 287(5):3147–3155
141. Gao L, Gao F, Wang L, Geng C, Chi L, Zhao J, Qu Y (2012) *N*-glycoform diversity of cellobiohydrolase I from *Penicillium decumbens* and synergism of nonhydrolytic glycoform in cellulose degradation. J Biol Chem 287(19):15906–15915
142. Xu L, Shen Y, Hou J, Peng B, Tang H, Bao X (2014) Secretory pathway engineering enhances secretion of cellobiohydrolase I from *Trichoderma reesei* in *Saccharomyces cerevisiae*. J Biosci Bioeng 117(1):45–52
143. Qin Y, Wei X, Liu X, Wang T, Qu Y (2008) Purification and characterization of recombinant endoglucanase of *Trichoderma reesei* expressed in *Saccharomyces cerevisiae* with higher glycosylation and stability. Protein Expr Purif 58(1):162–167
144. Mandels M, Weber J, Parizek R (1971) Enhanced cellulase production by a mutant of *Trichoderma viride*. Appl Microbiol 21(1):152–154
145. Montenecourt BS, Eveleigh DE (1977) Preparation of mutants of *Trichoderma reesei* with enhanced cellulase production. Appl Environ Microbiol 34(6):777–782
146. Ribeiro O, Magalhães F, Aguiar TQ, Wiebe MG, Penttilä M, Domingues L (2013) Random and direct mutagenesis to enhance protein secretion in *Ashbya gossypii*. Bioengineered 4 (5):322–331
147. Shafique S, Bajwa R, Shafique S (2010) Molecular characterisation of UV and chemically induced mutants of *Trichoderma reesei* FCBP-364. Nat Prod Res 24(15):1438–1448
148. Abdullah R, Zafar W, Nadeem M, Iqtedar M, Naz S, Syed Q, Butt ZA (2015) Random mutagenesis and media optimisation for hyperproduction of cellulase from *Bacillus* species using proximally analysed *Saccharum spontaneum* in submerged fermentation. Nat Prod Res 29(4):336–344
149. Iwakuma H, Koyama Y, Miyachi A, Nasukawa M, Matsumoto H, Yano S, Ogihara J, Kasumi T (2015) Generation of a glucose de-repressed mutant of *Trichoderma reesei* using disparity mutagenesis. Biosci Biotechnol Biochem 80(3):486–492
150. Cheng Y, Song X, Qin Y, Qu Y (2009) Genome shuffling improves production of cellulase by *Penicillium decumbens* JU-A10. J Appl Microbiol 107(6):1837–1846
151. Zi Z, Shi W, Li Y, Zhai Z, Han L, Wang M, Fang X (2013) Modification of celllulose-producing industrial filamentous fungi with plasma induced mutation. Chin Brew 32(S1):5–8
152. Qin Y, Wei X, Song X, Qu Y (2008) Engineering endoglucanase II from *Trichoderma reesei* to improve the catalytic efficiency at a higher pH optimum. J Biotechnol 135(2):190–195
153. Nakazawa H, Okada K, Onodera T, Ogasawara W, Okada H, Morikawa Y (2009) Directed evolution of endoglucanase III (Cel12A) from *Trichoderma reesei*. Appl Microbiol Biotechnol 83(4):649–657
154. Mahadevan SA, Wi SG, Lee DS, Bae HJ (2008) Site-directed mutagenesis and CBM engineering of Cel5A (*Thermotoga maritima*). FEMS Microbiol Lett 287(2):205–211
155. Becker D, Braet C, Brumer H 3rd, Claeyssens M, Divne C, Fagerstrom BR, Harris M, Jones TA, Kleywegt GJ, Koivula A, Mahdi S, Piens K, Sinnott ML, Stahlberg J, Teeri TT, Underwood M, Wohlfahrt G (2001) Engineering of a glycosidase Family 7 cellobiohydrolase to more alkaline pH optimum: the pH behaviour of *Trichoderma reesei* Cel7A and its E223S/ A224H/L225 V/T226A/D262G mutant. Biochem J 356(Pt 1):19–30
156. Chokhawala HA, Roche CM, Kim TW, Atreya ME, Vegesna N, Dana CM, Blanch HW, Clark DS (2015) Mutagenesis of *Trichoderma reesei* endoglucanase I: impact of expression host on activity and stability at elevated temperatures. BMC Biotechnol 15:11
157. Nordwald EM, Brunecky R, Himmel ME, Beckham GT, Kaar JL (2014) Charge engineering of cellulases improves ionic liquid tolerance and reduces lignin inhibition. Biotechnol Bioeng 111(8):1541–1549
158. Akcapinar GB, Venturini A, Martelli PL, Casadio R, Sezerman UO (2015) Modulating the thermostability of Endoglucanase I from *Trichoderma reesei* using computational approaches. Protein Eng Des Sel 28(5):127–135

159. Yao G, Li Z, Gao L, Wu R, Kan Q, Liu G, Qu Y (2015) Redesigning the regulatory pathway to enhance cellulase production in *Penicillium oxalicum*. Biotechnol Biofuels 8:71
160. Liu C, Cai M, Ma X (1993) Secreening of a β-glucosidase product repression resistant *Penicillium decumbens* strain. Microbiol Res Appl 613:5–8
161. Kim JJ, Kwon YK, Kim JH, Heo SJ, Lee Y, Lee SJ, Shim WB, Jung WK, Hyun JH, Kwon KK, Kang DH, Oh C (2014) Effective microwell plate-based screening method for microbes producing cellulase and xylanase and its application. J Microbiol Biotechnol 24 (11):1559–1565
162. Xiao Z, Storms R, Tsang A (2004) Microplate-based filter paper assay to measure total cellulase activity. Biotechnol Bioeng 88(7):832–837
163. Li ZH, Du CM, Zhong YH, Wang TH (2010) Development of a highly efficient gene targeting system allowing rapid genetic manipulations in *Penicillium decumbens*. Appl Microbiol Biotechnol 87(3):1065–1076
164. Zhang G, Hartl L, Schuster A, Polak S, Schmoll M, Wang T, Seidl V, Seiboth B (2009) Gene targeting in a nonhomologous end joining deficient *Hypocrea jecorina*. J Biotechnol 139(2):146–151
165. Schuster A, Bruno KS, Collett JR, Baker SE, Seiboth B, Kubicek CP, Schmoll M (2012) A versatile toolkit for high throughput functional genomics with *Trichoderma reesei*. Biotechnol Biofuels 5:1
166. Steiger MG, Vitikainen M, Uskonen P, Brunner K, Adam G, Pakula T, Penttilä M, Saloheimo M, Mach RL, Mach-Aigner AR (2011) Transformation system for *Hypocrea jecorina* (*Trichoderma reesei*) that favors homologous integration and employs reusable bidirectionally selectable markers. Appl Environ Microbiol 77(1):114–121
167. Derntl C, Kiesenhofer DP, Mach RL, Mach-Aigner AR (2015) Novel strategies for genomic manipulation of *Trichoderma reesei* with the purpose of strain engineering. Appl Environ Microbiol 81(18):6314–6323
168. Magaña-Ortiz D, Coconi-Linares N, Ortiz-Vazquez E, Fernandez F, Loske AM, Gómez-Lim MA (2013) A novel and highly efficient method for genetic transformation of fungi employing shock waves. Fungal Genet Biol 56:9–16
169. He R, Guo W, Wang L, Zhang D (2015) Construction of an efficient RNAi system in the cellulolytic fungus *Trichoderma reesei*. J Microbiol Methods 108:70–73
170. Nødvig CS, Nielsen JB, Kogle ME, Mortensen UH (2015) A CRISPR-Cas9 system for genetic engineering of filamentous fungi. PLoS ONE 10(7):e0133085
171. Fuller KK, Chen S, Loros JJ, Dunlap JC (2015) Development of the CRISPR/Cas9 system for targeted gene disruption in *Aspergillus fumigatus*. Eukaryot Cell 14(11):1073–1080
172. Tripathi SA, Olson DG, Argyros DA, Miller BB, Barrett TF, Murphy DM, McCool JD, Warner AK, Rajgarhia VB, Lynd LR, Hogsett DA, Caiazza NC (2010) Development of *pyrF*-based genetic system for targeted gene deletion in *Clostridium thermocellum* and creation of a *pta* mutant. Appl Environ Microbiol 76(19):6591–6599
173. Olson DG, Tripathi SA, Giannone RJ, Lo J, Caiazza NC, Hogsett DA, Hettich RL, Guss AM, Dubrovsky G, Lynd LR (2010) Deletion of the Cel48S cellulase from *Clostridium thermocellum*. Proc Natl Acad Sci USA 107(41):17727–17732
174. Olson DG, Lynd LR (2012) Transformation of *Clostridium thermocellum* by electroporation. Methods Enzymol 510:317–330
175. Dashtban M, Qin W (2012) Overexpression of an exotic thermotolerant β-glucosidase in *Trichoderma reesei* and its significant increase in cellulolytic activity and saccharification of barley straw. Microb Cell Fact 11:63
176. Zhang J, Zhong Y, Zhao X, Wang T (2010) Development of the cellulolytic fungus *Trichoderma reesei* strain with enhanced β-glucosidase and filter paper activity using strong artificial cellobiohydrolase 1 promoter. Bioresour Technol 101(24):9815–9818
177. Hu Y, Xue H, Liu G, Song X, Qu Y (2015) Efficient production and evaluation of lignocellulolytic enzymes using a constitutive protein expression system in *Penicillium oxalicum*. J Ind Microbiol Biotechnol 42(6):877–887

178. Fang H, Xia L (2013) High activity cellulase production by recombinant *Trichoderma reesei* ZU-02 with the enhanced cellobiohydrolase production. Bioresour Technol 144:693–697
179. Lv X, Zheng F, Li C, Zhang W, Chen G, Liu W (2015) Characterization of a copper responsive promoter and its mediated overexpression of the xylanase regulator 1 results in an induction-independent production of cellulases in *Trichoderma reesei*. Biotechnol Biofuels 8:67
180. Derntl C, Gudynaite-Savitch L, Calixte S, White T, Mach RL, Mach-Aigner AR (2013) Mutation of the Xylanase regulator 1 causes a glucose blind hydrolase expressing phenotype in industrially used *Trichoderma* strains. Biotechnol Biofuels 6:62
181. Mello-de-Sousa TM, Gorsche R, Rassinger A, Poças-Fonseca MJ, Mach RL, Mach-Aigner AR (2014) A truncated form of the Carbon catabolite repressor 1 increases cellulase production in *Trichoderma reesei*. Biotechnol Biofuels 7(1):129
182. Fujii T, Inoue H, Ishikawa K (2013) Enhancing cellulase and hemicellulase production by genetic modification of the carbon catabolite repressor gene, *creA*, in *Acremonium cellulolyticus*. AMB Express 3(1):73
183. Zou G, Shi S, Jiang Y, van den Brink J, de Vries RP, Chen L, Zhang J, Ma L, Wang C, Zhou Z (2012) Construction of a cellulase hyper-expression system in *Trichoderma reesei* by promoter and enzyme engineering. Microb Cell Fact 11:21
184. Wang S, Liu G, Wang J, Yu J, Huang B, Xing M (2013) Enhancing cellulase production in *Trichoderma reesei* RUT C30 through combined manipulation of activating and repressing genes. J Ind Microbiol Biotechnol 40(6):633–641
185. Chen F, Chen XZ, Su XY, Qin LN, Huang ZB, Tao Y, Dong ZY (2015) An Ime2-like mitogen-activated protein kinase is involved in cellulase expression in the filamentous fungus *Trichoderma reesei*. Biotechnol Lett 37(10):2055–2062
186. Reilly MC, Qin L, Craig JP, Starr TL, Glass NL (2015) Deletion of homologs of the SREBP pathway results in hyper-production of cellulases in *Neurospora crassa* and *Trichoderma reesei*. Biotechnol Biofuels 8:121
187. Denton JA, Kelly JM (2011) Disruption of *Trichoderma reesei cre2*, encoding an ubiquitin C-terminal hydrolase, results in increased cellulase activity. BMC Biotechnol 11:103
188. Wei H, Wang W, Alahuhta M, Vander Wall T, Baker JO, Taylor LE 2nd, Decker SR, Himmel ME, Zhang M (2014) Engineering towards a complete heterologous cellulase secretome in *Yarrowia lipolytica* reveals its potential for consolidated bioprocessing. Biotechnol Biofuels 7(1):148
189. Hong J, Yang H, Zhang K, Liu C, Zou S, Zhang M (2014) Development of a cellulolytic *Saccharomyces cerevisiae* strain with enhanced cellobiohydrolase activity. World J Microbiol Biotechnol 30(11):2985–2993
190. Fitzpatrick J, Kricka W, James TC, Bond U (2014) Expression of three *Trichoderma reesei* cellulase genes in *Saccharomyces pastorianus* for the development of a two-step process of hydrolysis and fermentation of cellulose. J Appl Microbiol 117(1):96–108
191. Kim S, Baek SH, Lee K, Hahn JS (2013) Cellulosic ethanol production using a yeast consortium displaying a minicellulosome and β-glucosidase. Microb Cell Fact 12:14
192. Wang D, Qu Y, Gao P (1995) Regulation of cellulase synthesis in mycelial fungi: Participation of ATP and cyclic AMP. Biotechnol Lett 17:593–598
193. Baba Y, Sumitani J, Tani S, Kawaguchi T (2015) Characterization of *Aspergillus aculeatus* β-glucosidase 1 accelerating cellulose hydrolysis with *Trichoderma* cellulase system. AMB Exp 5(1):3
194. Gao L, Gao F, Zhang D, Zhang C, Wu G, Chen S (2013) Purification and characterization of a new β-glucosidase from *Penicillium piceum* and its application in enzymatic degradation of delignified corn stover. Bioresour Technol 147:658–661
195. Bunterngsook B, Eurwilaichitr L, Thamchaipenet A, Champreda V (2015) Binding characteristics and synergistic effects of bacterial expansins on cellulosic and hemicellulosic substrates. Bioresour Technol 176:129–135

196. Kang K, Wang S, Lai G, Liu G, Xing M (2013) Characterization of a novel swollenin from *Penicillium oxalicum* in facilitating enzymatic saccharification of cellulose. BMC Biotechnol 13:42
197. da Silva Delabona P, Cota J, Hoffmam ZB, Paixão DA, Farinas CS, Cairo JP, Lima DJ, Squina FM, Ruller R, Pradella JG (2013) Understanding the cellulolytic system of *Trichoderma harzianum* P49P11 and enhancing saccharification of pretreated sugarcane bagasse by supplementation with pectinase and α-L-arabinofuranosidase. Bioresour Technol 131:500–507
198. Berlin A, Maximenko V, Gilkes N, Saddler J (2007) Optimization of enzyme complexes for lignocellulose hydrolysis. Biotechnol Bioeng 97(2):287–296
199. Gonçalves GA, Takasugi Y, Jia L, Mori Y, Noda S, Tanaka T, Ichinose H, Kamiya N (2015) Synergistic effect and application of xylanases as accessory enzymes to enhance the hydrolysis of pretreated bagasse. Enzyme Microb Technol 72:16–24
200. Callow NV, Ju LK (2012) Promoting pellet growth of *Trichoderma reesei* Rut C30 by surfactants for easy separation and enhanced cellulase production. Enzyme Microb Technol 50(6–7):311–317
201. Sipos B, Szilagyi M, Sebestyen Z, Perazzini R, Dienes D, Jakab E, Crestini C, Reczey K (2011) Mechanism of the positive effect of poly(ethylene glycol) addition in enzymatic hydrolysis of steam pretreated lignocelluloses. C R Biol 334(11):812–823
202. Bilgin R, Yalcin MS, Yildirim D (2015) Optimization of covalent immobilization of *Trichoderma reesei* cellulase onto modified ReliZyme HA403 and Sepabeads EC-EP supports for cellulose hydrolysis, in buffer and ionic liquids/buffer media. Artif Cells Nanomed Biotechnol. doi:10.3109/21691401.2015.1024842
203. Abraham RE, Verma ML, Barrow CJ, Puri M (2014) Suitability of magnetic nanoparticle immobilised cellulases in enhancing enzymatic saccharification of pretreated hemp biomass. Biotechnol Biofuels 7:90
204. Cavka A, Alriksson B, Rose SH, van Zyl WH, Jonsson LJ (2014) Production of cellulosic ethanol and enzyme from waste fiber sludge using SSF, recycling of hydrolytic enzymes and yeast, and recombinant cellulase-producing *Aspergillus niger*. J Ind Microbiol Biotechnol 41 (8):1191–1200
205. Maurya DP, Singh D, Pratap D, Maurya JP (2012) Optimization of solid state fermentation conditions for the production of cellulase by *Trichoderma reesei*. J Environ Biol 33(1):5–8
206. Ma L, Li C, Yang Z, Jia W, Zhang D, Chen S (2013) Kinetic studies on batch cultivation of *Trichoderma reesei* and application to enhance cellulase production by fed-batch fermentation. J Biotechnol 166(4):192–197
207. Wang M, Han L, Liu S, Zhao X, Yang J, Loh SK, Sun X, Zhang C, Fang X (2015) A Weibull statistics-based lignocellulose saccharification model and a built-in parameter accurately predict lignocellulose hydrolysis performance. Biotechnol J 10(9):1424–1433
208. Rana V, Eckard AD, Teller P, Ahring BK (2014) On-site enzymes produced from *Trichoderma reesei* RUT-C30 and *Aspergillus saccharolyticus* for hydrolysis of wet exploded corn stover and loblolly pine. Bioresour Technol 154:282–289
209. Huang GL, Anderson TD, Clubb RT (2014) Engineering microbial surfaces to degrade lignocellulosic biomass. Bioengineered 5(2):96–106
210. Hahn-Hagerdal B, Karhumaa K, Fonseca C, Spencer-Martins I, Gorwa-Grauslund MF (2007) Towards industrial pentose-fermenting yeast strains. Appl Microbiol Biotechnol 74 (5):937–953
211. Jeffries TW (2006) Engineering yeasts for xylose metabolism. Curr Opin Biotechnol 17 (3):320–326
212. Hou J, Suo F, Wang C, Li X, Shen Y, Bao X (2014) Fine-tuning of NADH oxidase decreases byproduct accumulation in respiration deficient xylose metabolic *Saccharomyces cerevisiae*. BMC Biotechnol 14:13
213. Scalcinati G, Otero JM, van Vleet JR, Jeffries TW, Olsson L, Nielsen J (2012) Evolutionary engineering of *Saccharomyces cerevisiae* for efficient aerobic xylose consumption. FEMS Yeast Res 12(5):582–597

214. Walfridsson M, Anderlund M, Bao X, Hahn-Hagerdal B (1997) Expression of different levels of enzymes from the *Pichia stipitisXYL1* and *XYL2* genes in *Saccharomyces cerevisiae* and its effects on product formation during xylose utilisation. Appl Microbiol Biotechnol 48(2):218–224
215. Kuyper M, Harhangi HR, Stave AK, Winkler AA, Jetten MS, de Laat WT, den Ridder JJ, Op den Camp HJ, van Dijken JP, Pronk JT (2003) High-level functional expression of a fungal xylose isomerase: the key to efficient ethanolic fermentation of xylose *by Saccharomyces cerevisiae*? FEMS Yeast Res 4(1):69–78
216. van Maris AJ, Winkler AA, Kuyper M, de Laat WT, van Dijken JP, Pronk JT (2007) Development of efficient xylose fermentation in *Saccharomyces cerevisiae*: xylose isomerase as a key component. Adv Biochem Eng Biotechnol 108:179–204
217. Zhou H, Cheng JS, Wang BL, Fink GR, Stephanopoulos G (2012) Xylose isomerase overexpression along with engineering of the pentose phosphate pathway and evolutionary engineering enable rapid xylose utilization and ethanol production by *Saccharomyces cerevisiae*. Metab Eng 14(6):611–622
218. Eliasson A, Christensson C, Wahlbom CF, Hahn-Hagerdal B (2000) Anaerobic xylose fermentation by recombinant *Saccharomyces cerevisiae* carrying *XYL1*, *XYL2*, and *XKS1* in mineral medium chemostat cultures. Appl Env Microbiol 66(8):3381–3386
219. Bengtsson O, Hahn-Hagerdal B, Gorwa-Grauslund MF (2009) Xylose reductase from *Pichia stipitis* with altered coenzyme preference improves ethanolic xylose fermentation by recombinant *Saccharomyces cerevisiae*. Biotechnol Biofuels 2:9
220. Hou J, Shen Y, Li XP, Bao XM (2007) Effect of the reversal of coenzyme specificity by expression of mutated *Pichia stipitis* xylitol dehydrogenase in recombinant *Saccharomyces cerevisiae*. Lett Appl Microbiol 45(2):184–189
221. Matsushika A, Watanabe S, Kodaki T, Makino K, Inoue H, Murakami K, Takimura O, Sawayama S (2008) Expression of protein engineered $NADP^+$-dependent xylitol dehydrogenase increases ethanol production from xylose in recombinant *Saccharomyces cerevisiae*. Appl Microbiol Biotechnol 81(2):243–255
222. Verho R, Londesborough J, Penttilä M, Richard P (2003) Engineering redox cofactor regeneration for improved pentose fermentation in *Saccharomyces cerevisiae*. Appl Env Microbiol 69(10):5892–5897
223. Zhang GC, Liu JJ, Ding WT (2012) Decreased xylitol formation during xylose fermentation in *Saccharomyces cerevisiae* due to overexpression of water-forming NADH oxidase. Appl Env Microbiol 78(4):1081–1086
224. Bro C, Regenberg B, Forster J, Nielsen J (2006) *In silico* aided metabolic engineering of *Saccharomyces cerevisiae* for improved bioethanol production. Metab Eng 8(2):102–111
225. Grotkjaer T, Christakopoulos P, Nielsen J, Olsson L (2005) Comparative metabolic network analysis of two xylose fermenting recombinant *Saccharomyces cerevisiae* strains. Metab Eng 7(5–6):437–444
226. Hou J, Vemuri GN, Bao X, Olsson L (2009) Impact of overexpressing NADH kinase on glucose and xylose metabolism in recombinant xylose-utilizing *Saccharomyces cerevisiae*. Appl Microbiol Biotechnol 82(5):909–919
227. Hou J, Shen Y, Jiao C, Ge R, Zhang X, Bao X (2015) Characterization and evolution of xylose isomerase screened from the bovine rumen metagenome in *Saccharomyces cerevisiae*. J Biosci Bioeng 121(2):160–165
228. Lee JY, Jang YS, Lee J, Papoutsakis ET, Lee SY (2009) Metabolic engineering of *Clostridium acetobutylicum* M5 for highly selective butanol production. Biotechnol J 4 (10):1432–1440
229. Madhavan A, Tamalampudi S, Ushida K, Kanai D, Katahira S, Srivastava A, Fukuda H, Bisaria VS, Kondo A (2009) Xylose isomerase from polycentric fungus *Orpinomyces*: gene sequencing, cloning, and expression in *Saccharomyces cerevisiae* for bioconversion of xylose to ethanol. Appl Microbiol Biotechnol 82(6):1067–1078
230. Brat D, Boles E, Wiedemann B (2009) Functional expression of a bacterial xylose isomerase in *Saccharomyces cerevisiae*. Appl Env Microbiol 75(8):2304–2311

231. Hector R, Dien B, Cotta M, Mertens J (2013) Growth and fermentation of D-xylose by *Saccharomyces cerevisiae* expressing a novel D-xylose isomerase originating from the bacterium *Prevotella ruminicola* TC2-24. Biotechnol Biofuels 6:84
232. Peng B, Shen Y, Li X, Chen X, Hou J, Bao X (2012) Improvement of xylose fermentation in respiratory-deficient xylose-fermenting *Saccharomyces cerevisiae*. Metab Eng 14(1):9–18
233. Shen Y, Chen X, Peng B, Chen L, Hou J, Bao X (2012) An efficient xylose-fermenting recombinant *Saccharomyces cerevisiae* strain obtained through adaptive evolution and its global transcription profile. Appl Microbiol Biotechnol 96(4):1079–1091
234. Diao L, Liu Y, Qian F, Yang J, Jiang Y, Yang S (2013) Construction of fast xylose-fermenting yeast based on industrial ethanol-producing diploid *Saccharomyces cerevisiae* by rational design and adaptive evolution. BMC Biotechnol 13:110
235. Hahn-Hagerdal B, Karhumaa K, Jeppsson M, Gorwa-Grauslund MF (2007) Metabolic engineering for pentose utilization in *Saccharomyces cerevisiae*. Adv Biochem Eng Biotechnol 108:147–177
236. Schleif R (2000) Regulation of the L-arabinose operon of *Escherichia coli*. Trend Genet 16 (12):559–565
237. Sedlak M, Ho NW (2001) Expression of *E. coli araBAD* operon encoding enzymes for metabolizing L-arabinose in *Saccharomyces cerevisiae*. Enzyme Microb Technol 28(1):16–24
238. Wang C, Shen Y, Zhang Y, Suo F, Hou J, Bao X (2013) Improvement of L-arabinose fermentation by modifying the metabolic pathway and transport in *Saccharomyces cerevisiae*. Biomed Res Int 2013:461204
239. Wright J, Bellissimi E, de Hulster E, Wagner A, Pronk JT, van Maris AJ (2011) Batch and continuous culture-based selection strategies for acetic acid tolerance in xylose-fermenting *Saccharomyces cerevisiae*. FEMS Yeast Res 11(3):299–306
240. Atsumi S, Liao JC (2008) Metabolic engineering for advanced biofuels production from *Escherichia coli*. Curr Opin Biotechnol 19(5):414–419
241. Ren C, Gu Y, Hu S, Wu Y, Wang P, Yang Y, Yang C, Yang S, Jiang W (2010) Identification and inactivation of pleiotropic regulator CcpA to eliminate glucose repression of xylose utilization in *Clostridium acetobutylicum*. Metab Eng 12(5):446–454
242. Xiao H, Gu Y, Ning Y, Yang Y, Mitchell WJ, Jiang W, Yang S (2011) Confirmation and elimination of xylose metabolism bottlenecks in glucose phosphoenolpyruvate-dependent phosphotransferase system-deficient *Clostridium acetobutylicum* for simultaneous utilization of glucose, xylose, and arabinose. Appl Env Microbiol 77(22):7886–7895
243. Jin L, Zhang H, Chen L, Yang C, Yang S, Jiang W, Gu Y (2014) Combined overexpression of genes involved in pentose phosphate pathway enables enhanced D-xylose utilization by *Clostridium acetobutylicum*. J Biotechnol 173:7–9
244. Baer SH, Blaschek HP, Smith TL (1987) Effect of butanol challenge and temperature on lipid composition and membrane fluidity of butanol-tolerant *Clostridium acetobutylicum*. Appl Env Microbiol 53(12):2854–2861
245. Jiang Y, Liu J, Jiang W, Yang Y, Yang S (2015) Current status and prospects of industrial bio-production of *n*-butanol in China. Biotechnol Adv 33(7):1493–1501
246. Hu S, Zheng H, Gu Y, Zhao J, Zhang W, Yang Y, Wang S, Zhao G, Yang S, Jiang W (2011) Comparative genomic and transcriptomic analysis revealed genetic characteristics related to solvent formation and xylose utilization in *Clostridium acetobutylicum* EA 2018. BMC Genom 12:93
247. Jiang Y, Xu C, Dong F, Yang Y, Jiang W, Yang S (2009) Disruption of the acetoacetate decarboxylase gene in solvent-producing *Clostridium acetobutylicum* increases the butanol ratio. Metab Eng 11(4–5):284–291
248. Jang YS, Malaviya A, Lee SY (2013) Acetone-butanol-ethanol production with high productivity using *Clostridium acetobutylicum* BKM19. Biotechnol Bioeng 110(6):1646–1653
249. Hong KK, Nielsen J (2012) Metabolic engineering of *Saccharomyces cerevisiae*: a key cell factory platform for future biorefineries. Cell Mol Life Sci 69(16):2671–2690

250. Shen CR, Lan EI, Dekishima Y, Baez A, Cho KM, Liao JC (2011) Driving forces enable high-titer anaerobic 1-butanol synthesis in *Escherichia coli.* Appl Environ Microbiol 77 (9):2905–2915
251. Schirmer A, Rude MA, Li X, Popova E, del Cardayre SB (2010) Microbial biosynthesis of alkanes. Science 329(5991):559–562
252. Steen EJ, Kang Y, Bokinsky G, Hu Z, Schirmer A, McClure A, Del Cardayre SB, Keasling JD (2010) Microbial production of fatty-acid-derived fuels and chemicals from plant biomass. Nature 463(7280):559–562
253. Bruschi F, Dundar M, Gahan PB, Gartland K, Szente M, Viola-Magni MP, Akbarova Y (2011) Biotechnology worldwide and the 'European Biotechnology Thematic Network' Association (EBTNA). Curr Opin Biotechnol 22(Suppl 1):S7–S14
254. Chen Y, Nielsen J (2013) Advances in metabolic pathway and strain engineering paving the way for sustainable production of chemical building blocks. Curr Opin Biotechnol 24 (6):965–972
255. Jantama K, Haupt MJ, Svoronos SA, Zhang X, Moore JC, Shanmugam KT, Ingram LO (2008) Combining metabolic engineering and metabolic evolution to develop nonrecombinant strains of *Escherichia coli* C that produce succinate and malate. Biotechnol Bioeng 99 (5):1140–1153
256. Henry CS, Broadbelt LJ, Hatzimanikatis V (2010) Discovery and analysis of novel metabolic pathways for the biosynthesis of industrial chemicals: 3-hydroxypropanoate. Biotechnol Bioeng 106(3):462–473
257. Rathnasingh C, Raj SM, Jo JE, Park S (2009) Development and evaluation of efficient recombinant *Escherichia coli* strains for the production of 3-hydroxypropionic acid from glycerol. Biotechnol Bioeng 104(4):729–739
258. Borodina I, Kildegaard KR, Jensen NB, Blicher TH, Maury J, Sherstyk S, Schneider K, Lamosa P, Herrgard MJ, Rosenstand I, Oberg F, Forster J, Nielsen J (2015) Establishing a synthetic pathway for high-level production of 3-hydroxypropionic acid in *Saccharomyces cerevisiae* via beta-alanine. Metab Eng 27:57–64
259. Chen Y, Bao J, Kim IK, Siewers V, Nielsen J (2014) Coupled incremental precursor and co-factor supply improves 3-hydroxypropionic acid production in *Saccharomyces cerevisiae.* Metab Eng 22:104–109
260. Maury J, Asadollahi MA, Moller K, Clark A, Nielsen J (2005) Microbial isoprenoid production: an example of green chemistry through metabolic engineering. Adv Biochem Eng Biotechnol 100:19–51
261. Withers ST, Keasling JD (2007) Biosynthesis and engineering of isoprenoid small molecules. Appl Microbiol Biotechnol 73(5):980–990
262. Paddon CJ, Westfall PJ, Pitera DJ, Benjamin K, Fisher K, McPhee D, Leavell MD, Tai A, Main A, Eng D, Polichuk DR, Teoh KH, Reed DW, Treynor T, Lenihan J, Fleck M, Bajad S, Dang G, Dengrove D, Diola D, Dorin G, Ellens KW, Fickes S, Galazzo J, Gaucher SP, Geistlinger T, Henry R, Hepp M, Horning T, Iqbal T, Jiang H, Kizer L, Lieu B, Melis D, Moss N, Regentin R, Secrest S, Tsuruta H, Vazquez R, Westblade LF, Xu L, Yu M, Zhang Y, Zhao L, Lievense J, Covello PS, Keasling JD, Reiling KK, Renninger NS, Newman JD (2013) High-level semi-synthetic production of the potent antimalarial artemisinin. Nature 496(7446):528–532
263. Westfall PJ, Pitera DJ, Lenihan JR, Eng D, Woolard FX, Regentin R, Horning T, Tsuruta H, Melis DJ, Owens A, Fickes S, Diola D, Benjamin KR, Keasling JD, Leavell MD, McPhee DJ, Renninger NS, Newman JD, Paddon CJ (2012) Production of amorphadiene in yeast, and its conversion to dihydroartemisinic acid, precursor to the antimalarial agent artemisinin. Proc Natl Acad Sci USA 109(3):E111–E118
264. Richter H, Qureshi N, Heger S, Dien B, Cotta MA, Angenent LT (2012) Prolonged conversion of *n*-butyrate to *n*-butanol with *Clostridium saccharoperbutylacetonicum* in a two-stage continuous culture with *in-situ* product removal. Biotechnol Bioeng 109(4):913–921

265. Atsumi S, Wu TY, Eckl EM, Hawkins SD, Buelter T, Liao JC (2010) Engineering the isobutanol biosynthetic pathway in *Escherichia coli* by comparison of three aldehyde reductase/alcohol dehydrogenase genes. Appl Microbiol Biotechnol 85(3):651–657
266. Connor MR, Liao JC (2008) Engineering of an *Escherichia coli* strain for the production of 3-methyl-1-butanol. Appl Environ Microbiol 74(18):5769–5775
267. Atsumi S, Liao JC (2008) Directed evolution of *Methanococcus jannaschii* citramalate synthase for biosynthesis of 1-propanol and 1-butanol by *Escherichia coli*. Appl Environ Microbiol 74(24):7802–7808
268. Zhang K, Sawaya MR, Eisenberg DS, Liao JC (2008) Expanding metabolism for biosynthesis of nonnatural alcohols. Proc Natl Acad Sci USA 105(52):20653–20658
269. Zhang F, Carothers JM, Keasling JD (2012) Design of a dynamic sensor-regulator system for production of chemicals and fuels derived from fatty acids. Nat Biotechnol 30(4):354–359
270. Liu A, Tan X, Yao L, Lu X (2013) Fatty alcohol production in engineered *E. coli* expressing *Marinobacter* fatty acyl-CoA reductases. Appl Microbiol Biotechnol 97(15):7061–7071
271. Choi YJ, Lee SY (2013) Microbial production of short-chain alkanes. Nature 502 (7472):571–574
272. Peralta-Yahya PP, Ouellet M, Chan R, Mukhopadhyay A, Keasling JD, Lee TS (2011) Identification and microbial production of a terpene-based advanced biofuel. Nat Comm 2:483
273. Hara KY, Araki M, Okai N, Wakai S, Hasunuma T, Kondo A (2014) Development of bio-based fine chemical production through synthetic bioengineering. Microb Cell Fact 13:173

Chapter 8
Biofuels Production from Renewable Feedstocks

Jerald A. Lalman, Wudneh A. Shewa, Joe Gallagher and Sreenivas Ravella

Abstract Predicted increases in greenhouse gas emissions, depleting fossil fuel supplies, global conflicts, and energy security are major factors driving the search for renewable energy supplies. Based on future energy demand projections, biofuels production is expected to increase. However, this increase represents a small fraction of this growing demand because the land area required to grow sufficient biofuels crops is unavailable. Hence, fulfilling the growing energy demand after attaining peak fossil fuel production will include using a combination of energy sources such as renewables, wind, geothermal, nuclear, hydroelectric, solar, and coal. Current and potential feedstocks include grains, grasses, root crops, oil seeds, algae, and lignocellulosics. Grains, sugar crops, and lignocellulosics are the main feedstocks used in full-scale first- and second-generation ethanol processes. While first-generation biodiesel is produced mainly from corn, soybeans, canola oil, rapeseed, palm oil, Jatropha, and coconut oil, second-generation fuels are produced from lignocellulosics. Third-generation technology employs several processes to produce a variety of biofuels from algae while fourth-generation technologies, a developing concept, is intended to employ genetically modified terrestrial or aquatic plants. In another concept, fourth-generation technologies can be configured with CO_2 sequestration and storage. First-generation biobutanol is produced from corn or molasses and from sugar beet as well as sugarcane, while second-generation production processes utilize lignocellulosics such as corn stover, rice straw, corn fiber, switchgrass, alfalfa, reed canary grass, sugarcane bagasse, *Miscanthus*, waste paper, dry distillers grain with solubles (DDGS), and soy molasses. A variety of technologies, based on the enzyme systems, are currently under investigation for producing biohydrogen. Biohydrogen production routes are divided into biophotolysis (direct/indirect), dark fermentation, and photofermentation. Increasing

J.A. Lalman (✉) · W.A. Shewa
Department of Civil and Environmental Engineering, University of Windsor, Windsor, Ontario N9B 3P4, Canada
e-mail: lalman@uwindsor.ca

J. Gallagher · S. Ravella
Institute of Biological, Environmental and Rural Sciences, Aberystwyth University, Aberystwyth, Ceredigion SY23 3EE, UK

P.C.K. Lau (ed.), *Quality Living Through Chemurgy and Green Chemistry*,
Green Chemistry and Sustainable Technology,
DOI 10.1007/978-3-662-53704-6_8

global demand is expected to drive increasing bioethanol and biobutanol production using food and nonfood feedstocks. At the same time, researchers are developing technologies to produce biohydrogen and biodiesel. Biohydrogen and biodiesel production technologies are in their developmental stages; however, with innovation, these technologies are expected to mature into economical processes.

Keywords Global energy supply and demand · Annual biofuel feedstock production · Biofuels production processes · first-, second-, third-, and fourth-generation processes

8.1 Introduction

Predicted increases in the global population to approximately 8.3 billion by 2030 is linked to issues such as climate change, depleting fossil fuel supplies, energy security, and affluence. Population growth and increasing wealth per capita are key drivers connected to growing energy demand. Developing renewable energy supplies offers a mechanism to reduce carbon emissions and combat issues associated with population growth. A primary driver for developing energy crops is the desire to decrease the quantity of greenhouse gases (GHG) associated with utilizing fossil fuels. Bioenergy production from renewable feedstocks could be a major mechanism in reducing carbon dioxide emissions.

Despite increasing energy efficiencies, consumption is increasing globally (Fig. 8.1). In 2012, the global biofuels industry was valued at approximately $95 billion U.S., a 14.4 % increase from 2010. The industry is expected to grow to $145.6 billion U.S. by 2023 [1]. Global primary energy consumption is projected to grow by 1.5 % per year from 2012 to 2035. This increase is expected to add 41 % to the global energy consumption by 2035. Renewables (including biofuels) are the fastest-growing fuels with growth averaging 6.4 % per year from 2012 to 2035. Nuclear (2.6 % per year) and hydro (2.0 % per year) are both expected to grow more than the total energy consumption [2]. As fossil fuel demand peaks around 2030, biofuels, hydroelectric, solar, wind, geothermal, and nuclear are expected to fill the gap created by depleting fossil fuel supplies (Fig. 8.1).

The deployment of bioenergy technologies, if carefully and strategically managed in a sustainable manner, could produce the following advantages [3]:

- provide a large contribution to the global energy supply;
- contribute to reducing GHG emissions;
- improve energy security and trade balances, by substituting imported fossil fuels with domestic biomass;
- provide opportunities for economic and social development in rural communities; and,
- allow for the utilization of wastes and residues which subsequently result in reducing waste disposal problems and allow for efficient utilization of global resources.

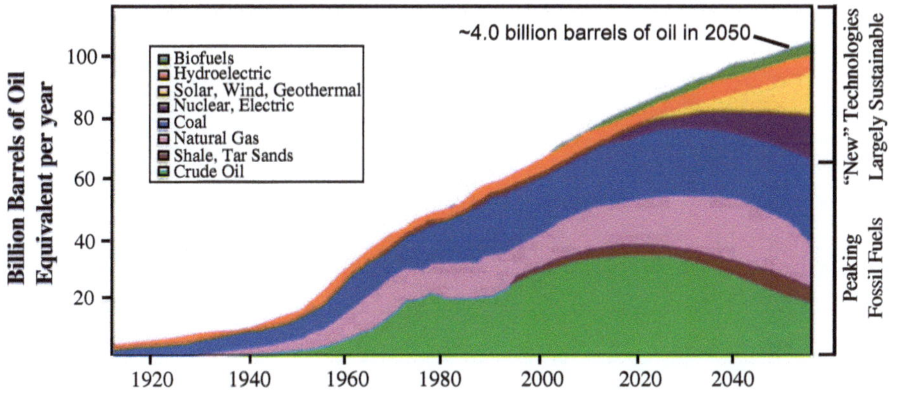

Fig. 8.1 World energy demand—long-term energy reserves. Data from Edwards [4] and Orr [5]

8.2 Biomass Feedstocks

Biomass generally refers to organic materials derived from living and nonliving matter. After coal, oil, and natural gas, biomass is the fourth largest energy source. Cultivation of terrestrial renewable energy crops across the globe is shown in Fig. 8.2. The global map accounts for crop cultivated on lands suitable and poorly suitable for first, second, and third-generation feedstocks.

First-generation biofuels are classified based on using commercially available technologies and conventional feedstocks such as oil seeds and grains [6]. Corn, sugarcane, wheat, and other grains plus rapeseed and palm oil are categorized as first-generation feedstocks because they are readily fermentable. Second-generation biofuels production processes utilize lignocellulosic materials feedstocks such as wheat stalks, corn stover, wood, and energy crops such as *Miscanthus* and switchgrass. Second-generation feedstocks (often referred to as lignocellulosics) require pretreatment including the use of complex enzyme mixtures prior to fermentation [6]. Second-generation biofuel technologies are used to produce hydrogen, ethanol, dimethylfuran (DMF), dimethyl ether (DME), Fischer–Tropsch (FT) diesel, and mixed alcohols. These technologies, including both biological and thermochemical, are not cost competitive with first-generation biofuels production. Second-generation technologies with substantial energy/environment benefits when compared to most first-generation biofuels is due primarily to greater biomass usability per unit land area. However, these technologies are characterized by greater capital intensity and lower feedstock costs when compared to first-generation technologies. Third-generation biofuels are a new category for classifying biofuels. Third-generation feedstocks include materials such as

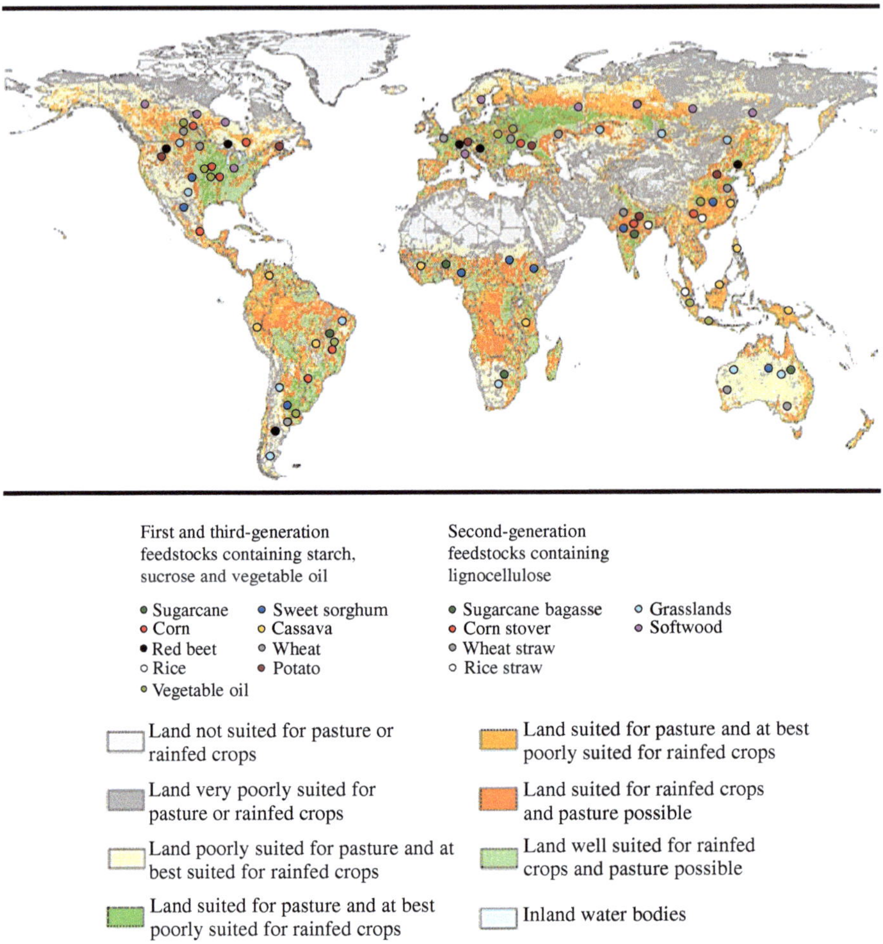

Fig. 8.2 Land suitability for pasture and rainfed crops. *Notes* Land suitability data map adapted from van Velthuizen et al. [7]. Crop cultivation data adapted from Stöcker and Tschentscher [8]. Vegetable oil production data accessed from http://www.indexmundi.com/agriculture/?commodity=soybean-oil&graph=production [9]. General cultivation locations are shown. The circle size is not representative of quantity produced

microalgae do not compete for land area [6]. Algae, the most significant feedstock for this technology, is cultivated to produce biodiesel from triglycerides, a metabolic product. Fourth-generation fuels have been classified based on synthetic biology of algae and cyanobacteria [10–12]. Synthetic biology includes the design and manufacture of biological components, devices and systems, and the reengineering of existing, natural biological systems for producing biofuels [13]. In an opposing concept, fourth-generation biofuel systems are proposed to be comprised of processing methods such as thermochemical coupled to carbon capture and storage technologies which divert the CO_2 generated into geological formations or mineral storage as carbonates [13]. In the opposing concept, the technology ability to sequester and store CO_2 leads to the carbon negative concept.

Crop cultivation in various areas is dependent on factors such as irradiation time and intensity, temperatures, rainfall, season length, and soil quality. Temperate regions are associated with the cultivation of softwood while in arid regions, sweet sorghum, and switchgrass are preferred crops due to the low water demand. The annual global energy consumption is expected to increase with population and economic growth. In 2013, approximately 4185 million tonnes of fuel was consumed globally. Based on this consumption and assuming varying biomass yields of 5–50 tonne biomass (ha year)$^{-1}$ [14] and conversion efficiencies ranging from 25 to 75 % tonne biomass per tonne fuel, the land area requirement is expected to range from approximately 110-3350 Mha (Table 8.1). Using biomass to produce biofuels will partially meet this demand; however, the exorbitant area required to grow sufficient terrestrial as well as aquatic plants is unavailable (Table 8.1). Hence, fulfilling this global growing energy demand after attaining peak fossil fuels production will include utilizing a combination of other energy sources such as renewables, wind, geothermal, nuclear, hydroelectric, solar, and coal (Fig. 8.1).

Biomass feedstocks are the largest renewable energy option available for producing energy and chemicals [15]. Major biomass feedstocks categories are shown in Table 8.2 [16]. Climate change is attributed as the leading cause for varying biomass yields. Depending on the models and input climate scenario as well as assumptions, many global areas may experience significant decreases as well as significant increases in crop yields [17]. The global primary biomass potential forecast by numerous studies estimate annually over a 100-year timeframe is

Table 8.1 Estimated biomass conversion and land area requirement [14]

Yield (tonne biomass (ha year)$^{-1}$)	Proportion of biomass converted to biofuel (%)	Land area (Mha)
5	25	3348
	50	1674
	75	1116
10	25	1674
	50	837
	75	558
25	25	837
	50	419
	75	279
50	25	335
	50	167
	75	112

1 Mha = million hectare
2 Based upon 4185 million tonnes / year fuel consumption as of 2013 [14]

Table 8.2 Major biomass feedstock categories [16]

Biomass category	Biomass feedstock
Forest products	Wood, logging residues, trees, shrubs, and wood residues, sawdust, bark, etc.
Biorenewable wastes	Agricultural wastes, mill wood wastes, urban wood wastes, urban organic wastes
Energy crops	Short-rotation woody crops, herbaceous woody crops, grasses, forage crops
Food crops	Residue from grains and oil crops
Sugar crops	Sugarcane, sugar beets, sorghum
Landfill	Municipal solid wastes
Industrial wastes	Food wastes, organic acid wastes, and vegetable oil wastes
Algae, kelps, lichens, and mosses	Water hyacinth, mushrooms, etc.
Aquatic plants	Algae, water weed, water hyacinth, reed, and rushes

expected to range from 50 to 1550 EJ [18] (equivalent to 8.2–253 billion barrels of oil equivalent (Bboe) per annum). Other studies have reported an annual global production potential of 30 EJ from forest and agricultural residues [19]. According to Demirbas [20], the global biomass production is estimated at 146 giga tonnes (Gt) per year. Klass [21] and Hall et al. [22] reported annual biomass yields range from 170 to 200 Gt (equivalent to 500–599 Bboe assuming an energy yield of 18 GJ per tonne biomass).

The annual global production of selected crops is shown in Table 8.3. Corn, sugarcane, and oilseeds are the major contributors for producing biofuels with corn and sugarcane accounting for approximately 76 % of the total bioethanol produced [23].

Table 8.3 Annual production of selected crops

Crop	Annual production (Million tonnes year^{-1})
Wheat	725[a]
Rice, milled	480[a]
Corn	1000[a]
Barley	145[a]
Oats	23[a]
Rye	13[a]
Cassava	250[b]
Sorghum	70[a]
Sugar beet	160[a]
Sugarcane	1880[c]
Soybeans	280[b]
Oilseed	530[a], 425[d]

[a]USDA [24]; [b]FAO [25]; [c]Koo and Taylor [26]; [d]OECD/FAO Agriculture Outlook [27]

The global total crop biomass annual production is estimated at approximately 8950 million tonnes, of which 5570 million tonnes (62 %) is categorized as crops and 3380 million tonnes (38 %) as above ground residues. Crops are produced for food (69 %), feed (19 %), and energy and materials (7 %). A small fraction is used for seed while another portion is waste (5 %). Of the estimated residues, 10 % is used for feed (corn and straw) and 6 % for bedding (in developed countries) [28]. Most of the residues remaining in the field are burned or grazed by livestock.

Approximately 1000 million tonnes of corn is produced annually (Table 8.3 and Fig. 8.3) with the U.S. and China accounting for approximately 60 % in 2014 [24]. Corn is utilized for feed (55 %), food (20 %), and also for producing biofuels while corn stover is used for livestock feed. Oil is extracted from corn kernels for consumption, cooking, and nonfood products. The average quantity of oil extracted is approximately 15 % of the total crop. The leaves and straw are recycled into the soil. Sugar beet is produced for food and biofuels. Annual wheat production as a major food source is estimated at 725 million tonnes. Cassava is produced (250 million tonnes year^{-1}) in tropical climates and mainly used for food. Global beet production, mainly in the European Union, is approximately 160 million tonnes year^{-1}. Brazil and India account for approximately 60 % of the global sugarcane production of 1880 million tonnes year^{-1} (Table 8.3). Sugarcane is cultivated for sugar (sucrose) and ethanol production. Brazil is the largest producer, followed by India and China. Sugarcane is a diversified crop used for producing food (sucrose, alcohol, molasses), energy, and biofuels (ethanol) [29]. Approximately 50 % of the

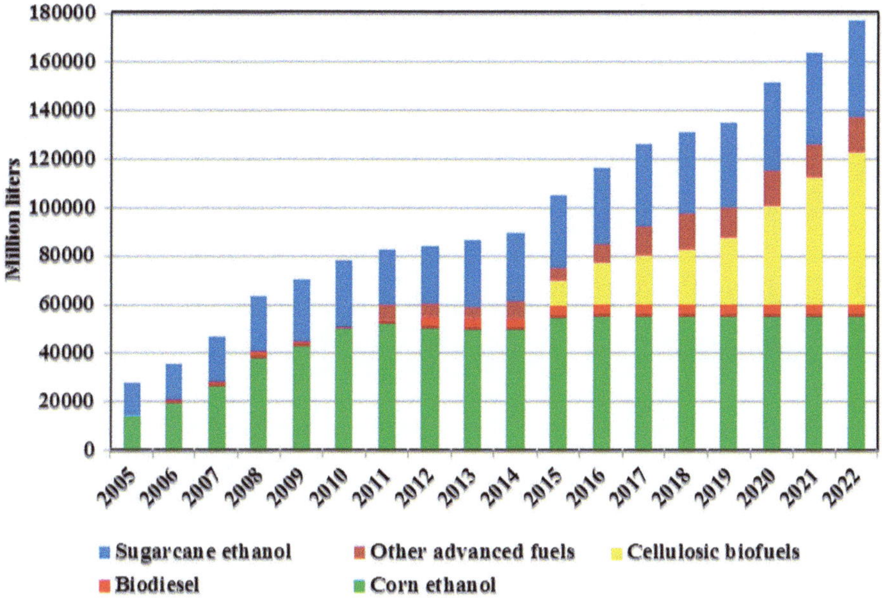

Fig. 8.3 Global biofuels production. *Source*: OECD/FAO [27]. United States Environmental Protection Agency [30]. United States Energy Information Administration [31]

Table 8.4 Chemical composition of selected agricultural grains, residues, and brown kelp

Feedstock	Cellulose	Hemicellulose	Mannitol	Algin	Starch	Fat	Ash	Protein	Lignin
Corn[a]	2.0	7.6			76.0	5.7	1.6	11.4	1.0
Soyabean[a]	2.0	5.0				18.8	5.5	42.8	
Wheat[a]	8.0	4.0			70.0	2.2	1.6	12.3	2.0
Sorghum[b]					73.0	4.6	1.2	10.9	
Switchgrass[a]	33.5	26.5					6.4	5.3	18.1
Brown rice[a]	1.0	2.0			74.4	2.6	1.6	8.5	
Giant brown kelp[c]	4.8		18.7	14.2			45.8	15.9	
Bermuda grass[c]	31.7	40.2					5.0	12.3	4.1
Poplar[c]	41.3	32.9					1.0	2.1	25.6

All data given in %w w^{-1}
[a]Abbas et al. [32]; [b]Léder [33]; [c]Klass [34]

bagasse produced is used for energy production and the rest remains unused in the environment [35]. Vegetable oils are produced (910 million tonnes $year^{-1}$) mainly for food products.

Biomass residues from fruits and vegetables, roots and tubers, and sugar beets are limited in quantity but have potential for bioenergy production when available in sufficient quantities. Their relatively high economic value makes these crops less favorable for uses other than food. Sugar beet is comparable to sugarcane, and the high sugar content makes the crop suitable for food, fuel, and energy.

Biomass sources include crops such as beet and sugarcane, woody and herbaceous species, wood wastes, agricultural residues, waste paper, municipal solid waste, biosolids, food processing waste, animal wastes, aquatic plants, bacteria, and algae. Cellulose, hemicelluloses, lignin, proteins, and fats are the major organic constituents of biomass. In the case of lignocellulosics biomass, the main constituents are cellulose (35–50 %), hemicellulose (20–35 %), and lignin (10–25 %) (Table 8.4).

8.3 Biofuels Production

8.3.1 Overview

On a global scale, biofuel production from renewable biomass is projected to continue increasing over the next decade [36]. Corn, sugarcane, beet and lignocellulosics are major feedstocks for producing bioethanol while vegetable oil plus oilseeds are utilized to produce biodiesel. The U.S., Brazil, the EU, and Argentina are the largest biofuel producers while the world's largest biofuel exporters include Argentina, Brazil, and the U.S., with Argentina, Brazil, and the U.S. specialize in producing soybean oil-based biodiesel, sugarcane-based ethanol, and corn-based ethanol, respectively [36].

Currently, the EU remains the world's largest biofuels importer over a projected period until 2024. Biodiesel accounts for the majority of the EU's biofuel imports with Brazil supplying a large fraction of the EU's ethanol imports. The EU is also projected to import, oilseeds and vegetable oils for biodiesel feedstocks mainly from Ukraine, Russia, and Indonesia. Global trade for wheat, coarse grains plus soybeans, and soybean products is expected to reach 175, 175, and 235 million tonnes, respectively, by 2024 [36].

Wheat, coarse grains, soybeans, sugarcane, and beet feedstocks are employed to produce a variety of chemical and fuels utilizing biological, physical, thermal, and thermochemical processes. Currently, the major biofuels produced include methane, ethanol, and biodiesel (Fig. 8.3). Other chemicals produced from biomass include benzene/toluene/xylene (BTX), furans, organic acids, diols, alkenes, alkanes, and polyhydroxyalkanoates (PHAs) (Fig. 8.4). Producing biofuels and chemicals from renewable crops and residues is important from an energy security and environmental perspectives. However, approximately 43 % of the total crop produced annually is used for food and only 5 % is converted into energy producing chemicals [28]. The small fraction of the total global annual crop produced which is converted into energy is due to limited factors such as the total annual production, crop yield, climatic conditions, and the quantity used to produce food products.

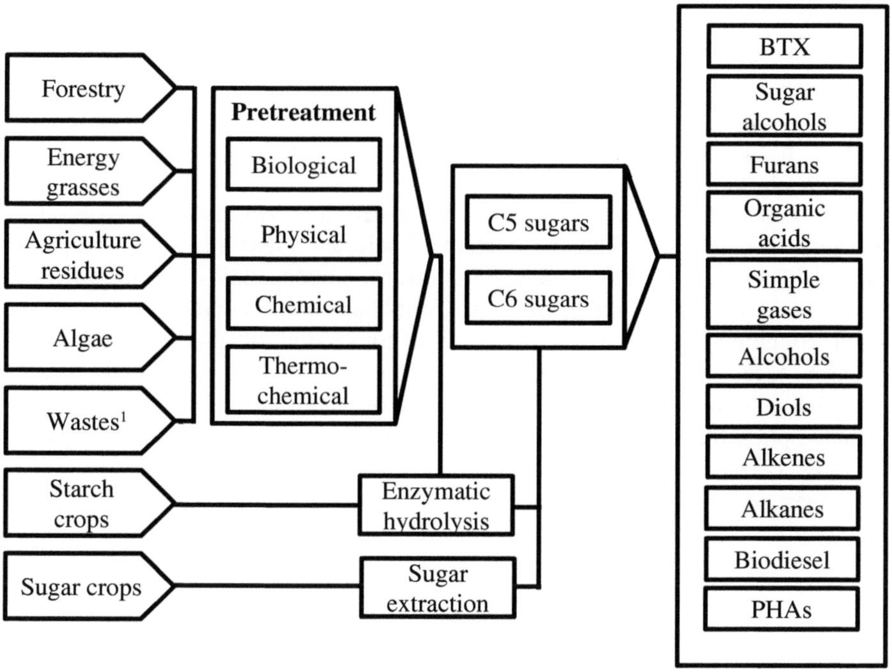

Fig. 8.4 Biofuels production pathways (adapted from European Commission [37]). [1]Municipal and industrial

8.3.2 Biofuels Production Technologies

Biomethane, bioethanol, and biodiesel are produced in large quantities from a variety of feedstocks. Production processes for these fuels are described in subsequent sections.

8.3.2.1 First- and Second-Generation Biomethane Production

Biomethane, a useful energy carrier, is derived from renewable feedstocks. Biomethane under the first-generation category is produced by anaerobic digestion while the second-generation production process is configured with biomass gasification, purification, and catalytic methanation. According to the European biogas association (EBA), the 2010 total biogas production in Europe was estimated at 14.7 billion m^3 [93.8 million boe (barrel of oil equivalent) or 20.6 billion m^3 biogas] based on natural gas equivalents. The EBA projects the level of biogas production will reach 28 billion m^3 (based on natural gas equivalents) in 2020 [38]. Contradictory data by EurObserv'ER [39] reported the quantity of biogas produced in 2013 reached approximately 25.7 billion m^3. Major biomethane producing countries in Europe include Germany, Sweden, the Netherlands, Austria, Finland and the UK [39]. Germany is Europe's largest biogas producer [40] with a total capacity from biomass of 12.8 billion kWh [85.8 million boe (barrel of oil equivalent)] in 2010 [41]. In the U.S., methane biogas from cow manure is sufficient to produce 100 billion kilowatt hours (equivalent to 16.7 billion m^3 biogas or 61.4 Mboe or 11.3 billion m^3 natural gas equivalent). Methane biogas can reduce 99 ± 59 million tonnes of GHG emissions or approximately 3.9 ± 2.3 % of the GHGs produced by the U.S. [42].

A vast selection of crops has been evaluated for their potential to produce methane. The methane yield per hectare of selected sugar and starch crops are shown in Table 8.5. Methane can be produced from lignocellulosic crops; however, the methane yield from these crops is lower than that obtained from sugar and starch containing crops. Even though the conversion of lignocellulosics into methane faces an initial barrier related to the enzyme access to readily biodegradable components, the process is more environmentally sound and sustainable option for renewable energy production because lignocellulosic crops can be cultivated on marginal and set-aside lands [43].

Converting agriculture feedstock into methane is accomplished sequentially by various microbial populations during anaerobic digestion. These organisms which include hydrolytic degraders, acidogens, acetogens, and methanogens, produce monomers, organic acids, acetic acid, and eventually methane, respectively. Sugars are easily degraded to methane by anaerobic microorganisms; however, the process reaction rate is impaired significantly when the feed contains substrates such as lignocellulosics and lipids.

Table 8.5 Biomethane yield in function of methane production and crop field yield [43]

Crops	Yield (m^3 STP CH_4 ha^{-1})
Cocksfoot	2390
Corn	5300–12,390
Festlolium	2800
Giant knotweed	3800
Hemp	2840
Jerusalem artichoke	3100–5400
Reed canary grass	3800–4200
Rhubarb	800–1700
Sugarbeets	5400
Sunflower	4695
Tall fescue	2749
Timothy	1840–2335
Timothy clover grass	2900–4000
Triticale	1112–6600

Biomethane production from a H_2 plus CO synthesis is classified as a second-generation biofuel. The production process involves four stages. In the first gasification stage, lignocellulosics are converted into a synthesis gas which is rich in CO and H_2. The synthesis gas is purified to eliminate tars and inorganic compounds in the second stage. Next, the synthesis gas is converted into biomethane using a heterogeneous catalytic conversion process. Finally, the biomethane stream is conditioned using gas separation processes in order to be compliant with standard specifications for natural gas [44].

8.3.2.2 First- and Second-Generation Bioethanol Production

First-Generation Ethanol Production from Starch- and Sucrose-Based Crops

Bioethanol production in 2015 from sugarcane and corn was estimated at approximately 84 billion liters (Fig. 8.3). By 2022, the global ethanol production is projected to reach approximately 157.2 billion liters (Fig. 8.1). Based on the 2022 data, corn, sugarcane, and lignocellulosics account for 35, 25, and 40 %, respectively, of the total first-generation biofuels produced. The 2022 ethanol production (157.2 billion liters) accounts for approximately less than 1 % of the total annual global energy utilized (Fig. 8.1). The 2022 data excludes production data for wheat, beet, and cassava because of the low annual ethanol production levels for these three feedstocks. Sugar beet bioethanol production accounted for approximately 0.83 billion liters in 2015 [45–47]. The annual low levels of bioethanol produced from wheat and cassava is due to the small fraction available for processing. In the

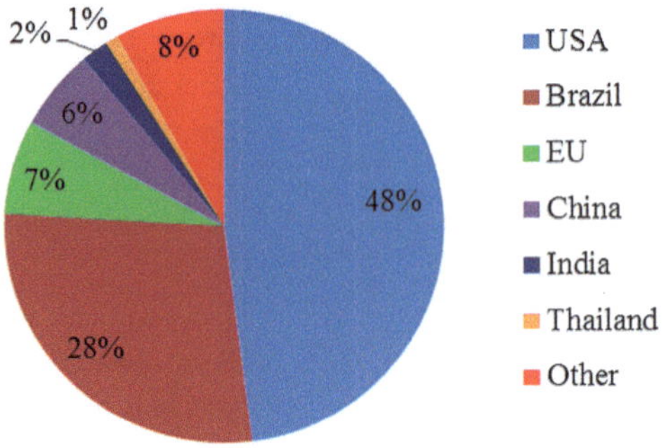

Fig. 8.5 Percent ethanol production based on countries [23]

case of wheat, only 1.0 % of global wheat production is available while for cassava, only 0.6 % is available for producing ethanol [46].

Approximately 76 % of the ethanol produced globally is derived from sugarcane and corn with the remaining produced from beet, wheat, and other crops (Fig. 8.3). The U.S. and Brazil are major ethanol producers with a market share of approximately 75 % in 2012 (Fig. 8.5). Ethanol yields for selected crops are shown in Table 8.6. Sugarcane and coarse grains are the largest yielding ethanol producing crops with lower yields derived from oats and cassava. Ethanol production based on crop type is shown in Fig. 8.6. Sugarcane to ethanol production facilities are located in Brazil while corn to ethanol are situated in North America and Asia. Beet to ethanol production facilities are stationed in Europe.

Hydrolysis, the first reaction step in producing ethanol from cereal grains, involves converting starch to glucose monomers using hydrolyzing enzymes. In the second step, glucose and other sugar monomers are converted into ethanol using

Table 8.6 First-generation ethanol yields for sucrose and starch with grains, terrestrial, and root crops

Feedstock	Moisture (% wt)	Starch (% wt)	Sugar (% wt)	Crop yield (tonnes ha^{-1})	Ethanol yield (L dry $tonne^{-1}$)
Sucrose					480[g] actual
Starch		100[a]			720[a]
Barley	9.7[a]	67.1[a]			399[a]
Beet	75[b]		17–18[b,f]		375[c]
Cassava	32[d]	35[d]			150[d]
Corn	13.8[a]	71.8[a]			407[a,c]
Oats	10.9[a]	44.7[a]			262[a]
Sweet Sorghum	70[e]		13[e]		400[g,] 520[g]
Sugarcane	82[f]		15–16[f]	74[g]	460–654[a,c]
Wheat	10.9[a]	63.8[a]			375[a]

[a]Saskatchewan Agriculture and Food [47]; [b]Asadim [48]; [c]Shapouri et al. [49]; [d]Kuiper et al. [50]; [e]Höfer [51]; [f]Cardona et al. [52]; [g]Bonin et al. [53]

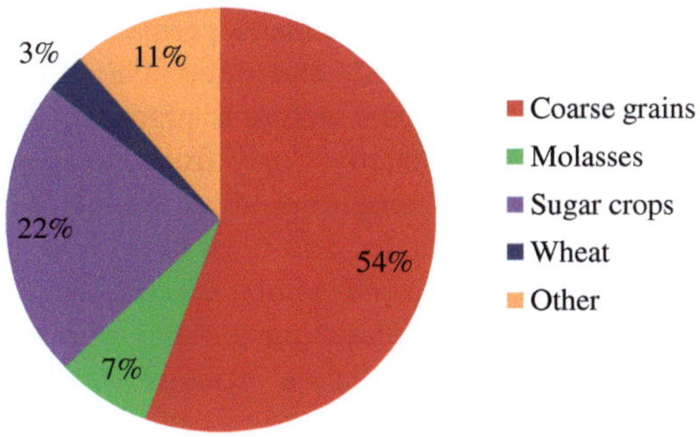

Fig. 8.6 Percent ethanol production based on crop type [23]

yeast. The fraction of biomass converted into fuels and chemicals depends on the availability of easily degradable sugars (Table 8.6). Starch, grain crops, root crops, and grasses are easily fermented into ethanol with less pretreatment when compared to ethanol production from lignocellulosics.

Ethanol production from grain crop such as corn and other starch rich grains is classified as wet or dry milling. Major differences between the two processes are the byproducts produced and the initial capital cost. In the wet milling process, corn is converted into fuel ethanol and dry distillers grain with solubles (DDGS). DDGS contains primarily protein and is sold as animal feed. The unit processes are configured in the following sequence: [54] 1. Grain grinding into a meal using hammer or roller mills; 2. Liquefying the meal using α-amylase at a 90 °C reaction temperature; 3. The liquefied mash is cooled and pH adjusted to 4–5. Glucoamylase is added to convert dextrins and oligosaccharides into glucose and maltose. Sugar monomers are fermented at approximately 30 °C for 40–60 h to produce ethanol and CO_2; and 4. The fermented mash is distilled to separate the ethanol from solids and water. After distillation, the residues are centrifuged into solid (grain residue) and liquid (thin stillage) fractions. The liquid fraction is further concentrated by evaporation to produce wet distillers grain (WDG). The WDG is sold directly as animal feed or concentrated to produce DDGS [16]. Almost all the industrially produced ethanol fermentation processes utilizes *Saccharomyces cerevisiae*. Other microorganisms include *S. uvarum*, *Schizosaccharomyces pombe*, and *Kluyveromyces* [55].

The wet milling process configuration is more complicated than the dry milling process. The wet milling of grains such as corn and wheat are different because of the gluten protein content. In general, the wet milling process consists of steeping, degerming, defibring, starch/gluten separation, liquefaction, saccharification and fermentation, distillation and dehydration [56]. Steeping involves a combination of biochemical, chemical, and mechanical processes. Maize kernels are conditioned to separate the germ, fiber, and gluten from starch. Separation is achieved by conditioning the kernels at 50–52 °C with 0.12–0.2 % sulfurous acid for 24–28 h [57]. Steeping water is concentrated to a liquor containing 35–45 % proteins and sold as fermentation nutrients or used to produce a gluten feed. During degerming, the steeped maize kernels are broken apart to produce a mixture containing maize germ plus a starchy slurry. The maize germs, which are separated from the starch slurry, are washed, dewatered, and dried. Oil is extracted from the maize germs and subsequently, the spent germs is processed into germ meal or used to produce a gluten feed. The starchy slurry from the degerming process undergoes washing, grinding, and screening to produce a mixture containing fibers, starch plus gluten. The fibers are separated and dried to produce a gluten feed. The stream containing the remaining starch plus gluten is processed to produce a gluten-rich stream which is concentrated to a gluten meal. After removing the gluten, the starch-rich process stream is used to produce a series of starch products which are used to produce ethanol [16].

Ethanol yields from corn fermentation are estimated at approximately 400–425 L $tonne^{-1}$ for first-generation production technologies [48, 58, 59] while for lignocellusoics, the yield is approximately 300 L $tonne^{-1}$ of sulfite pretreated Douglas-fir forest residue (Table 8.6) [60]. The ethanol yield is not only feedstock dependent but it is also affected by the process design configuration. For instance, the yields for dry and wet corn milling processes are approximately 460 and 440 L $tonne^{-1}$, respectively [61].

The sugarcane to ethanol industry in Brazil evolved out of the 1970s global energy crisis. Sugarcane cultivation in Brazil's economy generates 2.5 % of its gross domestic product with the ethanol industry contributing a further 1.1 % [62] or $60 billion U.S. [63]. The sugarcane to ethanol first-generation process is configured with a series of unit operations which involve physical and biochemical processes [64]. In the first step, crushing and pressing releases the juice while heating and clarification unit processes separate solids from the liquid fraction. Evaporation, crystallization, centrifugation are used to concentrate and purify the sugar stream. After producing ethanol by yeast fermentation, the liquid stream is distilled and dried. The bagasse byproduct is used as an energy source. Alternatively, after pretreating bagasse, glucose is recovered and subsequently used to produce ethanol. The solids fraction from pretreating bagasse is utilized to produce power and heat.

Beet sugar production facilities are configured with unit processes to produce sugar which is subsequently converted to ethanol. Screening and washing are employed to remove rocks and other unwanted fractions such as leaves and small roots. The beet roots are chopped into 3 mm size fractions and processed to extract the sugar using hot water fed to rotary drum screens, followed by dewatering. The remaining solids are sold as animal feed. Liquid from the hot water treatment and dewatering processes are combined for further processing. The liquid sugar stream is clarified before fermentation to ethanol. Typical ethanol yield for beet is approximately 375 L dry $tonne^{-1}$ (Table 8.6).

Second-Generation Ethanol Production from Lignocellulosics

Currently, corn stover, cereal straws, and sugarcane bagasse are the primary feedstocks for producing second-generation lignocellulosic ethanol (Table 8.7).

Table 8.7 Lignocellulosic biomass composition and ethanol yield

Crop	Cellulose plus hemicellulose content	Lignin	Ash	Ethanol yield (L dry tonne^{-1})
	dry wt%			
Bagasse	59–76[1], 70.8[2]	19–24[1,2]	4.5–9.0[1,2]	300[3]
Corn stover	63[4]	17.5[4]	6.6[4]	300[3],220–255[5,10], 238[6]
Wheat straw	59[4]	25[4]	6[4]	340[7]
Miscanthus	69[8]	13[8]	3[8]	165–295[9]
Switchgrass	59[8]	18[8]	5[8]	225[10]

[1]Soccol et al. [65]; [2]Canikha e al [66]; [3]Somerville et al. [67]; [4]Abbas et al. [32]; [5]Tumbala et al. [68]; [6]Aden et al. [69]; [7]Talebina [70]; [8]Abramson et al. [71]; [9]Lee and Kuan [72]; [10]Ewanick and Bura [73]

However, energy crops, such as *Miscanthus*, a relative of sugarcane which can grow in cooler climates and switchgrass are considered as potential future lignocellulosic feedstocks. *Miscanthus* yields in various European countries typically range from 4 to 44 tonne ha^{-1} while in the U.S., the yield (tonne ha^{-1}) is typically 11–14.5 [74–76]. Typical switchgrass yields in the U.S. range from 5 to 11 tonne ha^{-1} [77].

The process design configuration for second-generation lignocellulosic ethanol production generally includes pretreatment followed by fermentation [78]. The initial pretreatment step involves deconstruction of the biomass using physical, physicochemical, chemical, or biological treatments to render the complex carbohydrates accessible to hydrolysing enzymes. These complex polysaccharides (cellulose and hemicellulose) are then treated with enzyme cocktails which include cellulases, xylanases, ferulic acid esterases, etc., to produce both hexose and pentose sugars. The sugars are fermented to ethanol by a range of microorganisms [79]. Recent pretreatments methodologies have focused on the separation of the hemicellulose fraction prior to enzymic digestion enabling a harsher treatment to be used on the cellulose fraction. Following fermentation, the ethanol stream is concentrated by distillation while lignin is recovered and used for energy production. Typical second-generation lignocellulosic ethanol yields are shown in Table 8.7.

A lignocellulosic to ethanol process developed by Iogen was recently licensed to Raízen Energia Participacoes S/A, a major sugarcane processor, in Brazil. The Iogen process, which uses a bagasse feedstock, is configured with pretreatment, enzymatic hydrolysis, lignin separation, and processing plus fermentation and distillation. The high temperature pretreatment process is designed with a short residence time and mild acid hydrolysis. Acid hydrolysis prepares the feedstock for enzymatic hydrolysis. During enzymatic hydrolysis, cellulose is converted into six-carbon sugars. The hydrolysis products are divided into solid and liquid fractions. The solid portion is primarily lignin while the liquid fraction containing sugars is converted to ethanol by fermentation. Five- and six-carbon sugars are converted to ethanol using genetically modified yeasts. The dilute ethanol stream is concentrated into commercial-grade fuel ethanol. Other cellulosic ethanol facilities operating in the U.S. include Abengoa BioEnergy (Hugoton, Kansas) and Ineos Bio (Vero Beach, Florida) [80]. Facilities operational in Europe include Abengoa (Spain), Chempolis (Finland), Clariant/Sud-Chemie (Germany), Inbicon (Denmark), Mossi and Ghisolfi (Chemtex) (Italy), and St1 Biofuels (Finland) [81].

Saccharomyces cerevisiae and *Zymomonas mobilis*, genetically modified microorganisms are useful ethanologens for converting biomass into ethanol. These microorganisms have been used extensively in industry for producing ethanol using starch or sugar-based feedstocks [82, 83]. However, they only utilize hexose monomers and cannot consume pentose sugars [82, 83]. Bacteria such as *Escherichia coli* and *Klebsiella oxytoca* can consume a wide range of substrates including hexose and pentose sugars in cellulosic biomass hydrolysate. These excellent microorganisms have been genetically modified to ferment sugars in a hydrolysate from cellulosic biomass [82, 84, 85].

Second-Generation Ethanol Production Using a Hybrid Thermal/Fermentation Process

Thermal treatment of biomass to biofuels is divided between gasification and pyrolysis technologies. The Fischer–Tropsch three-step process combines coal gasification or natural gas reforming in the first step to produce a CO plus H_2 synthesis gas with a second-step heterogeneous catalytic step to produce a wide range of gaseous and liquid fuels. In the third step, longer chain, waxy synthetic hydrocarbons are hydrocracked to fuel grade fractions [86]. Coal gasification or alternatively methane reforming or partial oxidation produces a CO and H_2 rich gas under limited oxygen conditions [87, 88]. The technology was initially developed by Germany during World War II. The process was subsequently adopted by SASOL, a South African company, after the 1970s global energy crisis. The gasification and heterogenous catalytic processes are designed to operate under typical conditions of 2–20 MPa [89] plus 125–1600 °C [89–90] and 0.2–1.5 MPa [91] plus 125–325 °C [87], respectively.

Utilizing synthesis gas to produce bioethanol is a second-generation process under development. The process configuration for producing ethanol from biomass is simpler when compared to natural gas or coal processing to produce a CO plus H_2 rich synthesis gas. The hybrid biomass gasification plus fermentation process produces ethanol from synthesis gas. The fermentation reaction is mediated by anaerobic micoorganisms such as *Clostridium ljungdahlii* and *C. autoethanogenum* [92, 93]. The technology is licensed to Coskata Warrenville, Illinois. In a typical design configuration, the hybrid gasification fermentation pilot-scale technology is configured with a Westinghouse plasma gasifier [94].

8.3.2.3 Second-Generation Biofuel Production Using Thermal Processing of Lignocellulosics

Biomass pyrolysis is an evolving second-generation technology used to produce bio-oils, char, and gas. Pyrolysis is the thermal degradation of wood in the absence of oxygen. The technology is generally categorized as "fast" or "slow" depending on the reaction time for processing into byproducts. During fast pyrolysis, bio-oil yield can reach 80 % of the product on a dry fuel basis. Bio-oil can be used as a liquid fuel or as a feedstock for producing other chemicals. Biomass pyrolysis is typically performed at a relatively low temperature range (300–650 °C). During slow pyrolysis, the biomass is heated at a moderate rate to approximately 600 °C with a residence time 5–30 min. Typically for fast pyrolysis, the residence time is <2 s and the heating rate is high with a final temperature at 500 °C [95].

8.3.2.4 First-Generation Biodiesel Production from Vegetable Oils

Biodiesel production is mainly concentrated in the EU and the U.S. Approximately 60 % of the global total is produced in the EU. Biodiesel production is also emerging in Brazil China, India, Indonesia, and Malaysia. In 2007, the total biodiesel produced was estimated at approximately 10 billion liters [96].

Acid Hydrolysis

First-generation biodiesel production technology is well developed with feedstocks derived from corn, soybeans, canola oil, rapeseed, palm oil, Jatropha, waste frying oil, and coconut oil. Biodiesel (long-chain fatty acid methyl esters (LCFAME)) is produced by combining a vegetable oil with an alcohol (ethanol or methanol) and a catalyst (usually NaOH or KOH). The transesterification reaction involves converting the vegetable oil molecule into three LCFAME molecules plus one glycerol molecule [97]. The reaction mixture process stream, after mixing the catalyst, alcohol plus vegetable oil, is separated by centrifugation into methyl esters plus glycerol. In the next stage, fatty acid esters are neutralized with acid and glycerol is removed due to its low solubility. The centrate stream containing glycerol (50 %), fatty acid soaps, and methanol, is acidified and subsequently, the free fatty acids are phase separated and removed. Methanol is also removed and the glycerol stream is concentrated to 85 %. Methanol removed from the neutralization and acidification stages is combined and vacuum flash evaporated to separate methanol from water. The LCFAME stream is washed and subsequently dried to produce the biodiesel product [98].

Enzymatic Hydrolysis

Lipase-catalyzed hydrolysis of triacylglycerol is utilized to produce free fatty acids (FFA) plus glycerol. The enzymatic reaction is advantageous because glycerol is easily recovered without complex processing. The FFA are completely converted to methyl esters and wastewater treatment is not required [99]. Lipase-catalyzed transesterification is performed at low temperature and ambient pressure and hence, less energy intensive when compared to chemically catalyzed reactions. The technology is underdeveloped because of constraints such as enzyme cost and enzyme inhibition by methanol beyond a threshold level [99–102].

8.3.2.5 Third- and Fourth-Generation Biodiesel Production Using Algae and Yeast

Algae have been considered as third- and fourth-generation biofuel feedstock because existing processing routes can be designed to produce biodiesel. However, microalgae can be potentially processed into other fuels such as methane, alcohols, and distillate fuels [103]. Microalgae, which are high yielding, do not compete with land use and as food crops and hence, they are excellent feedstocks. Cultivation of any biomass requires water as well as nutrients. Terrestrial biomass cultivation generally requires water sources with relatively low salt levels because of problems associated with salt accumulation. Nutrients for terrestrial crop cultivation can be supplied from fertilizers, treated effluents as well as stabilized microbial cultures produced from wastewater treatment. In the case of algae cultivation, growth can be accomplished in seawater and nutrients supplied from wastewater streams.

Oleaginous yeast is another feasible source of third-generation feedstock. Glucose is the primary sugar used by yeast such as *Saccharomyces cerevisiae.* However, using glucose as a carbon source is not economically feasible and alternate electron donor carbon sources, especially waste material have received more attention to produce microbial lipid. A comprehensive review summarizing lipid analyses for 480 yeast strains has also reported 46 strains belonging to 14 species were oleaginous with yields reaching up to approximately 70 % oil (w w^{-1}) [104]. Many major species studied to date include *Yarrowia lipolytica, Rhodotorula glutinis, Rhodotorula graminis, Cryptococcus curvatus, Cryptococcus albidus, and Rhodosporidium toruloides* [105].

In the U.S., many algae strains have been identified by the Department of Energy (DOE) Aquatic Species Program (ASP). Several promising algal species such as *Botryococcus braunni*, *Chaetoceros calcitrans*, several *Chlorella* species, *Isochrysis galbana*, *Nanochloropsis*, *Schizochytrium limacinum,* and *Scenedesmus* species have been examined as potential biofuel feedstock sources [103, 106, 107]. The U.S. and Europe are the largest producers of algae biodiesel with the U.S. accounting for 87 % of the total global industrial production [103]. Typical lipid yields for a variety of algae and yeast species are provided in Tables 8.8 and 8.9, respectively.

Table 8.8 Biomass composition of microalgae expressed on a dry matter basis [108, 109]

Strain	Protein	Carbohydrates	Lipid
	% dry w w^{-1}		
Anabaena cylindrica	43–56	25–30	4–7
Botryococcus braunii	40	2	33
Chlamydomonas rheinhardii	48	17	21
Chlorella vulgaris	41–58	12–17	10–22
Dunaliella salina	57	32	6
Euglena gracilis	39–61	14–18	14–20
Porphyridium cruentum	28–39	40–57	9–14
Prymnesium parvum	28–45	25–33	22–39
Scenedesmus dimorphus	8–18	21–52	16–40
Spirogyra sp.	6–20	33–64	11–21

Table 8.9 Typical lipid composition for yeasts [104]

Species	Lipid content (% dry w w^{-1})	TG	FA	Sterol	SE	PL
		Percent composition				
Debaryomyces hanseni	7	27	4	6	1	60
Hansenula anomala	13	77	Trace	3	15	17
Lipomyces starkeyi	16	60	22	3	1	9
Saccharomyces cerevisiae (agar plate)	7	7	4	4	22	62
Saccharomyces cerevisiae (batch culture)	9	40	6	–	20	30
Yarrowia lipolytica (Candida lipolytica)	36	52	5	–	4	9
Candida utilis	11	55	1	2	1	38
Rhodotorula rubra	15	63	6	2	1	25

TG = triacylglycerols, FA = nonesterified fatty acids, SE = sterol esters, PL = phospholipids

Research and development activities as well as pilot-scale facilities development for third-generation algae to biodiesel facilities are primarily located in the U.S., Europe, and China. The U.S. DOE has invested heavily in developing the technology [110]. Many U.S. companies as well as research institutions are major players leading the technology development efforts.

The third-generation technology employs several processes to produce a variety of biofuels from algae. The unit process configuration utilizes oil extraction, transesterification, and subsequently, distillation. It is envisaged that future new fourth-generation technologies will be developed that advances the process design of third-generation technologies using a genetically modified feedstock, namely those terrestrial or aquatic plants which can capture CO_2 more efficiently. In addition, genetically modifying algae may be employed to increase CO_2 utilization with a subsequent increase in lipid production [103].

8.3.2.6 First- and Second-Generation Biobutanol Production

Microbially produced butanol can be utilized without blending with other fuels or at any blend ratio in combustion engines. Butanol is produced together with other solvents by Clostridia [111–114]. Acetone, butanol, and ethanol (ABE) are produced by fermentation of hexoses. *C. acetobutylicum, C. beijerinckii, C. saccharobutylicum,* and *C. saccharoperbutylacetonicum* are mainly utilized in industrial production processes [115–119]. These microorganisms utilize mainly starch or sugars. Corn (starch) or molasses from sugar beet and sugarcane (sugar) are used for industrial ABE fermentations [112]. In addition, other substrates in wastes such as microalgae-based biodiesel residues have been used with varying success [120]. Microbial butanol yield is affected by the cost for substrates such as glucose, corn, sugarcane molasses, and whey permeate. An alternate approach to produce second-generation biobutanol is reducing the production cost by utilizing lignocellulosics such as corn stover, rice straws, corn fiber, switchgrass, alfalfa, reed canary grass, sugarcane bagasse, *Miscanthus*, waste paper, DDGS, and soy molasses [121]. Waste agriculture residues and lignocellulosics are complex structures containing a mixture of cellulose, hemicellulose, and lignin. Because the structure complexity does not permit direct fermentative microbial activity, pretreatment and coupled with hydrolysis are required prior to fermentation. During pretreatment, inhibitor chemicals impair the reaction rates and hence, purifying the hydrolysate is necessary before fermentation [121]. Microbial cultures such as C. *sacchrobutylicum* P262 and *C. beijerinckii* P260 have been utilized to produce hydrolysates containing toxic chemicals. Strains capable of withstanding chemical toxicity will accelerate development of the technology [121]. Butanol is commercially produced by companies such as GreenBiologics Ltd in the UK.

8.3.2.7 Biohydrogen Production

A variety of technologies have been under investigation for producing biohydrogen. The technology type is based on the enzyme systems used to produce hydrogen. Biohydrogen production routes are divided into biophotolysis (direct/indirect), dark fermentation, and photofermentation. In the direct biophotolysis process, water is converted into hydrogen plus oxygen via photosynthetic reactions in the presence of solar energy (Eq. 1). A major disadvantage is that the Fe-hydrogenase activity is oxygen sensitive. *Chlamydomonas reinhardtii*, a green algae, has been shown to deplete the oxygen level during the oxidative respiration [122]. However, the reaction is short-lived and the hydrogen production rate is approximately one-tenth of other photosynthetic reactions [123].

$$2H_2O + \text{light energy} \rightarrow 2H_2 + O_2 \quad (1)$$

In the indirect process, enzymatic inhibition caused by O_2 is resolved by separating O_2 and H_2 evolution (Eqs. 2 and 3) [123]. Carbon dioxide is fixed and serves as the electron carrier between the O_2 producing (water splitting) reaction and the O_2 sensitive hydrogenase reactions. Many green algae and cyanobacteria are able to fix CO_2 via photosynthesis (Eq. 3) and they also have the ability to fix nitrogen and use enzymes which are able to catalyze the second H_2 generation step (Eq. 4) [123, 124].

$$12H_2O + 6CO_2 \rightarrow C_6H_{12}O_6 + 6O_2 \quad (2)$$

$$C_6H_{12}O_6 + 12H_2O \rightarrow 12H_2 + 6CO_2 \quad (3)$$

In the photofermentaton process, H_2 production by purple non-sulfur bacteria is attributed to nitrogenases under O_2 deficient conditions using light energy and electron donors (organic acids). *Rhodopseudomonas palustris* is able to oxidize organic electron donors in the presence of light energy to produce H_2 [125].

$$C_6H_{12}O_6 + 12H_2O + \text{light energy} \rightarrow 12H_2 + 6CO_2 \quad (4)$$

Dark fermentation is mediated by a consortium of microorganisms consisting of hydrolytic degraders, acetogens, acidogen plus methanogens. Carbohydrates, mainly glucose, are the preferred carbon sources used to produce acetic and butyric acids together with H_2 (Eqs. 5 and 6) [124].

$$C_6H_{12}O_6 + 2H_2O \rightarrow 2CH_3COOH + 2CO_2 + 4H_2 \quad (5)$$

$$C_6H_{12}O_6 + 2H_2O \rightarrow CH_3CH_2COOH + 2CO_2 + 2H_2 \quad (6)$$

Clostridiales and *Enterobacteriaceae* are recognized H_2-producers during dark fermentation. However, H_2-consumers such as hydrogenotrophic methanogens and homoacetogens reduce the H_2 yield by producing methane and acetic acid, respectively. In addition to environmental and engineering factors, chemical inhibition of H_2 consumers is able to increase the H_2 yield. In general, factor are selected to promote the growth of H_2 producers and reduce the growth of H_2 consumers. Heat, pH adjustment, ethylene, and 2-bromoethanesulfonate (BES) have been used to inhibit H_2 consumers in mixed anaerobic cultures [126, 127]. Even though BES and ethylene are effective inhibitors, storage as well as toxicity are major issues if these chemicals are discharged into the environment. Long chain fatty acids (LCFAs) are renewable chemicals and relative nontoxic because they are degradable. LCFAs have been used effectively to inhibit H_2 consumers and subsequently, increase the H_2 yield in mixed anaerobic cultures [125].

Researchers have also employed genetic modification of pure cultures to control the production of reduced products such as alcohols and selected organic acids [128]. Kumar et al. [128] demonstrated that improving the H_2 yield can be achieved through redirecting the metabolic fluxes by blocking formation of alcohol and some organic acids in *Enterobacter cloacae* IIT-BT 08. The double mutant with defects in both alcohol and organic acid formation pathways was shown to yield 3.8 H_2 mol mol^{-1} of glucose. This yield is close to the theoretical H_2 yield of 4 mol mol^{-1} glucose.

Microbial technologies for producing biohydrogen can be classified based on substrate utilization and genetic modification. A large amount of work has focused on optimizing the H_2 yield using hexoses and pentoses feedstocks fed to mixed microbial cultures. However, low H_2 yields coupled with the production of reduced carbon products (alcohols plus organic acids) and high substrate cost are major factors preventing the development of utilizing carbohydrate feedstocks. Employing lignocellulosics have been used to produce H_2 using pure and mixed cultures [129–132].

8.4 Conclusions and Perspectives

Ethanol, diesel, butanol, and methane are the major biofuels accounting for a large fraction of the current renewable energy market. Currently, biofuels are produced mainly from food-based crops such as corn, wheat, sugarcane, sugar beet, palm oil, rapeseed, and soy. "Nonfood" feedstocks available globally for biofuel production include energy crops (*Miscanthus*, switchgrass, Jatropha), wastes (waste oils, food processing wastes), agricultural residues (straw, corn stover), forestry residues as well as emerging feedstocks such as algae. Further technology development includes converting lignocellulosics to ethanol, butanol, methane, and hydrogen and algae to biodiesel and developing cost-competitive technologies is a large component of this effort.

Although biofuels offer numerous benefits to society, a global debate over the past several years has critically evaluated the impact on food production and prices. Increasing global population growth is linked to increasing competition for land and water, for woody biomass (e.g., timber for construction), and for bioenergy (heat and power) as well as liquid biofuels. Biofuels are part of a growing global industry which is driven by issues such as reducing our dependence on fossil fuels, decelerate climate change, and increase fuel security. However, growth of the industry will be limited by the availability of suitable land and water resources.

Acknowledgments Financial support for this work was provided by the University of Windsor and Aberystwyth University. Funding for Dr. Ravella was also provided by the Welsh European Funding Office (WEFO) and the UK Biotechnology and Biological Sciences Research Council (BBSRC) (BBS/E/W/10963A01).

References

1. Pernick R, Wilder C, Belcher J (2014) Energy Trends 2014, Clean Edge reports, The Clean Market Authority
2. BP Energy Outlook 2035 (2014) BP, London, United Kingdom. http://www.bp.com/content/dam/bp/pdf/energy-economics/energy-outlook-2015/bp-energy-outlook-2035-booklet.pdf
3. World Energy Council (2013) World Energy Resources 2013 Survey. World Energy Council, Registered in England and Wales, No. 4184478, Regency House, London
4. Edwards JD (2000) Twenty first century energy decline of fossil fuels increase of renewable non-polluting energy sources. In: Part II conference, petroleum provinces of the 21st century, January 12th–15th, 2000, San Diego, CA
5. Orr L (2006) Changing the world's energy systems. Stanford University Global and Energy Project (after John Edwards, American Association of Petroleum Geologists)
6. Haque SM, Bhat AH, Khan I (2015) Biomass: an ageless raw material for biofuels. In: Hakeem KR, Jawaid M, Alothman OY (eds) Agricultural biomass based potential materials. Springer, New York, p 446
7. vanVelthuizen H, Huddleston B, Fischer G, Salvatore M, Ataman E, Nachtergaele FO, Zanetti M, Bloise M (2006) Mapping biophysical factors that influence agricultural production and rural vulnerability, Environmental and Natural Resources, series 11. FAO and IIASA, Rome
8. Stöcker M, Tschentscher R (2015) Biomass to liquid biofuels via heterogeneous catalysis. In: Jacinto Sa (ed) Fuel production with heterogeneous catalysis. CRC Press, Taylor & Francis Group, Boca Raton, FL, p.219
9. Vegetable oil production data accessed from http://www.indexmundi.com/agriculture/?commodity=soybean-oil&graph=production
10. Berla BM, Saha R, Immethun CM, Maranas CD, Moon TS, Pakrasi HB (2013) Synthetic biology of cyanobacteria: unique challenges and opportunities. Front Microbiol 246(4):1–14
11. Hays SG, Ducat DC (2015) Engineering cyanobacteria as photosynthetic feedstock factories. Photosynth Res 123:285–295
12. Scaife MA, Nguyen GTDT, Rico J, Lambert D, Helliwell KE, Smith AG (2015) Establishing *Chlamydomonas reinhardtii* as an industrial biotechnology host. Plant J 82:532–546
13. Aro E-M (2016) From first generation biofuels to advanced solar biofuels. Ambio 45 (Suppl 1):24–31
14. Henry RJ (2010) Evaluation of plant biomass resources available for replacement of fossil oil. Plant Biotechnol J 8:288–293
15. Ladanai S, Vinterbäck J (2009) Global potential of sustainable biomass for energy, SLU, Institutionen förenergio chteknik, Swedish University of Agricultural Sciences, Department of Energy and Technology, Uppsala, Finland, Report 013, ISSN 1654-9406
16. Gupta RB, Demirbas A (2010) Biomass availability in the world. In: Gasoline, diesel and ethanol biofuels from grasses and plants. Cambridge University Press, New York, Chap. 4, pp 56–72
17. Haberl H, Erb KH, Krausmann F, Bondeau A, Lauk C, Müller C, Plutzar C, Steinberger JK (2011) Global bioenergy potentials from agricultural land in 2050: sensitivity to climate change, diets and yields. Biomass Bioenerg 35(12):4753–4769
18. Slade R, Saunders R, Gross R, Bauen A (2011) Energy from biomass: the size of the global resource. Imperial College Centre for Energy Policy and Technology and UK Energy Research Centre, London
19. McKendry P (2002) Energy production from biomass (part 1): overview of biomass. Bioresour Technol 83:37–46
20. Demirbas A (2000) Biomass resources for energy and chemical industry. Energy Edu Sci Technol 5:21–45

21. Klass DL (1998) Biomass for renewable energy, fuels, and chemicals. Academic Press, San Diego, CA
22. Hall DO, Rosillo-Calle F, Williams RH, Woods J (1993) Biomass for energy: supply prospects. In: Johansson TB, Kelly H, Reddy AKN, Williams RH (eds) Renewable energy: sources for fuels and electricity. Island Press, Washington, DC, pp 593–651
23. OECD/FAO (2013) Agricultural outlook 2013, Chapter 3 Biofuels, OECD
24. United States Department of Agriculture (USDA) (2015) World Agricultural Production, Foreign Agricultural Service Circular Series WAP 12–15, December 2015
25. FAO (2013) Food Outlook, Biannual report global food markets. www.fao.org/3/a-i4136e.pdf
26. Koo WW, Taylor RD (2013) Outlook of the U.S. and World Sugar Markets, 2012–2022. Agribusiness and Applied Economics Report No. 714 June 2013. Center for Agricultural Policy and Trade Studies, Department of Agribusiness and Applied Economics, North Dakota State University, Fargo, North Dakota 58108-6050
27. OECD, FAO (2015) Oilseeds and oilseed products, in OECD-FAO agricultural outlook 2015. OECD Publishing, Paris
28. van den Born GJ, van Minnen JG, Olivier JGJ, Ros JPM (2014) Integrated analysis of global biomass flows in search of the sustainable potential for bioenergy production, PBL Netherlands Environmental Assessment Agency, PBL publication number 1509, p 8
29. Pippo WA, Luengo CA (2013) Sugarcane energy use: accounting of feedstock energy considering current agro-industrial trends and their feasibility. Int J Energy Environ Eng 4 (1):10–13
30. United States Environmental Protection Agency—www3.epa.gov
31. United States Energy Information Administration—www.eia.gov
32. Abbas C, Binder TP, Beery KE, Cecava MJ, Doane PH, Holzgraefe DP, Solheim LP (2011) Process for the production of animal feed and ethanol and novel animal feed. U.S. patent US 7998511 B2
33. Léder I (2004) Sorghum and millets. Grains and cereals, in cultivated plants, primarily as food sources. In: Fulek G (ed) Encyclopedia of life support systems (EOLSS), developed under the auspices of the UNESCO, vol 1. EOLSS Publishers, Paris, p 66 http://www.eolss.net
34. Klass DL (2004) Biomass for renewable energy and fuels in Encyclopedia of Energy, Cutler CJ (Editor-in-Chief), Elsevier, San Diego, CA
35. Chandel AK, da Silva SS, Carvalho W, Singh OV (2012) Sugarcane bagasse and leaves: foreseeable biomass of biofuel and bio-products. J Chem Technol Biotechnol 87(1):11–20
36. United States Department of Agriculture (USDA) (2015) Agricultural projections to 2024. Office of the Chief Economist, World Agricultural Outlook Board, U.S. Department of Agriculture. Prepared by the Interagency Agricultural Projections Committee. Long-term Projections Report OCE-2015-1
37. European Commission (2015) From the sugar platform to biofuels and biochemical. Final report for the European Commission Directorate-General Energy No. ENER/C2/423-2012/SI2.673791
38. European Biogas Association. EBA's Biomethane fact sheet. Renewable Energy House, Rue d'Arlon 63–65, www.european-biogas.eu
39. EurObserv'ER (2014) Biogas Barometer accessed Feb. 10th, 2016 http://www.energies-renouvelables.org/observ-er/stat_baro/observ/baro224_Biogas_en.pdf
40. Voegele E (2014) EurObserv'ER annual report highlights EU bioenergy sector, Biomass Magazine accessed on Feb. 10th, 2016 http://biomassmagazine.com/articles/10208/eurobserver-annual-report-highlights-eu-bioenergy-sector
41. Böhme D, Dürrschmidt W, van Mark M (2011) General and fundamental aspects of renewable energy sources. Federal Ministry for the Environment, Nature Conservation and Nuclear Safety (BMU) Division KI III 1
42. Cuèllar AD, Webber ME (2008) Cow power: the energy and emissions benefits of converting manure to biogas. Environ Res Lett 3:034002
43. Frigon JC, Guiot SR (2010) Biomethane production from starch and lignocellulosic crops: a comparative review. Biofuels, Bioprod Biorefin 4(4):447–458

44. http://www.projetgaya.com/en/the-2nd-generation-biomethane-production-process
45. http://www.statista.com/statistics/249609/sugar-beet-production-worldwide/
46. Koizumi T (2015) Biofuels and food security: Biofuel impact on food security in Brazil, Asia and major producing countries. Springer, New York
47. Saskatchewan Agriculture and Food, Agriculture Development and Diversification Secretariat, Environment and Engineering Branch (1993) Establishing an ethanol production business. The Secretariat, Saskatchewan
48. Asadim M (2007) Beet-Sugar Handbook. Chapter 3, Sugarbeet processing. Wiley, New York, p 105
49. Shapouri H, Salassi M, Fairbanks JN (2006) The economic feasibility of ethanol production from sugar in the United States. United States Department of Agriculture report, Office of the Chief Economist, USDA report July 2006
50. Kuiper L, Ekmekci B, Hamelinck C, Hettinga W, Meyer S, Koop K (2007) Bio-ethanol from cassava. Ecofys Netherlands BV, NL-3503 RK Utrecht, The Netherlands, Project number: PBIONL062937
51. Höfer R (2015) Sugar- and starch-based biorefineries. In: Pandey A, Höfer R, Larroche C, Taherzadeh M, Nampoothiri KM (eds) Industrial biorefineries and white biotechnology. Elsevier, Waltham, MA
52. Cardona CA, Sanchez OJ, Gutierrez LF (2010) Process synthesis for fuel ethanol production. Chapter 3. Feedstock for ethanol production. CRC Press, Taylor and Francis, Boca Raton, p 45
53. Bonin CL, Heaton EA, Cogdill TJ, Moore KJ (2016) Management of sweet sorghum for biomass production. Sugar Tech 2:150–159
54. Brown RC, Wright MM (2008) Capturing solar energy through biomass. In: Kreith F, Krumdieck S (eds) Principles of sustainable energy systems, 2nd ed. CRC Press, Taylor and Francis Group, Boca Raton, FL
55. Lee S, Shah YT (2013) Biofuels and bioenergy: processes and technologies, CRC Press, Taylor and Francis Group, Boca Raton, FL
56. Licht S (2014) Fermentation for biofuels and bio-based chemicals. In: Vogel HC, Todaro CM (eds) Fermentation and biochemical engineering handbook principles, process design, and equipment, 3rd edn. Elsevier, Waltham, MA
57. Zhao R, Wu X, Bean S, Wang D (2010) Ethanol from grain crops. In: Singh BP (ed) Industrial crops and uses. CAB International, Oxfordshire
58. https://www.eia.gov/todayinenergy/detail.cfm?id=21212
59. Kelsall DR, Lyons TP (1999) Grain dry milling and cooking for alcohol production: designing for 23 % ethanol and maximum yield. Chapter 2. In: Jacques KA, Lyons TP, Kelsall DR (eds) The alcohol textbook. 3rd ed. Nottingham University Press. Nottingham
60. Cheng J, Leu SY, Zhu J, Gleisner R (2015) High titer and yield ethanol production from undetoxified whole slurry of Douglas-fir forest residue using pH profiling in SPORL. Biotechnol Biofuels 8:22
61. Overend RP, Wright LL (2008) Biomass energy in energy conversion. Taylor & Francis, Boca Raton, Chapter 3, pp 3–14
62. Anon (2010) Year one of the RTFO renewable fuels agency report on the renewable transport fuel obligation 2008/09. Printed in the UK for The Stationery Office Limited on behalf of the Controller of Her Majesty's Stationery Office
63. http://www.tradingeconomics.com/brazil/gdp
64. de Souza Dias MO, Filho RM, Mantelatto PE, Cavalett O, Rossell CEV, Bonomi A, Leal MRLV (2015) Sugarcane processing for ethanol and sugar in Brazil. Environ Dev 15:35–51
65. Soccol CR, de Souza Vandenberghe LP, Medeiros ABP, Karp SG, Buckeridge M, Ramos LP et al (2009) Production of bioethanol and other bio-based materials from sugarcane bagasse: integration to conventional bioethanol production process. Chem Eng Res Des 87(9):1206–1216

66. Canilha L, Santos V, Rocha G, Almeida e Silva J, Giulietti M, Silva S et al (2011) A study on the pretreatment of a sugarcane bagasse sample with dilute sulfuric acid. J Ind Microbiol Biotechnol 38(9):1467–1475
67. Somerville C, Youngs H, Taylor C, Davis SC, Long SP (2010) Feedstocks for lignocellulosic biofuels. Science 329:790–792
68. Tumbala P, Thelen, KD, Adkins A, Dale B, Balan V, Gunawan C, Gao J (2016) Corn stover ethanol yield as affected by grain yield, Bt trait, and environment. Biomass Bioenerg 119–125
69. Aden A, Ruth M, Ibsen K, Jechura J, Neeves K, Sheehan J et al (2002) Lignocellulosic biomass to ethanol process design and economics utilizing co-current dilute acid prehydrolysis and enzymatic hydrolysis for corn stover. National Renewable Energy Laboratory, Golden, Colorado. Report # NREL/TP-510-32438
70. Talebina F (2015) Bioethanol from lignocellulosics wastes: current status. In: Karimi K (ed) Lignocellulose-based bioproducts. Springer, New York
71. Abramson M, Shoseyov O, Hirsch S, Shani Z (2013) Genetic modifications of plant cell walls to increase biomass and bioethanol production. In: Lee JW (ed) Advanced biofuels and bioproducts. Springer, New York
72. Lee W-C, Kuan W-C (2015) *Miscanthus* as cellulosic biomass for bioethanol production. Biotechnol J 10(6):840–854
73. Ewanick S, Bura R (2011) The effect of biomass moisture content on bioethanol yields from steam pretreated switchgrass and sugarcane bagasse. Bioresourc Technol 102(3):2651–2658
74. Angelini LG, Ceccarini L, o Di Nasso NN, Bonari e (2009) Comparison of *Arundodonax* L and *Miscanthus* × *giganteus* in a long-term field experiment in Central Italy. Biomass Bioenergy 33(4):635–643
75. Lewandowski I, Clifton BJC, Scurlock JMO, Huisman W (2000) *Miscanthus*: European experience with a novel energy crop. Biomass Bioenerg 19:209–227
76. Hayes DJM (2013) Biomass composition and its relevance to biorefining. In: Triantafyllidis KS, Lappas AA, Stocker M (eds) The role of catalysis for the sustainable production of bio-fuels and bio-chemicals. Elsevier, Amsterdam
77. Schmer MR, Vogel P, Mitchell RB, Perrin RK (2008) Net energy of cellulosic ethanol from switchgrass. Proc Nat Acad Sci 105:464–469
78. Lewis RS, Datar RP, Huhnke RL (2005) Biomass to ethanol. In: Lee S (ed) Encyclopedia of chemical processing, 8th edn. Marcel Dekker, New York, NY
79. Kang Q, Appels L, Tan T, Dewil R (2014) Bioethanol from lignocellulosic biomass: Current findings. The Sci World J 1–13
80. Renewable Energy Association (2013) Battling for the barrel: 2013 ethanol industry outlook. Renewable Energy Association
81. https://dglassassociates.wordpress.com/2013/02/25/commercial-cellulosic-ethanol-projects-brazil-and-europe/
82. Zaldivar J, Nielsen J, Olsson L (2001) Fuel ethanol production from lignocellulose: a challenge for metabolic engineering and process integration. Appl Microbiol Biotechnol 56:17–34
83. Dien BS, Cotta MA, Jeffries TW (2003) Bacteria engineered for fuel ethanol production: current status. Appl Microbiol Biotechnol 63:258–266
84. Lynd LR, Wyman CE, Gerngross TU (1999) Biocommodity engineering. Biotechnol Prog 15:777–793
85. Ingram LO, Aldrich HC, Borges AC, Causey TB, Martinez A, Morales F et al (1999) Enteric bacterial catalysts for fuel ethanol production. Biotechnol Prog 15:855–866
86. Speigh JG (2014) The Fischer-Tropsch process in gasification of unconventional feedstocks. Gulf Professional Publishing, Wyman Street, Waltham, p 119
87. Urakawa A, Sá J (2015) CO_2 to Fuels. In: Sá J (ed) Fuel production with heterogeneous catalysis. CRC Press, Taylor & Francis, Boca Raton, p 93
88. Lee H-J (2010) Optimization of Fischer–Tropsch plant. Doctor of Philosophy, Faculty of Engineering and Physical Sciences, University of Manchester, United Kingdom

89. Higman C, van der Burgt M (2008) Gasification. 2nd Edition, Gulf Professional Publishing (Imprint of Elsevier), Atlanta, GA
90. McKendry P (2002) Energy production from biomass (part 3): gasification technologies. Bioresour Technol 83(1):55–63
91. Mirzaei AA, Vahid S, Feyz, M (2009) Fischer–Tropsch synthesis over iron manganese catalysts: effect of preparation and operating conditions on catalyst performance. Adv Phy Chem Article 12 p ID 151489. doi:10.1155/2009/151489
92. Abubackar HN, Veiga MC, Kennes C (2015) Carbon monoxide fermentation ethanol by *Clostridium autoethanogenum* in a bioreactor with no accumulation of acetic acid. Bioresour Technol 186:122–127
93. Younesi H, Najafpour G, Mohamed AR (2005) Ethanol and acetate production from synthesis gas via fermentation processes using anaerobic bacterium, *Clostridium ljungdahlii*. Biochem Eng J 27:110–119
94. Runge W (2013) Technology entrepreneurship: a treatise on entrepreneurs and entrepreneurship for and in technology ventures, vol 1. KIT Scientific Publishing, Karlsruhe
95. Basu P (2010) Pyrolysis and torrefaction in biomass gasification and pyrolysis: practical design and theory. Elsevier, Waltham, MA
96. FAO (2008) The state of food and agriculture—biofuels: prospects, risks and opportunities. FAO, Rome
97. Issariyakul T, Dalai AK (2014) Biodiesel from vegetable oils (review). Renew Sustain Energy Rev 31:446–471
98. Hoogendoorn A, van Kasteren H (2011) Transportation biofuels: novel pathways for the production of ethanol. Chapter 4 Enzymatic biodiesel, Royal Society of Chemistry, p 133
99. Al-Zuhair S, Fan WL, Lim SJ (2007) Proposed kinetic mechanism of the production of biodiesel from palm oil using lipase. Proc Biochem 42:951–960
100. Al-Zuhair S (2008) Enzymatic production of bio-diesel from waste cooking oil using lipase. Open Chem Eng J 2:84–88
101. Shimada Y, Watanabe Y, Samukawa T, Sugihara A, Noda H, Fukuda H, Tominaga Y (1999) Conversion of vegetable oil biodiesel using immobilized *Candida antarctica* lipase. J Am Oil Chem Soc 76:789–793
102. Pourzolfaghar H, Abnisa F, Daud WMAW, Aroua MK (2016) A review of the enzymatic hydroesterification process for biodiesel production. Renew Sustain Energy Rev 61:245–257
103. Singh J, Gu S (2010) Commercialization potential of microalgae for biofuels production. Renew Sustain Energy Rev 14:2596–2610
104. Rattray JBM (1988) Yeasts. In: Ratledge C, Wilkinson SG (eds) Microbial lipids, vol 1. Academic Press, London, pp 555–697
105. Sitepu IR, Garay LA, Sestric R, Levin D, Block DE, German JB, Boundy-Mills KL (2014) Oleaginous yeasts for biodiesel: current and future trends in biology and production. Biotechnol Adv 32(7):1336–1360
106. Rodolfi L, Zittelli GC, Bassi N, Padovani G, Biondi N, Bonini G (2008) Microalgae for oil: strain selection, induction of lipid synthesis and outdoor mass cultivation in a low-cost photobioreactor. Biotechnol Bioeng 102(1):100–112
107. Chisti Y (2007) Biodiesel from microalgae. Biotechnol Adv 25(3):294–306
108. Um B-H, Kim Y-S (2009) Review: a chance for Korea to advance algal-biodiesel technology. J Ind Eng Chem 15:1–7
109. Sydney EB, Sturm W, de Carvalho JC, Thomaz-Soccol V, Larroche C, Pandey A, Soccol CR (2010) Potential carbon dioxide fixation by industrially important microalgae. Bioresourc Technol 101:5892–5896
110. Han SF, Jin WB, Tu RJ, Wu WM (2015) Biofuel production from microalgae as feedstock: current status and potential. Crit Rev Biotechnol 35(2):255–268
111. Dürre P (2008) Fermentative butanol production: Bulk chemical and biofuel. Ann NY Acad Sci 1125:353–362
112. Ezeji TC, Qureshi N, Blaschek HP (2005) Industrially relevant fermentations. In: Dürre P (ed) Handbook on clostridia. CRC Press, Boca Raton, pp 797–812

113. Dürre P, Bahl H, Gottschalk G (1988) Membrane processes and product formation in anaerobes. In: Erickson LE, Fung DY-C (eds) Handbook on anaerobic fermentations. Marcel Dekker, New York, pp 187–206
114. Jones DT, Woods DR (1986) Acetone-butanol fermentation revisited. Microbiol Rev 50:484–524
115. Dolejš I, Krasňan V, Stloukal R, Rosenberg M, Rebroš M (2014) Butanol production by immobilised *Clostridium acetobutylicum* in repeated batch, fed-batch, and continuous modes of fermentation. Bioresour Technol 169:723–730
116. Jones DT, Keis S (1995) Origins and relationships of industrial solvent-producing clostridial strains. FEMS Microbiol Rev 17:223–232
117. Keis S, Shaheen R, Jones DT (2001) Emended descriptions of *Clostridium acetobutylicum* and *Clostridium beijerinckii*, and descriptions of *Clostridium saccharoperbutylacetonicum* sp. nov. and *Clostridium saccharobutylicum* sp. nov. Int J Syst Evol Microbiol 51:2095–2103
118. Qureshi N, Blaschek HP (2001) Recent advances in ABE fermentation: hyper-butanol producing *Clostridium beijerinckii* BA101. J Ind Microbiol Biotechnol 27(5):287–291
119. Shamsudin S, MohdSahaid HK, Wan Mohtar WY (2006) Production of acetone, butanol and ethanol (ABE) by *C. saccharoperbutylacetonicum* N1-4 with different immobilization systems. Pakistan. J Biol Sci 9(10):1923–1928
120. Cheng HH, Whang LM, Chan KC, Chung MC, Wu SH, Liu CP, Tien SY, Chen SY, Chang JS, Lee WJ (2015) Biological butanol production from microalgae-based biodiesel residues by *Clostridium acetobutylicum*. Bioresour Technol 184:379–385
121. Qureshi N, Liu S, Ezeji TC (2012) Cellulosic butanol production from agricultural biomass and residues: recent advances in technology. Adv Biofuels Bioprod 247–265
122. Melis A, Zhang LP, Forestier M, Ghirardi ML, Seibert M (2000) Sustained photobiological hydrogen gas production upon reversible inactivation of oxygen evolution in the green alga *Chlamydomonas reinhardtii*. Plant Physiol 122:127–135
123. Benemann JR (2000) Hydrogen production by microalgae. J Appl Phycol 12:291–300
124. Benemann JR (1996) Hydrogen biotechnology: Progress and prospects. Nat Biotechnol 14:1101–1103
125. Nath K, Das D (2005) Hydrogen production by *Rhodobactersphaeroides* strain O.U.0011 using spent media of *Enterobacter cloacae* strain DM1. Appl Microbiol Biotechnol 68:533–541
126. Philpott J (2010) Bio-hydrogen production from glucose degradation using a mixed anaerobic culture in the presence of natural and synthetic inhibitors. MASc. Thesis, University of Windsor, Windsor, Ontario, Canada
127. Sparling R, Risbey D, Poggi-Varaldo HM (1997) Hydrogen production from inhibited anaerobic composters. Int J Hydrogen Energy 22:563–566
128. Kumar N, Ghosh A, Das D (2001) Redirection of biochemical pathways for the enhancement of H_2 production by *Enterobacter cloacae*. Biotechnol Lett 23(7):537–541
129. Shanmugam SR, Chaganti SR, Lalman JA, Heath DD (2014) Effect of inhibitors on hydrogen consumption and microbial population dynamics in mixed anaerobic cultures. Int J Hydrogen Energy 39(1,2):249–257
130. Veeravalli S, Chaganti S, Lalman J, Heath D (2014) Optimizing hydrogen production from a switchgrass steam exploded liquor using a mixed anaerobic culture in an upflow anaerobic sludge blanket reactor. Inter J Hydrogen Energy 39(7):3160–3175
131. Lai Z, Zhu M, Yang X, Wang J, Li S (2014) Optimization of key factors affecting hydrogen production from sugarcane bagasse by a thermophilic anaerobic pure culture. Biotechnol Biofuels 7:119
132. Chen R, Wang YZ, Liao Q, Zhu X, Xu TF (2013) Hydrolysates of lignocellulosic materials for biohydrogen production. BMB Rep 46(5):244–251

Chapter 9
Lignin: A Platform for Renewable Aromatic Polymeric Materials

Jairo H. Lora

Abstract Lignin is the most abundant renewable aromatic material on Earth and is a logical link in the sustainable supply chain of the chemicals and materials that today we derive mostly from fossil-based benzene. In this chapter, we review lignin's industrial sources and the recent progress in developing polymeric materials that use lignin as a macromonomer, polymer blend, functional additive, and precursor for advanced carbon materials. Several high-purity lignin production facilities were installed recently at pulp mills in three continents. Additional lignin availability is expected as second generation biofuel biorefineries realize that it makes sense to add value beyond the energy content of the lignin-rich residues that they generate. In the last decade, many lignin applications have seen significant progress, including lignin use in thermoset resins, polyurethanes, polyesters, polymer blends, elastomers, smart memory materials, antioxidants, functional packaging, electrically conductive materials for energy storage, coatings, advanced carbon materials, engineering plastics, ingredients for inks and varnishes, UV light blockers, coatings, compostable plastics, foams, rubber products, insulation, sealants, fiber composites, etc. Several of the applications highlighted here have already entered commercialization, while many others are expected to do so within the next few years. Increasing lignin use will result in environmental benefits and products with reduced carbon footprint. Application to lignin of emerging technologies may open new routes for lignin valorization in the near future.

Keywords Lignin · Lignin producers · Lignin applications · Thermoset resins · Elastomers · Polymer blends · Macromonomer · Bioaromatics · Antioxidants · UV blockers · Carbon materials · Flame retardants · Smart memory materials · Energy storage materials

J.H. Lora (✉)
Lora Consulting LLC, Media, PA 19063, USA
e-mail: jhlorac@gmail.com

P.C.K. Lau (ed.), *Quality Living Through Chemurgy and Green Chemistry*,
Green Chemistry and Sustainable Technology,
DOI 10.1007/978-3-662-53704-6_9

9.1 Introduction

The generation of energy from fossil fuels is the main contributor to the increase in carbon dioxide concentration in the atmosphere and the resulting damaging climate change trends that planet Earth is currently experiencing and mankind is trying to reverse. A smaller, but nevertheless important source of greenhouse gases is the production of chemicals and materials used in virtually every aspect of daily life, given that today's chemical industry is primarily based on fossil feedstocks and many of the goods that mankind uses have a significant carbon footprint. Alternative sustainable approaches based on renewable feedstocks are required not only for energy generation but also for the production of chemicals and materials.

There are already multiple renewable feedstocks and sustainable pathways to generate the key aliphatic chemicals as exemplified by ethylene production from bioethanol, a strategy that has already been implemented in Brazil on an industrial scale [1]. Methane from anaerobic digestion may also be used to generate important aliphatic intermediates which today are produced from natural gas. The aromatic chemicals are primarily derived from benzene, a chemical with a 40 million ton/year market for which sustainable production options are fewer and less advanced.

Lignin—the biopolymer that holds fibers together in trees and other biomass—is nature's most plentiful renewable aromatic chemical. Given its abundance and ubiquity, lignin appears to be the main logical sustainable platform for aromatic chemicals. Progress in the development of renewable aromatic products from lignin has been slow due to various factors including its natural complexity and limited industrial supply chain of high purity, well-understood lignins. In addition, volatile oil prices when low are a disincentive to renewable alternatives, particularly when society has no established mechanisms to give financial credit for the value of renewable carbon.

In this chapter, we provide an update on current and emerging industrial sources of lignin. Then we review the state of the art of industrial applications of lignin in polymeric materials as a macromonomer, as a blend with other polymers, as a functional additive contributing specialized properties to other polymers and as a precursor for advanced materials.

9.2 Industrial Lignin Sources

In nature, the main lignin building blocks are coniferyl, synapyl, and *p*-coumaryl alcohols [2] (Fig. 9.1), which are formed from photosynthesis glucose, and through a series of biochemical transformations in the plant cell become a complex polymer of phenylpropane units functionalized with hydroxyl, methoxyl, carboxyl, carbonyl, and double bonds interconnected by ether, ester, and carbon-to-carbon bonds. Guaiacyl (G) units derived from coniferyl alcohol dominate softwoods lignins,

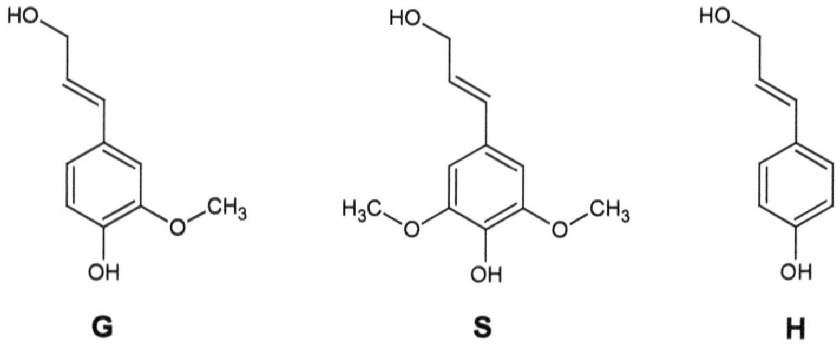

Fig. 9.1 Lignin building blocks: guaiacyl (G), syringyl (S), *p*-coumaryl (H) alcohols

while hardwoods in addition to G units contain syringyl (S) units derived from synapyl alcohol, and annual plants have G, S as well as *p*-coumaryl units (H).

Traditionally, industrial lignins have been available in limited quantities as co-products of the pulp and paper industry. Lignins are solubilized during the manufacture of cellulose pulp from wood and other biomass. The vast majority of this solubilized lignin is burned in a highly capital intensive recovery process, which produces steam and regenerates pulping active chemicals. Out of more than 70 million tons/year of lignin separated in pulping operations worldwide, only less than 2 % is recovered and used for non-energy applications.

Kraft, soda, and sulfite are the main types of chemical pulping processes practiced industrially at this time. The soda process is mostly used on annual fiber feedstocks such as sugarcane bagasse, flax, or wheat straw, while kraft, and sulfite pulping are predominantly used on wood. Lignin characteristics are affected by several factors, including pulping process, intensity of pulping, and feedstock.

Sulfite pulping introduces sulfonate groups covalently bound to the aliphatic chain of the lignin molecule. The resulting sulfite lignins (or lignosulfonates) are highly water soluble under acid or alkaline conditions. Commercial lignosulfonates, in general, may contain significant levels of impurities, such as hemicellulose sugars and inorganic spent pulping chemicals. Such lignosulfonates may be enriched in lignin content by fermenting the hemicellulose sugars or by membrane filtration technologies. As illustrated in Table 9.1, sulfite lignins are the largest type of lignin currently marketed for non-energy applications going principally into the dispersants, surfactants, and binders markets [3].

During kraft and soda pulping, lignins are solubilized under alkaline conditions as the corresponding ionized aromatic hydroxyl salts to form the stream conventionally known as black liquor. Kraft and soda lignins may be recovered in relatively high purity by controlled acidification of the black liquor, since these lignins' water solubility decreases as the pH is lowered while hemicellulose sugars and ash remain largely in solution. The isolated kraft and soda lignins have much lower molecular weight and hydrophillicity than lignosulfonates (Table 9.2). Their principal functional groups are aromatic and aliphatic hydroxyl groups, as well as

Table 9.1 Major industrial producers of various types of lignin

Company	Capacity[a], tons/year	Lignin type	Country of production
Borregaard Lignotech	470,000	Sulfite	Various
Tembec	570,000[b]	Sulfite	Canada/France
Aditya Birla Domsjo	120,000	Sulfite	Sweden
Nippon Paper	70,000	Sulfite	Japan
Stora Enso	50,000	Kraft	Finland
Cartiere Burgo	40,000	Sulfite	Italy
Ingevity	35,000	Kraft, sulfonated kraft	USA
Domtar	25,000	Kraft	USA
West Fraser	10,000	Kraft	Canada
GreenValue	5,000	Soda	India

[a]Solids content may vary according to producer
[b]Tembec sells a lot less than this stated capacity, as it can process a significant portion of their spent liquors through its recovery boilers

Table 9.2 Typical properties of industrially available lignins

	Soda (Straw)	Kraft (Spruce)	Sulfite
Ash (%)	<2	<1	
Carbohydrates (%)	<4	<2	Up to ~35
Sulfur (%)	<1	~2	4–8
Sulfonate	Nil	Nil	0.9–3.3
Weight average molecular weight (M_W) (dalton)	4000–12,000	4000–12,000	~150,000
Number average molecular weight(M_N) (dalton)	1000–2000	1000–2000	Not available
Softening temperature (°C)	200	150-200	Decomposes
Water solubility	Very low	Very low	Very high

carboxylic acid groups. Lignins obtained from annual plants (typically via soda pulping) have higher amounts of carboxylic acid groups than lignins from wood. Kraft lignins have higher sulfur content than soda lignins. Kraft pulping forms mercaptans and introduces thiol groups in the lignin molecule, which also contains elemental sulfur as an impurity.

For many decades one company (Ingenity, a recent spin-off from WestRock and previously part of MeadWestVaco) was the only industrial producer of alkaline lignin (kraft-type) and some derivatives including sulfonated and aminated products. During recent years alkaline lignin capacity has tripled. In 2005, a straw-based soda lignin recovery plant was installed in India [3]. By 2016 installed annual capacity of alkaline lignin had further increased by about 85,000 tons, as three softwood kraft pulp producers in US, Canada, and Europe implemented lignin production systems [4–6]. This recent acceleration in alkaline lignin recovery installations is a result of a debottlenecking strategy in which by diverting part of the spent liquor from combustion to lignin extraction these pulp mills are able to incrementally increase cellulose production. This option is claimed to be more economically attractive than having to install a larger and very expensive recovery boiler and associated systems used to generate energy and regenerate pulping chemicals.

Potential industrial additions to the future supply of lignin could emerge along with the large-scale implementation of biorefinery technologies in which ligno-cellulosic biomass is used to produce biofuels or other chemicals and materials. In the current technology practiced in the biorefineries already in operation in USA, Brazil, and Europe biomass is pretreated with water, steam, or dilute mineral acid at high temperature and pressure to facilitate enzymatic hydrolysis. Fermentation of the enzymatic hydrolyzate produces a beer that upon distillation to recover the alcohol leaves behind a lignin-rich residue. This is usually burned for steam and power generation, but should be considered a potential source of higher purity lignin products. As illustrated in Fig. 9.2, this can be accomplished by an extraction and precipitation process, since the pretreatment normally generates significant amounts of a highly reactive lignin fraction, which is soluble in aqueous alkali or certain solvents [7]. We estimate that such a process could generate high purity lignin equivalent to about 7–11 % by weight of typical non-wood feedstocks, equivalent to more than 1/3 of the alcohol output.

Other potential source of lignin is its extraction from biomass with organic solvents, which has been proposed as an environmentally superior way to produce hardwood cellulose pulp or pretreat biomass for biofuels production. An ethanol-based organosolv pulping technology (industrially known as the Alcell

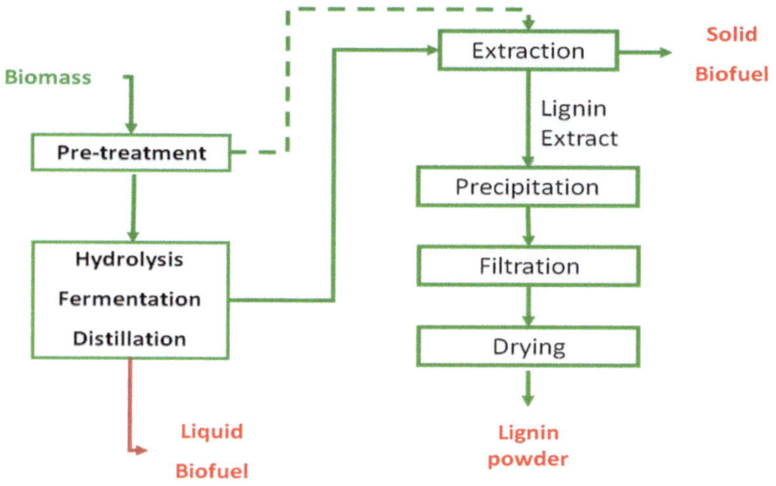

Fig. 9.2 Scheme for lignin production integrated with a cellulosic alcohol biorefinery

Table 9.3 Typical properties of lignins obtained by autocatalyzed alcohol organosolv process and by extraction from the residue after fermentation when the pretreatment is an aqueous acid step

	Alcohol organosolv	Aqueous pretreatment
Feedstock	Hardwood	Straw
Lignin recovery	Precipitation from alcohol extract by dilution and acidification	Alkaline extraction from fermentation residue followed by precipitation from black liquor by acidification
Ash (%)	<0.2	<2
Carbohydrates (%)	<1	<2
M_W (dalton)	~3000	10,000–40,000
M_N (dalton)	~1000	1000–3000
Softening temperature (°C)	~120	>220
Water solubility	Very low	Low

process) advanced up to semi-industrial scale in Canada in the 1990s [8, 9] and permitted the production of high quality lignin with very high aromatic hydroxyl content, low molecular weight, low polydispersity, low glass transition temperature, and low levels of contaminants. Recently, this technology was acquired by one of the largest Brazilian producers of hardwood pulp, but no plans for industrial implementation are known. Organosolv processes based on other solvents such as formic acid or glycerol are in pilot or pre-pilot scale [10–12]. Organosolv technologies, in general, offer the promise of exceptionally high purity lignin amenable to utilization in a wide range of applications, but is not practiced on a full commercial scale at this time. Other emerging techniques not discussed further here that may lead to novel routes for lignin production and modification in the not too distant future include the use of ionic liquids (ILs) and deep eutectic mixtures [13]. See Chap. 10 for a brief description on ILs.

As shown in Table 9.3, lignins from biorefinery processes that include aqueous pretreatments prior to the production of bioethanol can also be obtained in high purity, but relative to alcohol based organosolv lignins they normally have a higher molecular weight, broader molecular weight distribution and higher softening temperatures.

9.3 Lignin Applications in Polymeric Materials

The field of lignin applications is an evolving one that has been reviewed recently by several authors [3, 14–16]. In this section, we give an update on lignin use as polymer for industrial purposes. We pay significant attention to information on applications that have recently entered or could soon be entering the commercial arena. Traditional applications based on kraft or sulfite lignins are mentioned briefly as they exemplify a variety of current uses for the limited range of lignins that have been available. New applications that have become industrially established in the last few years (in some cases as a consequence of the availability of new lignin materials in industrial and semi-industrial amounts) are highlighted, as well as novel promising uses that have been recently proposed. The subject of degradation of lignin to simpler derivatives that may be used as platform aromatic chemicals to produce other chemicals or polymers is not within the scope of this chapter.

9.3.1 Tensioactive Agents

Historically, the most important traditional lignin application has been the use of lignosulfonates obtained directly from the sulfite pulping process as dispersants, competing effectively against synthetic, petrochemical based alternatives. Provided they meet purity and certain broad composition parameters, lignosulfonates perform in a cost-effective and versatile manner to increase fluidity in concrete, gypsum, and drilling muds for oil recovery, and as dispersants for agricultural chemicals, among others [17]. Other applications, such as dispersion of dyes for textiles and of carbon black in battery electrodes are more demanding and need to meet targets for purity, molecular weight, and/or degree of sulfonation. Such specifications are best satisfied by kraft lignins that have been sulfonated after isolation, rather than by lignosulfonates produced by sulfite pulping. The starting kraft lignin normally already has very low levels of sugar and ash impurities and properties such as molecular weight and content of sulfonate groups may be adjusted during the modification process which is actually done in a sulfomethylation reaction. A few new applications for lignins modified in this way have been recently proposed. High purity sulfomethylated kraft lignin with average molecular weight above 25,000 has been claimed to be a suitable dispersant for the exacting requirements of inkjet formulations [18] and sulfomethylated organosolv lignin has been proposed as superplasticizer for concrete, capable to perform competitively against petroleum-based naphthalene sulfonate-based products [19].

Lignin products are also used as emulsifiers for asphalt. Sodium salts of kraft lignin are used as anionic emulsifiers [20] and aminated lignins as cationic ones [21]. Alcohol organosolv lignin in combination with polymerized rosin and their sodium salts have been used industrially as asphalt emulsifiers and as asphalt and concrete air entrainers to increase durability under freezing-thawing conditions [22–24].

A promising alternative approach to tensioactive agents from lignin involves modification by treatment with oxygen under alkaline conditions. Depending on the conditions used, this treatment leads to oxidation (which introduces carboxylic acid groups bound to the lignin, as well as lower molecular weight oxidized products) and also to lignin condensation. The product obtained was superior to a commercial lignosulfonate-based concrete plasticizer and approached the performance of a petroleum-based concrete superplasticizer [25]. This development is significant, as oxidation is expected to have a much better environmental footprint than sulfonation.

9.3.2 Applications in Asphalt

Most paved roads use asphalt, in which oil-derived bitumen is the binder. Annual world consumption of bitumen is more than 100 million tons, with about 80 % going to paving, followed by roofing products at about 10 %. Asphalt is a material prone to gradual degradation not only from mechanical wear, but also through bitumen oxidation.

Lignin could favorably impact the sustainability of roads and roofs by reducing bitumen demand by at least two routes: Lignin hindered phenolic structure gives it antioxidant properties and therefore can slow down asphalt oxidative deterioration making replacement less frequent. In addition, certain lignins have similar adhesion properties and dimensional stability as bitumen, which makes possible substitution of substantial portions of fossil-derived bitumen with renewable lignin. Applied research concerning inclusion of lignin and lignin derivatives in asphalt formulations has recently advanced to larger scale testing. For instance, in The Netherlands sections of a highway were paved in mid-2015 with asphalt in which 50 % of conventional bitumen had been replaced with a modified wheat straw soda lignin [26]. In the US, corn stover lignin pyrolysis oil is under testing as a component of asphalt in a bike trail at 3 % inclusion level [27–29].

9.3.3 Macromonomer

The concept of using lignin as a reactive macromonomer that is present during a polymerization reaction and gets incorporated in the architecture of the resulting copolymer has been explored in the past and some progress has been made toward its industrial implementation (Fig. 9.3).

Lignin functionality includes being a polyphenol and a polyol, and therefore the polymer types in which the macromonomer concept has been explored most include phenol formaldehyde (PF) thermoset resins and use as a polyol, particularly in polyurethanes (PUs) and polyesters. This field has been reviewed recently [3, 30] and here we give an update on developments of the last few years.

Lignin polyphenolic structure can be incorporated in phenolic resins either during the synthesis of the resin or (depending on type and properties of the lignin and on resin specifications) as a partial replacement for the final product. When the lignin is incorporated during resin synthesis it is often desirable to preactivate the lignin, for instance by pre-reacting with formaldehyde or with phenol [31]. Such a preactivation is designed to get around steric hindrances that may cause the lignin to

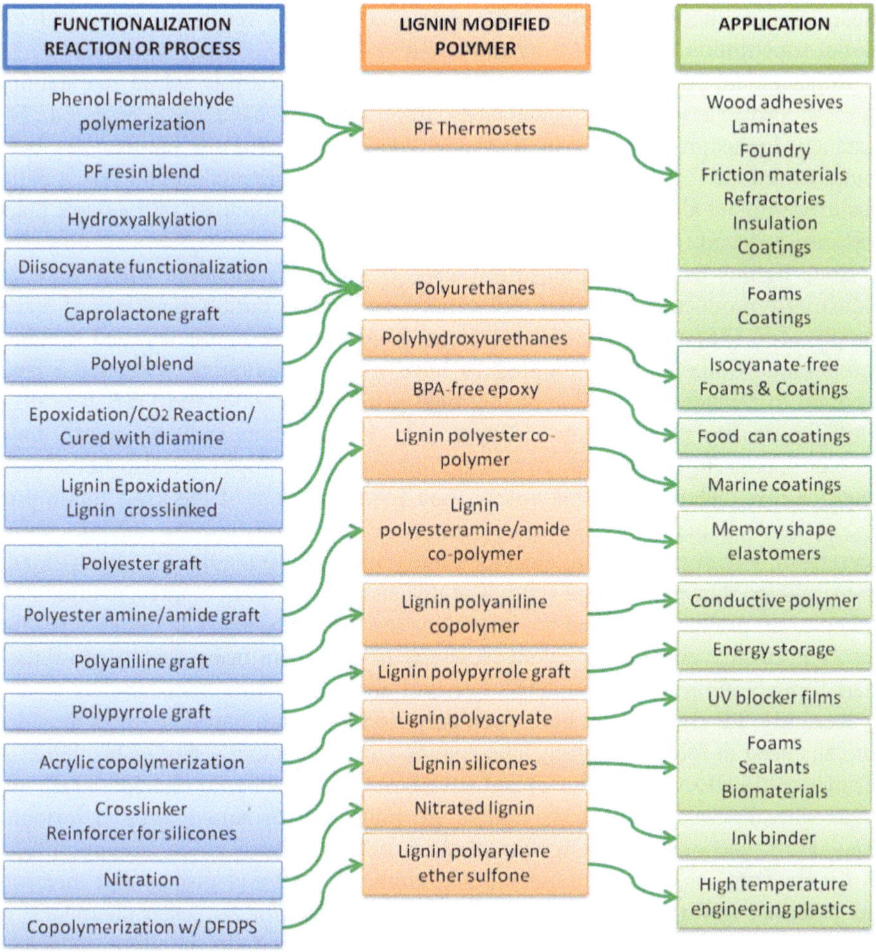

Fig. 9.3 Lignin reactions as a macromonomer, the resulting lignin-modified polymers and their areas of application

be a sluggish co-reactant. Lignin activation by enzymatic routes has also been proposed [32, 33]. Other activation techniques that are under initial development include the use of ILs and deep eutectic solvents. The latter have shown some promise when applied to a wheat straw soda lignin, but a lot of work is still needed before they become industrial realities [34].

In the 1990s, hardwood organosolv lignin from the Alcell demonstration plant mentioned above was routinely used in industry as partial replacement of PF binders used for oriented strand board (OSB) panels, foundry sands, and brake pads. The low molecular weight and low softening temperature of this organosolv lignin made it particularly compatible with certain powder PF resins, permitting direct partial substitution of the resin by the end user [8, 15].

Replacement levels up to about 25 % resin were possible while maintaining strength properties, and in some cases improvements have been reported. For instance, inclusion of lignin in formaldehyde-based thermoset binders has been found to reduce formaldehyde press emissions during oriented strand board panel manufacture and also during the life of the panel [35]. Particulate in the work place was also reduced.

Soda lignins have been industrially used (particularly in India, where they are industrially produced) as partial replacement for phenol during synthesis of resins for plywood [36], foundry shell moldings, high pressure laminates, and molding compounds. Soda lignins modified to have a lower softening point have been reported to replace 20 % of a conventional petrochemical-derived OSB face resin while improving resistance to moisture [37]. Kraft lignins (which recently have become increasingly available in North America and Europe) are also being marketed as raw materials for thermoset wood binders. Some of the kraft lignin producers and/or marketers are also involved in wood panels manufacturing operations, which may lead to enhanced integration and synergies [6, 38]. It should be mentioned that commercial success in this and other similar markets is intimately linked to oil price. At times in the last few years, volatility in oil and phenol pricing has significantly curbed the extent to which lignins have been able to grow their market penetration as replacement of phenol and other petrochemicals.

Recently, PF crosslinkers have been proposed for lignin-based coatings intended for cans used in food packaging. The coatings are substantially free of bisphenol-A (a suspected endocrine disruptor) and have been prepared based on soda, kraft, and organosolv lignins. While soda lignins are preferred for aqueous-based formulations, kraft, and organosolv lignins function well in solvent based coatings [39].

Lignin contains multiple aliphatic and aromatic hydroxyl groups and therefore has potential to function as a polyol to react in PU and/or polyester chemistry, as illustrated in previous reviews. Earlier work focused in alkaline catalyzed hydroxypropylation with propylene oxide under pressure as a way to improve the reactivity of the polyol [15, 30]. Recent studies have emphasized green chemistry approaches such as the use of glycerol and its di- and tri-glycidyl ethers as reactive solvents (that may also graft on the lignin), polypropylene glycol triol or castor oil as chain extenders, and water as the blowing agent [40, 41].

Among other modifications, acid-catalyzed reaction of kraft lignin with cardanol (a biobased product extracted from cashew nut shells) has been proposed as a way to obtain a polyol that when reacted with 2,4 toluene-diisocyanate results in elastomeric PUs with a range of glass transition temperatures and crosslink densities that may be tuned by varying the cardanol to lignin ratio [42]. Solvent fractionation followed by crosslinking with formaldehyde in a solvent system was used to convert organosolv lignin into a high molecular weight polyol that has been used in polyesters and PUs. Thus, base-catalyzed reaction of crosslinked organosolv lignin with dicarboxy terminated polybutadiene (Mn 4200) in dioxane gave polyester with hard and soft segments, capable of forming a free-standing film [43]. Similarly, reacting 65–80 % crosslinked lignin with telechelic isocyanate-terminated polybutadiene (molecular weight about 2500) in anhydrous dioxane gave thermoplastic PU products capable of forming films and with properties acceptable for many applications, in spite of high lignin content [44]. Another promising approach involves direct reaction of the 2-methyltetrahydrofuran soluble fraction of kraft lignin with toluene-diisocyanate to yield highly hydrophobic, thermal stable thermoset coatings [45].

Pre-modifying lignin or fractionating it to improve its polyol characteristics for inclusion in PU and polyesters has not been practiced industrially, yet. Alternative approaches in which unmodified lignin is pre-dissolved in a polyether or polyester polyol have been proposed [46] and may offer a more ready route to industrial implementation. For instance, a 30 % by weight solution of organosolv lignin in a commercial polyether polyol was used to make rigid foams intended for automotive applications when reacted with a commercial isocyanate [47]. More recently, polyester resins have been prepared by the reaction at 200 C of an annual fiber derived soda lignin with 2 methyl 1,3 propane diol, a mixture of polyacids, and additives. When used in marine coatings these resins have been claimed to impart superior corrosion resistance relative to polyester coatings without lignin [48]. Work on isocyanate binders for OSB-production carried out by ICI Polyurethanes (now part of Huntsman Inc.) showed that the use of lignin in combination with specific lignin solvents results in improved performance relative to lignin-free isocyanates used at the same binder-level. As a result it was possible to reduce isocyanate requirements to achieve equal or better mechanical properties. The use of lignin also resulted in improved release of the product from the press platens, alleviating a problem that has hindered the more widespread use of isocyanate binders [49]. Functionalization of lignin with isocyanate has been used to improve compatibility in PU chemistry. Thus, Kessler et al. have recently reported successful incorporation of up to 30 % of lignin modified with octadecyl isocyanate in vegetable oil based PU composites [50]. Similarly, blends of an isocyanate prepolymer and lignin functionalized with diphenyl methane diisocyanate have been proposed as binders for lignocellulosic composites [51, 52]. Introduction of lignin and hydrophobically treated nanocellulose in PU formulations based on 100 % soy polyol resulted in foams with about 6 % higher density, lower open cell content, higher mechanical properties, and less odor relative to formulations without lignin and nanofibers [53].

Grafting polycaprolactone chains on lignin's hydroxyl groups is a strategy to modify lignin by chain extension while using a biocompatible and biodegradable ingredient. A couple of decades ago, Oliveira and Glasser reported on star-like polymers with a hydroxypropylated lignin core and polycaprolactone arms [54]. More recently, the reaction has been extensively studied by other authors. Hatakeyama et al. [55] reported on the graft copolymerization of caprolactone and hardwood organosolv or softwood kraft lignins and in the subsequent reaction with isocyanate to form PUs. When the caprolactone/OH ratio was 5, the corresponding graft copolymer was a flexible elastomer. At higher ratios crystalline and brittle materials were obtained. Crystallinity was observed for the derived PU when the caprolactone to OH ratio was 10. Averous et al. [56, 57] produced copolymers of straw soda lignin and caprolactone by ring opening polymerization in the presence of stannous octoate and confirmed Hatakeyama's reagents ratios required for elastomeric and brittle lignin–caprolactone graft polymers.

Grafting acrylic acid groups on a lignin obtained as co-product of bioalcohol fuel production and copolymerizing with butyl methacrylate and methyl methacrylate resulted in a copolymer that in film form had excellent UV light absorbing properties, good transparency in the visible range, good mechanical properties, and chemical resistance [58].

A recent development in lignin valorization is its proposed use in advanced elastomeric copolymers by crosslinking low molecular weight lignin fractions with a poly(ester amine) or poly(ester amine amide) prepolymer. The methanol soluble fraction of industrially produced wheat straw soda or softwood kraft lignins has been crosslinked in a one-pot melt copolymerization with prepolymers obtained from the reaction between adipic acid (or other dicarboxylic acid such as dodecanedioic acid) and triethanolamine and/or tris(hydroxy methyl) amino methane. The resulting highly branched poly(ester amine) or poly(ester amine amide) have high functionality, high molecular weight, and low viscosity/molecular weight ratio. By taking advantage of the reactivity differences between dicarboxylic acids and hydroxyl and amine groups, polymers with a tertiary amine core and amide linkages along the backbone of the polymer were obtained. Lignin was used up to 50 % content level and the resulting copolymers were elastomers with a single glass transition temperature, and improved young's modulus, tensile strength and ultimate strain. When adipic acid was the diacid, toughness was particularly enhanced at lignin content around 40 % by weight [59, 60]. Use of dodecanedioic acid improved the copolymers elasticity by increasing the distance between the branch points. Materials with dual and triple shape memory effects triggered by temperature can be prepared and their properties can be tuned by changing the prepolymer branching structure and the lignin content. These smart polymer systems could be used in medical applications and consumer products where body temperature actuation of the material is required [61–63]. The concept has been extended to the preparation of shape memory elastomers with a high renewable content by starting with a glycerol-based polyester prepolymer [64].

Another recent new development in the field of elastomers is the inclusion of lignin in polysiloxane (silicone) formulations. Silicone is used in many applications including foams, sealants, printing, and biomaterials. Silicones are expensive and require fillers (typically fumed silica or other minerals) for enhancing some of their mechanical properties. Stiubianu et al. first reported in 2009 on blends of soda non-wood lignin and silicone and found comparable properties to silica-filled silicone and projected that lignin inclusion could make silicone formulations more cost competitive and expand the uses for silicone-based products [65, 66]. More recently, Brook et al. have reported that softwood lignin can be incorporated as a crosslinker as well as a reinforcer in silicone elastomers. Their approach is based on the reduction of various functional groups in lignin by hydrosilane groups in the silicone polymer in the presence of silicon hydride, and with $B(C_6F_5)_3$

(tris(pentafluorophenyl)borane) as catalyst [67]. The products have mechanical properties tunable according to lignin content and molecular weight of the silicones and within the range of commercial products. The elastomer exhibited excellent resistance to high temperature aging and acceptable resistance to solvent extraction. The principle has been used for the production of biocomposite foams and elastomers [68].

Inherently conductive polymers (ICPs) such as polyaniline have significant potential in antistatic fabric coatings, electrostatic dissipation, corrosion prevention, shielding of electronic equipment and aircraft guiding systems, smart windows, and radar invisible coatings. In their conductive form ICPs are water insoluble, which makes them difficult to process. This has led to the use of soluble polymeric acids (such as lignosulfonic acids) as templates for the polymerization of the ICP. When a lignosulfonic acid derived from a commercially available, highly sulfonated, ethoxylated kraft lignin was present during the aniline polymerization reaction, ICP chains grafted on lignosulfonic acid were obtained. The graft copolymer is cross-linkable and the pendant sulfonic acid groups not only make it highly water soluble, but also act as dopants to enhance conductivity [69–74]. Ligno-pani can be used as an ingredient in corrosion protection products [75, 76]. Applications for Ligno-pani have also been reported in xerographic printing [77]. Highly conductive composites of Ligno-pani with carbon black or graphite have been prepared and could have use in electrodes manufacture, as primers for plastics in electrospray coatings and as plastic additives [78]. The promising lignin polyaniline copolymer originally developed by NASA was produced industrially in the US by Polyone in the early part of this century [79] but was discontinued for reasons unknown to us.

Other recent developments in electrochemical applications of lignin include its use in batteries for energy storage. It has been reported that a 50:50 lignosulfonate polypyrrole interpenetrating composite obtained by the electrochemical polymerization of pyrrole in lignosulfonate aqueous solution has been proposed as a low cost energy storage material [80]. Testing of carbon nanotubes surface functionalized with kraft lignin has shown that the reversible redox activity of the lignin adsorbate can be utilized for charge storage, providing more than a 100 % increase in the capacitance of lignin-modified carbon nanotubes compared with unmodified ones [81]. The use of carbon nanomaterials derived from lignin in electrochemistry is discussed in the Sect. 9.3.8 "Carbon Fibers and other carbon materials".

The objective to incorporate lignin in high temperature stable engineering plastics appears to get closer: softwood kraft lignin was recently copolymerized with 4, 4′-diflourodiphenyl sulfone (DFDPS) to form lignin poly-(arylene ether) sulfones, which exhibited high temperature resistant thermoplastic behavior. The process requires solvent fractionation of the lignin and careful protection of the aromatic hydroxyl groups in lignin in order to avoid gelation [82].

Lignin could potentially help the printing industry develop alternative strategies to achieve products with a lower carbon footprint. Recently, it has been reported that derivatives of wheat straw soda lignin were nitrated with concentrated nitric acid and that the resulting nitrated lignin was used as a binder component in flexographic ink and overprint varnishes, giving better heat resistance than the control [83]. Esterification prior to nitration allowed the use of ester solvents during the nitration and the final esterified-nitrated lignin had enhanced solubility in alcohol/ester blends, which are the solvent mixtures typically used in flexographic or gravure inks. The nitrated lignin esters could partly or completely replace more costly nitrocellulose [84].

It has been reported that lignin inclusion in inks may bring about significant property and environmental improvements in the case of alkyd resins. For instance, addition of hardwood organosolv lignin to alkyd resin-based commercial offset inks, varnishes, and paints modifies their rheology in such a way that a very significant and desirable reduction of misting in the printing press takes place, without impact on drying or other properties [85]. Kraft lignin from a mix of hardwood and bamboo has been reported to replace about 35 % of a specially formulated alkyd resin for use in screen ink formulations. It is claimed that the combination uses significantly less petroleum-based products (including lower volatile organic compounds), at a lower cost and without negatively impacting printing properties. In fact, properties such as hiding power improved with the lignin formulation [86].

Past examples of lignin use in epoxy formulations have been discussed elsewhere [15, 87]. Recently, steam explosion bamboo lignin (SEBL) was epoxidized with epichlorohydrin and cured with unmodified SEBL, i.e., a completely bisphenol-A-free system with a high renewable content. Flexural strength and thermal properties were better than for epoxidized SEBL cured with the conventional epoxy curing compound 1-(2-cyanoethyl)-2-ethyl-4-methylimidazole (2E4MZ-CN) but fell short relative to the control, which was diglycidyl ether of bisphenol-A cured with 2E4MZ-CN [88].

Epoxidation of soda herbaceous lignin or hardwood kraft lignin with epichlorohydrin followed by catalyzed addition of CO_2 to the oxirane ring in an imidazodium-based, reusable ionic liquid medium has been used to form cyclic lignin carbonates [89]. Cyclic carbonates can be reacted with diamines to form polyhydroxyurethanes, which offers an isocyanate-free route to urethane products [90].

9.3.4 Lignin in Blends with Thermoplastic Polymers

Incorporation of lignin in thermoplastics has been seen for years as a potential strategy to produce materials with modified properties and lower cost. This topic has been reviewed in the past by others [87, 91–93] and here we just give an update of some recent developments. Figure 9.4 summarizes some of the most significant developments in this field in the last few years. The discussion in this section is limited to blends with polymeric materials other than rubber and biopolymers, which are discussed in separate sections.

When developing successful polymer blends a factor of paramount importance is the compatibility between the components of the blend. Blends need to be processed at temperatures sufficiently high for good flow but below the degradation

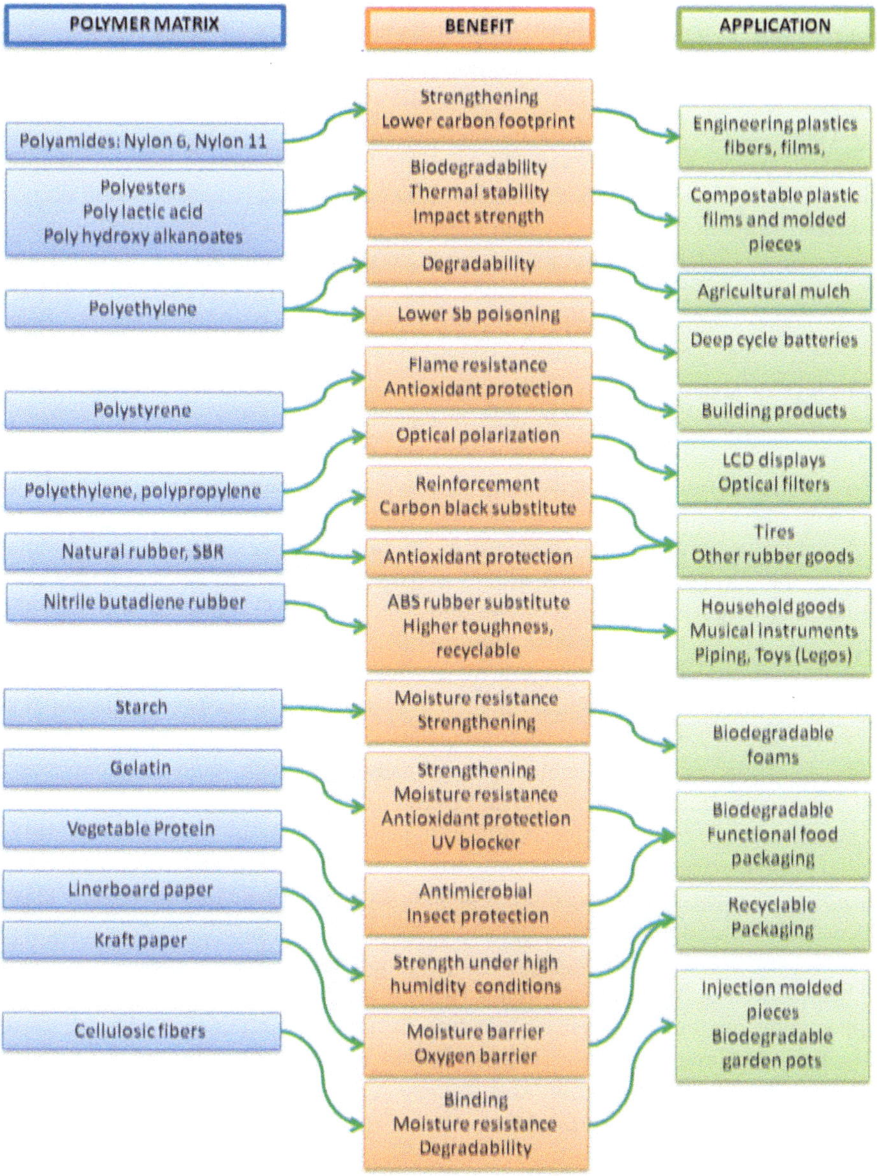

Fig. 9.4 Examples of significant lignin-based polymer blends illustrating end products and highlighting benefits derived from the use of lignin

point of the components or the point at which lignin may self condense. Good solubility of the lignin in the other polymer(s) and favorable interactions (such as hydrogen bonding or reactions) between the functional groups are conducive to compatible blends. Pouteau et al. conducted a systematic image analysis of extruded binary blends of 1 % softwood kraft lignin and various polymers, which revealed that best compatibility was achieved with polyester amide, polystyrene, unplasticized polyvinyl chloride, and poly(butylene succinate adipate). Blends with low density polyethylene (PE), polypropylene (PP), and polyvinyl acetate exhibited less compatibility as evidenced by higher amounts of lignin aggregates. Lower molecular weight fractions obtained by solvent fractionation were more compatible with PP [94]. It is also possible that the higher content of non-polar wood derived resinous components in the dichloromethane soluble fraction improved compatibility with polyolefins. Kadla and Kubo have shown compatibility of lignin with polyethylene oxide (PEO) but immiscibility with polyvinyl alcohol and PP [95]. These authors have illustrated the crucial role of hydrogen bonding for better miscibility [96–98]. The quality of polymer blends may also be profoundly impacted by the size and morphology of the lignin particles. A recent patent application by the Fraunhoffer Institute in Germany claims that a high proportion of particles of 5 microns or less with a round or slightly elliptical shape is desirable [99]. Lignin particle size reduction to less than 100 nm was also effective for making polyvinyl alcohol-lignin films with enhanced thermal stability [100].

To improve compatibility in blends of lignin with other polymers one common strategy that has been tried is the use of compatibilizers such as an esterified version of the matrix polymer. For instance, maleated PP [101], ethylene vinyl acetate copolymer [102, 103], anhydride modified PE [104], and maleated polycaprolactone [105] have shown to result in blends with better properties than when no compatibilizer is used. Feldman et al. have shown that plasticizing lignin may improve dispersibility during processing and lead to enhanced blend properties [106]. Grafting on lignin chemical groups compatible with the matrix polymer is another useful strategy. Alkylation and/or esterification have been extensively used to increase lignin content significantly in blends with various polymers [107–118]. Graft copolymerization of 2-ethyl-2-oxazoline with lignin extracted with phenol has also been proposed to increase thermal stability and improve compatibility [119]. Recently, sylanation has been proposed to make the lignin more hydrophobic, and therefore more compatible with typical petrochemical polymers [68].

Unmodified softwood kraft lignin (which was the most industrially available lignin with potential for polymer blending in the mid-20th century) has relatively poor thermal flow properties, which presents significant challenges for polymer blends. When hardwood organosolv lignins with high purity and better thermal flow characteristics became available in pilot and semi-industrial quantities in the late 20th century some lignin polymer blend applications advanced beyond the laboratory. For instance, blends of PE and Alcell organosolv lignin were processed in commercial film blowing equipment and the products were tested as degradable plastic mulch. Furthermore, it was found that the presence of lignin also facilitated the recyclability of the plastic, since the degradation temperature of the blend was higher than for PE by itself [120, 121]. In a specialized application, separators for lead acid batteries in deep cycle uses (such as those used in electric golf carts and forklifts) were manufactured from blends of ultra high molecular weight PE and hardwood organosolv lignin and resulted in improved battery life relative to conventional rubber leaf separators as antimony poisoning (a common problem in batteries) was significantly reduced when lignin was present [122]. This technology to prepare battery separators was later applied to soda herbaceous lignins [123].

The enhanced compatibility of lignins with polyester and polyamides has been confirmed in extrusion studies. Thus, Nitz et al. [124] melt blended polyamide 11 (derived from castor oil), biodegradable polyester amide, and biodegradable aromatic polyester with hardwood organosolv lignin and sisal and abaca soda lignins up to 50 % lignin contents. The blends of the polyester and any of the lignins at 30 % or lower exhibited elongation at break above 500 % and very high impact strength. At higher contents sisal lignin was somewhat lower in strain and impact strength relative to the other two lignins. Up to about 40 % lignin content the materials had about the same or slightly higher Young's modulus and yield stress as the neat polyester. At 50 % lignin content modulus and yield stress were significantly higher, particularly in the case of sisal lignin. For the polyester amide based blends the Young's modulus and yield stress generally decreased as lignin content increased with the exception of slightly higher Young's modulus for organosolv lignin below 20 % content and higher yield stress for organosolv and abaca soda lignin above 20 %. Blends of polyamide 11 with all lignins showed increase in modulus up to 50 % lignin content and increase in stress up to 30 %, followed by a decrease. Electron microscopy showed that better dispersions were achieved with hardwood organosolv and abaca soda lignins within the polyester and polyester amide matrices. A separate study by the same authors using polycaprolactone and hardwood organosolv lignin showed an increase in Young's modulus (particularly when lignin content was 60–70 %), a moderate decrease in yield stress, a significant decrease in impact stress, while maintaining high strain [105]. Recently, homogeneous single phase blends of up to 30 % wheat straw soda lignin and polyamide 6 (PA6) were prepared in a twin-screw extruder and the blends properties generally followed the trends reported above for the PA11-lignin blends. FTIR analyses revealed hydrogen bonding interactions between the two polymers. Young's modulus and yield stress of the blends were close to those for 100 % PA6. Elongation at break was not negatively affected [125].

The morphology of single screw extruder prepared blends of polyethylene terephthalate (PET) and up to 20 % lignin recovered from bagasse after steam hydrolysis indicated good dispersibility and a significant increase of crystallinity and crystal size of PET [126]. Lignin nanoparticles prepared by precipitation of soda lignin from wheat straw from an ethylene glycol solution were added at a rate of 1.5 % by weight to bio-poly(trimethylene terephthalate) by melt blending in a rheomix and resulted in improvement in tensile and flexural properties (strength and modulus) in the 8–16 % range, impact strength improved by 30 %. The presence of the lignin nanoparticles enhanced biodegradability. Inclusion of vapor-grown carbon fibers at a level of 7 % resulted in further strength and biodegradability improvements [127].

The US company Cyclewood Solutions appears to be approaching commercialization of its blends of esterified hydroxypropyl lignin with thermoplastic polymers (such as the biodegradable polyester Ecoflex) to produce compostable film and injection molded products for use in packaging, garbage bags, pet waste bags, etc. [128–133].

Optical properties are affected when lignin is present in polymer blends. For instance, UV absorbance, and light reflectance (gloss) increase [134, 135]. Recently it was reported by Kimberly Clark that polyolefin blends with up to 30 % wheat straw soda lignin cast into films with a twin extruder had optical polarization (OP) properties. OP filters are conventionally made using iodine. They are used for a range of applications from sunglasses to liquid crystal displays and the growth in their demand is putting significant pressure on the supply of iodine, a valuable micronutrient. Lignin was found as lamellae oriented in the machine direction of the polyolefin films and this lamellar structure gave the films OP properties. With additional optimization, lignin could become a sustainable alternative to iodine for the production of OP films [104].

Among the most sought after properties when incorporating lignin in plastic blends is protection against oxidative, ultraviolet, and/or thermal degradation. Demonstration of the in vitro antioxidant properties of lignins is supported by a large amount of research [136–143]. The antioxidant performance of lignin-containing blends in plastic products has been studied for PP [144, 145]. Lignin UV protection effects have been demonstrated for blends with polyolefins [146–148]. Similarly, effects on thermal stability have been shown for lignin blends with PU [149, 150], virgin and recycled polystyrene [151] polyvinyl alcohol [152, 153] and polyolefins [146]. Furthermore, improvement on flame retardancy (attributed to lignin's char forming ability) has been shown for lignin blends with polyolefins [117, 154–156] and polystyrene [157]. In the majority of studies, lignin use led to improvements in all of the above properties. Some studies have shown the existence of optimum inclusion rates and the importance of the nature of the polymer being protected. As an example, Ciobanu et al. found deterioration of thermal stability for PU-flax soda lignin films, particularly above 4.2 % lignin [150] while Alexy et al. have reported stabilizing effects at low concentration in lignin-PP blends, but not in lignin-PE blends [146].

9.3.5 Lignin in Rubber

Rubber reinforcement with kraft lignins and lignosulfonates instead of carbon black was intensively researched at the time of World War II. As reviewed elsewhere, earlier efforts showed little evidence of reinforcement by these lignins when used in dry compounding, which was attributed to their tendency to form self-aggregates during rubber processing [87]. Coprecipitation of kraft lignin and rubber from a lignin latex liquid blend gave better results, but clearly not good enough to warrant implementation on a significant scale at that time.

While recent attempts to use dry compounding to incorporate lignin as carbon black replacement indicate some progress relative to efforts more than 60 years ago

[158], most recent work has been directed toward finding ways to improve compatibility with rubber via functionalization and/or by reducing the particle size and also toward the evaluation of lignins other than kraft and sulfite. Lignin modified by treatment with benzoyl peroxide was evaluated as a replacement for carbon black in nitrile rubber vulcanizates and gave better thermal stability and resistance to oil and fuel, but the reinforcement characteristics were inferior to carbon black [159]. Modification of a commercial lignosulfonate (a very polar type of lignin) either by esterification or silylation resulted in materials that compared well with carbon black in rubbers according to a patent recently granted to tire company Goodyear [160]. A patent application by Pirelli claimed that sulfur-free lignins functionalized by etherification and esterification were successful partial replacements of carbon black [161]. The tensile and strain at break of styrene-butadiene rubber (SBR) compositions containing 30 parts per hundred rubber (phr) of commercial calcium lignosulfonate modified with cyclohexylamine increased proportionally to the quantity of cyclohexylamine used in the modification. This technique appears to be promising, but no comparison with carbon black was published [162]. Evaluation of the effect of lignin nanoparticles prepared in the presence of poly-(diallyldimethylammoniumchloride) on the properties of natural rubber confirmed that particle size reduction enhances the efficiency of lignin as a reinforcing agent. Much work, however, is needed on practical and economical protocols for the preparation of such nanoparticles [163].

Among lignins other than kraft and lignosulfonate that have been evaluated in this century, a sulfur-free lignin extracted from autohydrolyzed beech wood was tested in SBR and was well dispersed and appeared to have reacted with sulfur and other compounds in the vulcanization system. Properties such as stress at 100 % elongation and tensile and elongation at break improved with increasing lignin up to 50 phr, while curing behavior was little changed up to 20 phr lignin, but slowed down at higher levels. The study did not provide a comparison with carbon black [164]. A purified lignin extracted from a biorefinery residue did not perform well as a 30 % replacement for carbon black in a natural rubber/SBR formulation. Pre-reaction of this lignin with hexamethylenetetramine (HMT) brought the properties close to the formulation with the carbon black control. A modified commercial soda lignin also pretreated with HMT was close in performance to the control, but all lignin products showed poor dispersion, which may be improved by particle size reduction [165].

Researchers at Oak Ridge National Laboratory in the US have recently studied 50–50 blends of various lignins (mostly softwood kraft and its methanol soluble fraction) with nitrile butadiene rubbers having nitrile contents of 33–51 %. The authors proposed that the new material is a compositional analog of acrylonitrile-butadiene-styrene rubber (ABS), where the styrene fraction is completely replaced with lignin. The elastomers obtained exhibit a range of properties that depends on the type of rubber, lignin, and processing conditions. The compounding method used leads to the formation of 5–100 nm lignin lamellae, which are smaller than previously reported for lignin in a rubber matrix. The material has significantly higher toughness than nonrenewable ABS copolymers and is highly recyclable [166].

Recently a novel reinforcement with lignin-epoxy networks has been applied to SBR and nitrile butadiene rubber. Initially, 50:100 lignin:latex coprecipitates were produced and then were melt compounded with a novolac–epoxy resin (up to 20 phr). FTIR data substantiated ring opening reaction of the epoxy groups with the OH groups of lignin. Dynamic material analysis and scanning electron microscopy were consistent with the occurrence of a reaction. Hardness, modulus, tear strength, crosslinking density, and thermal stability were significantly enhanced for the lignin-novolac epoxy containing formulations [167, 168].

Carbon black replacement by a renewable product could lead to a very significant reduction of the carbon footprint of tires and other rubber goods given the large quantities of carbon black that are used. The relatively low cost of carbon black, however, may make it difficult for the higher purity lignins to compete economically as reinforcing fillers, particularly if they have to be functionalized and their particle size reduced. There are other examples of opportunities for lignin in rubber formulations which are smaller markets in terms of volume, but may contribute higher value functional properties. For example, kraft lignin coprecipitated with tall oil pitch has been proposed as a rubber tackifier [169]. Hardwood organosolv lignin has been reported to be a dual purpose (tackifier and antioxidant) additive [170]. The antioxidant, photo-, and thermal-stabilizer effects of lignins in various types of elastomers has been documented in several studies [171–174] Flame retardancy effects for ABS rubber have been reported and are attributed to the strong char forming ability of the lignin [175].

Lignosulfonates have been reported to be effective emulsifiers in water based adhesive formulations that replace solvent-based adhesives used to join two different rubber tire portions together [176], and also as components of a non-aqueous adhesive for the textile ply in tires [177]. Lignosulfonate has also been claimed to be an effective dispersant for silica used as filler in tire rubber compounds [178].

9.3.6 Lignin Blends with Biopolymers

Biopolymers are alternative raw materials to conventional petrochemical polymers that offer the promise of a lower carbon footprint, increase in renewable content and potential for degradability and compostability. They include naturally occurring polymers such as polysaccharides and proteins, and polymers such as polylactic acid and polyhydroxyalkanoates, which are manufactured from biobased monomers. Addition of lignin has been proposed as a way to improve water resistance, modulus, antioxidant protection or other properties that the primary biopolymer may be deficient in. Starch is a polysaccharide that may be processed by casting or extrusion in the presence of a plasticizer to give films, and molded articles that are biodegradable and compostable. Starch, however, is hygroscopic, has poor moisture barrier properties and low tensile strength. Incorporation of lignin into the starch matrix generally permits the production of materials with better water resistance, antioxidant properties, thermal resistance, and tensile strength than starch on its own [179–185]. Electron beam irradiation appears to improve the water resistance of starch–lignin films especially when the lignin is relatively rich in carbohydrates and low molecular weight bioaromatic fractions [186, 187] Lignin esterification improves further the effect on water resistance and strength [188]. Starch films plasticized with glycerol had maximum stress and modulus of elasticity improvements of 256–265 % and 1274–1478 %, respectively, when 5 % soda lignin and 1 % nano-crystalline cellulose were added [189, 190].

Starch foams are well known as biodegradable alternatives to foamed polystyrene. It has been reported that the addition of 20 % commercial kraft lignin did not interfere with foam density or morphology, but resulted in improvements such as reduced water absorption, thermally stabilized the starch and brought the strength properties to a comparable range with polystyrene foams [191].

Starch–lignin, gelatin–lignin and starch–gelatin–lignin films have been tested as materials for preparation of drug delivery systems [192, 193] Gelatin–lignin films plasticized with glycerol have also been studied in the context of food packing. Recent studies involving two herbaceous soda lignins (Protobind 1000 and Protobind 2400) have shown that the resulting composite films were more elastic than gelatin alone and gained in light barrier properties, which could be useful for preventing ultraviolet-induced lipid oxidation of packaged foods [194]. Three Eucalyptus lignosulfonates were also evaluated and relative to soda lignin had lower film elasticity and antioxidant activity [195]. The films that were obtained with soda lignin Protobind 1000 were tested in a high pressure packing system for ready-to-eat cooked salmon which was used as an alternative to conventional thermal processing. High pressure processing in combination with gelatin–lignin film showed that salmon's fresh appearance was preserved, protein quality was improved, levels of carbonyl groups formed immediately after processing were reduced, and lipid oxidation was prevented at advanced stages of chilled storage [196].

Lignin has also shown to improve moisture resistance, UV light and water vapor barrier properties, thermal stability, processability and strength characteristics of films made with other natural biopolymers that may be candidates for food packaging and edible films such as agar [197], wheat gluten [198–200], and zein protein [201]. Films based on lignin and natural polymers have been claimed to perform well as controlled release matrices for bioactive compounds. For instance, bilayer films containing 3 % essential oils (such as clove and citronella) in a matrix made of soy protein isolate, herbaceous soda lignin, and formaldehyde exhibited sustained antifungal and insecticidal properties. If an alternative to formaldehyde is demonstrated, such films could be the basis of packaging systems capable of protecting fruit and other food, while at the same time exhibiting UV light blocking properties and being biodegradable [202, 203]. A recent patent application claims that the incorporation of suitable bioactives in formulations based on chitin nanofibrils and lignin have antioxidant and anti-inflammatory properties that are useful in a wide range of consumer products [204].

Among polymers based on biobased monomers, polylactic acid (PLA) has reached significant industrial scale. PLA is a crystalline polyester with relatively good tensile strength but poor impact properties that has received significant attention in recent years as a potential renewable plastic that could replace petrochemical-derived polymers used in packaging and other applications. Blends of lignin and PLA prepared in a twin-screw extruder exhibited good compatibility. Antioxidant and oxygen barrier properties were enhanced. PLA plasticized with PEG was more compatible with softwood kraft lignin than unplasticized PLA [205]. It has also been reported that impact strength of PLA–lignin blends after accelerated weathering is highly improved relative to PLA [206]. Furthermore, PLA-graft lignin polymer blends and copolymers have been prepared by lignin butyration followed by hot blending with PLA [207]. Lignin along with compatibilizers such as an oligomeric styrene-acrylic epoxide or polymeric methylene diphenyl diisocyanate have been incorporated with some success in compostable and biodegradables blends of PLA with other polyesters such as poly butylene succinate, polyhydroxyalkanoates, or poly (butylene adipate-co-terephthalate) [208–211]. Compatibility of lignin and PLA has also been improved by lignin acetylation. Thus, blending PLA and acetylated softwood kraft lignin or acetylated organosolv almonds shell lignin maintained properties close to those of neat PLA, with the exception of elongation at break, which was increased significantly [212]. While preserving biodegradability is often a target when introducing lignin in formulations based on degradable polymers, there have been reports of significant slowing down of biodegradability for melt extrusion blends of sugarcane bagasse soda lignin with a polyhydroxyalkanoate [213] and blends of hardwood organosolv lignin with caprolactone [105].

9.3.7 Lignin as a Component in Cellulosic Fiber Composite Products

Cellulose fibers are the basis of highly sustainable and recyclable products including paper, specialty celluloses, and composites. In the case of paper goods, additives may be added to improve performance characteristics. Often such additives are of nonrenewable origin and frequently interfere with the efficient recycling of paper. There is scope for incorporating industrial lignins in the manufacture of paper, particularly packaging grades made with weaker recycled fiber, where lignin may enhance properties such as stiffness and resistance to humidity.

In an effort to improve the stiffness of paper under high humidity conditions, linerboard was treated in a size press with a blend of styrene-acrylic and either softwood kraft or hardwood organosolv lignin (ratio of lignin to synthetic polymer 2:1) resulting in a pickup of 13-20 % formulation solids based on paper. The treated linerboard exhibited a 103–120 % increase in the Cross Direction Short Column Test (an indication of stiffness) at 50 % relative humidity (RH), and 94–112 % increase at 95 % RH [214]. It has also been proposed to saturate paperboard with a lignosulfonate solution and to follow with a treatment with an amine polymer epichlorohydrin adduct that insolubilizes the lignosulfonate on the paper and reduces lignosulfonate bleeding under humid conditions [215].

Lignin has also been tested to enhance the moisture barrier characteristics of paper. Effective barrier properties require film forming formulations and one strategy that has been proposed relies on the use of low molecular weight, low glass transition temperature hardwood organosolv lignin. In an application that was demonstrated using industrial equipment in the 1990s, this lignin was formulated as a reduced particle size ammonia-based aqueous dispersion containing also a latex binder and mica (lignin:latex:mica 55:17:28) and was used to coat paper. At a coat weight of 44-54 g/m^2 a water vapor transmission rate (WVTR) of 7.3 g/m^2/day at 25 C 50 % RH was reported. Coated paper thus prepared was laminated and used to wrap paper rolls produced in a large coated paper mill. The lignin-based wrapping performed comparably to wax laminates and effectively prevented changes of moisture and brightness of the paper rolls. The wrapping could be recycled by a number of methods and was expected to be highly compostable [216]. In an alternative approach studied more recently, kraft lignin reacted with octylamine in the Mannich reaction had reduced particle size and was used to prevent excessive stickiness in barrier coatings based on prevulcanized natural rubber [217].

Ammonia-based lignin liquid formulations have also been used in conjunction with crosslinkers, sizing agents, plasticizers, and deodorizers followed by an acid treatment of the coated paper to provide water resistant products [218, 219].

Another approach to implementation of barrier coatings has been to esterify kraft lignin with long chain organic acids, such as lauric, palmitic, or tall oil fatty acid, which increases lignin's hydrophobicity and lowers its softening point. Best results reported are for coatings based on lignin palmitate solutions in acetone, which at 10.4 g/m^2 coating gave MVTR of 40 g/m^2/24 h and oxygen transmission rates of 1750 cm^3/m^2/24 h. While the work discussed makes a case for the usefulness of the lignin esters as moisture barrier products, significant additional work is needed before implementation beyond the laboratory [220, 221].

In addition to the paper related applications mentioned above, lignin can be used in fiber composite products, area in which commercial products have already been introduced. One example are the products of German company Tecnaro, which are made of lignin, cellulosic fibers, and natural additives, such as plasticizers, resins, processing aids, and antioxidants. These composites are processed like plastic and the end products are wood-like. They are used to make a large and diverse range of injection molded products including enclosures for high performance acoustic electronic equipment, coffee capsules (pods), coat hangers, and designer footwear, among many others [222–224]. Another example of lignin-containing fiber composites are biodegradable horticultural plant pots developed at University of Toronto in which the lignin acts as a binder for recycled paper fibers. The plant is planted without removal of the pot, which degrades in the soil without causing the waste associated with gardening pots made of plastic or other materials. They have been sold in Canadian garden centers since 2013 [225].

The recombination of cellulose and lignin to bring about composite products with enhanced properties is bound to benefit from the use of the latest advanced techniques. For instance, in an application of nanotechnology, coatings made with lignin and cellulose nanocrystals were shown to have excellent UV absorbing properties while being transparent in the visible range [226]. From solutions of lignin, starch and cellulose in the IL 1-allyl-3-methylimidazolium chloride, biocomposite films were prepared by aqueous precipitation. The films were amorphous, had good transparency and a range of gas permeability rates, which could be tuned according to composition [227]. Nanocellulose was made hydrophobic by coating with lignin and has been proposed as a component for rubber tires [228].

9.3.8 Carbon Fibers and Other Carbon Materials

Having cost competitive carbon fibers would allow their increased use in motor vehicles, leading to decrease in vehicle weight (without compromising safety), fuel economy improvement, and greenhouse gas emissions reduction. Most of the cost associated with the production of carbon fiber is associated with the cost of the precursor, which in most cases is polyacrylonitrile (PAN). Significant attention and government funding has been given to lignin as a precursor, because of lignin's abundance, ubiquity, and anticipated high carbon yield.

Manufacture of carbon fiber involves spinning the precursor into filaments, followed by oxidative thermostabilization and carbonization in an inert atmosphere [229]. Different types of spinning systems are used, including melt spinning or spinning from a solution, either into a coagulating solvent, or into a drying atmosphere, or into an electric field (electrospinning). For utilization in carbon fibers, lignins need to meet very exacting purity specifications. Impurities result in defects and may interfere with processing. The purest lignins are generally organosolv lignins, but they are not truly commercially available, yet. Kraft and soda lignins from existing pulping operations may be purified to the extent required for carbon fiber by acid washing, solvent extraction [230] or membrane fractionation [231].

Early attempts to develop lignin-based carbon fiber were primarily focused on melt spinning and on the ambitious goal of having very high lignin levels. It was found that thermostabilization of extruded lignin fibers was necessary in order to maintain a fiber morphology during carbonization. During thermostabilization crosslinking occurs and lignin transforms from a thermoplastic to a thermoset-type material. If lignin fibers are heated faster than the rate at which the softening temperature increases due to crosslinking, they tend to soften and fuse [232–234]. Thus, for melt spinning, the thermal behavior of the precursor has to satisfy the requirement to flow with temperature during spinning and the need for an acceptable rate of thermostabilization. Various strategies have been proposed for lignin to meet such demands, including:

- Chemical modification such as esterification [235–237]
- Increase of glass transition temperature by treatments such as thermal [238] or solvent purification [239]

- Blending with polymers such as PEO [233, 240], PET [241], or PLA [239, 242]
- Blending with other lignins or their fractions (for instance, blending hardwood lignin as plasticizer for softwood lignin) [243]
- Addition of nanofillers to improve spinnability, such as carbon nanotubes [244] or clays [245].

Lignin-based carbon fibers produced by melt spinning processes have made significant progress. While they still have strength properties below the targets for structural automotive use, in the short term they may find utilization in other functional products.

Wet spinning is conventionally used for industrial PAN processing and introduction of lignin in such a system could be a faster route to commercialization relative to melt spinning. Bissett and Herriott claimed that adding acetylated softwood kraft lignin at an inclusion rate of 10–45 % of the solids to a solution of PAN in dimethyl formamide (DMF), dimethyl sulfoxide or dimethylacetamide (DMAc) reduced the viscosity and permitted an increase in solids and as a consequence a faster throughput rate [246]. Other researchers have shown that addition of softwood kraft lignin to PAN has a favorable effect on the stabilization kinetics and promotes PAN cyclization and crosslinking [247].

Since 2011, commercial manufacturer of carbon fiber Zoltek and its subcontractor pulp company Weyerhaeuser have been collaborating under a US Department of Energy sponsored project to demonstrate partial replacement of PAN with softwood kraft lignin on commercial scale equipment. As of the last available project update the partnership had achieved demonstration in commercial equipment of spinning and carbonization at lignin contents of up to 25 % (Fig. 9.5). Macrovoids in the carbonized fiber appeared at higher lignin content. These were attributed to low solution viscosities. Attempts to use PAN with higher molecular weight to increase the viscosities, have not given the desired results. Even at 25 % lignin mechanical properties did not meet targets [248].

Fig. 9.5 Softwood kraft lignin:PAN (25:75) based precursor (*left*) and carbonized fiber (*right*) produced in commercial spinning and carbonization equipment by Zelcor–Weyerhaeuser collaboration. Reproduced from Husman [248]

Lignin copolymerization with acrylonitrile has been proposed in order to improve results in wet spinning. Products of copolymerization of hardwood kraft lignin with acrylonitrile were processed by wet spinning and showed good carbon fiber morphology, but no strength properties were reported [249]. Desalted and esterified lignosulfonate was copolymerized with acrylonitrile and processed in wet spinning system. Good morphology carbon fibers were produced and strength properties were better for the copolymer relative to a solution blend of lignosulfonate and PAN [250] but were below targets achievable with PAN by itself.

Seydibeyoğlu [251] reported that uniform blends of PAN with several non-acetylated annual fiber soda lignins can be prepared from a DMAc solution. Cast films prepared with all of the lignins tested did not show phase separation. One of the blends (70:30 PAN:Protobind 2400) showed significant increases in storage and loss modulus relative to PAN precursor, suggesting that these blends could be feasible carbon fiber precursors. Further to Seydibeyoğlu's publication, Liu et al. [252] successfully fabricated composite fibers by wet-spinning starting with 70:30 PAN:Protobind 2400 solutions in DMAc coagulated in a methanol bath. When carbonized at 1100 °C, PAN/lignin carbon fiber exhibited at least as good mechanical properties as PAN carbon fiber and exceeded U.S. Department of Energy target mechanical properties for automotive composites, i.e., tensile strength 1.75 GPa, and tensile modulus 175 GPa. Addition of carbon nanotubes (CNT) at 3 wt% on lignin-PAN solids resulted in somewhat lower properties relative to Protobind 2400:PAN due to the presence of microvoids and may improve if smaller CNT are used.

Electrospinning is another technique to generate carbon fiber precursors that has received renewed attention in recent years. Lallave et al. in 2007 were the first to report successful production of nanofibers and nanotubes by electrospinning of an ethanol solution of 50 % Alcell organosolv lignin without the use of added binders. The surface chemistry of the fibers was characterized, but no mechanical properties were presented [253]. Other researchers have reported that inclusion of very small amounts of PEO facilitates electrospinning [254] and addition of multiwalled carbon nanotubes leads to higher conductivity of the final product [255]. Electrospinning has been used to prepare precursors with up to 50:50 PAN:kraft lignin which if treated with electron beam radiation could be thermostabilized faster and could result in carbon fiber with increased tensile strength [256].

In another example of a beneficial treatment to improve thermostabilization rate, partially acetylated softwood kraft lignin processed by dry spinning from a 75 % solids solution in acetone needed thermostabilization time as long as 40 h [257], but if the precursor fibers are UV irradiated the thermostabilization time went down to 4 h [258].

Lignins have also been used in conjunction with carbon fibers precursors other than PAN, such as cellulose and pitch. For example, solutions containing 20–60 % lignin and 40–80 % cellulose or cellulose derivatives were dissolved in DMF. After wet spinning and stabilization, the product was carbonized to produce carbon fibers (no mechanical properties disclosed) or treated with steam to produce activated carbon [259]. Hardwood organosolv lignin and cellulose acetate were electrospun and treated with iodine to improve the carbon structure [260]. Pitch with softening point 70–105 °C was added at 0.1–1 % by weight to phenolated lignin and was processed by melt spinning [261]. A blend of hardwood kraft lignin and pyrolyzed fuel oil (70:30) was melt spun. The resulting fibers ultimately produced consisted of an amorphous core derived from lignin and a crystalline surface originating from pyrolyzed fuel oil. Mechanical properties were below automotive targets [262].

Since satisfying automotive mechanical targets has proven challenging for lignin-based carbon fibers, parallel attention has been given to functional, non-structural applications. Carbon-based materials are highly versatile and in addition to composites applications may be used in diverse areas such as energy storage and generation, filtration, and catalysts support, among others. Interest in sustainable carbon products is increasing and there are already several examples illustrating that lignin may play a central role. For instance, wet spinning has been used to prepare hollow fibers from PAN/lignin solutions in DMSO by manipulation of the composition of the coagulation bath. Reports concerning processing these materials to carbon fibers are not available yet, but it is anticipated that they could be used as adsorbents, microreactors, and in microfluidic applications [263]. Dallmeyer et al. fractionated softwood kraft lignin by sequential solvent extractions and prepared precursors with different ratios of F_{1-3} (methanol soluble fraction) and F_4 (fraction insoluble in methanol, but soluble in 70:30 methanol:methylene chloride) in the presence of 0.2 % PEO. After electrospinning and thermostabilization different morphologies were obtained depending on the ratios of the fractions used. These included single fibers, bonded non-woven mats, porous films and eventually smooth films. Furthermore, the film products were smart materials with moisture memory shape properties [264]. Stabilized and carbonized bonded non-woven products made with 70:30 F_4:F_{1-3} exhibited higher electrical conductivity than products prepared with PAN and are candidates for further study as potential materials for flexible carbon electrodes [234]. Electrospinning of hardwood organosolv lignin was used to produce submicron carbon nanofibers supports for platinum catalysts and have also been proposed for hydrogen storage for fuel cells [265]. Soda herbaceous lignin appears to be a good candidate for the production of electrochemically active nanoshell-containing carbon materials which have been proposed as surrogates for platinum to catalyze the oxygen reduction reaction in proton exchange fuel cell membranes. The intrinsically somewhat higher nitrogen content of the herbaceous lignin appears to be responsible for the higher oxygen reduction reaction activity of the lignin-based product [266]. Recently, Choi et al. demonstrated that carbon nanofibers prepared from DMF solutions of PAN and a 10,000 molecular weight lignin by electrospinning followed by stabilization and carbonization had potential to be used as lithium-ion battery anodes due to their high cycling performance. It was reported that increasing the lignin content resulted in thinner fibers, but a practical limit for processability was found at 50 % lignin [267]. Glass fibers functionalized with carbon product obtained from lignosulfonate carbonization have been patented as filters for the removal of microorganisms by Proctor and Gamble [268].

The pursue of nonstructural applications has been the focus of the current cooperation between the company Graftech and Oak Ridge National Laboratory in the US. They already demonstrated that lignin-based carbon fibers from hardwood organosolv lignin were a drop-in replacement for pitch used in a commercial line of GrafTech's rigid insulation products. In fact, lignin tends to form an isotropic fiber, which is better suited for insulation and filtration applications than higher strength fibers normally used. Ongoing work under the auspices of a US Department of Energy cooperative agreement is focused on qualifying multiple lignin sources and eliminating bottlenecks in melt spinning and stabilization. This work will also evaluate the lignin in prototypes and commercial products [233].

9.4 Concluding Remarks

Being the most abundant aromatic biopolymer, lignin is the logical link in a green supply chain aimed at the partial or complete replacement of petrochemical-derived aromatics. The first years of the twenty-first century have seen the installation and start-up of a few new industrial sources of high purity lignin, generated as a co-product of cellulose pulp. The emerging biorefinery industry may eventually add volume and diversity to the portfolio of available lignins, as side streams from biofuels production may get upgraded to produce lignin. These new industrial sources and the drive to find sustainable routes to produce the chemicals, materials, and fuels we use today have catalyzed recent R&D work on lignin utilization. Interestingly enough, this work has been done not just in the academic milieu, but with industry and including not only big and small lignin producers but also lignin users. Knowledge has significantly advanced in many polymer applications where lignin may play different roles such as macromonomer, polymer blend ingredient, additive with functional properties, or precursor for advanced materials. The partial replacement of phenol and phenolic resins by lignin in thermoset resin applications has long been established as technically feasible. Progress in commercial utilization has been made in the last decade, but economic considerations, volatile oil prices, and absence of a robust and diverse supply chain have curtailed the extent of wider market adoption. New developments in polymeric uses of lignin are predicated by lignin's sustainability, ubiquity, and cost but more and more take advantage of its unique properties, such as natural moisture resistance, antioxidant performance, and thermal resistance, Furthermore, lignin use often brings about other benefits such as reduction in use of and exposure to harmful compounds such as formaldehyde or bisphenol-A. We discussed a multitude of examples of lignin industrial polymeric products and applications that have been developed over the last decade in diverse areas including:

- Fiber composites
- Component of bioasphalt
- Additive for conventional asphalt

- Component in polymer blends, particularly with polyesters for compostable products
- Polymer blends with polyamides in which lignin inclusion causes significant mechanical properties improvement
- Renewable functional additive for rubber, petrochemical polymers, and bio-based polymers, with antioxidant and flame retardant properties
- UV light blocker
- Component of moisture vapor barrier coatings
- Binder and rheology modifier in inks
- Precursor for carbon fiber and other carbon materials for structural as well as nonstructural applications such as insulation, filtration, catalysts supports, and energy storage
- Ingredient in anticorrosion coatings and specialty varnishes
- Ingredient in coatings for cans free of bisphenol-A
- Smart memory shape elastomers and flexible films that are triggered by temperature or moisture
- Materials with enhanced optical properties such as optical polarization attributes and UV absorbance
- Component of high sustainable content biobased functional packaging
- Lignin elastomers, such as copolymers with polyester amines, polyester amine amides and silicones.

Several of the polymeric applications highlighted have already entered the commercial phase, while many others are expected to do so within the next few years. Environmental considerations will continue favoring lignin utilization, while application to lignin of nanotechnology, ILs, green chemistry, biocatalysis, and other emergent technologies may open new routes for lignin valorization.

References

1. http://www.braskem.com/site.aspx/green-products-USA. Accessed 16 May 2015
2. Gellerstedt G, Henriksson G (2008) Lignins: major sources, structure and properties, Chapter 9. In: Belgacem M, Gandini A (eds) Monomers, polymers and composites from renewable materials. Elsevier, Amsterdam
3. Lora JH (2008) Industrial commercial lignins: sources, properties and applications, Chapter 10. In: Belgacem M, Gandini A (eds) Monomers, Polymers and Composites from Renewable Materials. Elsevier, Amsterdam
4. http://www.compositesworld.com/news/domtar-inaugurates-commerical-lignin-production. Accessed 14 May 2016
5. http://www.valmet.com/media/news/press-releases/2015/valmet-supplied-lignoboost-plant-now-handed-over-to-stora-ensos-sunila-mill-in-finland/. Accessed 14 May 2016
6. http://www.bioplasticsmagazine.com/en/news/meldungen/2016-04-08-Innovative-linin-extraction-plant-opened-in-Canada.php. Accessed 14 May 2016
7. Lora JH, Wayman M (1978) Delignification of hardwoods by autohydrolysis and extraction. Tappi 61(6):47–50

8. Pye EK (2006) Industrial lignin production and applications, Chapter 5. In: Kamm B, Gruber PR, Kamm M (eds) Biorefineries—industrial processes and products. Status quo and future directions. 2006 Wiley-VCH Verlag GmbH & Co. KGaA, Weinheim
9. Pye EK, Lora JH (1991) The Alcell Process: a proven alternative to kraft pulping. Tappi J 74 (3):113–118
10. http://www.chempolis.com/. Accessed 14 May 2016
11. http://cimv.fr/research-development/cimv-research-development.html. Accessed 14 May 2016
12. http://www.leafresources.com.au/. Accessed 14 May 2016
13. Francisco M, van den Bruinhorst A, Kroon MC (2012) New natural and renewable low transition temperature mixtures (LTTMs): screening as solvents for lignocellulosic biomass processing. Green Chem 14:2153–2157
14. Laurichesse S, Avérous L (2014) Chemical modification of lignins: Towards biobased polymers. Progr Polym Sci 39(7):1266–1290
15. Lora JH, Glasser WG (2002) Recent industrial applications of lignin: a sustainable alternative to nonrenewable materials. J Polym Environ 10(1/2):39–48
16. Kai D, Tan MJ, Chee PL, Chua YK, Yap YL, Loh XY (2016) Towards lignin-based functional materials in a sustainable world. Green Chem 18:1175–1200
17. Gargulak JD, Lebo SE (1999) Commercial use of lignin-based materials, Chapter 15. In: Glasser WG, Northey RA, Schultz TP (eds) Lignin: historical, biological and materials perspectives. ACS Symposium Series 742, American Chemical Society, Washington, DC
18. Bullock J, Krapp M, Siemensmeyer K (2003) Aqueous colorant preparation for inkjet printing. US Patent Application 2003/0193554 A1
19. Zhor J, Bremner TW, Lora JH (1994) Effect of chemical characteristics of Alcell® lignin based methylsulphonates on their performance as water-reducing admixtures. In: Proceedings fourth Canmet-ACI international conference on superplasticizers and other chemical admixtures in concrete, Montreal
20. http://www.mwvspecialtychemicals.com/content/product-guides/specialty-chemicals/asphalt/emulsionindulinc.pdf. Accessed 28 Dec 2015
21. Borgfeldt MJ (1964) Anionic bituminous emulsions. US Patent 3,123,569
22. Suchanec RR (1997) Asphalt emulsion with lignin containing emulsifier. US Patent 5683497
23. Suchanec RR (1997) Lignin-containing resinous compositions. US Patent 5,656,733
24. Suchanec RR (1997) Air entrained concrete with lignin-containing air entraining agent. US Patent 5,702,521
25. Kalliola A, Vehmas T, Liitiä T, Tamminen T (2015) Alkali-O_2 oxidized lignin—a bio-based concrete plasticizer. Ind Crops Prod 74:150–157
26. http://www.bioplasticsmagazine.com/en/news/meldungen/20151002-Bio-asphalt-trial-in-Zeeland.php. Accessed 25 Oct 2015
27. Bourzac K (2015) Inner workings: paving with plants. In: Proceedings national academy of sciences, September 22, vol. 112, no. 38, pp 11743–11744
28. Williams RC, Brown RC, Tang S (2014) Asphalt materials containing bio-oil and methods for production thereof. US8696806 B2
29. Williams RC, Raouf MA, Metwally M, Brown RC (2015) Bio-oil formulation as an asphalt substitute. US Patent 9200161 B2
30. Gandini A, Belgacem MN (2008) Lignins as components of macromolecular materials, Chapter 11. In: Belgacem M, Gandini A (eds) Monomers, polymers and composites from renewable materials. Elsevier, Amsterdam
31. Podschun J, Stücker A, Saake B, Lehnen R (2015) Structure–function relationships in the phenolation of lignins from different sources. ACS Sustain Chem Eng 3(10):2526–2532
32. Zhang Y, Wang D-Q, Wang X-M, Feng M, Brunette G (2014) Process for fungal modification of lignin and preparing wood adhesives with the modified lignin and wood composites made from such adhesives. US Patent Application 2014/0163142 A1
33. Zhang Y, Wang D-Q, Wang X-M, Feng M, He G (2015) Fungus-modified lignin and its use in wood adhesive for manufacturing wood composites. For Prod J 65(1–2):43–47

34. Lian H, Hong S, Carranza A, Mota-Morales JD, Pojman JA (2015) Processing of lignin in urea–zinc chloride deep-eutectic solvent and its use as a filler in a phenol-formaldehyde resin. RSC Adv 5(36):28778–28785
35. Senyo WC, Creamer AW, Wu CF, Lora JH (1996) The use of organosolv lignin to reduce press vent formaldehyde emissions in the manufacture of wood composites. For Prod J 46 (6):73–77
36. Khan MA, Lora JH (2006) Protobind 1075—an indigenous economical and eco-friendly renewable raw material for the plywood industry. Ply Gazette pp 68–77
37. Lora JH, Wu Q (2007) Performance of two non-wood soda lignin derivatives in oriented strand board powder phenolic adhesives. In: Proceedings 8th International Lignin Institute Forum, Rome
38. http://bioplasticsnews.com/2014/07/23/upm-to-distribute-biochoice-lignin/. Accessed 27 Dec 2015
39. Bode D, Wilson P, Craun GP (2015) Lignin based coating compositions. US Patent Application 2015/0344737 A1
40. Bernardini J, Anguillesi I, Coltelli M-B, Cinelli P, Lazzeri A (2015) Optimizing the lignin based synthesis of flexible polyurethane foams employing reactive liquefying agents. Polym Int 64:1235–1244
41. Bernardini J, Cinelli P, Anguillesi I, Coltelli M-B, Lazzeri A (2015) Flexible polyurethane foams green production employing lignin or oxypropylated lignin. Eur Polym J 64:147–156
42. Ton TTM (1996) Cardanol–lignin-based polyurethanes. Polym Int 41:13–16
43. Saito T, Brown RH, Hunt MA, Pickel DL, Pickel JM, Messman JM, Baker FS, Keller M, Naskar AK (2012) Turning renewable resources into value-added polymer: development of lignin-based thermoplastic. Green Chem 14:3295–3303
44. Saito T, Perkins JH, Jackson DC, Trammel NE, Hunt MA, Naskar AK (2013) Development of lignin-based polyurethane thermoplastics. RSC Adv 3:21832
45. Griffini G, Passoni V, Suriano R, Levi M, Turri S (2015) Polyurethane coatings based on chemically unmodified fractionated lignin. ACS Sustain Chem Eng 3(6):1145–1154
46. Hatakeyama H (2002) Polyurethanes containing lignin. In: Hu TQ (ed) Chemical modification, properties and usage of lignin. Kluwer Academic/Plenum Publishers, New York, pp 41–56
47. Kurple KR (2000) Lignin based polyols. US Patent 6025452
48. Bowman MP, Conley CA, Schwendeman IG, Hibbert MM (2015) Polyester polymers comprising lignin. US Patent Application 2015/0197667 A1
49. Phanopoulos C, Vanden Ecker JMV (1998) Process for binding lignocellulosic material. US Patent 5,750,201
50. Zhang C, Wu H, Kessler MR (2015) High bio-content polyurethane composites with urethane modified lignin as filler. Polymer 69:52–57
51. Chauhan M, Gupta M, Singh B, Singh AK, Gupta VK (2014) Effect of functionalized lignin on the properties of lignin–isocyanate prepolymer blends and composites. Eur Polym J 52:32–43
52. Gupta M, Chauhan M, Naseeba Khatoon N, Singh B (2010) Studies on biocomposites based on pine needles and isocyanate. Adhes J Biobased Mater Bioenergy 4:353–362
53. Faruk O, Sain M, Farnood R, Pan Y, Xiao H (2014) Development of lignin and nanocellulose enhanced Bio PU foams for automotive parts. J Polym Environ 22:279–288
54. de Oliveira W, Glasser WG (1994) Multiphase materials with lignin. 11. Starlike copolymers with caprolactone. Macromolecules 27(1):5–11
55. Hatakeyama T, Izuta Y, Hirose S, Hatakeyama H (2002) Phase transitions of lignin based caprolactones and their polyurethanes derivatives. Polymer 43:1177–1182
56. Laurichesse S, Avérous L (2013) Synthesis, thermal properties, rheological and mechanical behaviors of lignins-grafted-poly (ε-caprolactone). Polymer 54(15):3882–3890
57. Pérez-Camargo RA, Saenz G, Laurichesse S, Casas MT, Puiggalí J, Avérous L, Müller AJ (2015) Nucleation, crystallization, and thermal fractionation of poly (ε-caprolactone)-

grafted-lignin: effects of grafted chains length and lignin content. J Polym Sci, Part B: Polym Phys 53(24):1736–1750
58. Liu X, Wang J, Li S, Zhuang X, Xu Y, Wang C, Fuxiang C (2014) Preparation and properties of UV-absorbent lignin graft copolymer films from lignocellulosic butanol residue. Ind Crops Prod 52:633–641
59. Sivasankarapillai G, McDonald AG (2011) Synthesis and properties of lignin-highly branched poly (ester-amine) polymeric systems. Biomass Bioenergy 35:919–931
60. Sivasankarapillai G, McDonald AG, Li H (2012) Lignin valorization by forming toughened lignin-co-polymers: development of hyperbranched prepolymers for cross-linking. Biomass Bioenergy 47:99–108
61. Li H, Sivasankarapillai G, McDonald AG (2015) Lignin valorization by forming toughened thermally stimulated shape memory copolymeric elastomers: evaluation of different fractionated industrial lignins. J Appl Polym Sci 132(5):41389–41400
62. Li H, Sivasankarapillai G, McDonald AG (2014) Lignin valorization by forming thermally stimulated shape memory copolymeric elastomers—partially crystalline hyperbranched polymer as crosslinks. J Appl Polym Sci 131(22):41103–41112
63. Sivasankarapillai G, Li H, McDonald AG (2015) Lignin-based triple shape memory polymers. Biomacromolecules 16:2735–2742
64. Li H, Sivasankarapillai G, McDonald AG (2015) Highly biobased thermally-stimulated shape memory copolymeric elastomers derived from lignin and glycerol-adipic acid based hyperbranched prepolymer. Ind Crops Prod 67:143–154
65. Stiubianu G, Cazacu M, Cristea M, Vlad A (2009) Polysiloxane-lignin composites. J Appl Polym Sci 113:2313–2321
66. Stiubianu G, Nistor A, Vlad A, Cazacu M (2011) Modification of water sorption capacity of polydimethylsiloxane based composites by incorporation of lignin. Materiale Plastice 48 (4):289–294
67. Zhang J, Cheng Y, Sewell P, Brook MA (2015) Utilization of softwood lignin as both crosslinker and reinforcing agent in silicone elastomers. Green Chem 17:1811–1819
68. Zhang J, Fleury E, Brook MA (2015) Foamed lignin–silicone bio-composites by extrusion and then compression molding. Green Chem 17(9):4647–4656
69. Viswanathan T (2001) Conducting compositions of matter. US Patent 6299800 B1
70. Viswanathan T (2005) Synthesis of lignosulfonic acid-doped polyaniline using transition metal ion catalysts. US Patent 6,977,050 B1
71. Viswanathan T (2007) Corrosion prevention of cold rolled steel using water dispersible lignosulfonic acid doped polyaniline. US Patent 7,179,404 B1
72. Viswanathan T (2005) Catalytic effects of transition metal ions in the synthesis of polyaniline grafted lignosulfonic acid. US 6,916,900 B1
73. Berry BC, Viswanathan T (2002) Lignosulfonic acid-doped polyaniline (Ligno-Pani)—a versatile conducting polymer. In: Hu TQ (ed) Chemical modification, properties and usage of lignin. Kluwer Academic/Plenum Publishers, New York, pp 21–40
74. Taylor KK, Cole CV, Soora R, Dilday JC, Hill AM, Berry B, Viswanathan T (2008) The use of lignosulfonic acid in the synthesis of water-dispersible polyaniline. J Appl Polym Sci 108:1496–1500
75. Horton SD (2010) Cathodic protection compounds. US Patent 7,794,626 B2
76. Viswanathan T (2005) Corrosion prevention of cold rolled steel using water dispersible lignosulfonic acid doped polyaniline. US Patent 6,972,098 B1
77. Mishra S, Yu RCU, Post RL, Horgan AM, Grabowski EF, Carmichael KM, Parikh SR, Hsieh BR, VonHoene DC (2006) Imageable seamed belts with lignin sulfonic acid doped polyaniline. US Patent 7,031,647 B2
78. Visnawathan T (2006) Lignosulfonic acid-doped polyaniline composites with carbon allotropes. US Patent 7,063,808 B1
79. http://www.plasticsnews.com/article/20030929/NEWS/309299974/polyone-licensing-conductive-polymers. Accessed 5 May 2016

80. Milkzarek G, Inganäs O (2012) Renewable cathode materials from biopolymer/conjugated polymer interpenetrating networks. Science 335:1468–1471
81. Milczarek G, Nowicki M (2013) Carbon nanotubes/kraft lignin composite: characterization and charge storage properties. Mater Res Bull 48:4032–4038
82. Argyropoulos DS, Sadeghifar H, Cui C, Sen S (2013) Synthesis and characterization of poly (arylene ether sulfone) kraft lignin heat stable copolymers. ACS Sustain Chem Eng 2:264–271
83. Mohapatra N, Noirot PA, Magrez M (2015) Printing ink or overprint varnish with renewable binder component. US Patent application 2015/0000559 A1
84. Kessler MR, Xia Y, Caes B, Rivas MJ, Bergman J (2015) Nitrated lignin ester and process of making the same. US Patent application 20150126716
85. Belgacem MN, Blayo A, Gandini A (2003) Organosolv lignin as a filler in inks, varnishes and paints. Ind Crops Prod 18:145–153
86. Mandavgane SA, Gogte BB, Subramanian D (2007) Sulphonated lignin based screen ink formulations. Indian J Chem Technol 14:321–324
87. Feldman D (2002) Lignin and its polyblends. In: Hu TQ (ed) Chemical modification, properties and usage of lignin. Kluwer Academic/Plenum Publishers, New York, pp 81–99
88. Sasaki M, Wanaka H, Takagi S, Tamura C Asada, Nakamura Y (2013) Evaluation of epoxy resins synthesized from steam-exploded bamboo lignin. Ind Crops Prod 43:757–761
89. Salanti A, Zoia L, Orlandi M (2016) Chemical modifications of lignin for the preparation of macromers containing cyclic carbonates. Green Chem 18:4063–4072
90. Nohra B, Candy L, Blanco JF, Guerin C, Raoul Y, Mouloungui Z (2013) From petrochemical polyurethanes to biobased polyhydroxyurethanes. Macromolecules 46 (10):3771–3792
91. Kumar MNS, Mohanty AK, Erickson L, Misra M (2009) Lignin and its applications with polymers. J Biobased Mater Bioenergy 3:1–24
92. Doherty W, Mousavioun P, Fellows C (2011) Value-adding to cellulosic ethanol: lignin polymers. Ind Crops Prod 33(2):259–276
93. Thakur VK, Thakur MK, Raghavan P, Kessler MR (2014) Progress in green polymer composites from lignin for multifunctional applications: a review. ACS Sustain Chem Eng 2 (5):1072–1092
94. Pouteau C, Baumberger S, Cathala B, Dole P (2004) Lignin–polymer blends: evaluation of compatibility by image analysis. C R Biol 327:935–943
95. Kadla JF, Kubo S (2004) Lignin-based polymer blends: analysis of intermolecular interactions in lignin–synthetic polymer blends. Compos Part A 35:395–400
96. Kadla JF, Kubo S (2003) Miscibility and hydrogen bonding in blends of poly(ethylene oxide) and kraft lignin. Macromolecules 36:7803–7811
97. Kubo S, Kadla JF (2005) Kraft lignin/poly(ethylene oxide) blends: effect of lignin structure on miscibility and hydrogen bonding. J Appl Polym Sci 98:1437–1444
98. Kubo S, Kadla JF (2004) Poly(ethylene oxide)/organosolv lignin blends: relationship between thermal properties, chemical structure, and blend behavior. Macromolecules 37:6904–6911
99. Erdmann J, Ganster J, Engelmann G (2016) Thermoplastic polymer compounds with low molecular lignins, method for the production thereof, moulded articles and also uses. US patent application 2016/0002467 A10
100. Nair SS, Sharma S, Pu Y, Sun Q, Pan S, Zhu SY, Deng Y, Ragauskas AJ (2014) High shear homogenization of lignin to nanolignin and thermal stability of nanolignin-polyvinyl alcohol blends. ChemSusChem 7:3513–3520
101. Toriz G, Denes F, Young RA (2002) Lignin-polypropylene composites. Part 1: composites from unmodified lignin and polypropylene. Polym Compos 23(5):806–813
102. Alexy P, Kosikova B, Crkonova G, Gregorova A, Martis P (2004) Modification of lignin-polyethylene blends with high lignin content using ethylene-vinylacetate copolymer as modifier. J Appl Polym Sci 94:1855–1860

103. Samal SK, Fernandes EG, Corti A, Chiellini E (2014) Bio-based polyethylene-lignin composites containing a pro-oxidant/pro-degradant additive: preparation and characterization. J Polym Environ 22:58–68
104. Shi B, Shlepr M (2016) "Thermoplastic films containing lignin and their optical polarization properties. J Polym Eng 36:521–528
105. Nitz H, Semke H, Landers R, Mulhaupt R (2001) Reactive extrusion of polycaprolactone compounds containing wood flour and lignin. J Appl Polym Sci 81:1972–1984
106. Feldman D, Banu D, Campanelli J, Zhu H (2001) Blends of vinylic copolymer with plasticized lignin: thermal and mechanical properties. J Appl Polym Sci 81:861–874
107. Li Y, Sarkanen S (2002) Alkylated kraft lignin-based thermoplastic blends with aliphatic polyesters. Macromolecules 35:9707–9715
108. Li Y, Sarkanen S (2005) Miscible blends of kraft lignin derivatives with low-Tg polymers. Macromolecules 38:2296–2306
109. Rosu L, Cascaval CN, Rosu D (2009) Effect of UV radiation on some polymeric networks based on vinyl ester resin and modified lignin. Polym Testing 28:296–300
110. Thielemans W, Wool RP (2004) Butyrated kraft lignin as compatibilizing agent for natural fiber reinforced thermoset composites. Composites A 35:327–338
111. Zhong M, Dai H, Yao H, Dai D, Zhou Y, Yang J, Chen F (2011) Strong, flexible high-lignin polypropylene blends. Polym Res Online. doi:10.2417/spepro.003642
112. Sailaja RRN, Deepthi MV (2010) Mechanical and thermal properties of compatibilized composites of polyethylene and esterified lignin. Mater Des 31:4369–4379
113. Acha BA, Marcovich NE, Reboredo MM (2009) Lignin in jute fabric-polypropylene composites. J Appl Polym Sci 113:1480–1487
114. Maldhure AV, Ekhe JD, Deenadayalan E (2012) Mechanical properties of polypropylene blended with esterified and alkylated lignin. J Appl Polym Sci 125:1701–1712
115. Mariotti N, Wang X-M, Rodrigue D, Stevanovic T (2014) Combination of esterified kraft lignin and MAPE as coupling agent for bark/HDPE composites. J Mater Sci Res 3(2):8–22
116. Teramoto Y, Lee SH, Endo T (2009) Phase structure and mechanical property of blends of organosolv lignin alkyl esters with poly(-caprolactone). Polym J 41:219–227
117. Maldure AV, Chaudhari AR, Ekhe JD (2011) Thermal and structural studies of polypropylene blended with esterified industrial waste lignin. J Therm Anal Calorim 103:625–632
118. Chen F, Dai H, Dong X, Yang J, Zhong MP (2011) Physical properties of lignin-based polypropylene blends. Polym Compos 32:1019–1025
119. Nemoto T, Konishi G, Tojo Y, An YC, Funaoka M (2012) Functionalization of lignin: Synthesis of lignophenol-graft-poly(2-ethyl-2-oxazoline) and its application to polymer blends with commodity polymers. J Appl Polym Sci 123:2636–2642
120. Bono P, Lambert C (1994) Degradable plastics film including lignin as active vegetable filler. US Patent 5,321,065
121. Bono P, Feldman D, Banu D, Lora JH (1995) Degradable polymer blends and polymer products. PCT Int Appl 95/34604
122. Navarrete J, Chapman AJ, Parikh CJ, Toomey RT (2002) Use of lignins in thermoplastics. US Patent 6,485,867 B1
123. Navarrete J (2005) Use of grass lignins in thermoplast ics. US Patent Application 2005/0058909 A1
124. Nitz H, Semke H, Mulhaupt R (2001) Influence of lignin type on the mechanical properties of lignin based compounds. Macromol Mater Eng 286:737–743
125. Sallem-Idrissi N, Sclavons M, Debecker DP, Devaux J (2016) Miscible raw lignin/nylon 6 blends: Thermal and mechanical performances. J Appl Polym Sci 133(6):42963
126. Canetti M, Bertini F (2007) Supermolecular structure and thermal properties of poly (ethylene terephthalate)/lignin composites. Compos Sci Technol 67:3151–3157
127. Gupta AK, Mohanty S, Nayak SK (2015) Influence of addition of vapor grown carbon fibers on mechanical, thermal and biodegradation properties of lignin nanoparticle filled bio-poly (trimethylene terephthalate) hybrid nanocomposites. RSC Adv 5(69):56028–56036

128. Cao N, Oden K, Glasser WG (2014) Chemical modification of lignin and lignin derivatives. US patent 8,809,426 B2
129. Cao N, Oden K, Glasser WG (2014) Chemical modification of lignin and lignin derivatives. US patent 8,865,802 B2
130. Cao N, Oden K (2014) Blending lignin with thermoplastic and a coupling agent or compatibilizer. US Patent Application 2014/ 0121307 A1
131. Cao N, Oden K (2014) Injection of a chemical reagent in a process stream that contains lignin. US Patent Application 2014/ 0121360 A1
132. Cao N, Oden K, Glasser WG (2013) Chemical modification of lignin and lignin derivatives. US Patent Application 2013/ 0338347 A1
133. Cao N, Oden K, Glasser WG (2013) Chemical modification of lignin and lignin derivatives. US Patent Application 2015/0126685 A1
134. Ling S, Guizhen F (2014) Light transmittance and gas permeability of alkaline lignin-poly vinyl alcohol film with formaldehyde crosslinker. Trans Chin Soc Agric Eng 30(4):239–246
135. Mishra S, Mishra A, Kaushik N, Khan M (2007) Study of performance properties of lignin-based polyblends with polyvinyl chloride. J Mater Process Tech 183:273–276
136. Aguié-Béghin V, Foulon L, Soto P, Crônier D, Corti E, Legée F, Baumberger S (2015) Use of food and packaging model matrices to investigate the antioxidant properties of biorefinery grass lignins. J Agric Food Chem 63(45):10022–10031
137. Garcia A, Toledano A, Andres MA, Labidi J (2010) Study of the antioxidant capacity of Miscanthus sinensis lignins. Process Biochem 45:935–940
138. Zoia L, Salanti A, Frigerio P, Orlandi M (2014) Exploring allylation and Claisen rearrangement as a novel chemical modification of lignin. BioResources 9(4):6540–6561
139. Hussin MH, Rahim AA, Nasir M, Ibrahim M, Yemloul M, Perrin D, Brosse N (2014) Investigation on the structure and antioxidant properties of modified lignin obtained by different combinative processes of oil palm fronds (OPF) biomass. Ind Crops Prod 52:544–551
140. Boeriu C, Bravo D, Gosselink R, Dam J (2004) Characterisation of structure-dependent functional properties of lignin with infrared spectroscopy. Ind Crop Prod 20:205–218
141. Arshanitsa A, Ponomarenko J, Dizhbite T, Andersone A, Gosselink RJA, van der Putten J, Lauberts M, Telysheva G (2013) Fractionation of technical lignins as a tool for improvement of their antioxidant properties. J Anal Appl Pyrol 103(2013):78–85
142. Ponomarenko J, Dizhbite T, Lauberts M, Viksna A, Dobele G, Bikovens O, Telysheva G (2014) Characterization of softwood and hardwood lignoboost kraft lignins with emphasis on their antioxidant activity. BioResources 9(2):2051–2068
143. Dizhbite Tatiana, Jurkjane Galina Telysheva Vilhelmina, Viesturs Uldis (2004) Characterization of the radical scavenging activity of lignins—natural antioxidants. Bioresour Technol 95:309–317
144. Pouteau C, Dole P, Cathala B, Averous L, Boquillon N (2003) Antioxidant properties of lignin in polypropylene. Polym Degrad Stab 81:9–18
145. Gregorova A, Cibulkova Z, Kosikova B, Simon P (2005) Stabilization effect of lignin in polypropylene and recycled polypropylene. Polym Degrad Stab 89:553–558
146. Alexy P, Kosıkova B, Podstranska G (2000) The effect of blending lignin with polyethylene and polypropylene on physical properties. Polymer 41:4901–4908
147. Toh K, Nakano S, Yokoyama H, Ebe K, Gotoh K, Noda H (2005) Anti-deterioration effect of lignin as an ultraviolet absorbent in polypropylene and polyethylene. Polym J 37:633–635
148. Pucciariello R, Villani V, Bonini C, Auria MD, Vetere T (2004) Physical properties of straw lignin-based polymer blends. Polymer 45:4159–4169
149. Faruk O, Sain M (2013) Continuous extrusion foaming of lignin enhanced thermoplastic polyurethane (TPU). J Biobased Mater Bioenergy 7(3):309–314
150. Ciobanu C, Ungureanu M, Ignat L, Ungureanu D, Popa V (2004) Properties of lignin polyurethane films prepared by casting method. Ind Crop Prod 20:231–241
151. Lisperguer J, Nuñez C, Perez-Guerrero P (2013) Structure and thermal properties of maleated lignin recycled polystyrene composites. J Chil Chem Soc 58(4) Concepción dic. http://dx.doi.org/10.4067/S0717-97072013000400005

152. Fernandes D, Hechenleitner A, Pineda E (2006) Kinetic study of the thermal decomposition of poly (vinyl alcohol)/kraft lignin derivative blends. Thermochim Acta 441:101–109
153. Corradini E, Pineda EAG, Hechenleitner AAW (1999) Lignin-poly (vinyl alcohol) blends studied by thermal analysis. Polym Degrad Stab 66:199–208
154. Chirico A, Armanini M, Chini P, Cioccolo G, Provasoli F, Audisio G (2003) Flame retardants for polypropylene based on lignin. Polym Degrad Stab 79:139–145
155. Canetti M, Bertini F, De Chirico A, Audisio G (2006) Thermal degradation behaviour of isotactic polypropylene blended with lignin. Polym Degrad Stab 91:494–498
156. Morandim-Giannetti A, Agnelli J, Lancas B, Magnabosco R, Casarin S, Bettini S (2012) Lignin as additive in polypropylene/coir composites: thermal, mechanical and morphological properties. Carbohyd Polym 87:2563–2568
157. Barzegari MR, Alemdar A, Zhang Y, Rodrigue D (2013) Thermal analysis of highly filled composites of polystyrene with lignin. Polym Polym Compos 21:357
158. Veas C, Hotaling EL (2009) Lignin in tire components. WO 2009/145784
159. Setua DK, Shukla MK, Nigam V, Singh H (2000) Lignin reinforced rubber composites. Polym Compos 21(6):988–995
160. Benko D, Hahn BR, Cohen MP, Dirk SM, Cicotte KN (2014) Functionalized lignin, rubber containing functionalized lignin and products containing such rubber composition. U.S. Pat. 8664305B2
161. Hanel T, Castellani L, Orlandi M, Frigeiro P, Zoia L (2014) Tyre for vehicle wheels. WO 2014/097108A1
162. Bahl K, Jana S (2014) Surface modification of lignosulfonates for reinforcement of styrene–butadiene rubber compounds. J Appl Polym Sci 131:40123
163. Jiang C, He H, Jiang H, Ma L, Jia DM (2013) Nano-lignin filled natural rubber composites: Preparation and characterization. Express Polym Lett 7(5):480–493
164. Kosikova B, Gregorova A (2005) Sulfur-free lignin as reinforcing component of styrene-butadiene rubber. J Appl Polym Sci 97:924–929
165. Frigerio P, Zoia L, Orlandi M, Hanel T, Castellani L (2014) Application of sulphur-free lignins as a filler for elastomers: effect of hexamethylenetetramine treatment. BioResources 9 (1):1387–1401
166. Tran Chau D, Chen Jihua, Keum Jong K, Naskar Amit K (2016) A new class of renewable thermoplastics with extraordinary performance from nanostructured lignin-elastomers. Adv Funct Mater. doi:10.1002/adfm.201504990
167. Yu P, He H, Jiang C, Wang D, Jia Y, Zhou L, Jia DM (2015) Reinforcing styrene butadiene rubber with lignin-novolac epoxy resin networks. Express Polym Lett 9(1):36–48
168. Yu P, He H, Jiang C, Jia Y, Wang D, Yao X, Jia D, Luo Y (2016) Enhanced oil resistance and mechanical properties of nitrile butadiene rubber/lignin composites modified by epoxy resin. J Appl Polym Sci 133:42922
169. Dimitri MS, Force CG (1981) Tackifying composition for rubber. US Patent 4,287,104
170. Lora JH, Trojan M, Klingensmith W (1993) Rubber compositions containing high purity lignin derivatives. US Patent 5,196,460
171. Gregorová A, Košíková B, Moravčík R (2006) Stabilization effect of lignin in natural rubber. Polym Degrad Stab 91(2):229–233
172. Furlan LT, Rodrigues MA, De Paoli MA (1985) Sugar cane bagasse-lignin a stabilizer for rubbers, Part III styrene/butadiene rubber and natural rubber. Polym Degrad Stab 13:337–350
173. De Paoli MA, Furlan LT (1985) Sugar cane bagasse-lignin a stabilizer for rubbers, Part II butadiene rubber. Polym Degrad Stab 13:129–138
174. De Paoli MA, Furlan LT (1985) Sugar cane bagasse-lignin as photo-stabilizer for butadiene rubber. Polym Degrad Stab 11:327–333
175. Song P, Cao Z, Fu S, Fang Z, Wu Q, Ye J (2011) Thermal degradation and flame retardancy properties of ABS/lignin: effects of lignin content and reactive compatibilization. Thermochim Acta 518:59–65

176. Cotugno S, De Luca F, Caliano L (2015) Method of joining two rubber tyre portions. WO 2015033285 A1
177. Cotugno S, Caliano L Tyre body ply skim. US Pat Application 2016/0009906 A1
178. Kamada S (2015) Rubber composition for tire tread and pneumatic tire. US Patent 9012540 B2
179. Baumberger S (2002) Starch-lignin films. In: Hu TQ (ed) Chemical modification, properties, and usage of lignin. Kluwer Academic/Plenum Publishers, New York, pp 1–19
180. Spiridon I, Teaca CA, Bodirlau R (2011) Preparation and characterization of adipic acid-modified starch microparticles/plasticized starch composite films reinforced by lignin. J Mater Sci 46:3241–3251
181. Bodirlau R, Teaca C, Spiridon I (2013) Influence of natural fillers on the properties of starch-based biocomposite films. Compos B 44:575–583
182. Espinoza Acosta JL, Torres Chávez PI, Ramírez-Wong B, Bello-Pérez LA, Vega Ríos A, Carvajal Millán E, Plascencia Jatomea M, Ledesma Osuna AI (2015) Mechanical, thermal, and antioxidant properties of composite films prepared from durum wheat starch and lignin. Starch Stärke 67:502–511
183. Bhat R, Abdullah N, Din RH, Tay GS (2013) Producing novel sago starch based food packaging films by incorporating lignin isolated from oil palm black liquor waste. J Food Eng 119(4):707–713
184. Stevens ES, Willett JL, Shogren RL (2007) Thermoplastic starch-kraft lignin-glycerol blends. J Biobased Mater Bioenergy 1(3):351–359
185. Ban W, Song J, Lucia LA (2007) Influence of natural biomaterials on the absorbency and transparency of starch-derived films: an optimization study. Ind Eng Chem Res 46 (20):6480–6485
186. Lepifre S, Baumberger S, Pollet B, Cazaux F, Coqueret X, Lapierre C (2004) Reactivity of sulphur-free alkali lignins within starch films. Ind Crops Prod 20:219–230
187. Lepifre S, Froment M, Cazaux F, Houot S, Lourdin D, Coqueret X, Lapierre C, Baumberger S (2004) Lignin incorporation combined with electron-beam irradiation improves the surface water resistance of starch films. Biomacromolecules 5:1678–1686
188. Kaewtatip K, Thongmee J (2013) Effect of kraft lignin and esterified lignin on the properties of thermoplastic starch. Mater Des 49:701–704
189. Miranda CS, Ferreira MS, Magalhães MT, Bispo APG, Oliveira JC, Silva JBA, José NM (2015) Starch-based films plasticized with glycerol and lignin from piassava fiber reinforced with nanocrystals from eucalyptus. Mater Today Proc 2(1):134–140
190. Miranda CS, Ferreira MS, Magalhães MT, Santos WJ, Oliveira JC, Silva JBA, José NM (2015) Mechanical, thermal and barrier properties of starch-based films plasticized with glycerol and lignin and reinforced with cellulose nanocrystals. Mater Today Proc 2(1):63–69
191. Stevens ES, Klamczynski A, Glenn GM (2010) Starch-lignin foams. Express Polym Lett 4 (5):311–320
192. Çalgeris İ, Çakmakçı E, Ogan A, Kahraman MV, Kayaman-Apohan N (2012) Preparation and drug release properties of lignin–starch biodegradable films. Starch/Stärke 64:399–407
193. Vengal JC, Srikumar M (2005) Processing and study of novel lignin-starch and lignin-gelatin biodegradable polymeric films. Trends Biomater Artif Organs 18(2):237–241
194. Núñez-Flores R, Giménez B, Fernández-Martín F, López-Caballero ME, Montero MP, Gómez-Guillén MC (2013) Physical and functional characterization of active fish gelatin films incorporated with lignin. Food Hydrocoll 30:163–172
195. Núñez-Flores R, Giménez B, Fernández-Martín F, López-Caballero ME, Montero MP, Gómez-Guillén MC (2012) Role of lignosulphonate in properties of fish gelatin films. Food Hydrocolloids 27:60–71
196. Ojagh SM, Núñez-Flores R, López-Caballero ME, Montero MP, Gómez-Guillén MC (2011) Lessening of high-pressure-induced changes in Atlantic salmon muscle by the combined use of a fish gelatin–lignin film. Food Chem 125:595–606
197. Shankar S, Reddy JP, Rhim JW (2015) Effect of lignin on water vapor barrier, mechanical, and structural properties of agar/lignin composite films. Int J Biol Macromol 81(81):267–273

198. Kaewtatip K, Menut P, Auvergne R, Tanrattanakul V, Morel MH, Guilbert S (2010) Interactions of kraft lignin and wheat gluten during biomaterial processing: evidence for the role of phenolic groups. J Agric Food Chem 58:4185–4192
199. Kunanopparat T, Menut P, Morel MH, Guilbert S (2009) Modification of the wheat gluten network by kraft lignin addition. J Agric Food Chem 57:8526–8533
200. Kunanopparat T, Menut P, Morel MH, Guilbert S (2012) Improving wheat gluten materials properties by kraft lignin addition. J Appl Polym Sci 125:1391–1399
201. Oliviero M, Verdolotti L, Nedi I, Docimo F, Di Maio E, Iannace S (2012) Effect of two kinds of lignins, alkaline lignin and sodium lignosulfonate, on the foamability of thermoplastic zein-based bionanocomposites. J Cell Plast 48(6):516–525
202. Arancibia MY, López-Caballero ME, Gómez-Guillén C, Montero P (2014) Release of volatile compounds and biodegradability of active soy protein lignin blend films with added citronella essential oil. Food Control 44:7–15
203. Arancibia M, Rabossi A, Bochicchio PA, Moreno S, López-Caballero ME, Gómez-Guillén C, Montero P (2013) Biodegradable films containing clove or citronella essential oils against the mediterranean fruit fly ceratitis capitata (diptera tephritidae). J Agric Food Tech 3(3):1–7
204. Massey-brooker AD, Mauro Vaccaro M, Scialla S, Walker SJ, Morganti P, Carezzi F, Benjelloun-mylayah B, Crestini C, Lange H, Bartzoka E (2016) Consumer goods product comprising chitin, lignin and a polymer or co-polymer. US Patent Application 2016/0074311 A1
205. Rahman MA, De Santis D, Spagnoli G, Ramorino G, Penco M, Phuong VT, Lazzeri A (2013) Biocomposites based on lignin and plasticized poly(L-lactic acid). J Appl Polym Sci 129:202–214
206. Spiridon I, Leluk K, Resmerita AM, Darie RN (2015) Evaluation of PLA–lignin bioplastics properties before and after accelerated weathering. Compos B 69(2015):342–349
207. Dorgan JR, Eyser MP, Perbix C (2013) Polylactide-graft lignin blends and co-polymers. US Patent Application 2013/0281582 A1
208. Mohanty AK, Misra M, Sahoo S (2012) Lignin based materials and methods of making those. US Patent Application 20120071591Al
209. Abdelwahab MA, Taylor S, Misra M, Mohanty AK (2015) Thermo-mechanical characterization of bioblends from polylactide and lignin. Macromol Mater Eng 300(3):299–311
210. Sahoo S, Misra M, Mohanty AK (2014) Biocomposites from switchgrass and lignin hybrid and poly (butylene succinate) bioplastic: studies on reactive compatibilization and performance evaluation. Macromol Mater Eng 299(2):178–189
211. Mousavioun P, Doherty WOS, George G (2010) Thermal stability and miscibility of poly (hydroxybutyrate) and soda lignin blends. Ind Crops Prod 32:656–661
212. Gordobil O, Egüés I, Llano-Ponte R, Labidi J (2014) Physicochemical properties of PLA lignin blends. Polym Degrad Stab 108:330–338
213. Mousavioun P, George GA, Doherty WOS (2012) Environmental degradation of lignin/poly (hydroxybutyrate) blends. Polym Degrad Stab 97:1114–1122
214. Jopson RN (1993) Saturation technology for corrugated containers. Tappi J 76(4):207–214
215. Owens BA, Collias DI, Wnuk AJ (2003) Methods for the reduction of bleeding of lignosulfonates from lignosulfonate-treated substrates. US Patent 6,458,419 B2
216. Lyons AV, Berry E (2000) Lignin-based vapor barrier formulations. EP 0904330 A4
217. Wang H, Easteal AJ, Edmonds N (2008) Prevulcanized natural rubber latex/modified lignin dispersion for water vapour barrier coatings on paperboard packaging. Adv Mater Res 47–50:93–96
218. Doherty WOS, Halley P, Cronin D, Edye LA (2010) Method for treating paper product. US Patent Application 2010/0166968 A1
219. Edye LA, Tietz AJ (2015) Lignin-based waterproof coating. WO2015054736A1
220. Tamminen T, Ropponen J, Hult E-L, Poppius-Levlin K (2014) Functionalized lignin and method of producing the same. US Patent application 2014/0243511 A1

221. Hult E-L, Koivu K, Asikkala J, Ropponen J, Wrigstedt P, Sipilä J, Poppius-Levlin K (2013) Esterified lignin coating as water vapor and oxygen barrier for fiber-based packaging. Holzforschung 67(8):899–905
222. Nägele H, Pfitzer J, Eisenreich N, Eyerer P, Elsner P, Eckl W (2003) Plastic material made from a polymer blend. US Patent 6,509,397 B1
223. Nägele H, Pfitzer J, Inone E, Eyerer P, Eisenreich N, Eckl W (2004) Composition for the production n of shaped bodies and methods for the production of shaped bodies made of said composition. US Patent 6,706,106 B1
224. http://www.tecnaro.de/english/aktuell.htm?section=we. Accessed 18 Feb 2016
225. Goldberg B (2013) Revolutionary fibre gardening pots developed at U of T to hit shelves soon. U of T News, http://www.news.utoronto.ca/revolutionary-fibre-gardening-pots-developed-u-t-hit-shelves-soon. Accessed 18 Feb 2016
226. Hambardzumyan A, Foulon L, Chabbert B, Aguie-Beghin V (2012) Natural organic UV-absorbent coatings based on cellulose and lignin: designed effects on spectroscopic properties. Biomacromolecules 13:4081–4088
227. Wu R-L, Wang X-L, Li F, Li H-Z, Wang Y-Z (2009) Green composite films prepared from cellulose, starch and lignin in room-temperature ionic liquid. Biores Technol 100:2569–2574
228. Karampelas BE (2016) Automotive tires containing hydrophobic nanocellulose. US Patent Application 2016/01222515
229. Kadla JF, Kubo S, Gilbert RD, Venditti RA Lignin based carbon fibers, Chapter 7. In: Hu TQ (ed) Chemical modification, properties and usage of lignin. Kluwer Academic/Plenum Publishers, New York, pp 121–137
230. Saito T, Perkins JH, Vautard F, Meyer HM, Messman JM, Tolnai B, Naskar AK (2014) Methanol fractionation of softwood kraft lignin: impact on the lignin properties. ChemSusChem 7:221–228
231. Norberg I, Nordstrom Y, Drougge R, Gellerstedt G, Sjoholm E (2013) A new method for stabilizing softwood kraft lignin fibers for carbon fiber production. J Appl Polym Sci 128 (6):3824–3830
232. Baker DA, Rials TG (2013) Recent advances in low-cost carbon fiber manufacture from lignin. J Appl Polym Sci 130:713–728
233. Paul R, Burwell D, Dai X, Naskar A, Gallego N, Akato K (2015) Recent progress in producing low-cost and renewable lignin-based carbon fibers. CAMX Conference Proceedings. Dallas, TX, October 26–29. CAMX—The Composites and Advanced Materials Expo. http://www.osti.gov/scitech/servlets/purl/1224680/. Accessed 12 May 2016
234. Dallmeyer I, Lin LT, Li Y, Ko F, Kadla JF (2014) Preparation and characterization of interconnected, kraft lignin-based carbon fibrous materials by electrospinning. Macromol Mater Eng 299(5):540–551
235. Eckert RC, Abdullah Z (2010) Method for the production of lignin-containing precursor fibres and also carbon fibres. US Patent 7,678,358 B2
236. Thunga M, Chen K, Grewell D, Kessler MR (2014) Bio-renewable precursor fibers from lignin/polylactide blends for conversion to carbon fibers. Carbon 68:159–166
237. Chatterjee S, Jones EB, Clingenpeel AC, McKenna AM, Rios O, McNutt NW, Keffer DJ, Johs A (2014) Conversion of lignin precursors to carbon fibers with nanoscale graphitic domains. ACS Sustain Chem Eng 2(8):2002–2010
238. Baker DA, Gallego NC, Baker FS (2012) On the characterization and spinning of an organic-purified lignin toward the manufacture of low-cost carbon fiber. J Appl Polym Sci 124:227
239. Baker DA, Hosseinaei O (2014) High glass transition lignins and lignin derivatives for the manufacture of carbon and graphite fibers. U.S. Patent Application No. 20140271443
240. Awal A, Sain M (2013) Characterization of soda hardwood lignin and the formation of lignin fibers by melt spinning. J Appl Polym Sci 129(5):2765–2771
241. Kubo S, Kadla JF (2005) Lignin-based carbon fibers: effect of synthetic polymer blending on fiber properties. J Polym Environ 13(2):97–105

242. Thunga M, Chen K, Kessler MR (2016) Process of making carbon fibers from compositions including esterified lignin and poly(lactic acid). US Patent 9,340,425
243. Nordstrom Y, Norberg I, Sjoholm E, Drougge R (2013) A new softening agent for melt spinning of softwood kraft lignin. J Appl Polym Sci 129(3):1274–1279
244. Baker FS, Baker DA, Menchhofer PA (2011) Carbon nanotube enhanced precursor for carbon fiber production and method of making a CNT-enhanced continuous lignin fiber. US Patent Application 2011285049
245. Sevastyanova O, Qin W, Kadla JF (2010) Effect of nanofillers as reinforcement agents for lignin composite fibers. J Appl Polym Sci 117(5):2877–2881
246. Bissett PJ, Herriott CW (2014) Lignin/polyacrylonitrile-containing dopes, fibers, and methods of making same. U.S. Patent 8,771,832 B2
247. Liu HC, Chien AT, Newcomb BA, Davijani AAB, Kumar S (2016) Stabilization kinetics of gel spun polyacrylonitrile/lignin blend fiber. Carbon 101:382–389
248. Husman G (2014) Development and commercialization of a novel low cost carbon fiber. http://energy.gov/sites/prod/files/2014/07/f17/lm048_husman_2014_o.pdf. Accessed 12 May 2016
249. Maradur SP, Kim CH, Kim SY, Kim B-H, Kim WC, Yang KS (2012) Preparation of carbon fibers from a lignin copolymer with polyacrylonitrile. Synth Metals 162:453–459
250. Xia K, Ouyang Q, Chen Y, Wang X, Qian X, Wang L (2016) Preparation and characterization of lignosulfonate-acrylonitrile copolymer as a novel carbon fiber precursor. ACS Sustain Chem Eng 4:159–168
251. Seydibeyoğlu MÖ (2012) A novel partially biobased PAN-lignin blend as a potential carbon fiber precursor. J Biomed Biotechnol 2012:1–8
252. Liu HC, Chien AT, Newcomb BA, Liu Y, Kumar S (2015) Processing, structure, and properties of lignin-and CNT-incorporated polyacrylonitrile-based carbon fibers. ACS Sustain Chem Eng 3(9):1943–1954
253. Lallave M, Bedia J, Ruiz-Rosas R, Rodrıguez-Mirasol J, Cordero T, Otero JC, Marquez M, Barrero A, Loscertales IG (2007) Filled and hollow carbon nanofibers by coaxial electrospinning of alcell lignin without binder polymers. Adv Mater 19:4292–4296
254. Dallmeyer I, Ko F, Kadla JF (2010) Electrospinning of technical lignins for the production of fibrous networks. J Wood Chem Technol 30:315–329
255. Teng N-Y, Dallmeyer I, Kadla JF (2013) Incorporation of multiwalled carbon nanotubes into electrospun softwood Kraft lignin-based fibers. J Wood Chem Technol 33(4):299–316
256. Seo DK, Jeun JP, Kim HB, Kang PH (2011) Preparation and characterization of the carbon nanofiber mat produced from electrospun PAN/lignin precursors by electrobeam radiation. Rev Adv Mater Sci 28:31
257. Zhang M, Ogale A (2014) Carbon fibers from dry-spinning of acetylated softwood kraft lignin. Carbon 69:626–629
258. Zhang M, Jin J, Ogale AA (2015) Carbon fibers from UV-assisted stabilization of lignin-based precursors. Fibers 3:184–196
259. Lehmann A, Horst Ebeling H, Fink H-P (2014) Method for the production of lignin-containing precursor fibres and also carbon fibres. US Patent Application 2014/0194603A1
260. Schreiber M, Vivekanandhan S, Mohanty AK, Misra M (2015) Iodine treatment of lignin-cellulose acetate electrospun fibers: enhancement of green fiber carbonization. ACS Sustain Chem Eng 3(1):33–41
261. Ichikawa H, Yokoyama A, Nankjima N (1992) JP Pat. 4,194,029
262. Kim MS, Lee DH, Kim CH, Lee YJ, Hwang JY, Yang C-M, Kim YA, Yang KS (2015) Shell-core structured carbon fibers via melt spinning of petroleum-and wood-processing waste blends. Carbon 85:194–200
263. Jia Z, Lu C, Liu Y, Zhou P, Wang L (2016) Lignin/polyacrylonitrile composite hollow fibers prepared by wet-spinning method. ACS Sustain Chem Eng 4(5):2838–2842

264. Dallmeyer I, Chowdhury S, Kadla JF (2013) Preparation and characterization of kraft lignin-based moisture responsive films with reversible shape-change capability. Biomacromolecules 14:2354–2363
265. Ruiz-Rosas P, Bedia J, Lallave M, Loscertales IG, Barrero A, Rodrıguez-Mirasol J, Cordero T (2010) The production of submicron diameter carbon fibers by the electrospinning of lignin. Carbon 48:696–705
266. Kannari N, Takigami M, Maie T, Honda H, Kusadokoro S, Ozaki J (2013) Nanoshell-containing carbon cathode catalyst for proton exchange membrane fuel cell from herbaceous plants lignin. Smart Grid Renew Energy 4:10–15
267. Choi DI, Lee J-N, Song J, Kang PH, Park J-K, Lee YM (2013) Fabrication of polyacrylonitrile/lignin-based carbon nanofibers for high-power lithium ion battery anodes. J Solid State Electrochem 17:2471–2475
268. Mitchell MD, Owens BA, Collias DI, Wnuk AJ (2004) Filters and filter materials for the removal of microorganisms and processes for making the same. US Patent 6,827,854 B2

Chapter 10
Green Processes for Lignin Conversion

Fanny Monteil-Rivera

Abstract Lignin is a complex heterogeneous aromatic polymer consisting of up to 30 % of plant material. Its aromatic structure suggests that it is a possible renewable source for aromatic chemicals. However, the natural complexity and high stability of lignin makes its depolymerization a highly challenging task. Many efforts have been directed toward a better understanding of the structure and composition of lignin in order to design more efficient and greener deconstruction paths. This chapter aims at providing an overview of key advances in the field of lignin depolymerization, with special emphasis on chemical catalysis, ionic liquids, and biocatalysis. The various technologies are discussed and critically evaluated in terms of potential for further industrial implementation. Research gaps that still need to be addressed and the most promising approaches are highlighted.

Keywords Lignin structure and properties · Lignin depolymerization · Chemical catalysis · Ionic liquids · Biocatalysis

10.1 Introduction

Together with cellulose and hemicellulose, lignin is an important component of plant material constituting up to 30 % of the weight and 40 % of the fuel value of biomass [1]. Lignin is a complex three-dimensional amorphous polymer consisting of phenylpropanoid units of various types, which makes it the widest natural source of bio-based aromatic chemicals.

Despite this singularly attractive chemical property, only few commercial examples of lignin application exist besides its use as energy source [2]. Most of the lignin markets are limited to low value products, e.g., dispersing, binding, or stabilizing agents. Efforts have also been done to develop higher value applications of

F. Monteil-Rivera (✉)
National Research Council of Canada, Aquatic and Crop Resource Development,
6100 Royalmount Ave., Montreal, QC H4P 2R2, Canada
e-mail: fanny.monteil@cnrc-nrc.gc.ca

P.C.K. Lau (ed.), *Quality Living Through Chemurgy and Green Chemistry*,
Green Chemistry and Sustainable Technology,
DOI 10.1007/978-3-662-53704-6_10

lignin in its macromolecular form, which includes using lignin in thermoplastics, phenolic and epoxy resins, polyurethane foams, or carbon fibers. These polymeric applications have been well summarized in several reviews and chapters [3, 4] and are not discussed herein.

In this chapter we focus on the production of value-added low molecular weight chemicals from lignin. Except for the production of vanillin and DMSO from lignin sulfonate by Borregaard in Norway [5, 6], most commercial lignin depolymerization routes are still in their infancy. Several reasons explain the limited commercial maturity of these reactions. Among them are: (i) the recalcitrance of lignin toward both chemical and biochemical attacks [7] and, (ii) the complexity of the reaction media produced which imply high separation costs. Today, the valorization of lignin into biochemicals represents a real challenge in terms of both sustainability and environmental protection [5, 8, 9]. The need to develop high-volume production of aromatics from lignin is an urgent matter considering the depletion of fossil resources. To be advantageous against the currently available production of petroleum-based aromatics, the new processes have to be efficient, economical, and environmentally friendly. The present chapter aims at summarizing new promising green approaches for the depolymerization of lignin.

10.2 Lignin Structure and Properties

Lignin is biosynthesized in plants by free-radical polymerization from three phenolic monomers, namely *p*-coumaryl alcohol, coniferyl alcohol, or sinapyl alcohol which correspond to the *p*-hydroxyphenyl (H), guaiacyl (G), or syringyl (S) units of lignin, respectively (Fig. 10.1). The proportion of each unit varies with the type of biomass, the part of the plant, the growing conditions, the geographical growing area, the extraction process, and even sometimes with the analytical methods used for characterization. Despite these heterogeneities, general trends have been drawn for various families of lignin. Softwood lignin consists almost exclusively of G units, while hardwood lignin contains a mixture of S and G units, and grass lignin contains minor amounts of H units in addition to G and S units [1]. The additional methoxyl groups on the aromatic rings of hardwood and grass lignins prevent formation of resistant 5-5′ (aryl aryl) linkages, and thus cause hardwood and grass lignins to form more linear and less condensed structures compared to softwood.

The phenylpropanoid units of lignin are bound together by ether bonds or C–C bonds which play a major role in determining the reactivity of lignin. In native lignin, ether bonds represent two thirds of the linkages or more, while the rest corresponds to C–C bonds [10]. A generic schematic representation of lignin is provided in Fig. 10.2 that includes the major functional groups and linkages known to arise in softwood, hardwood, or grass lignins. Major linkages occurring between the structural units of lignin are β-O-4 (β-aryl ether), 5-5′ (aryl aryl), β-5 (phenylcoumaran), α-O-4 (α-aryl ether), and 4-O-5 (diaryl ether) (Table 10.1), with

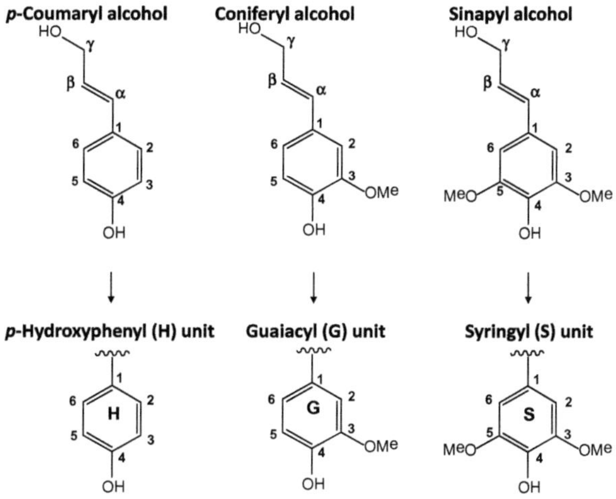

Fig. 10.1 Lignin building blocks and their corresponding units

the β-O-4 linkage being, by far, the most abundant. Other linkages, such as α-O-γ (aliphatic ether), β–β (resinol), and β-1 (spirodienone) can also be found in lignin [10–12]. The most abundant and "easy" to break β-O-4 linkage has attracted the most attention in attempts to depolymerize lignin.

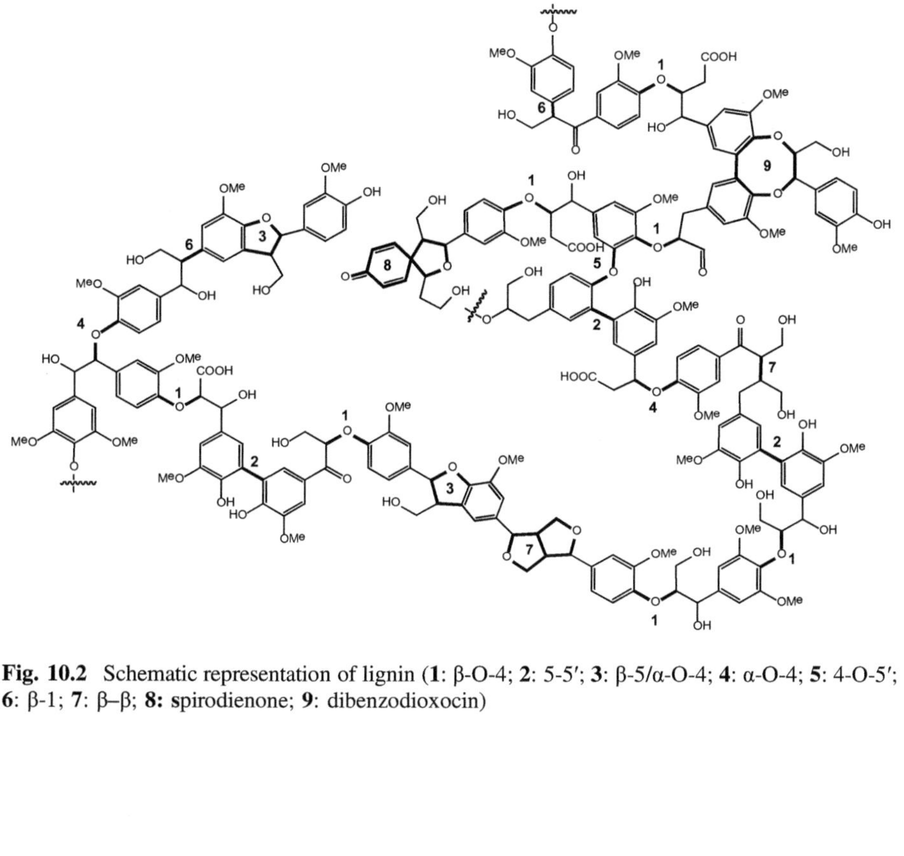

Fig. 10.2 Schematic representation of lignin (**1**: β-O-4; **2**: 5-5′; **3**: β-5/α-O-4; **4**: α-O-4; **5**: 4-O-5′; **6**: β-1; **7**: β–β; **8:** spirodienone; **9**: dibenzodioxocin)

Table 10.1 Frequency of inter-unit linkages in different native lignins (number of linkage per 100 phenylpropanoid units, ppu) [11–14]

Linkage type (# in Fig. 10.2)	Dimer structure	Linkages/100 ppu	
		Softwood	Hardwood
β-O-4 (**1**)	Phenylpropane β-aryl ether	43–50	47–65
5-5′ (**2**)	Biphenyl and dibenzodioxocin	18–25	2–12
β-5/α-O-4 (**3**)	Phenyl coumaran	9–12	4–6
α-O-4 (**4**)	Phenylpropane α-aryl ether	6–8	4–8
4-O-5′(**5**)	Diaryl ether	4–8	6–9
β-1 (**6**)	1,2-Diaryl propane	3–10	1–7
β–β (**7**)	β-β-Linked structures, resinol	2–4	3–12
Spirodienone (**8**)	–	2	3–5
Dibenzodioxocin (**9**)	Diaryl propane diether	5–7	1–2

The chemical structure of grass lignins is much less understood than that of wood lignins, likely due to their more complex structural features, their wider variability from species to species, and their more recent entry in the world of biorefineries as byproduct of cellulosic ethanol production [15–17]. Nevertheless, it has been recurrently observed that lignin in non-woody plants is naturally acetylated, coumarylated, and/or *p*-hydroxybenzoylated. Grass lignins are acylated by *p*-coumaric acid at the γ-position of lignin side chains, usually on the S units [15, 18]. Ferulic acid can also be present in smaller amounts with linkage occurring through either ether or ester bonds [19]. The content of ester groups in grass lignin is strongly affected by the pulping conditions, with acid or alkaline reagents acting as hydrolysis promoters.

Analyses of lignin chemical structure including its monolignol ratio, inter-unit linkages and functional groups have advanced considerably over the two last decades owing to the substantial progress of NMR technology. In particular, two-dimensional (2D) NMR correlation experiments, such as heteronuclear single quantum coherence (HSQC) spectroscopy, have contributed enormously to the elucidation of lignin subunits and lignin carbohydrate complexes [20, 21]. 2D NMR, however, has its limitations since it is not quantitative and it can lead to various overlapping signals. A complementary, more quantitative, approach was developed by Argyropoulos and his team, which consists of derivatizing all OH groups of lignin with an appropriate phosphorous reagent and analyzing quantitatively the phosphorylated lignin by ^{31}P NMR [18, 22]. Since the pioneering work of Argyropoulos, quantitative ^{31}P NMR has been widely used to quantify the aliphatic, phenolic, and carboxylic OH groups of numerous wood and grass lignins (Table 10.2). Knowledge of the content of various G, S, or H groups in lignin is a predeterminant of the type of target chemicals that can be obtained by depolymerization of a specific lignin.

While the original distribution of phenolic groups in lignin is mainly governed by the biomass, the absolute content of aliphatic, carboxylic, and in lesser extent

Table 10.2 Hydroxyl group contents (mmol g^{-1}) of various lignins as determined by ^{31}P NMR

Lignin type[a]	Aliphatic OH	Phenolic OH				COOH	References
		C5-substituted	Syringyl	Guaiacyl	p-Hydroxy phenyl		
Softwood, MWL	4.13–4.57	0.08–0.50	–	0.57–1.18	0.03–0.12	0.02–0.22	[21]
Softwood, EMAL	n.d.	0.41–0.63	–	0.79–1.06	0.10–0.16	0.11–0.19	[21]
Softwood, EAL	4.92	0.30	–	0.72	0.06	0.09	[21]
Softwood, OSL	4.70	1.80	–	1.20	0.10	n.d.	[10]
Softwood, Kraft	2.34–2.41	1.37–1.91	–	1.95–1.96	0.20–0.26	0.32–0.39	[23, 24]
Hardwood, MWL	4.05–5.72	0.13–0.29	0.16–0.23	0.19–0.37	0.03–0.20	0.06–0.14	[21]
Hardwood, EMAL	n.d.	n.d.	0.62	0.35	0.02	0.15	[21]
Hardwood, OSL	1.10	1.18	1.63	0.80	0.13	0.23	[23]
Non-wood, MWL	2.83–5.54	0.18–0.35	0.09–0.22	0.38–0.46	0.10–0.64	0.07–0.33	[10]
Non-wood, EMAL	n.d.	0.18	0.21	0.57	0.30	n.d.	[25]
Non-wood, OSL	1.19–2.95	0.11–0.49	0.39–1.53	0.79–1.20	0.02–0.94	0.01–0.33	[17, 26]
Non-wood, Soda	2.12–3.1	0.03–0.52	0.24–0.82	0.35–0.79	0.07–0.55	0.07–0.97	[27–29]

[a]*MWL* milled wood lignin, *EMAL* enzymatic mild acidolysis lignin, *EAL* enzymatic acidolysis lignin, *OSL* organosolv lignin

phenolic OH groups is largely dependent on the process applied for its isolation. A full description of the production process, properties, and applications of various commercial lignins, including sulfite, kraft, soda, or organosolv lignin has been provided by Lora [2] and reader are referred to the preceding chapter of this book as well. Milled wood lignin obtained by milling in water/dioxane is the lignin that is the closest to native lignin [30]. It is generally rich in aliphatic OHs and poorer in phenolic and carboxylic OHs. Heated alkaline or acid processes such as soda and kraft processes or organosolv processes, respectively, induce a decrease in aliphatic OH groups, and an increase in either carboxylic or phenolic OH groups.

In summary, with its unique structure and chemical properties, lignin represents the only viable renewable source for the production of aromatic compounds. However, its high complexity and variability imply that lignin needs to be well characterized in order to target the desired chemicals from an appropriate source. In addition, lignin heterogeneity also implies that transformation processes need to be selective enough to limit the cost of product purification.

10.3 Chemical Catalysis for Depolymerization of Lignin

Lignin can be depolymerized using technologies including pyrolysis [31–33], homogenous acid/base-catalysis [12, 34–41], and thermal hydrogenolysis [32, 42, 43]. However due to harsh conditions used and a lack of selectivity, most of these techniques give rise to a large number of monomeric products in addition to substantial amounts of unwanted side products such as char and light gases. Such a broad product array requires dedicated separation efforts and a wide variety of catalytic reforming in order to upgrade all the products into useful chemicals. Chemical catalysis has thus been regarded as a key technology to tackle, in a more efficient and selective way, the chemical and biological resistance of lignin. Many catalytic pathways have been reported to depolymerise lignin or its model compounds, e.g., β-ether dimers and oligomers, under oxidative or reducing conditions, and several comprehensive reviews are available on the subject [7, 10, 11, 44].

10.3.1 Reductive Lignin Depolymerization

Most studies reported on reductive depolymerization of lignin have been focused on the production and upgrading of bio-oils and fuels [11] that fall beyond the scope of this chapter. Non-catalyzed thermochemical conversion of lignin into monomeric phenols has also been well documented, as reviewed by Roy and coworkers [42] and more recently by Pandey and Kim [32]. In this section, we focus on the production of phenols or arenes from lignin using catalytic reductive reactions as opposed to pure thermal processes.

Catalytic reduction of lignin or its model compounds normally occurs at temperatures within the range 100–400 °C and gives rise to aromatic compounds, such as phenols, benzene, toluene, xylene, and even alkanes if hydrogen upgrading is involved [10]. Reactions of a substrate with hydrogen, from H_2 or from a hydrogen donor, can be of three main types: hydrogenation, i.e., saturation of C=O, C=C, or C≡C bonds; hydrogenolysis, i.e., cleavage of carbon–carbon or carbon–heteroatom bonds; or hydrodeoxygenation, i.e., removal of oxygen from oxygenated substrates. Hydrogenation generally occurs simultaneously with hydrogenolysis or hydrodeoxygenation, and is largely dependent on the catalyst system and reaction conditions. Since ether bonds constitute the most frequent linkages in lignin, many examples of lignin hydrogenolysis can also be considered as hydrodeoxygenation reactions. Table 10.3 gathers reported examples of hydrogenation and hydrodeoxygenation (or hydrogenolysis) applied to real lignins or whole biomass.

The first attempts to reductively depolymerize lignin have been conducted in the 1930s. Since then, many studies have been reported that consist of pressurized molecular hydrogen and a monometallic or bimetallic catalyst either in a solubilized or supported form (Table 10.3). The use of catalysts enhances hydrodeoxygenation by suppressing the formation of char and increasing the yield of oil [32]. Nickel

Table 10.3 Catalyzed reductive lignin depolymerisation

Lignin type[a]	Catalyst	Support	H source	Solvent	Conditions			Major products	Conv. (%)	References
					Temp (°C)	Pressure (MPa)	Time (h)			
Hardwood OSL	Cu-CrO	None	H_2	Dioxane	260	40	20	Mix of alcohols (methanol,4-propylcyclohexanol, …)	70	[45]
Lignins and lignosulfonates	Ni-Co-catalysts[b]	Al_2O_3, SiO_2…	H_2	Protic, e.g. H_2O, MeOH,…	150–300	2–20	3–20	Monophenols, benzene	50	[46]
Kraft	Ni-W	Al_2O_3-SiO_2	H_2	Phenol	300–450	3.5–24	NS	Monophenols	High	[47]
Softwood Kraft	FeS-CuS-SnS	None	H_2	H_2O, Phenol, MeOH	300–450	>15	1–2	Monophenols C_6–C_9	NS[c]	[48]
Aspen OSL	Co-Mo	Al_2O_3 sulfided	H_2	1-Methyl-naphthalene	400–450	16	1	Monophenols, solid residue	NS	[49]
Softwood OSL	Sulfided Ni-Mo	Al_2O_3, Zeolite A	H_2	Oils	375–450	10	2	Monophenols	>13	[50]
Softwood OSL	Pd	Activated C	H_2	None	380	18	0.25	Monophenols, oils	80	[51]
Softwood OSL	Fe_2O_3	None	H_2	None	380	30	0.25	Monophenols, oils	17	
Softwood OSL	Raney Ni	None	H_2	None	380	22	0.25	Monophenols, oils	54	
Softwood OSL	Ni-Mo	Al_2O_3-SiO_2	H_2	None	420	22	0.25	Monophenols, oils	66	
Softwood OSL	Ni-Mo	Zeolite	H_2	None	420	30.5	0.25	Monophenols, oils	17	
OSL or Kraft	Ni-Mo	Al_2O_3-SiO_2	H_2	None	395–430	~20	0.5–0.7	Monophenols, oils	49–71	[52]
Pine Kraft	Mo	None	H_2	None, H_2O	430	22	2	Phenols, cyclohexanes, benzenes, naphthalenes	61	[53]
Pine EMAL	Pd	C	H_2	Dioxane/H_2O	195	3.45	4	Oils, dihydroconiferyl alcohol, 4-propylguaiacol	89	[54]
Hydrolysis lignin	Ni-Mo-P_2O_5	Al_2O_3, sulfided or not	H_2	None	320–380	4–7	4	Org. phase (hydrocarbons), aq. phase (phenols), gas, solids	100	[55]

(continued)

Table 10.3 (continued)

Lignin type[a]	Catalyst	Support	H source	Solvent	Conditions			Major products	Conv. (%)	References
					Temp (°C)	Pressure (MPa)	Time (h)			
Lignosulfonates	Ni	Activated C	H_2	Ethylene glycol	200	5		4-ethyl-guaiacol, 4-propyl-guaiacol	91	[56]
Hardwood OSL	Ni_7Au_3	None	H_2	H_2O	170	1	12	Guaiacylpropanol, syringylpropanol, …	14	[57]
Hardwood OSL	$Ni_{85}Ru_{15}$	None	H_2	H_2O	130	1	?	?	42	[58]
	$Ni_{85}Rh_{15}$	None	H_2	H_2O	130	1			50	
	$Ni_{85}Pd_{15}$	None	H_2	H_2O	130	1			48	
Alkaline lignin	WP	Activated C	H_2	H_2O + EtOH	280	2	2	4-propylguaiacol, 4-ethylguaiacol, guaiacol,…	NS	[59]
Switchgrass OSL	Pt	C	HCOOH	EtOH	350	0.1 (Air)	4	4-propylguaiacol, 4-methylguaiacol, guaiacol,…	NS	[60]
Candlenut OSL	Cu	PMO[d]	H_2 + MeOH	Supercritical MeOH	180	4	14	C_9 catechols	93	[61]
O. Eurpoea OSL	Ni	SBA-15	HCOOH	HCOOH	150	0.1	0.5	Desaspidinol, syringaldehyde, syringol…	35–45	[62]
O. Eurpoea OSL	Ni, Pt, Pd, Ru	SBA-15	Tetralin, HCOOH	Tetralin, HCOOH	140	0.1	0.5	Biool (mesitol, syringaldehyde), biochar…	27–70	[63]
Birch sawdust	Ni, magnetic	Activated C Al_2O_3 SBA-15	MeOH, EtOH, or Ethyleneglycol	MeOH, EtOH, or Ethyleneglycol	200	0.1 (Ar)	0	Propylguaiacol, propylsyringol	50	[64]
Hardwood biomass	Zn/Pd	C	H_2	MeOH	225	3.4	12	2,6-dimethoxy-4-propyl phenol, dihydroeugenol	19–53	[65]

[a]*OSL* organosolv lignin, *EMAL* enzymatic mild acidolysis lignin
[b]Co-catalysts include Al, Co, Sn, Fe, Ce, Zn, Cu, Mo, Cr
[c]*NS* not specified
[d]*PMO* porous metal oxide

(especially nickel-molybdenum) catalysts have been the most studied catalysts in lignin hydrotreatment. High partial hydrogen pressure (>1 MPa) and high temperature (often >300 °C) have both been shown to favor higher lignin conversion rates [51]. However, although conversion rates higher than 60 % have been repeatedly reported in lignin hydrotreatment (Table 10.3) the actual yields of individual phenols or arenes present in the resulting oils rarely exceed few tens of mg per g of lignin (<1 wt% of lignin). In addition, in most of these studies the stability and reusability of the catalyst have not been well documented.

A patent by Engel and Steigleder claimed that hydrocracking of Kraft lignin gave rise to high yields of phenols and cresols using a supported Ni-tungsten catalyst [47]. Nevertheless, the individual yields of phenolics were not provided, and the reaction was carried out in phenol, which constitutes a major environmental barrier to the industrial development of this process.

The use of fossil-fuel derived H_2 for deoxygenation of lignin affects the carbon footprint of the processes. Finding greener sources of H_2 could therefore contribute to reducing the negative environmental impact of lignin depolymerization. Tetralin has been used as a source of hydrogen in various hydrocracking processes but it is not renewable and it is often associated with phenol or cresol used as solvent [66]. A renewable source of hydrogen that has recently attracted growing attention for depolymerization of lignin into phenolics is formic acid, obtainable from cellulose. This first non-catalyzed method involves the treatment of lignin in a high pressure reactor using formic acid as active H_2 donor and alcohol as solvent [67, 68]. Upon heating at 300 °C, formic acid decomposes into CO_2 and H_2, and the latter combines with oxygen from the methoxyl groups of lignin to form water, thus reducing the number of possible phenolic products. While promising, the method still requires optimization in terms of yields, product isolation, and solvent recovery. A second approach combines the use of formic acid with that of catalysts [7, 60, 62, 63]. Ni-based catalyst with formic acid was found to provide optimum depolymerization results as compared to noble metals including Rh, Ru, Pd, and Pt. The use of aluminium-substituted mesoporus material (Al-SBA-15) as support was advantageous due to its good (hydro) thermal stability, large surface area, and high acidity which, like for formic acid, helps quenching aromatic radicals. Importantly, Ni-catalysts supported on Al-SBA-15 were shown to be fairly stable and reusable under the investigated reaction conditions [62, 63]. Formic acid thus appears as a promising H_2 replacement for the reductive depolymerization of lignin. In spite of these positive results, the separation of a complex product mixture and the utilization of an expensive formic acid reagent are key issues to be addressed in order to bring this technology to industrial applications [7].

An alternative solution to the complex downstream processing of mixtures of high boiling phenols is to proceed to the reduction of the resulting phenols into arenes or diarenes, expected to be more volatile and hence more easily valorizable using conventional refinery processes. Recently, Rinaldi and his group managed to deconstruct organosolv lignin by combining transfer hydrogenation (using isopropanol), hydrogenolysis with a Raney-Nickel catalyst, and an acidic beta-zeolite [69]. The proposed approach was conducted at relatively mild temperatures (150–

240 °C) and preliminary results indicated the cleavage of ether linkages, demethoxylation of phenol units, and dehydroxylation of phenol intermediates resulting in 40 % conversion of lignin into oil mainly composed of arenes and alkanes. Jongerius et al. [70] also applied a two-step approach consisting of (i) lignin depolymerization over a 1 % Pt/Al_2O_3 catalyst at 225 °C in alkaline ethanol–water, and (ii) subsequent hydrodeoxygenation of the lignin-oil obtained using $CoMo/Al_2O_3$ and Mo_2/Carbon nanofiber at 300 °C. Although encouraging, these complex two-step processes still require an increase of yields and selectivity before they could move to industrial scale.

In view of the difficulty to convert polymeric lignin into simple phenols in high yield, researchers have started to investigate the possible deconstruction of lignin directly from the lignocellulose matrix [64, 65, 71]. In a remarkable study, Song et al. designed a Ni-based magnetically separable catalyst able to depolymerise native birch wood lignin into monomeric phenols [64]. The proposed methodology conducted in methanol, ethanol, or ethylene glycol, at moderate temperatures (200 ° C), and under argon (active H species were provided by the alcohol), achieved conversions of ca. 50 % and lignin was selectively cleaved into propylguaiacol and propylsyringol at >90 %. This is one of the first methodologies that allow converting lignin selectively, in green solvent, under relatively mild conditions, and with a catalyst that can be recovered and reused. Parsell et al. also reported a system applicable to whole lignocellulosics, i.e., a Zn/Pd/C catalyst that convert lignin in intact hardwood biomass directly into dihydroeugenol and 2,6-dimethoxy-4-propyl phenol, leaving behind the carbohydrates as a solid residue [65]. Interestingly, the system worked better for intact lignocellulose than for organosolv lignin and the leftover carbohydrate residue was easily hydrolysable by cellulases. Finally, Van den Bosch et al. reported a biorefinery process where a Ru/C catalyst was able to yield carbohydrate pulp and lignin oil (50 % of phenolic monomers, mainly 4-n-propylguaiacol and 4-n-propylsyringol) from hardwood with a nearly complete retention of cellulose in the pulp [72]. These "lignin-first" strategies demonstrate that deconstruction of lignin directly from the lignocellulosic matrix could be done more selectively and efficiently than from the isolated lignin. The new "lignin-first" approach could possibly constitute the basis of new industrial developments for lignin depolymerization.

10.3.2 Oxidative Lignin Depolymerization

In comparison to reductive depolymerization, oxidation of lignin-like material occurs at milder temperatures (20–200 °C) and gives rise to polyfunctionalized aromatic chemicals including aldehydes, ketones, and acids. Oxygen, air, hydrogen peroxide, or peracids are oxidative agents that have been most investigated for this purpose; nitrobenzene has also been used but mainly for analytical purposes. Catalysts used for lignin oxidation comprise organometallics, metal oxides, metal-free organics, acid/base, or metal salts [10].

Among the reported oxidation methods, many use simple lignin model compounds that lack key lignin structural features and properties; only few have been devoted to lignin itself. Oxidative processes applied to actual lignin are summarized in Table 10.4. When targeting aromatic fine chemicals, hydrogen peroxide is normally avoided to prevent ring opening and retain the aromatic features of products. Oxygen is almost always used as oxidative agent due to its abundance, low cost, and environmental friendliness. It is often coupled with alkaline conditions that are used to enhance lignin dissolution and to favor oxidation reactions of ether bonds at the Cα and Cβ positions, C–C bond cleavage in the side chains, quinone methide formation, and/or nucleophilic addition of hydroxide ion on quinone methide [34].

Vanillin (or syringaldehyde) is generally obtained as the major product of oxidation, depending on the lignin origin, but at individual yield that rarely exceeds 10 wt% (Table 10.4). Vanillate (syringate) or acetovanillone (acetosyringone) are also formed in significant amount. At higher pH, the formation of aldehydes is favored versus that of acids [86]. Compounds, such as guaiacol and syringol, lactones, dimers, or organic acids are also formed although their miniscule amounts are usually not provided in the literature.

The main drawback of oxidative processes resides in the production of radicals that lead to partial re-polymerization and multiple reaction pathways associated to low selectivity [84]. Industrially, the only aromatic chemical that has ever been produced by oxidation of lignin is vanillin. Until the 1980s, vanillin was produced via alkaline oxidation of lignosulfonates, the byproduct resulting from sulfite pulping processes [5, 87]. However, with 160 kg of caustic liquids generated for each kg of vanillin produced, these processes raised environmental concerns and were largely abandoned in North America by the late 80's. Today, Borregaard is the only company that produces vanillin from Norway spruce lignosulfonates. Although yields around 8 % have been reported in the literature for alkaline oxidation of lignin [78], Borregaard claimed a vanillin yield of 1 % [6], thus leaving plenty of room for the development of more efficient green processes that are more selective, economically viable, and bypass potentially harmful byproducts.

Kraft processes are currently the most widely used pulping processes in the world. Developing a depolymerization process that works well for Kraft lignin could thus be the key to success. New oxidation processes of Kraft lignin involving the use of polyoxometalates as reversible oxidants in combination with radical scavengers to prevent lignin fragments from repolymerizing have been reported [74, 75]. These new approaches that take place in alcohols and do not involve high concentrations of alkali processes are very promising in terms of environmental footprint. Another process was recently reported by Stahl and his group that allowed fairly high yields of syringyl compounds and their guaiacyl analogues, using a two-step process. This consists of (i) chemoselective aerobic organocatalytic oxidation of secondarybenzylic alcohols of lignin, and (ii) mild depolymerization of oxidized lignin in aqueous formic acid [84, 85] (Table 10.4). The metal-free process worked effectively for native hardwood (aspen) lignin. If applicable to Kraft lignins, and more specifically to the more condensed softwood

Table 10.4 Catalyzed oxidative lignin depolymerisation

Lignin type[a]	Catalyst	Oxidant	Solvent	Conditions			Major product	Yield (%)	References
				Temp (°C)	Pressure (MPa)	Time (h)			
Spruce Kraft Hardwood OSL	$MeReO_3$	H_2O_2	Glacial CH_3COOH	RT	0.1	NS	Side-chain oxidized and demethylated fragments	NA	[73]
Kraft Aldrich	$H_3PMo_{12}O_{40}$	O_2	$H_2O + CH_3OH$	170	0.5	0.33	Vanillin/methyl vamillate	5	[74, 75]
Yellow poplar wood chips	$CuSO_4/FeCl_3$	O_2	H_2O + Alkali	190	1.38	0.33	Syringaldehyde	8.4	[76]
							Vanillin	4.3	
Poplar Alkali	$CuSO_4/FeCl_3$	O_2	H_2O + Alkali	170		0.16	Syringaldehyde	9.5	[77]
							Vanillin	4.7	
Lignosulfonates	$CuSO_4$	O_2	H_2O + Alkali	190	1.1	1.25	Vanillin	7.2	[78]
Loblolly Pine MWL	$CuSO_4$/ Phenanthroline	O_2	H_2O + Alkali	80	0.27	24	Vanillin	Low	[79]
							Vanillic acid		
Hardwood Kraft	CuO	O_2	H_2O + Alkali	180	1.52	2	Syringaldehyde	3.5	[80]
Hardwood Kraft	CoO	O_2	H_2O + Alkali	180	1.52	2	Syringaldehyde	2.5	[80]
Sugarcane Alkali	Pd)-Al_2O_3	O_2	H_2O + Alkali	140	0.5	0.5	Syringaldehyde	3.0	[81]
							Vanillin	4.4	
							p-hydroxybenzaldehyde	7.1	
Cornstalk EMAL	$LaMnO_3$	N_2/O_2	H_2O + Alkali	120	1.5/0.5	3	Syringaldehyde	9.0	[82]
							Vanillin	4.0	
							p-hydroxybenzaldehyde	1.0	
Eucalyptus Kraft	Co(salen) w or w/o NaY	$CH_3C(O)OOH$	$H_2O + CH_3OH$	70	0.1	3	Guaiacol	5.2	[83]

(continued)

Table 10.4 (continued)

Lignin type[a]	Catalyst	Oxidant	Solvent	Conditions			Major product	Yield (%)	References
				Temp (°C)	Pressure (MPa)	Time (h)			
Native Aspen	4-acetamido-TEMPO	O_2	$CH_3CN + H_2O + HNO_3 + HCl$	45	0.1	24	Aspen lignin oxidized at benzylic positions	NA	[84]
Oxidized native Aspen	–	–	$H_2O + HCOOH$	110	0.1	24	Syringyl diketone	13.1	[85]
							Syringaldehyde	8.5	
							syringol	7.9	

[a] *OSL* organosolv lignin, *MWL* milled wood lignin, *TEMPO* 2,2,6,6-tetramethylpiperidine-N-oxyl, *EMAL* Enzymatic mild acidolysis lignin, *RT* room temperature, *NA* non applicable

Kraft lignins, it could open the door to a highly effective and selective system for the production of guaiacyl diketone and vanillin.

Oxidative depolymerization of lignin so far leads mainly to vanillin or syringaldehyde, along with many other phenolic compounds and organic acids. None of the metal-catalyzed lignin oxidation processes have been scaled up to an industrial level due in part to high amounts of alkali needed, low yields and selectivities, complex, and costly downstream treatments, as well as catalyst poisoning and reusability issues. Current processes lead to complex mixtures, which are economically and technically hard to fractionate into pure compounds [88]. Developing new non-alkaline chemical processes for oxidative lignin depolymerization as well as new technologies for vanillin/syringaldehyde isolation from the reaction medium therefore appears to be prerequisites to a successful industrial scale up of chemical production of vanillin or syringaldehyde from lignin. Alternatively, a way to make the new processes economically viable is to co-produce, along with vanillin, lignin fragments that are usable as macromonomers for the production of bio-based polymers. Such an approach was reported by Borges da Silva et al. where vanillin was co-produced with a polymer that is suitable for the production of polyurethane foams [86].

10.3.3 Summary

Although many homogeneous and heterogeneous catalysts have been investigated at the laboratory scale for lignin depolymerization, the industrial implementation of these catalysts is still hampered by the associated harsh conditions, low yields, and reusability issues. The usual products of lignin being either alcohols or aromatic compounds, solid catalysts remain very susceptible to surface saturation and deactivation by the coordinating products. New approaches aimed at depolymerizing lignin directly from lignocellulosic biomass appear to work under milder and greener conditions and might just be the way to go.

10.4 Photocatalysis

Photocatalysis has been used as a clean oxidative technology leading to the total mineralization of various organic pollutants including phenols and other aromatics [89, 90]. In particular, photocatalysis was proven to permit decolourization of paper mill effluents by completely degrading lignins and phenols into CO_2 [91–98]. Titanium dioxide (TiO_2) is generally considered to be the best photocatalyst for this reaction due to its high photosensitizing power, chemical stability, and commercial availability. In some cases, the photocatalytic activity of TiO_2 can be enhanced by doping it with a noble metal such as platinum [95] or nonmetal ions such as sulfur or boron [99], or by adding ferrous iron in the medium [100]. Efforts have also been made to improve the separation of catalyst from the medium by immobilizing TiO_2 on a support, such as sepiolite [97], carbon fibers [94], glass plates [99], or (CeO_2, La_2O_3, or C) nanotubes [101]. Zinc oxide (ZnO) is another semiconductor material that was tested for lignin depolymerization and that appeared to be an even better photocatalyst than TiO_2 for degradation of Kraft wheat straw lignin [102].

Since most of these photocatalytic systems were developed to decolorize pulping effluents, emphasis was usually put on the ability of the catalyst to decrease the color or chemical oxygen demand (COD) in the solution. Various parameters (pH, load of catalyst, temperature) effects were investigated but little effort was dedicated to the identification of lignin products. Ksibi et al. first identified phenolics by gas chromatography–mass spectrometry (GC-MS) when photodegrading soluble alfalfa lignin using a TiO_2/UV photocatalytic technique under aerobic conditions [93]. When replacing air with H_2O_2, Kamwilaisak and Wright confirmed the formation of organic acids, namely acetic acid, malonic acid, and succinic acid, as main products of lignin TiO_2-catalyzed photodegradation [103]. The photodegradation of phenolic compounds by TiO_2 was suggested to occur via (i) $^{\cdot}OH$ radical attack on the phenyl rings producing catechol, resorcinol, and hydroquinone; (ii) phenyl ring opening to give malonic acid; (iii) degradation of malonic acid (and lignin side chains) to short-chain organic acids, such as maleic, oxalic, acetic, and formic acids; and (iv) release of CO_2 by decarboxylation of formic acid. Interestingly, under certain conditions, succinic, and malonic acids were produced in significant amounts and photocatalysis appeared as a potential process to produce organic acids from lignin. The challenge for this reaction was to get a highly active photocatalyst that does not favor complete degradation of products into CO_2. Keeping this in mind, Tonucci et al. designed a mild photocatalytic TiO_2 system working under aerobic conditions, that showed low carbon consumption, good preservation of aromatic rings, and that greatly reduced mineralization [104]. The system was tested with lignosulfonates only. However, it would be interesting to see whether positive results can also be obtained with water-insoluble lignin.

10.5 Electrocatalysis

Electrochemical oxidation has also been considered as a potential alternative in lignin oxidative degradation. Various anodes ranging from simple ones (e.g., Pb, Ni [105, 106]) to more complex ones (Ti/Sb-SnO_2, Ti/PbO_2, Ti/Ta_2O_5-IrO_2, Ti/SnO_2-IrO_2, Ti/RuO_2-IrO_2, and Ti/TiO_2-IrO_2) [107–109] have been used or designed for the purpose. Among the various IrO_2-based electrodes tested, Ti/RuO_2-IrO_2 electrode exhibited the highest activity and stability [107]. Depending on the systems and the advancement of reaction, vanillin/vanillic acid or organic acids were identified as the main products. Vanillin and other low molecular weight phenolic intermediates can undergo condensation by phenolic dimerization or condensation of quinonic radicals [110]. Quinonic radicals can be decarbonylated and lead to cyclopentadienes; quinones can undergo ring opening to form various dicarboxylic acids that further evolve to simpler diacids, such as maleic or oxalic acid [110]. The production of phenolic compounds thus requires limiting secondary reactions. It can be enhanced by stopping the reaction at the early stage of the reaction [106] or by conducting continuous extraction to avoid further decomposition reactions of the primary intermediate products [111].

Although electrochemical oxidation is a potentially promising technology for lignin oxidation/modification, the high cost and the electrode fouling caused by condensation of intermediate products limit its application [107]. Electrocatalysts with high activity, long lifetime, and low cost have thus been the focus of the newest research for lignin electrooxidation [107]. Nevertheless, the yields of the products are still too low to make the overall process economically viable [10].

Various synergistic approaches aiming at improving the yield of depolymerization have been explored. An electrochemical approach combining anode oxidation and electrogenerated H_2O_2 oxidation has been recently developed for converting lignin into aromatic chemicals [109]. Using GC-MS for the analysis of reaction products, the authors confirmed that C–C and C–O–C bonds were cleaved synergistically by direct anodic oxidation and indirect H_2O_2 oxidation, and the macromolecules were gradually depolymerized into monomers and dimers. In another synergistic example, a combination of photoelectrocatalysis using a Ta_2O_5-IrO_2 thin film as electrocatalyst, and TiO_2 nanotube arrays as photocatalyst provided ca. 92 % lignin oxidation as compared to 66 % using a single electrochemical approach under similar conditions [112]. Finally, the electro-oxidative cleavage of alkali lignin was successfully conducted in a protic ionic liquid (IL), namely triethylammonium methanesulfonate, using a highly active $Ru_{0.25}V_{0.05}Ti_{0.7}O_x$ anode [113]. Vanadium was identified as a crucial element in the electrode due to its ability to promote single electron transfer. The protic IL, with its ability for proton transfer from the acid to the base, provided a suitable medium for dissolution of lignin, ensured electrolysis at higher potentials and promoted the oxidative lignin cleavage mechanism. Interestingly, the product distribution was strongly affected by the applied potential. That higher potentials resulted in the formation of molecules of smaller molecular weights suggested the possibility of tuning the system (electrode and IL) toward specific selected products.

10.6 Ionic Liquids

Ionic liquids (ILs) are salts composed of large organic cations and inorganic or organic anions that exist as liquids at a relatively low temperature (<100 °C) [114, 115]. ILs tend to have good thermal and electrochemical stability, very low volatility, implying ease of recycling with virtually no VOCs (volatile organic compounds) release, and nonflammability, implying low or no risk of explosion [116–118]. The vast number of possible ion combinations in ILs results in a wide range of physical properties to choose from when selecting an IL for use as solvent or catalyst. The greener properties of ILs compared to conventional solvents associated with the extensive diversity of tunable properties made ILs the center of interest over the last few decades. Although most commercial developments involving ILs have been dedicated to petroleum-based reagents [119], new technologies are emerging where ILs are applied to pretreat, fractionate, or convert lignocellulosic materials.

10.6.1 Dissolution of Lignocellulosic Biomass in ILs

The transformation of lignocellulosic feedstocks has always been challenging due to their low solubility in almost any solvent [116]. Manufacturing of cellulose and its derivatives are normally conducted in environmentally undesirable media consisting of polar organic solvents mixed with charged compounds, e.g., dimethyl sulfoxide/tetrabutylammonium fluoride (DMSO/TBAF) or LiCl/dimethylacetamide (LiCl/DMAc) [116]. With their unique chemical inertness and solvation properties [120], ILs thus offer a promising greener alternative for the processing of biomass.

Interest in ILs for biomass dissolution started with the contribution of Rogers' group who found that 1-butyl-3-methylimidazolium chloride, [BMIM][Cl], could dissolve cellulose at concentrations as high as 25 % and who then demonstrated that whole wood could be fully solubilized in 1-ethyl-3-methylimidazolium acetate, [EMIM][OAc] [121, 122]. A high number of studies reporting the processing of biomass in ILs then followed this pioneering work, with most studies being focused on cellulose dissolution and processing.

The easy solubilization of polysaccharides in ILs has been exploited for the development of pretreatment technologies aimed at increasing the enzymatic hydrolysis of cellulose [116, 118, 123, 124]. Indeed, it appears that cellulose obtained after reconstitution from IL solutions is less crystalline and therefore more accessible to cellulase than the original material [123].

Several reviews are now available on the combination of ILs with cellulose and lignocellulosic material that covered the dissolution mechanisms, the effect of ions on solubility, the constituent regeneration or reconstitution, and the use of ILs in pretreatment before cellulose hydrolysis [116–118, 125–129].

10.6.2 Dissolution of Lignin in ILs

Compared to the wide efforts dedicated to the understanding of dissolution of polysaccharides in ILs, much less is known about the potential of ILs to dissolve lignin. Several studies investigated the dissolution of whole biomass in ILs followed by the addition of antisolvent, e.g.,water or water/acetone, to precipitate cellulose-rich or lignin-rich fractions [122, 124, 130–132]. Solvents such as [BMIM][Cl] and 1-allyl-3-methylimidazolium chloride, [AMIM][Cl], gave the best results for both hardwood and softwood [123, 133]. [EMIM][OAc] also led to excellent results with complete dissolution of hardwood at 90 °C [122, 133] or softwood at >150 °C [131]. Significantly more rapid dissolution of biomass and more effective lignin removal were observed with increasing temperature, but higher temperatures were found to degrade the ILs and therefore decrease their recyclability [124, 131, 134].

Another approach makes use of ILs to allow dissolution of lignin while leaving behind the undissolved polysaccharide material. Excellent lignin extraction yields (>93 %) were achieved by Tan et al. using the aromatic alkylbenzenesulfonate (ABS) anion with BMIM cation [135]. Unfortunately, few drawbacks arose when using [BMIM][ABS] that include: (i) the relatively high extraction temperatures (190 °C), (ii) the necessity to pretreat the biomass with steam, (iii) a considerable loss of carbohydrate, and (iv) difficulties to recover the IL. Inspired by the initial results of Tan et al. [135], Pinkert et al. studied the dissolution of lignin using imidazolium acesulfamate ILs [136]. The authors's choice was driven by three reasons: (i) the low cost and nontoxicity of potassium acesulfamate ([K][ace] is used as sugar substitute), (ii) previous studies suggested that large and bulky IL anions with delocalized charge do not dissolve cellulose, and (iii) the aromatic character of acesulfamate should allow a good interaction with lignin. Applying imidazolium [ace] to lignin dissolution from wood led to a lignin extraction yield of 43 % under gentle extraction conditions (T = 100 °C, t = 2 h), a steadiness of cellulose crystallinity, and a constancy of extraction yields over multiple sequential recycling of the IL (up to 6 runs). However, before any commercial plan could be realized for using [ace] IL to extract lignins from biomass some technical issues such as the sensitivity of the system to water present in biomass and the need to increase the recycling runs above 100 times, need to be addressed.

ILs were also used to dissolve isolated lignins. Different cations and anions screened for this purpose are summarized in Table 10.5. The role of cation consists in generating interactions between the IL and the biopolymer, whereas anion is primarily responsible for the initial disruption of the inter- and intramolecular hydrogen bonding that exist in the biopolymer. Imidazolium-based cations appeared to be efficient for the dissolution of lignin. Kilpelainen et al. suggested that π–π interactions exist between the imidazolium-based cations and the aromatic compounds of lignin [141], a hypothesis that was confirmed by the increased dissolution of softwood lignin when an allyl group replaced the butyl group of [BMIM][Cl] [123]. When dissolving softwood lignin with the cation $[BMIM]^+$ and various anions, the solubility varies with the anions following the order: $[OTf]^-$ ~ $[OMs]^-$ ~ $[MeSO_4]^-$ > > $[OAc]^-$ > $[HCOO]^-$ > $[Cl]^-$ ~ $[Br]^-$> > $[BF_4]^-$ ~ $[PF_6]^-$, suggesting that the strongly hydrogen-bonding anions such as $[OTf]^-$ or $[OMs]^-$ are efficient solvents for lignin, while the large non-coordinating anions, such as $[BF_4]^-$ and $[PF_6]^-$ are inefficient [118, 137, 138]. Chloride anion was used recently in combination with alkyl-diazabicyclo[5.4.0]undec-7-enium [$DBUC_n$] cation to dissolve softwood Kraft lignin but compared to imidazolium, these poorly unsaturated cations exhibited a modest dissolution power [140].

Table 10.5 Dissolution of isolated lignin in ILs

Ionic liquid	Lignin type	Solubility	Conditions	Reference
[AMIM][Cl]	Softwood Kraft lignin	>300 g/kg	90 °C, 24 h	[123]
[BMIM][Br]	Softwood Kraft lignin	17.5 g/L	75 °C	[137]
[BMIM][Br]/H_2O	Alcell lignin	225 g/kg	60 °C	[138]
[BMIM][BF_4]	Softwood Kraft lignin	40 g/kg	90 °C, 24 h	[123]
[BMIM][BF_4]/ H_2O	Alcell lignin	20 g/kg	60 °C	[138]
[BMIM][Cl]	Softwood Kraft lignin	13.9 g/L	75 °C	[137]
	Alkali lignin	88 g/kg	80–90 °C, 20 min	[139]
	Softwood Kraft lignin	>100 g/kg	90 °C, 24 h	[123]
	Softwood Kraft lignin	100 g/kg	90 °C, 24 h	[124]
[BMIM][OTf]	Softwood Kraft lignin	>500 g/kg	90 °C, 24 h	[123]
[BMIM][OMs]/ H_2O	Alcell lignin	475 g/kg	60 °C	[138]
[BMIM][$MeSO_4$]	Softwood Kraft lignin	312 g/L	50 °C	[137]
[BMIM][OAc]/ H_2O	Alcell lignin	450 g/kg	60 °C	[138]
[BMIM][OTf]	Softwood Kraft lignin	>500 g/kg	90 °C, 24 h	[123]
[BMIM][PF_6]	Softwood Kraft lignin	1 g/kg	90 °C, 24 h	[123]
[BMMIM][BF_4]	Softwood Kraft lignin	14.5 g/L	70–100 °C	[137]
[DMEA][HCOO]	Softwood Kraft lignin	280 g/kg	90 °C, 24 h	[124]
[DMEA][OAc]	Softwood Kraft lignin	190 g/kg	90 °C, 24 h	[124]
[CMIM][Br]	Alkali lignin	95 g/kg	80–90 C, 20 min	[139]
[EMIM][OAc]	Softwood Kraft lignin	>300 g/kg	90 °C, 24 h	[123]
	Softwood Kraft lignin	300 g/kg	90 °C, 24 h	[124]
[EMIM][OAc]/ H_2O	Alcell lignin	300 g/kg	60 °C	[138]
[HMIM][OAc]/ H_2O	Alcell lignin	400 g/kg	60 °C	[138]

(continued)

Table 10.5 (continued)

Ionic liquid	Lignin type	Solubility	Conditions	Reference
[HMIM][OTf]	Softwood Kraft lignin	275 g/L	70 °C	[137]
[PrMIM][Br]	Alkali lignin	62 g/kg	80–90 °C, 20 min	[139]
[$DBUC_4$][Cl]	Softwood Kraft lignin	~200 g/kg	105 °C, 9 h	[140]
[DBUC6][Cl]	Softwood Kraft lignin	~200 g/kg	105 °C, 11 h	[140]
[DBUC8][Cl]	Softwood Kraft lignin	~200 g/kg	105 °C, 13 h	[140]

$[AMIM]^+$, 1-allyl-3-methylimidazolium; $[BMIM]^+$, 1-butyl-3-methylimidazolium; $[BMMIM]^+$, 1-butyl-2,3-dimethylimidazolium; $[CMIM]^+$, 1-cyano-3-methylimidazolium; $[DMEA]^+$, N, N-dimethylethanolammonium; $[EMIM]^+$, 1-ethyl-3-methylimidazolium; $[HMIM]^+$, 1-hexyl-3-methylimidazolium; $[PrMIM]^+$, 1-Propyl-3-methylimidazolium; $[HCOO]^-$, formate; $[OAc]^-$, acetate; $[OMs]^-$, mesylate, $CH_3SO_3^-$; $[OTf]^-$, trifluoromethylsulfonate $CF_3SO_3^-$

Recently, it was demonstrated that adding an appropriate amount of water to ILs (optimum ratio IL/H_2O = 70/30) facilitated lignin dissolution, likely due to the increased mobility and diffusion constant of ions in the presence of water [138]. The addition of water promoted release of more ions from the IL stack, thus increasing the interaction probability between lignin and ILs.

10.6.3 Conversion of Lignin in ILs

A number of studies reporting the modification of lignin during IL biomass deconstruction have been published and recently reviewed by Brandt et al. [132]. Tan et al. reported that lignin extracted with [EMIM][ABS] had a lower molecular weight and a narrower polydispersity than one obtained by aqueous auto-catalyzed pretreatment [135]. Similarly, Kim et al. compared the characteristics of a lignin extracted from poplar wood using [EMIM][OAc] with the properties of milled wood lignin extracted from the same biomass and concluded that IL extraction led to a less polydisperse lignin of smaller molecular weight [130]. George et al. studied the impact of a range of ILs on several commercial lignins and demonstrated an intense effect of anion on the fragmentation mechanism and degree of depolymerization, while the cation did not play any significant role [142]. ILs containing alkyl sulfate anions appeared to have the greatest ability to fragment lignin and shorten polymer length. The order of molecular weight reduction was sulfate > lactate > acetate > chloride > phosphate. The anion was believed to undergo nucleophilic attack on the β-O-4 lignin linkages, thus reducing the average molecular weight of lignin.

Besides the observations on lignin chemical properties after IL treatment, ILs were also used as solvent for chemical transformation of lignin. Main research efforts addressing the conversion of lignin in ILs have been dedicated to either its acidolytic cleavage or catalytic oxidation. Moreover, most of the reported studies involved homogeneous or heterogeneous catalysis [118].

ILs can be used for lignin acidolytic depolymerization due to their ability to act as both an acidic catalyst and a solvent [7, 143]. For instance, 1-H-3-methylimidazolium chloride was shown to promote acidolytic cleavage of β-O-4 linkages in lignin over relatively mild temperatures (110–150 °C) [144]. The reaction was proposed to occur *via* protonation (or coordination) of the ether linkages, followed by attack of water (or any other nucleophile present in the system), similarly to the reactions taking place in conventional solvents.

ILs can also be used as solvent for the oxidative depolymerization of lignin, as reviewed recently [9]. A few good examples that use ILs as solvents for the production of aromatic chemicals from lignin are listed in Table 6 of Ref. [9]. Most of the oxidative reactions in ILs were applied to either organosolv or alkali lignin and were conducted using ILs based on phosphate or sulfonate anions due to (i) their stability against oxidation and, (ii) their ability to dissolve lignin. The oxidation of organosolv lignin in ILs in the presence of transition metals and molecular oxygen or air has been demonstrated by Stärk et al. [145], Weckhuysen and his team [146, 147], and Liu et al. [148]. Under optimal conditions, 66.3 % of the lignin could be converted into several monomeric units using $Mn(NO_3)_2$ in 1-ethyl-3-methylimidazolium trifluoromethanesulfonate [EMIM][CF_3SO_3] [145]. The predominant product was either dimethoxy-1,4-benzoquinone or syringaldehyde, depending upon the conditions applied. Interestingly, the potential antitumor agent, dimethoxy-1,4-benzoquinone, could be isolated in 11.5 wt% overall yield using a simple extraction/crystallization process [145]. Similarly, Alcell and soda lignins were oxidized under mild conditions (0.5 MPa O_2, 80 °C) using $CoCl_2{\cdot}6H_2O$ and NaOH in 1-ethyl-3-methylimidazolium diethylphosphate [EMIM][Et_2PO_4]. The catalyst rapidly oxidized benzyl and other alcohol functionalities in lignin, but left phenolic functionality, and 5-5′, β-*O*-4 and phenylcoumaran linkages intact [146]. In situ ATR-FT-IR, Raman and UV-Vis spectroscopy allowed demonstrating that the reaction proceeded via the coordination of alcohol-containing substrates to cobalt, followed by formation of a Co-superoxo species [147]. Finally, in another very interesting study, under optimal conditions (2.5 MPa O_2, 1.5 h, 175 °C), up to 100 % of the lignin could be converted into several monomeric units using $CuSO_4$ in dimethyl phosphate-based ILs [148]. Using [MMIM][Me_2PO_4], a total yield of 29.7 % aldehydes (vanillin, syringaldehyde, *p*-hydroxybenzaldehyde) was obtained, which was significantly higher than the yields usually obtained in aqueous NaOH systems (see Table 10.4 for oxidation yields in aqueous systems). In contrast to the previous studies which did not address the recycling of IL, Liu et al. confirmed that the products and IL phase were easily separated by solvent extraction, allowing the recycling and reuse of the IL-$CuSO_4$ phase up to six times without loss of efficiency [148].

10.6.4 Summary

Recent studies involving lignin and ILs demonstrate that ILs can be used (i) to extract selectively lignin from lignocellulosic biomass, and (ii) to oxidatively depolymerise lignin more selectively than in aqueous media. The possible control of selectivity by changing the experimental conditions (IL nature, reaction temperature, catalyst loading, extracting solvent, etc.) opens the door to new ILs-based processes for the production of value-added chemicals from lignin [9]. Although the conversion and depolymerization of lignin can be achieved in ILs, and even controlled by changing the structure of the IL or conditions, bulk separation of the products from ILs remains a formidable challenge that needs to be overcome [118].

10.7 Biocatalysis

The high selectivity and efficiency of enzymatic catalysts, the mild operating conditions required, the broad range of substrates, and the ability of some enzymes to react under adverse conditions (e.g., high temperatures, extreme pH values) are all advantageous properties when considering the application of biocatalysis for industrial processes [149]. Given the lack of selectivity observed when subjecting lignin to chemical treatments, one could see biocatalysis as a suitable means to depolymerise lignin in a selective, cleaner, and more environmental friendly way. However to date, no biocatalytic system has been shown to be efficient enough to access aromatics or non-aromatics from lignin in yields that could be commercialized.

In a desire to combine both an easier access of enzymes to cellulose for biofuel production and the valorization of lignin into value-added chemicals, there has been recently a renewed interest in the microbial breakdown of lignin [150]. While microbial degradation of lignin had been widely studied in white-rot and brown-rot fungi [151–154], emphasis has recently been placed on lignin-degrading bacteria which offer more opportunities for protein expression and genetic manipulation [155–159]. The present section is focused on recent progress in ligninolytic green biotechnology—either as microbes or enzymes—for the production of aromatic and non-aromatic chemicals.

10.7.1 Microbial Lignin Degradation

Lignin was reported to be biodegraded only under aerobic conditions [151]. The initial steps of lignin biodegradation consist in introducing new functional groups into its macromolecular structure by oxidative enzymes, which render lignin more susceptible towards its subsequent degradation by other enzymes [149]. Studies on the microbial degradation of lignin have focused primarily on white-rot and brown-rot fungi, with *Phanerochaete chrysosporium* being by far the most studied of all white-rot fungi [151, 152, 160]. White-rot fungi are much more active lignin-degrader than brown-rot fungi. Although white-rot fungi (basidiomycetes) do not use lignin as a carbon source for their growth, they have developed nonspecific methods for the degradation of lignin [161]. The initial depolymerization of lignin is thought to be promoted by extracellular oxidative enzymes (oxidoreductases) whereas subsequent transformation of smaller molecular weight lignin fragments is assumed to occur intracellularly [154]. The extracellular enzymes involved in lignin depolymerization are lignin peroxidases (LiP), manganese peroxidases (MnP), versatile peroxidases (VP) and phenol oxidases, also known as laccases.

Despite the extensive study of fungal lignin degradation since the mid-1980s, there is still no commercial biocatalytic process for lignin depolymerization, in part due to the practical challenges of fungal protein expression and fungal genetic manipulation [157]. In the last decade, efforts have thus been focused toward the breakdown of lignin by bacteria as indicated by the recent outbreak of reports on the subject [157, 159]. Various bacterial strains identified to have activity for lignin breakdown have been isolated from soils [158, 162–164] or from guts of termites [165, 166]. Most of these strains fall within three classes: actinomycetes, α-proteobacteria, and γ-proteobacteria [157]. Their activity is often less than that of white-rot fungus *P. chrysosporium*, but comparable to other lignin-degrading fungi [153]. Although the metabolic pathway of lignin-degrading bacteria is much less understood than that of white-rot and brown-rot fungi, there are indications that bacteria use similar types of extracellular enzymes found in fungi, i.e., peroxidases and laccases [157]. The next section provides more details on the enzymatic pathways involved in lignin degradation and transformation.

10.7.2 Enzymatic Pathways

10.7.2.1 Lignin Peroxidases (LiP)

LiP were the first lignolytic enzymes to be isolated from *P. chrysosporium* in the mid 80's. LiP are characterized by their high redox potential ($\sim$1.2 V), low pH optima and molecular mass varying from 37 to 50 kDa for different white-rot fungus strains [161]. Relatively non-specific to their substrates, LiP have been known to oxidize phenolic and non-phenolic aromatic substrates by abstracting one electron via a mechanism involving cation radicals. Its catalytic cycle is similar to other peroxidase enzymes, in which the resting state of the enzyme contains ferric heme, which reacts with H_2O_2 to form a compound I oxo-ferryl intermediate (two-electron oxidized form), and subsequently a compound II intermediate (one-electron oxidized form). Veratryl alcohol (VA), produced by ligninolytic fungi, has been proposed to act as a redox mediator, or electron shuttle. LiP differ from other peroxidases by the fact that the heme environment provides high redox potential to the oxo-ferryl complex, which allows oxidation of unusually high potential sites, such as aromatic rings [150, 159]. LiP oxidize the substrates in multistep electron transfers by forming intermediate radicals such as phenoxy radicals and VA radical cations. Phenoxy radicals can undergo nonenzymatic reactions such as repolymerization on the lignin polymer and/or Cα–Cβ breakdown, yielding *p*-quinones [161]. LiP are active on a wide range of aromatic compounds, such as VA, methoxybenzenes, and nonphenolic β-O-4 linked arylglycerol β-aryl ethers with a redox potential up to 1.4 V in the presence of H_2O_2 [167].

The main drawback in using purified LiP for oxidative depolymerization of lignin is the repolymerization reaction that occurs when the small phenolic products are not removed or consumed [168, 169]. The preference of LiP for phenolic lignin units favors the coupling of phenoxy radicals and increases the propensity to polymerize rather than depolymerise lignin samples under in vitro conditions [167]. A second drawback comes from the difficulties associated with the fast decay of H_2O_2, indispensable for LiP (and MnP) catalysis, and the inactivation of LiP (and MnP) for certain amounts of H_2O_2 [170].

10.7.2.2 Manganese Peroxidase (MnP)

MnP is a heme-containing glycoprotein of ~40–60 kDa, optimum pH of 4–7, and temperature of 40–60 °C, which was also first discovered in *P. chrysosporium*. It is produced and secreted by almost all basidiomycetes, including white-rot and various soil colonizing fungi [167]. Like LiP, MnP requires H_2O_2 as an oxidant, but in contrast to LiP, it also requires the presence of Mn^{2+} and chelators (organic acids) such as oxalate or malate. MnP catalyses the oxidation of Mn^{2+} to Mn^{3+}, which in turn can oxidize a large number of phenolic substrates. Complexed Mn^{3+} is widely accepted as a diffusible oxidant, able to oxidize secondary substrates at a distance away from the active site of MnP [150]. The Mn^{3+}-chelate complex oxidizes phenolic compounds (such as 2,6-dimethoxyphenol, guaiacol, 4-methoxyphenol and phenolic lignin residues), but is inactive on VA or nonphenolic substrates [167]. It is also small enough to diffuse into lignin or analogous structures, which are not necessarily available to the enzymes [161].

The drawbacks associated to the broader usage of LiP, i.e., repolymerization and inconvenient use of H_2O_2 apply to MnP.

10.7.2.3 Versatile Peroxidases (VP)

VP are glycoproteins which act in a bifunctional way by sharing typical features of the MnP and LiP fungal peroxidase families. Like MnP they oxidize Mn (II), and similar to LiP they oxidize both phenolic and non-phenolic aromatic compounds, including VA, methoxybenzenes, and lignin model compounds [167].

10.7.2.4 Laccases

Laccases (or benzenediol:oxygen oxidoreductase) are glycosylated blue multi-copper oxidoreductases, produced by many fungi, bacteria, plants, and insect. A comprehensive review on the ability of laccases to breakdown lignin is available elsewhere [171].

In plant, laccases are thought to participate in lignin biosynthesis, while in fungi, they are suspected to contribute to lignin depolymerization, though this last role is still unclear and controversial [150]. Laccases subtract one electron from phenolic–OH groups thus forming phenoxy radicals, which undergo polymerization via radical coupling. As a consequence, the polymerizing or depolymerizing effect of laccase treatment on native or technical lignin is ambiguous [171].

Laccase contains four copper ions of three different types: one type 1 (T1) copper, which gives laccase its characteristic blue color; one type 2 (T2) and two type 3 (T3) copper ions forming a trinuclear cluster. T1 is the site where substrate oxidation takes place, whereas O_2 reduction occurs at the trinuclear cluster site [171].

Laccase alone is not able to break down nonphenolic units of lignin [171]. Indeed, laccases have low redox potentials (0.5–0.8 V) that restrict their action to the oxidation of the phenolic lignin fragments [167]. The restriction of laccase action to phenolic subunits can, however, be overcome by use of a small molecular weight mediator that acts as electron shuttle [172]. Mediators are believed to expand the reach of laccase thanks to their comparatively smaller size that helps them diffuse into the plant cell wall or polymeric structures. In addition, they also broaden the range of oxidizable substrates including non-phenolic units by having higher redox potential than laccase. Natural compounds such as 4-hydroxybenzylic alcohol, *p*-cinnamic acid, sinapic acid, or syringaldehyde can act as laccase mediators. Synthetic compounds such as 2,2′-azinobis (3-ethylbenzothiazoline-6-sulfonic acid) (ABTS), 1-hydroxybenzotriazole (HBT), 2,2,6,6-tetramethylpiperidine 1-oxyl (TEMPO), or violuric acid can also be co-added with laccases to increase the activity of the latter [171].

In contrast to ligninolytic enzymes LiP, MnP, and VP, phenol oxidases such as laccases use O_2 as the final electron acceptor rather than H_2O_2, which could be a great advantage for industrial applications. The main drawbacks, however, in using laccases for oxidative transformation of lignin are the non-fully elucidated action mode of laccases on lignin, especially the ability of laccase to polymerise and depolymerise lignin, and the low extracellular production of laccases in basidiomycete fungi.

10.7.2.5 Other Enzymes

The above ligninolytic enzymes are characterized by their technically unresolved lack of specificity and their high predisposition to provoke the repolymerization of previously released monolignols [173]. Bacterial enzymes may be superior to their fungal counterparts with regard to specificity, thermostability, halotolerance, and mediator dependency [161]. New types of enzymes are therefore being investigated to favor depolymerization of lignin into monomers. Various β-O-4 aryl ether cleaving enzymes or enzyme systems have been isolated from microbes including fungus 2BW-1 [174] and bacteria *Sphingobium paucimobilis* SYK-6 and *Novosphingobium* sp. [173, 175–178], and biochemically characterized. Biphenyl bond cleavage enzyme systems involving demethylation on one ring, followed by dioxygenase-catalyzed ring opening of the resulting catechol, and C–C hydrolase at the benzylic ketone site have also been reported [150]. A tetrahydrofolate-dependent O-demethylase gene was also isolated from *S. paucimobilis* SYK-6 and found to convert syringate into 3-O-methylgallate [179]. A review recently published by Bugg and Rahmanpour describes recent developments in the understanding of bacterial enzymes for lignin breakdown [180]. The enzymes covered in this review include dye-decolorizing peroxidases (DyP), bacterial laccases, and beta-etherase enzymes. While the knowledge of microbial lignin degradation pathways is still incomplete, the use of pathway engineering methods to construct genetically modified microbes to convert lignin to renewable chemicals (e.g., vanillin, adipic acid) via fermentation seems a possible way to go [180].

10.7.3 Small Chemicals Obtained from Lignin Using Biocatalysis

Oxidative breakdown of spruce wood lignin by *P. chrysosporium* was found to lead to 28 low molecular weight products, 10 of which were aromatic carboxylic acids obtained by Cα–Cβ cleavage, and 13 others were acyclic 2,4-hexadiene-1,6-dioic acids (e.g., muconic acid) resulting from oxidative ring cleavage [181]. Many similar phenolic compounds have been detected when treating lignin with bacterial lignin degraders such as *S. paucimobilis* or *Bacillus* sp. [150, 157]. However, the natural microbial systems tend to produce the aromatic chemicals as intermediates which are further degraded into smaller molecules, and even down to CO_2 in the case of *P. chrysosporium*.

Recently, with the aim to produce vanillin in larger amounts and higher selectivity compared to chemical catalysis, a targeted pathway engineering strategy was applied to *Rhodococcus jostii* strain RHA1 [182]. When grown on minimal medium containing 2.5 % wheat straw lignocellulose and 0.05 % glucose, the strain in which the vanillin dehydrogenase gene had been deleted was found to accumulate vanillin with yields of up to 96 mg/L after 144 h, together with smaller amounts of ferulic acid and 4-hydroxybenzaldehyde. This pioneering work established that vanillin could be produced from lignin using a predictive gene deletion and demonstrated that lignin breakdown pathways could in principle be engineered for the production of aromatics.

Pathway engineering has also been used in *Pseudomonas putida* KT2440 to accumulate *cis,cis*-muconic acid from degradation of aromatic ring, via blockage of the protocatechuate cleavage pathway, and rerouting via catechol cleavage [183]. Using an alkaline pretreated liquor, muconic acid was obtained at a yield of 0.7 g/L after 24 h and subsequently reduced into adipic acid by Pd/C hydrogenation [183]. Interestingly, muconic acid could be obtained at high purity (>97 %) after a two-step purification. Given these two recent examples, metabolic engineering and synthetic biology seem to offer a tangible way for obtaining reasonable yields of chemicals from lignin *via* the reconstructed catabolic pathways.

10.7.4 Scale-up of Enzyme Production

Despite the potential industrial use of fungal peroxidases, the actual application of these enzymes in industrial processes is hampered by (i) the limited availability of the proteins in the natural hosts, and (ii) their rather low stability [170]. Efforts have been made to produce LiP and MnP recombinant proteins, using either homologous or heterologous host systems but those led to limited success [170]. Given the lack of efficient microbial expression systems for peroxidases, a significant number of studies have been carried out to express laccases genes in various fungal hosts. In spite of the active secretion of recombinant laccase in several heterologous systems, the levels of recombinant laccase were too low for industrial purposes [184]. The lack of efficient microbial expression systems together with the large amount of enzymes requirement constitute a serious bottleneck in the industrial application of fungal enzymes.

Until recently, the enzymology of bacterial lignin degradation was not well understood [180]. However, genome sequences of few lignin degraders, e.g., *R. jostii* RHA1 and *P. putida* KT2440, have recently become available. A better understanding of the metabolic pathway and their gene regulation should facilitate the goals of synthetic biology to address the bottlenecks encountered with fungal enzymes and hence engineer new routes for large scale production of lignin-derived chemicals from renewable feedstocks.

10.7.5 Summary

A wide variety of lignin-degrading fungi and bacteria have been identified, among which the white-rot fungi are the most active and the only ones that can completely break down lignin to CO_2 and H_2O. Aromatic carboxylic acids and aldehydes obtained by Cα–Cβ cleavage as well as acyclic 2,4-hexadiene-1,6-dioic acids resulting from ring cleavage are the products that have been the most often observed, usually as intermediates. The main extracellular enzymes participating in lignin degradation are the LiP, MnP, and laccases. Attempts to produce these enzymes using either homologous or heterologous host systems led to limited success. Efforts are currently centered on the elucidation of the enzymology of bacterial lignin breakdown. Metabolic engineering has been successfully applied recently to lignin degraders, allowing the production of chemicals like vanillin or muconic acid from lignin at concentration of 0.096 and 0.7 g/L, respectively. These promising results are encouraging signs for a great future for the biological depolymerization of lignin.

10.8 Computational Approaches

At first sight, the natural structural complexity of lignin seems rather incompatible with first-principles, computational approaches, which are limited with regard to the size and complexity of molecules they can be used for. Significant computational effort has been dedicated to understanding lignin model compounds reactivity. Few research groups have used β-O-4 lignin models to investigate homolytic bond dissociation energies [185–188], kinetic parameters [189, 190], and free-energy pathways occurring under various reaction conditions [191–193]. However, computational modeling of the heterogeneous lignin using the information gathered with only dimers would not have been representative.

In a recent computational screening approach, Kulik and coworkers have developed a chemical discovery technique to identify the chemically relevant putative fragments in eight of the known polymeric linkages of lignin [194]. From the identified cleavage pathways and resulting fragments the authors concluded that (i) ether bonds cleave readily in lignin polymers; (ii) biphenyl linkages are more recalcitrant; (iii) spirodienone linkages can fragment into many possible products with a high frequency of cleavage; and (iv) breaking an "easy" to break ether bond could also result in the spontaneous cleavage of harder to break biphenyl-like bonds [194]. While the two first observations had been previously made by others, the latter two were newly revealed by this theoretical approach. After the pioneering work conducted by Kulik's team with lignin oligomers consisting of up to six monomers, one hopes to see environmentally friendly theoretical approaches unraveling more mysteries of lignin in the near future.

10.9 Conclusions and Future Outlook

Currently, all aromatic chemicals are produced from petroleum-based sources. Lignin is a complex heterogeneous aromatic polymer consisting of up to 30 % of plant material. Its aromatic structure suggests that it is a possible renewable source for aromatic chemicals. However, the natural complexity and high stability of lignin makes its depolymerization a highly challenging task. Despite a strong renewed interest for lignin depolymerization over the last decade, a very limited number of lignin reactions, e.g., production of vanillin from lignosulfonates, have been actually scaled up and commercialized.

On a general point of view, the majority of efforts dedicated to lignin conversion have been done using simple model compounds that lack key lignin structural features and properties. Efforts are therefore necessary to confirm the applicability of promising processes to actual lignins which themselves offer another level of challenge by being largely a heterogenous and poorly soluble material.

This chapter presents recent advances and challenges to be addressed in the field of lignin depolymerization, with emphasis on chemical catalysis, ILs, or microbial and enzymatic approaches.

Although many homogeneous and heterogeneous catalysts have been investigated at the laboratory scale for lignin depolymerization, the industrial implementation of these catalysts was hampered by the harsh conditions, low selectivity, low yields, and issues with the reusability of catalysts. Efforts have still to be done to (i) increase the activity and selectivity of chemical catalysts; (ii) increase the tolerance of catalysts to impurities, especially sulfur or water; (iii) increase the separability and reusability of catalysts; and (iv) develop new technologies able to isolate phenolic compounds efficiently and economically. The promising results (50 % depolymerization, >90 % selectivity) recently obtained using reductive catalyzed depolymerization of lignin directly from lignocellulosic biomass opened the door to a new approach for lignin depolymerization, i.e., the "lignin-first" processes. By involving milder and greener conditions than the depolymerization of isolated lignin, this new approach gives rise to more environmentally friendly reactions that are more selective and therefore less costly in product recovery. Though all "lignin-first" experiments have been conducted at the lab-scale so far, they are encouraging to demonstrate the technology on a larger scale, and more efforts should be pursued towards the scale-up of "lignin-first" biorefinery processes. Under oxidative conditions, continuous separation of products is an effective way to increase the reaction yields by limiting radical condensation of products. Dedicating more effort to the engineering needs of the oxidative depolymerization of lignin should allow producing aromatic aldehyde and acids in higher yields and higher purity.

ILs are seen as another new green option for lignin conversion. Indeed, the use of ILs is becoming more and more accepted in the chemical industry and recent studies involving lignin and ILs demonstrate that ILs can be applied (i) to extract selectively lignin from lignocellulosic biomass and (ii) to oxidatively depolymerise lignin more selectively than in aqueous media. Despite the tremendous progress done in the last few years on lignin depolymerization in ILs, bulk separation of the products from ILs remains a major challenge that researchers should continue dedicating their efforts to.

Finally, lignin can also be deconstructed biologically. Regardless of the strong activity of fungal peroxidases for lignin depolymerization, the actual application of these enzymes in industrial processes is yet to take place. Efforts are now being pursued to elucidate the enzymology of bacterial lignin breakdown, and metabolic engineering is underway to construct genetically modified microbes that can convert lignin into renewable chemicals. The recent case studies of vanillin production (0.096 g/L) from lignin or muconic acid (0.7 g/L) from an alkaline pretreated liquor blaze the trail for a potentially great future for the biological depolymerization of lignins.

References

1. Holladay J, White JF, Bozell JJ, Johnson D (2007) Top value-added chemicals from biomass. Vol. II: results of screening for potential candidates from biorefinery lignin; Report PNNL-16983. U.S. Department of Commerce: Springfield, VA, pp 1–79
2. Lora J (2008) Industrial commercial lignins: Sources, properties and application. In: Belgacem MN, Gandini A (eds) Monomers, polymers and composites from renewable resources. Elsevier, Amsterdam, pp 225–241
3. Gandini A, Belgacem MN (2008) Lignins as components of macromolecular materials. In: Belgacem MN, Gandini A (eds) Monomers, polymers and composites from renewable resources. Elsevier, Amsterdam, pp 243–271
4. Doherty WOS, Mousavioun P, Fellows CW (2011) Value-adding to cellulosic ethanol: lignin polymers. Ind Crop Prod 33:259–276
5. Hocking MB (1997) Vanillin: synthetic flavoring from spent sulfite liquor. J Chem Educ 74:1055–1059
6. Borregaard http://www.borregaard.com/Business-Areas/Other-Businesses/Borregaard-Ingredients/ (14 Dec 2015)
7. Xu C, Arancon RAD, Labidi J, Luque R (2014) Lignin depolymerisation strategies: towards valuable chemicals and fuels. Chem Soc Rev 43:7485–7500
8. Ragauskas AJ, Beckham GT, Biddy MJ, Chandra R, Chen F, Davis MF, Davison BH, Dixon RA, Gilna P, Keller M, Langan P, Naskar AK, Saddler JN, Tschaplinski TJ, Tuskan GA, Wyman CE (2014) Lignin valorization: improving lignin processing in the biorefinery. Science 344:1246843
9. Chatel G, Rogers RD (2014) Review: oxidation of lignin using ionic liquids-an innovative strategy to produce renewable chemicals. ACS Sustain Chem Eng 2:322–339
10. Li C, Zhao X, Wang A, Huber GW, Zhang T (2015) Catalytic transformation of lignin for the production of chemicals and fuels. Chem Rev 115:11559–11624
11. Zakzeski J, Bruijnincx PCA, Jongerius AL, Weckhuysen BM (2010) The catalytic valorization of lignin for the production of renewable chemicals. Chem Rev 110:3552–3599
12. Rodrigues Pinto PC, da Silva EAB, Rodrigues AE (2011) Insights into oxidative conversion of lignin to high-added-value phenolic aldehydes. Ind Eng Chem Res 50:741–748
13. Chakar FS, Ragauskas AJ (2004) Review of current and future softwood kraft lignin process chemistry. Ind Crop Prod 20:131–141
14. Capanema EA, Balakshin MY, Kadla JF (2005) Quantitative characterization of a hardwood milled wood lignin by nuclear magnetic resonance spectroscopy. J Agric Food Chem 53:9639–9649
15. Buranov AU, Mazza G (2008) Lignin in straw of herbaceous crops. Ind Crop Prod 28:237–259
16. Zhang AP, Lu FC, Sun RC, Ralph J (2009) Ferulate-coniferyl alcohol cross-coupled products formed by radical coupling reactions. Planta 229:1099–1108
17. Monteil-Rivera F, Phuong M, Ye M, Halasz A, Hawari J (2013) Isolation and characterization of herbaceous lignins for applications in biomaterials. Ind Crop Prod 41:356–364

18. Crestini C, Argyropoulos DS (1997) Structural analysis of wheat straw lignin by quantitative P-31 and 2D NMR spectroscopy. The occurrence of ester bonds and alpha-O-4 substructures. J Agric Food Chem 45:1212–1219
19. Scalbert A, Monties B, Lallemand J-Y, Guittet E, Rolando C (1985) Ether linkage between phenolic acids and lignin fractions from wheat straw. Phytochemistry 24:1359–1362
20. Ralph J, Marita J, Ralph S, Hatfield RD, Lu F, Ede RM, Peng J, Quideau S, Helm R, Grabber J, Kim H, Jimenez-Monteon G, Zhang Y, Jung H-JG, Landucci L, MacKay J, Sederoff R, Chapple C, Boudet A (1999) Solution-state NMR of lignins. In: Argyropoulos DS (ed) Advances in lignocellulosics characterization. Tappi Press, Atlanta, pp 55–108
21. Pu Y, Cao S, Ragauskas AJ (2011) Application of quantitative ^{31}P NMR in biomass lignin and biofuel precursors characterization. Energy Environ Sci 4:3154–3166
22. Argyropoulos DS (1994) Quantitative phosphorus-31 NMR analysis of lignins, a new tool for the lignin chemist. J Wood Chem Technol 14:45–63
23. Cateto CA, Barreiro MF, Rodrigues AE, Brochier-Salon MC, Thielemans W, Belgacem MN (2008) Lignins as macromonomers for polyurethane synthesis: a comparative study on hydroxyl group determination. J Appl Polym Sci 109:3008–3017
24. Monteil-Rivera F, Paquet L (2015) Solvent-free catalyst-free microwave-assisted acylation of lignin. Ind Crop Prod 65:446–453
25. Yang Q, Wu SB, Lou R, Lv GJ (2011) Structural characterization of lignin from wheat straw. Wood Sci Technol 45:419–431
26. Hu G, Cateto C, Pu Y, Samuel R, Ragauskas AJ (2012) Structural characterization of switchgrass lignin after ethanol organosolv pretreatment. Energy Fuels 26:740–745
27. Hattalli S, Benaboura A, Ham-Pichavant F, Nourmamode A, Castellan A (2002) Adding value to Alfa grass (*Stipa tenacissima* L.) soda lignin as phenolic resins. 1. Lignin characterization. Polym Degrad Stab 75:259–264
28. Ahvazi B, Wojciechowicz O, Ton-That T-M, Hawari J (2011) Preparation of lignopolyols from wheat straw soda lignin. J Agric Food Chem 59:10505–10516
29. Moghaddam L, Zhang Z, Wellard RM, Bartley JP, Hara IMO, Doherty WOS (2014) Characterisation of lignins isolated from sugarcane bagasse pretreated with acidified ethylene glycol and ionic liquids. Biomass Bioenergy 70:498–512
30. Fengel D, Wegener G (1984) Chemical composition and analysis of wood. In: Fengel D, Wegener G (eds) Wood chemistry ultrastructure reactions. Walter de Gruyter, Berlin, pp 26–65
31. Jae J, Tompsett GA, Lin YC, Carlson TR, Shen J, Zhang T, Yang B, Wyman CE, Conner WC, Huber GW (2010) Depolymerization of lignocellulosic biomass to fuel precursors: maximizing carbon efficiency by combining hydrolysis with pyrolysis. Energy Environ Sci 3:358–365
32. Pandey MP, Kim CS (2011) Lignin depolymerization and conversion: a review of thermochemical methods. Chem Eng Technol 34:29–41
33. Wang K, Kim KH, Brown RC (2014) Catalytic pyrolysis of individual components of lignocellulosic biomass. Green Chem 16:727–735
34. Tarabanko VE, Petukhov DV, Selyutin GE (2004) New mechanism for the catalytic oxidation of lignin to vanillin. Kinet Catal 45:569–577
35. Wong Z, Chen K, Li J (2010) Formation of vanillin and syringaldehyde in an oxygen delignification process. Bioresoure 5:1509–1516
36. Araújo JDP, Grande CA, Rodrigues AE (2010) Vanillin production from lignin oxidation in a batch reactor. Chem Eng Res Des 88:1024–1032
37. Roberts VM, Stein V, Reiner T, Lemonidou A, Li X, Lercher JA (2011) Towards quantitative catalytic lignin depolymerization. Chem Eur J 17:5939–5948
38. Beauchet R, Monteil-Rivera F, Lavoie JM (2012) Conversion of lignin to aromatic-based chemicals (L-chems) and biofuels (L-fuels). Bioresour Technol 121:328–334

39. Gasson JR, Forchheim D, Sutter T, Hornung U, Kruse A, Barth T (2012) Modeling the lignin degradation kinetics in an ethanol/formic acid solvolysis approach. Part 1. Kinetic model development. Ind Eng Chem Res 51:10595–10606
40. Forchheim D, Gasson JR, Hornung U, Kruse A, Barth T (2012) Modeling the lignin degradation kinetics in an ethanol/formic acid solvolysis approach. Part 2. Validation and transfer to variable conditions. Ind Eng Chem Res 51:15053–15063
41. Toledano A, Serrano L, Labidi J (2012) Organosolv lignin depolymerisation with different base catalysts. J Chem Technol Biotechnol 87:1593–1599
42. Amen-Chen C, Pakdel H, Roy C (2001) Production of monomeric phenols by thermochemical conversion of biomass: a review. Bioresour Technol 79:277–299
43. Zakzeski J, Jongerius AL, Bruijnincx PCA, Weckhuysen BM (2012) Catalytic lignin valorization process for the production of aromatic chemicals and hydrogen. ChemSusChem 5:1602–1609
44. Bozell J (2014) Approaches to the selective catalytic conversion of lignin: a grand challenge for biorefinery development. In: Nicholas KM (ed) Selective catalysis for renewable feedstocks and chemicals. Springer, Berlin, pp 229–255
45. Harris EE, D'Ianni J, Adkins H (1938) Reaction of hardwood lignin with hydrogen. J Am Chem Soc 60:1467–1470
46. Kashima K, Maeda Y, Oshima M (1964) Method for liquefying lignin. Canadian Patent 700210
47. Engel DJ, Steigleder KZ (1987) Hydrocracking process for liquefaction of lignin. U.S. Patent 4,647,704
48. Urban P, Engel DJ (1988) Process for liquefaction of lignin U.S. Patent 4,731,491
49. Ratcliff MA, Johnson DK, Posey FL, Chum HL (1988) Hydrodeoxygenation of lignins and model compounds—scientific note. Appl Biochem Biotechnol 17:151–160
50. Meier D, Berns J, Faix O, Balfanz U, Baldauf W (1994) Hydrocracking of organocell lignin for phenol production. Biomass Bioenergy 7:99–105
51. Meier D, Ante R, Faix O (1992) Catalytic hydropyrolysis of lignin: influence of reaction conditions on the formation and composition of liquid products. Bioresour Technol 40:171–177
52. Oasmaa A, Alen R, Meier D (1993) Catalytic hydrotreatment of some technical lignins. Bioresour Technol 45:189–194
53. Oasmaa A, Johansson A (1993) Catalytic hydrotreating of lignin with water-soluble molybdenum catalyst. Energy Fuels 7:426–429
54. Torr KM, van de Pas DJ, Cazeils E, Suckling ID (2011) Mild hydrogenolysis of in-situ and isolated *Pinus* radiate lignins. Bioresour Technol 102:7608–7611
55. Horácek J, Homola F, Kubicková I, Kubicka D (2012) Lignin to liquids over sulfide catalysts. Catal Today 179:191–198
56. Song Q, Wang F, Xu J (2012) Hydrogenolysis of lignosulfonate into phenols over heterogeneous nickel catalysts. Chem Commun 48:7019–7021
57. Zhang JG, Asakura H, van Rijn J, Yang J, Duchesne P, Zhang B, Chen X, Zhang P, Saeys M, Yan N (2014) Highly efficient, NiAu-catalyzed hydrogenolysis of lignin into phenolic chemicals. Green Chem 16:2432–2437
58. Zhang JG, Teo J, Chen X, Asakura H, Tanaka T, Teramura K, Yan N (2014) A series of NiM (M = Ru, Rh, and Pd) bimetallic catalysts for effective lignin hydrogenolysis in water. ACS Catal 4:1574–1583
59. Ma XL, Tian Y, Hao WY, Ma R, Li YD (2014) Production of phenols from catalytic conversion of lignin over a tungsten phosphide catalyst. Appl Catal A 481:64–70
60. Xu W, Miller SJ, Agrawal PK, Jones CW (2012) Depolymerization and hydrodeoxygenation of switchgrass lignin with formic acid. ChemSusChem 5:667–675
61. Barta K, Warner GR, Beach ES, Anastas PT (2014) Depolymerization of organosolv lignin to aromatic compounds over Cu-doped porous metal oxides. Green Chem 16:191–196

62. Toledano A, Serrano L, Balu AM, Luque R, Pineda A, Labidi J (2013) Fractionation of organosolv lignin from olive tree clippings and its valorization to simple phenolic compounds. ChemSusChem 6:529–536
63. Toledano A, Serrano L, Pineda A, Romero AA, Luque R, Labidi J (2014) Microwave-assisted depolymerisation of organosolv lignin via mild hydrogen-free hydrogenolysis: catalyst screening. Appl Catal B Environ 145:43–55
64. Song Q, Wang F, Cai J, Wang Y, Zhang J, Yua W, Xu J (2013) Lignin depolymerization (LDP) in alcohol over nickel-based catalysts via a fragmentation–hydrogenolysis process. Energy Environ Sci 6:994–1007
65. Parsell T, Yohe S, Degenstein J, Jarrell T, Klein I, Gencer E, Hewetson B, Hurt M, Kim JI, ChoudhariH Saha B, Meilan R, Mosier N, Ribeiro F, Delgass WN, Chapple C, KenttamaaHI Agrawal R, Abu-Omar MM (2015) A synergistic biorefinery based on catalytic conversion of lignin prior to cellulose starting from lignocellulosic biomass. Green Chem 17:1492–1499
66. Vuori A, Bredenberg JB-S (1988) Liquefaction of Kraft lignin: 1. Primary reactions under mild thermolysis conditions. Holzforshung 42:155–161
67. Kleinert M, Barth T (2008) Phenols from lignin. Chem Eng Technol 31:736–745
68. Kleinert M, Barth T (2008) Towards a lignincellulosic biorefinery: direct one-step conversion of lignin to hydrogen-enriched biofuel. Energy Fuels 22:1371–1379
69. Wang X, Rinaldi R (2013) A route for lignin and bio-oil conversion: dehydroxylation of phenols into arenes by catalytic tandem reactions. Angew Chem Int Ed 52:11499–11503
70. Jongerius AL, Bruijnincx PCA, Weckhuysen BM (2013) Liquid-phase reforming and hydrodeoxygenation as a two-step route to aromatics from lignin. Green Chem 15:3049–3056
71. Yan N, Zhao C, Dyson PJ, Wang C, L-t Liu, Kou Y (2008) Selective degradation of wood lignin over noble-metal catalysts in a two-step process. ChemSusChem 1:626–629
72. Van den Bosch S, Schutyser W, Vanholme R, Driessen T, Koelewijn S-K, Renders T, De Meester B, Huijgen WJJ, Dehaen W, Courtin CM, Lagrain B, Boerjan W, Sels BF (2015) Reductive lignocellulose fractionation into soluble lignin-derived phenolic monomers and dimers and processable carbohydrate pulps. Energy Environ Sci 8:1748–1763
73. Crestini C, Pro P, Neri V, Saladino R (2005) Methyltrioxorhenium: a new catalyst for the activation of hydrogen peroxide to the oxidation of lignin and lignin model compounds. Bioorg Med Chem 13:2569–2578
74. Voitl T, von Rohr PR (2008) Oxidation of lignin using aqueous polyoxometalates in the presence of alcohols. ChemSusChem 1:763–769
75. Voitl T, von Rohr PR (2010) Demonstration of a process for the conversion of Kraft lignin into vanillin and methyl vanillate by acidic oxidation in aqueous methanol. Ind Eng Chem Res 49:520–525
76. Xiang Q, Lee YY (2001) Production of oxychemicals from precipitated hardwood lignin. Appl Biochem Biotechnol 91–93:71–80
77. Wu G, Heitz M (1995) Catalytic mechanism of Cu^{2+} and Fe^{3+} in alkaline O_2 oxidation of lignin. J Wood Chem Technol 15:189–202
78. Bjørsvik H-R, Minisci F (1999) Fine chemicals from lignosulfonates. 1. Synthesis of vanillin by oxidation of lignosulfonates. Org Proc Res Dev 3:330–340
79. Azarpira A, Ralph J, Lu FC (2014) Catalytic alkaline oxidation of lignin and its model compounds: a pathway to aromatic biochemicals. BioEnergy Res 7:78–86
80. Villar JC, Caperos A, Garcia-Ochoa F (2001) Oxidation of hardwood kraft-lignin to phenolic derivatives with oxygen as oxidant. Wood Sci Technol 35:245–255
81. Sales FG, Maranhão LCA, Lima Filho NM, Abreu CAM (2007) Experimental evaluation and continuous catalytic process for fine aldehyde production from lignin. Chem Eng Sci 62:5386–5391
82. Deng HB, Lin L, Sun Y, Pang CS, Zhuang JP, Ouyang PK, Li ZJ, Liu SJ (2008) Perovskite-type oxide $LaMnO_3$: an efficient and recyclable heterogeneous catalyst for the wet aerobic oxidation of lignin to aromatic aldehydes. Catal Lett 126:106–111

83. Zhou XF (2014) Selective oxidation of Kraft lignin over zeolite-encapsulated Co(II) $[H_4]$ salen and $[H_2]$salen complexes. J Appl Polym Sci 131:40809(1-9)
84. Rahimi A, Azarpira A, Kim H, Ralph J, Stahl SS (2013) Chemoselective metal-free aerobic oxidation in lignin. J Am Chem Soc 135:6415–6418
85. Rahimi A, Ulbrich A, Coon JJ, Stahl SS (2014) Formic-acid-induced depolymerization of oxidized lignin to aromatics. Nature 515:249–252
86. Borges da Silva EA, Zabkova M, Araújo JD, Cateto CA, Belgacem MF, Barreiro MN, Rodrigues AE (2009) An integrated process to produce vanillin and lignin-based polyurethanes from Kraft lignin. Chem Eng Res Des 87:1276–1292
87. Wong JT (2012) Technological, commercial, organizational, and social uncertainties of a novel process for vanillin production from lignin. MBA Project, Simon Fraser University
88. Fache M, Boutevin B, Caillol S (2016) Vanillin production from lignin and its use as a renewable chemical. ACS Sustain Chem Eng 4:35–46
89. Fox MA, Dulay MT (1993) Heterogeneous photocatalysis. Chem Rev 93:341–357
90. Lathasree S, Nageswara R, Sivasankar B, Sadasivam V, Rengaraj K (2004) Heterogeneous photocatalytic mineralization of phenols in aqueous solutions. J Mol Catal A Chem 223:101–105
91. Kobayakawa K, Sato Y, Nakamura S, Fujishima A (1989) Photodecomposition of kraft lignin catalyzed by TiO_2. Bull Chem Soc Jpn 62:3433–3436
92. Tanaka K, Calanag CR, Hisanaga T (1999) Photocatalyzed degradation of lignin on TiO_2. J Mol Catal A Chem 138:287–294
93. Ksibi M, Ben Amor S, Cherif S, Elaloui E, Houas A, Elaloui M (2003) Photodegradation of lignin from black liquor using a UV/TiO_2 system. J Photochem Photobiol A 154:211–218
94. Yuan RS, Guan RB, Liu P, Zheng JT (2007) Photocatalytic treatment of wastewater from paper mill by TiO_2 loaded on activated carbon fibers. Colloids Surf A 293:80–86
95. Ma YS, Chang CN, Chiang YP, Sung HF, Chao AC (2008) Photocatalytic degradation of lignin using Pt/TiO_2 as the catalyst. Chemosphere 71:998–1004
96. Ugurlu M, Karaoglu MH (2009) Removal of AOX, total nitrogen and chlorinated lignin from bleached Kraft mill effluents by UV oxidation in the presence of hydrogen peroxide utilizing TiO_2 as photocatalyst. Environ Sci Pollut Res 16:265–273
97. Ugurlu M, Karaoglu MH (2011) TiO_2 supported on sepiolite: preparation, structural and thermal characterization and catalytic behaviour in photocatalytic treatment of phenol and lignin from olive mill wastewater. Chem Eng J 166:859–867
98. Ugurlu M, Karaoglu MH, Vaizogullar AI (2013) Decolourization and removal of phenol compounds from olive mill wastewater byO_2/UV/$NaBO_2H_2O_2$-SH_2O. Fresenius Environ Bull 22:752–763
99. Portjanskaja E, Stepanova K, Klauson D, Preis S (2009) The influence of titanium dioxide modifications on photocatalytic oxidation of lignin and humic acids. Catal Today 144:26–30
100. Portjanskaja E, Preis S (2007) Aqueous photocatalytic oxidation of lignin: the influence of mineral admixtures. Int J Photoenergy 76730:1–7
101. Rangel R, Mercado GJL, Bartolo-Perez P, Garcia R (2012) Nanostructured-[CeO_2, La_2O_3, C]/TiO_2catalysts for lignin photodegradation. Sci Adv Mater 4:573–578
102. Kansal SK, Singh M, Sud D (2008) Studies on TiO_2/ZnO photocatalysed degradation of lignin. J Hazard Mater 153:412–417
103. Kamwilaisak K, Wright PC (2012) Investigating laccase and titanium dioxide for lignin degradation. Energy Fuels 26:2400–2406
104. Tonucci L, Coccia F, Bressan M, Alessandro N (2012) Mild photocatalysed and catalysed green oxidation of lignin: A useful pathway to low-molecular-weight derivatives. Waste Biomass Valorization 3:165–174
105. Bailey A, Brooks HM (1946) Electrolytic oxidation of lignin. J Am Chem Soc 68:445–446
106. Moodley B, Mulholland DA, Brookes HC (2011) The electrooxidation of lignin in Sappi Saiccor dissolving pulp mill effluent. Water SA 37:33–40
107. Tolba R, Tian M, Wen J, Jiang Z-H, Chen A (2010) Electrochemical oxidation of lignin at IrO 2-based oxide electrodes. J Electroanal Chem 649:9–15

108. Shao D, Liang JD, Cui XM, Xu H, Yan W (2014) Electrochemical oxidation of lignin by two typical electrodes: Ti/Sb-SnO_2 and Ti/PbO_2. Chem Eng J 244:288–295
109. Zhu HB, Chen YM, Qin TF, Wang L, Tang Y, SunYZ Wan PY (2014) Lignin depolymerization via an integrated approach of anode oxidation and electro-generated H_2O_2 oxidation. RSC Adv 4:6232–6238
110. Constant S, Robitzer M, Quignard F, Di Renzo F (2012) Vanillin oligomerization as a model of side reactions in lignin fragmentation. Catal Today 189:123–128
111. Parpot P, Bettencourt AP, Carvalho AM, Belgsir EM (2000) Biomass conversion: attempted electrooxidation of lignin for vanillin production. J Appl Electrochem 30:727–731
112. Tian M, Wen J, MacDonald D, Asmussen RM, Chen A (2010) A novel approach for lignin modification and degradation. Electrochem Commun 12:527–530
113. Reichert E, Wintringer R, Volmer DA, Hempelmann R (2012) Electro-catalytic oxidative cleavage of lignin in a protic ionic liquid. Phys Chem Chem Phys 14:5214–5221
114. Wasserscheid P, Stark A (eds) (2010) Handbook of green chemistry: green solvents, vol 6—ionic liquids, Wiley-VCH
115. Peleteiro S, Rivas S, Alonso JL, Santos V, Parajo JC (2015) Utilization of ionic liquids in lignocellulose biorefineries as agents for separation, derivatization, fractionation, or pretreatment. J Agric Food Chem 63:8093–8102
116. Olivier-Bourbigou H, Magna L, Morvan D (2010) Ionic liquids and catalysis: Recent progress from knowledge to applications. Appl Catal A 373:1–56
117. Stark A (2011) Ionic liquids in the biorefinery: a critical assessment of their potential. Energy Environ Sci 4:19–32
118. Hossain MM, Aldous L (2012) Ionic liquids for lignin processing: dissolution, isolation, and conversion. Aust J Chem 65:1465–1477
119. Plechkova NV, Seddon KR (2008) Applications of ionic liquids in the chemical industry. Chem Soc Rev 37:123–150
120. Parvulescu VI, Hardacre C (2007) Catalysis in ionic liquids. Chem Rev 107:2615–2665
121. Swatloski RP, Spear SK, Holbrey JD, Rogers RD (2002) Dissolution of cellulose with ionic liquids. J Am Chem Soc 124:4974–4975
122. Sun N, Rahman M, Qin Y, Maxim ML, Rodriguez H, Rogers RD (2009) Complete dissolution and partial delignification of wood in the ionic liquid 1-ethyl-3-methylimidazolium acetate. Green Chem 11:646–655
123. Lee SH, Doherty TV, Linhardt JS (2009) Ionic liquid-mediated selective extraction of lignin from wood leading to enhanced enzymatic cellulose hydrolysis. Biotechnol Bioeng 102:1368–1376
124. Fu D, Mazza G, Tamaki Y (2010) Lignin extraction from straw by ionic liquids and enzymatic hydrolysis of the cellulosic residues. J Agric Food Chem 58:2915–2922
125. Zhu S, Wu Y, Chen Q, Yu Z, Wang C, Jin S, Ding Y, Wu G (2006) Dissolution of cellulose with ionic liquids and its application: a mini-review. Green Chem 8:325–327
126. Pinkert A, Marsh KN, Pang SS, Staiger MP (2009) Ionic liquids and their interaction with cellulose. Chem Rev 109:6712–6728
127. Maki-Arvela P, Anugwom I, Virtanen P, Sjoholm R, Mikkola JP (2010) Dissolution of lignocellulosic materials and its constituents using ionic liquids—a review. Ind Crop Prod 32:175–201
128. Mora-Pale M, Meli L, Doherty TV, Linhardt RJ, Dordick JS (2011) Room temperature ionic liquids as emerging solvents for the pretreatment of lignocellulosic biomass. Biotechnol Bioeng 108:1229–1245
129. Wang H, Gurau G, Rogers RD (2012) Ionic liquid processing of cellulose. Chem Soc Rev 41:1519–1537
130. Kim J-Y, Shin E-J, Eoma I-Y, Wonb K, Kim YH, Choi D, Choi IG, Choi JW (2011) Structural features of lignin macromolecules extracted with ionic liquid from poplar wood. Bioresour Technol 102:9020–9025

131. Li W, Sun N, Stoner B, Jiang X, Lu X, Rogers RD (2011) Rapid dissolution of lignocellulosic biomass in ionic liquids using temperatures above the glass transition of lignin. Green Chem 13:2038–2047
132. Brandt A, Grasvik J, Hallett JP, Welton T (2013) Deconstruction of lignocellulosic biomass with ionic liquids. Green Chem 15:550–583
133. Zavrel M, Bross D, Funke M, Buchs J, Spiess AC (2009) High-throughput screening for ionic liquids dissolving (ligno-)cellulose. Bioresour Technol 100:2580–2587
134. Tan SSY, MacFarlane DR (2009) Ionic liquids in biomass processing. Top Curr Chem 290:311–339
135. Tan SSY, MacFarlane DR, Upfal J, Edye LA, Doherty WOS, Patti AF, Pringle JM, Scott JL (2009) Extraction of lignin from lignocellulose at atmospheric pressure using alkylbenzenesulfonate ionic liquid. Green Chem 11:339–345
136. Pinkert A, Goeke DF, Marsh KN, Pang S (2011) Extracting wood lignin without dissolving or degrading cellulose: investigations on the use of food additive-derived ionic liquids. Green Chem 13:3124–3136
137. Pu Y, Jiang N, Ragauskas AJ (2007) Ionic liquid as a green solvent for lignin. J Wood Chem Technol 27:23–33
138. Wang Y, Wei L, Li K, Ma Y, Ma N, Ding S, Wang L, Zhao D, Yan B, Wan W, Zhang Q, Wang X, Wang J, Li H (2014) Lignin dissolution in dialkylimidazolium-based ionic liquid–water mixtures. Bioresour Technol 170:499–505
139. Lateef H, Grimes S, Kewcharoenwong P, Feinberg B (2009) Separation and recovery of cellulose and lignin using ionic liquids: a process for recovery from paper-based waste. J Chem Technol Biotechnol 84:1818–1827
140. Diop A, Bouazza AH, Daneault C, Montplaisir D (2013) New ionic liquid for the dissolution of lignin. Bioresoure 8:4270–4282
141. Kilpelainen I, Xie H, King A, Granstrom M, Heikkinen S, Argyropoulos DS (2007) Dissolution of wood in ionic liquids. J Agric Food Chem 55:9142–9148
142. George A, Tran K, Morgan TJ, Benke PI, Berrueco C, Lorente E, Wu BC, Keasling JD, Simmons BA, Holmes BM (2011) The effect of ionic liquid cation and anion combinations on the macromolecular structure of lignins. Green Chem 13:3375–3385
143. Li B, Filpponen I, Argyropoulos DS (2010) Acidolysis of wood in ionic liquids. Ind Eng Chem Res 49:3126–3136
144. Cox BJ, Ekerdt JG (2012) Depolymerization of oak wood lignin under mild conditions using the acidic ionic liquid 1-H-3-methylimidazolium chloride as both solvent and catalyst. Bioresour Technol 118:584–588
145. Stärk K, Taccardi N, Bösmann A, Wasserscheid P (2010) Oxidative depolymerization of lignin in ionic liquids. ChemSusChem 3:719–723
146. Zakzeski J, Jongerius AL, Weckhuysen BM (2010) Transition metal catalyzed oxidation of Alcell lignin, soda lignin, and lignin model compounds in ionic liquids. Green Chem 12:1225–1236
147. Zakzeski J, Bruijnincx PCA, Weckhuysen BM (2011) In situ spectroscopic investigation of the cobalt-catalyzed oxidation of lignin model compounds in ionic liquids. Green Chem 13:671–680
148. Liu S, Shi ZL, Li L, Yu ST, Xie C, Song Z (2013) Process of lignin oxidation in an ionic liquid coupled with separation. RSC Adv 3:5789–5793
149. Sena-Martins G, Almeida-Vara E, Duarte JC (2008) Eco-friendly new products from enzymatically modified industrial lignins. Ind Crop Prod 27:189–195
150. Bugg TDH, Ahmad M, Hardiman EM, Rahmanpour R (2011) Pathways for degradation of lignin in bacteria and fungi. Nat Prod Rep 28:1883–1896
151. Kirk TK, Farrell RL (1987) Enzymatic "combustion": the microbial degradation of lignin. Annu Rev Microbiol 41:465–505
152. Ten Have R, Teunissen PJM (2001) Oxidative mechanisms involved in lignin degradation by white-rot fungi. Chem Rev 101:3397–3413

153. Ahmad M, Taylor CR, Pink D, Burton K, Eastwood D, Bending GD, Bugg TDH (2010) Development of novel assays for lignin degradation: comparative analysis of bacterial and fungal lignin degraders. Mol BioSyst 6:815–821
154. Hatakka A, Hammel KE (2010) Fungal biodegradation of lignocelluloses. In: The Mycota. Industrial Applications, 10, pp 319–340
155. Vicuna R (1988) Bacterial degradation of lignin. Enzyme Microb Technol 10:646–655
156. Zimmermann W (1990) Degradation of lignin by bacteria. J Biotechnol 13:119–130
157. Bugg TDH, Ahmad M, Hardiman EM, Singh R (2011) The emerging role for bacteria in lignin degradation and bio-product formation. Curr Opin Biotechnol 22:394–400
158. Bandounas L, Wierckx NJP, de Winde JH, Ruijssenaars HJ (2011) Isolation and characterization of novel bacterial strains exhibiting ligninolytic potential. BMC Biotechnol 11:94
159. Brown ME, Chang MCY (2014) Exploring bacterial lignin degradation. Curr Opin Chem Biol 19:1–7
160. Dey S, Maiti TK, Bhattacharyya BC (1994) Production of some extracellular enzymes by a lignin peroxidase-producing brown rot fungus, Polyporus ostreiformis, and its comparative abilities for lignin degradation and dye decolorization. Appl Environ Microbiol 60:4216–4218
161. Paliwal R, Rawat AP, Rawat M, Rai JPN (2012) Bioligninolysis: recent updates for biotechnological solution. Appl Biochem Biotechnol 167:1865–1889
162. DeAngelis KM, Allgaier M, Chavarria Y, Fortney JL, Hugenholtz P, Simmons B, Sublette K, Silver WL, Hazen TC (2011) Characterization of trapped lignin-degrading microbes in tropical forest soil. PLoS ONE 6:e19306
163. Taylor CR, Hardiman EM, Ahmad M, Sainsbury PD, Norris PR, Bugg TDH (2012) Isolation of bacterial strains able to metabolize lignin from screening of environmental samples. J Appl Microbiol 113:521–530
164. Huang X-F, Santhanam N, Badri DV, Hunter WJ, Manter DK, Decker SR, Vivanco JM, Reardon KF (2013) Isolation and characterization of lignin-degrading bacteria from rainforest soils. Biotechnol Bioeng 110:1616–1626
165. Pasti MB, Pometto AL, Nuti MP, Crawford DL (1990) Lignin-solubilizing ability of actinomycetes isolated from termite (Termitidae) gut. Appl Environ Microbiol 56:2213–2218
166. Kato K, Kosaki S, Sakuranaga M (1998) Degradation of lignin compounds by bacteria from termite guts. Biotechnol Lett 20:459–462
167. Pollegioni L, Tonin F, Rosini E (2015) Lignin-degrading enzymes. FEBS J 282:1190–1213
168. Haemmerli SD, Leisola MSA, Fiechter A (1986) Polymerisation of lignins by ligninases from *Phanerochaete chrysosporium*. FEMS Microbiol Lett 35:33–36
169. Hammel KE, Jensen KA, Mozuch MD, Landucci LL, Tien M, Pease EA (1993) Ligninolysis by a purified lignin peroxidase. J Biol Chem 268:12274–12281
170. Conesa A, Punt PJ, van den Hondel CAMJJ (2002) Fungal peroxidases: molecular aspects and applications. J Biotechnol 93:143–158
171. Munk L, Sitarz AK, Kalyani DC, Mikkelsen JD, Meyer AS (2015) Can laccases catalyze bond cleavage in lignin? Biotechnol Adv 33:13–24
172. Rochefort D, Leech D, Bourbonnais R (2004) Electron transfer mediator systems for bleaching of paper pulp. Green Chem 6:14–24
173. Reiter J, Strittmatter H, Wiemann LO, Schieder D, Sieber V (2013) Enzymatic cleavage of lignin beta-O-4-aryl ether bonds via net internal hydrogen transfer. Green Chem 15:1373–1381
174. Otsuka Y, Sonoki T, Ikeda S, Kajita S, Nakamura M, Katayama Y (2003) Detection and characterization of a novel extracellular fungal enzyme that catalyzes the specific and hydrolytic cleavage of lignin guaiacyl glycerol beta-aryl ether linkages. Eur J Biochem 270:2353–2362

175. Masai E, Ichimura A, Sato Y, Miyauchi K, Katayama Y, Fukuda M (2003) Roles of the enantioselective glutathione *S*-transferases in cleavage of β-aryl ether. J Bacteriol 185:1768–1775
176. Allocati N, Federici L, Masulli M, DiIlio C (2009) Glutathione transferases in bacteria. FEBS J 276:58–75
177. Gall DL, Kim H, Lu F, Donohoe TJ, Noguera DR, Ralph J (2014) Stereochemical features of glutathione-dependent enzymes in the *Sphingobium* sp. strain SYK-6 β-aryl etherase pathway. J Biol Chem 289:8656–8667
178. Picart P, Müller C, Mottweiler J, Wiermans L, Bolm C, Dominguez de Maria P, Schallmey A (2014) From gene towards selective biomass valorization: bacterial β-etherases with catalytic activity on lignin-like polymers. ChemSusChem 7:3164–3171
179. Masai E, Sasaki M, Minakawa Y, Abe T, Sonoki T, Miyauchi K, Katayama Y, Fukuda M (2004) A novel tetrahydrofolate-dependent O-demethylase gene is essential for growth of *Sphingomonas paucimobilis* SYK-6 with syringate. J Bacteriol 186:2757–2765
180. Bugg TDH, Rahmanpour R (2015) Enzymatic conversion of lignin into renewable chemicals. Curr Opin Chem Biol 29:10–17
181. Chen CL, Chang HM, Kirk TK (1983) Carboxylic acids produced through oxidative cleavage of aromatic rings during degradation of lignin in spruce wood by *Phanerochaete chrysosporium*. J Wood Chem Technol 3:35–57
182. Sainsbury PD, Hardiman EM, Ahmad M, Otani H, Seghezzi N, Eltis LD, Bugg TDH (2013) Breaking down lignin to high-value chemicals: the conversion of lignocellulose to vanillin in a gene deletion mutant of *Rhodococcus jostii* RHA1. ACS Chem Biol 8:2151–2156
183. Vardon DR, Franden MA, Johnson CW, Karp EM, Guarneri MT, Linger JG, Salm MJ, Strathman TJ, Beckham GT (2015) Adipic acid production from lignin. Energy Environ Sci 8:617–628
184. Kiiskinen LL, Saloheimo M (2004) Molecular cloning and expression in *Saccharomyces cerevisiae* of a laccase gene from the ascomycete *Melanocarpus albomyces*. Appl Environ Microbiol 70:137–144
185. Kim S, Chmely SC, Nimos MR, Bomble YJ, Foust TD, Paton RS, Beckham GT (2011) Computational study of bond dissociation enthalpies for a large range of native and modified lignins. J Phys Chem Lett 2:2846–2852
186. Parthasarathi R, Romero RA, Redondo A, Gnanakaran S (2011) Theoretical study of the remarkably diverse linkages in lignin. J Phys Chem Lett 2:2660–2666
187. Beste A, Buchanan AC III (2009) Computational study of bond dissociation enthalpies for lignin model compounds. Substituent effects in phenethyl phenyl ethers. J Org Chem 74:2837–2841
188. Younker JM, Beste A, Buchanan AC III (2011) Computational study of bond dissociation enthalpies for substituted β-O-4 lignin model compounds. ChemPhysChem 12:3556–3565
189. Beste A, Buchanan AC III (2011) Kinetic analysis of the phenyl-shift reaction in β-O-4 lignin model compounds: a computational study. J Org Chem 76:2195–2203
190. Beste A, Buchanan AC III (2012) Role of carbon–carbon phenyl migration in the pyrolysis mechanism of β-O-4 lignin model compounds: phenethyl phenyl ether and α-hydroxy phenethyl phenyl ether. J Phys Chem A 116:12242–12248
191. Beste A, Buchanan AC III, Harrison RJ (2008) Computational prediction of α/β selectivities in the pyrolysis of oxygen-substituted phenethyl phenyl ethers. J Phys Chem A 112:4982–4988
192. Beste A, Buchanan AC III (2012) Kinetic simulation of the thermal degradation of phenethyl phenyl ether, a model compound for the β-O-4 linkage in lignin. Chem Phys Lett 550:19–24
193. Beste A, Buchanan AC III (2013) Computational investigation of the pyrolysis product selectivity for α-hydroxy phenethyl phenyl ether and phenethyl phenyl ether: analysis of substituent effects and reactant conformer selection. J Phys Chem A 117:3235–3242
194. Mar BD, Qi HW, Liu F, Kulik HJ (2015) Ab initio screening approach for the discovery of lignin polymer breaking pathways. J Phys Chem A 119:6551–6562

Chapter 11
Production of Sialic Acid and Its Derivatives by Metabolic Engineering of *Escherichia coli*

Baixue Lin and Yong Tao

Abstract Sialic acid and its derivatives have attracted considerable interest due to their important biological functions, and are valuable resources with increasing demand in many fields. Significant efforts have been made for the efficient production of sialic acids. In this chapter, we review the achievements that have been obtained in the development of microbial processes for the production of sialic acids and its derivatives with an emphasis on the strategies of metabolic engineering for the enhanced production of N-acetylneuraminic acid, polysialic acid, and sialylated oligosaccharides, such as 3′ sialyllactose.

Keywords Metabolic engineering · *Escherichia coli* · N-Acetylneuraminic acid · Polysialic acid · 3′ sialyllactose

11.1 Introduction

Sialic acids are a family of nine-carbon amino sugars that encompass over 50 naturally occurring and structurally distinct acidic sugars and they play a wide variety of roles in nature [1, 2]. Sialic acids are frequently found as a terminal sugar in cell surface complex carbohydrates and are known to play important roles in biological, pathological, and immunological processes including cell adhesion, cell signaling, glycoprotein stability, bacterial virulence, and tumor metastasis [1, 3–6].

N-Acetylneuraminic acid (Neu5Ac) is the most abundant and widely studied sialic acid that undergoes extensive modifications to generate the diversity of the sialic acid family [7, 8]. Neu5Ac and its analogs can effectively inhibit sialidase and

B. Lin · Y. Tao (✉)
CAS Key Laboratory of Microbial Physiological and Metabolic Engineering,
Institute of Microbiology, Chinese Academy of Sciences, Beijing 100101, China
e-mail: taoyong@im.ac.cn

P.C.K. Lau (ed.), *Quality Living Through Chemurgy and Green Chemistry*,
Green Chemistry and Sustainable Technology,
DOI 10.1007/978-3-662-53704-6_11

therefore can be used as antiviral drugs (e.g., Relenza) to prevent virus infection and release [9–11]. Neu5Ac is a major component of gangliosides that are notably abundant in brain tissues [12]. It was reported that dietary Neu5Ac could increase its concentration in brain gangliosides and glycoproteins, and enhance learning during early development [13–15]. Therefore, Neu5Ac may be considered as a potential nutraceutical. Neu5Ac is also a useful tool in bacterial pathogenesis research, immunobiology and in the development of cancer diagnostic imaging techniques [7, 9, 10].

In bacteria, sialic acids have been identified in extracellular capsular polysaccharides and lipopolysaccharides as linear homopolymers of Neu5Ac. Polysialic acid (PSA), a polymer of α-(2,8)- and/or α-(2,9)-linked Neu5Ac, has many excellent characteristics, such as poor immunogenicity, biodegradability, and biocompatibility, and is regarded as an ideal material for use in the controlled drug release and for scaffolds in biomedical applications [16]. Sialylated oligosaccharides, especially sialyllactoses, are found in most mammalian milks from mice to humans, and are believed to protect breast-fed infants from infections [17, 18]. By serving sialylated oligosaccharides as an exogenous source of sialic acid, the concentration of brain gangliosides and glycoprotein sialic acid increase thus promoting the brain maturation in breast-fed infants [14]. Sialic acid and its derivatives, which have a variety of biofunctions have attracted considerable interest, and have been a valuable resource with an increasing demand in both medicine and biotechnology.

The availability of Neu5Ac is limited. Neu5Ac is traditionally isolated from natural sources, such as birds' nests [19], milk whey, and egg yolk [20] or by hydrolysis of colominic acid [21–23]. However, the yields are typically low and the procedures are not suitable for large-scale production. Chemical synthesis of Neu5Ac is challenging because of the complexity of the molecule [6]. The enzymatic synthesis of Neu5Ac from N-acetyl-D-glucosamine (GlcNAc) and pyruvate has attracted significant interest [24, 25]. Although a high efficiency can be achieved in an enzymatic process, the tedious preparation of enzymes and the requirement of adenosine triphosphate (ATP) make the process complicated and costly.

The above stated methods for producing sialic acid are insufficient to meet the large demand of medical and biotechnological industries. It is envisaged that microbial processes can overcome this problem and serve as a viable alternative to conventional methods. Based on the comprehensive understanding of the metabolism of sialic acids and metabolic engineering strategies have been applied to empower *Escherichia coli* into efficient microbial factories for sialic acid production. In this chapter, we focus on the metabolism and the current state of biotechnological production of sialic acid and its derivatives with a special emphasis on metabolic engineering strategies for the enhanced production of sialic acids using *E. coli* as a workhorse.

11.2 N-Acetylneuraminic acid (Neu5Ac)

11.2.1 Biosynthesis and Metabolism of Neu5Ac

The metabolism of Neu5Ac in bacteria has been well characterized [2, 26]. The *nan* operon (Fig. 11.1), which encodes the proteins for sialic acid catabolism is found in hundreds of bacterial species [2, 26]. In *E. coli*, the sialic acid catabolic pathway comprises sialic acid transporter (NanT), Neu5Ac aldolase (NanA), ManNAc-6-phosphate epimerase (NanE), ManNAc kinase (NanK), and a transcription factor NanR. NanT is responsible for Neu5Ac uptake [27]. NanA cleaves Neu5Ac to pyruvate and ManNAc. ManNAc is phosphorylated by NanK and subsequently epimerized by NanE to generate GlcNAc-6-P, which enters the amino sugar metabolic pathways [2, 8, 28]. The *nanR* gene encodes a transcription factor that represses the *nan* operon in the absence of sialic acid by interfering the binding

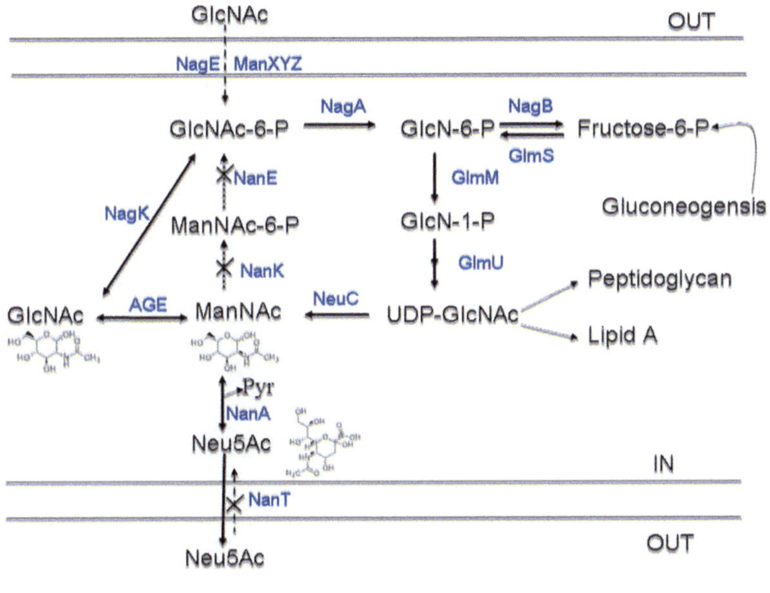

Fig. 11.1 Schematic presentation of the Neu5Ac metabolic pathway in engineered *E. coli*. Reprinted from Ref. [34], with permission from Springer Science+Business Media. The genes for AGE and NanA are overexpressed. Discontinued arrows represent the enzymatic activities that have been eliminated. AGE, GlcNAc 2-epimerase, EC5.1.3.8; NanA, Neu5Ac aldolase, EC4.1.3.3; NanT, sialic acid transporter; NanK, ManNAc kinase; NanE, ManNAc-6-P epimerase; NeuC, UDP-GlcNAc2-epimerase; GlmS, L-glutamine:D-fructose-6-phosphate aminotransferase (isomerizing); GlmM, phosphoglucosamine mutase; GlmU, bifunctional UDP-N-acetylglucosamine pyrophosphorylase/Glucosamine-1-phosphate N-acetyltransferase; NagA, N-acetylglucosamine-6-phosphate deacetylase; NagB, glucosamine-6-phosphate deaminase; NagK, GlcNAc kinase; NagE, GlcNAc-specific transporter; ManXYZ: mannose transporter

of RNA polymerase to the *nanA* promoter [29]. The products of sialic acid catabolism are GlcNAc and pyruvate, which are primary molecules of metabolism.

For the biosynthesis of sialic acid, Neu5Ac aldolase (NanA) (EC. 4.1.3.3) or Neu5Ac synthase (NeuB) (EC. 4.1.3.19) are used to catalyze the conversion of ManNAc to Neu5Ac [30–32]. NeuB catalyzes the energy-dependent reaction of ManNAc and phosphoenolpyruvate (PEP) to form Neu5Ac [7, 31]. NanA catalyzes ManNAc and pyruvate to produce Neu5Ac [33, 34]. Pyruvate is more readily available compared to PEP, which makes NanA a promising target in the engineered pathway for Neu5Ac synthesis.

Based on the comprehensive understanding of the sialic acid metabolic pathway, metabolic engineering strategies can be applied to improve the microbial production of sialic acid and its derivatives.

11.2.2 Microbial Production of Neu5Ac by Metabolic Engineering

11.2.2.1 Whole-Cell Biocatalysis of Neu5Ac by Coupling Two Strains

Whole-cell biocatalysis is widely applied as an alternative to conventional chemical methods for pharmaceutical synthesis because it is more efficient, cheaper, and environmentally friendlier. Whole-cell biocatalysis is preferred because ATP that is required for the activation of GlcNAc 2-epimerase can be supplied through the cellular activities. Otherwise ATP is expensive, easily degraded, and cannot be recycled in vitro [32, 35–37].

Whole-cell biocatalysis for production of Neu5Ac from GlcNAc was carried out in *E. coli* strains expressing GlcNAc 2-epimerase (AGE) and NeuB/or NanA [32, 35–37]. Tabata et al. (2002) first developed this process by a combination of three strains: recombinant *E. coli* expressing GlcNAc 2-epimerase (Slr1975), *E. coli* expressing Neu5Ac synthetase (NeuB) and *Corynebacterium ammoniagenes* [36]. In this system, GlcNAc was transformed to ManNAc by GlcNAc 2-epimerase, and then ManNAc and PEP were condensed to form Neu5Ac by NeuB. ATP and PEP were supplied by the cellular activities of *E. coli* and *C. ammoniagenes.* After reaction for 22 h, the Neu5Ac concentration reached 12.3 g/L with a GlcNAc conversion rate of 5 % [36].

Lee et al. (2007) coupled two recombinant *E. coli* strains (one strain expressing AGE from *Anabaena* sp. CH1 and the other strain expressing NanA from *E. coli*) for Neu5Ac synthesis from pyruvate and GlcNAc [35, 38]. In the process, a high concentration (122.3 g/L) and a high productivity (10.2 g/L/h) of Neu5Ac was reported, which might be attributed to the high concentration of cells (calculated to be 95.3 g cell/L, about 300 OD), and the high concentration of substrates (1.2 M of GlcNAc and 1.2 M of pyruvate) used in the system.

The process of whole-cell biocatalysis by combining two or three strains is complex because of the necessity to culture and concentrate the cells separately, and that the intermediates has to pass through the cell membranes twice (once to exit and a second time to enter). Although ATP is supplied by normal cell activity, mass transfer resistance is another major obstacle. This translated to a low catalytic efficiency of the process making it unsuitable for use on an industrial scale [30, 32, 37, 39].

11.2.2.2 Fermentative Production of Neu5Ac

Assembling a synthetic pathway for Neu5Ac production in one strain would reduce the mass transfer shortcoming and simplifying its operations. Metabolically engineered *E. coli* strains were constructed for the production of Neu5Ac from glucose by fermentation [7, 33]. Boddy and Lundgren (2009) assembled a six-step pathway for Neu5Ac production from fructose-6-phosphate by overexpressing genes of glucosamine synthase (*glmS*), Neu5Ac synthase (EC 4.1.3.19, *neuB*), and UDP-GlcNAc 2-epimerase (*neuC*) from *Neisseria meningitides*. The *nanA* and *nanT* genes were then knocked out to abolish sialic acid catabolism. In this engineered pathway, NeuB was responsible for converting ManNAc and PEP to Neu5Ac. As a result, 1.5 g/L Neu5Ac was obtained with glucose as the carbon source in a 98 h fermentation process [31].

In another study, the endogenous NanA of *E. coli* was used for converting ManNAc and pyruvate to Neu5Ac in which GlcNAc-2-epimerase (Slr1975) and glucosamine-6-phosphate acetyltransferase (GNA1) were introduced [33]. In this case, accumulation of GlcNAc and pyruvate was increased by gene deletion of Δ*nagAB* and Δ*ackA*Δ*poxB*Δ*ldhA*, and the catabolism of Neu5Ac was blocked by Δ*nanT*. As a result, 7.85 g/L of Neu5Ac was obtained using glucose as the carbon source in a 96 h fed-batch fermentation with the productivity of 0.08 g Neu5Ac/L/h [33]. Samain (2008) overexpressed NeuB and NeuC from *Campylobacter jejuni* in *neuA, nanA, nanK* and *nanT* knocked out *E. coli* strain, and 39 g/L of Neu5Ac was produced in high cell density fermentation with glycerol as the carbon source [40].

Fermentative production of Neu5Ac simplified the procedure and reduced the production cost. However, the pathway from glucose to Neu5Ac involved multistep reactions, and complex gene regulation. Moreover, the resulting titers and productivities of these strains are below values that are desired for cost-effective industrial production.

11.2.2.3 Whole-Cell Biocatalysis for Production of Neu5Ac by One Strain

Recently, the production of Neu5Ac has focused on whole-cell biocatalysis, where the requisite enzymes were packaged in an efficient manner in one engineered strain (Table 11.1). A two-step biosynthetic pathway of Neu5Ac consisting of AGE and

Table 11.1 Production of Neu5Ac by metabolic engineering in *E. coli*

Strains	Approaches (enzymes overexpression)	Pathway(s) blocked	Substrate	Titer (g/L)	References
E. coli K12	NeuB and NeuC from *N. meningitidis*, GlmS from *E. coli*	Catabolism of Neu5Ac: NanA and NanT encoding genes deleted	Glucose	1.7	[7, 31]
E. coli DH5α	Slr1975 from *Synechocystis* sp., GNA1 from *Saccharomyces cerevisiae*	Catabolism of GlcNAc: *nagA, nagB;* pyruvate accumulation: *ackA, poxB, ldhA* genes; catabolism of Neu5Ac: knockout of *nanTEK* operon	Glucose	7.85	[33]
E. coli SI2	NeuB and NeuC from *Campylobacter jejuni*	Catabolism of Neu5Ac: genes encoding NeuA, NanA, NanK and NanT deleted	Glycerol	39	[40]
E. coli N18-14 (acetate-resistant mutant)	Slr1975 from *Synechocystis* sp. and NeuB from *E. coli* K-1	Catabolism of Neu5Ac: *nanA* deleted	GlcNAc, glucose	53	[32]
E. coli K12	Slr1975 from *Synechocystis* sp. and NanA from *E. coli*	Transportation of GlcNAc: *nagE* gene deleted	GlcNAc, pyruvate	59	[30]
E. coli Rosetta (DE3) pLysS	AGE from *Anabaena* sp. and NanA from *E. coli* K12	–	GlcNAc, pyruvate	61.3	[41]
E. coli K12	AGE from *Anabaena* sp. and NanA from *E. coli* K12	Catabolism of Neu5Ac: knockout of *nanTEK* genes	GlcNAc, pyruvate	74.2	[34]

NanA/or NeuB was assembled in *E. coli* (Fig. 11.1) [30, 34, 41]. A two-step pathway would be simpler and easier to regulate.

Ishikawa and coworkers (2010) constructed a recombinant *E. coli* N18-14 strain by overexpressing genes of GlcNAc 2-epimerase (*slr1975*) and NeuB resulting in a yield of 53 g/Lof Neu5Ac (2.41 g/L/h) after 22 h [32]. Tao et al. (2011) and Sun et al. (2013) expressed AGE and NanA-encoding genes in *E. coli* that resulted in 59 g/L yield of Neu5Ac in 36 h (1.64 g/L/h) [30] and 61.3 g/L yield of Neu5Ac in 60 h (1.02 g/L/h) [41], respectively.

Although the various engineered strains produced relatively high titers of Neu5Ac and showed great promise, optimization of the heterologous pathways and host strains was required to obtain higher titers. Lin et al. (2013) developed a system for Neu5Ac production by assembling a heterologous biosynthetic pathway in *E. coli* consisting of AGE from *Anabaena* sp. PCC7120 and NanA from *E. coli* (Fig. 11.1) [34]. NanA was identified as the rate-controlling enzyme of the engineered pathway [34]. Efforts had been made to alleviate the bottleneck of catalysis by NanA in Neu5Ac biosynthesis by manipulating the amount of recombinant NanA. With increased expression of NanA, a ninefold increase in Neu5Ac production (65 mM) was observed [34].

Debugging the pathway by strain engineering usually resulted in the expected flux and high titer of the product. It was reported that the removal of the *nanA* and *nanT* genes abolished sialic acid catabolism [31]. The knockout of the *nanK* and *nanA* genes improved the efficiency of sialylation by preventing ManNAc and Neu5Ac from being diverted from the biosynthesis pathway [42]. The *nanTEK* genes *E. coli* encoding sialic acid transport, ManNAc-6-phosphate epimerase, and ManNAc kinase, respectively, were knocked out to block Neu5Ac uptake and the competing pathway [34]. This enhanced the Neu5Ac production by threefold, resulting in 173.8 mM [34]. Knocking out *nanTEK* genes improved the efficiency of the system significantly due to the combined effects of the following: (1) the competing catabolic pathway of ManNAc was disrupted by the deletion of *nanK* and *nanE*, so ManNAc was channeled to Neu5Ac synthetic pathway. (2) Neu5Ac transport into the cell was blocked by the *nanT* gene knockout. As a result, Neu5Ac was sequestered outside of the cell, which contributed to shifting the equilibrium toward Neu5Ac synthesis [34].

Multiple approaches were used to improve the efficiency of an engineered strain, including enhancing production of rate-controlling enzymes, deletion of competing pathways and promoting the substrate transport in Lin's system [34]. The maximum titer of 240 mM Neu5Ac (74.2 g/L) was obtained in 12 h, with a volumetric productivity of 6.2 g Neu5Ac/L/h and conversion yield of 40 % from GlcNAc [34]. Besides, the engineered strain could be reused at least five times with a productivity of >6 g/L/h [34].

11.2.2.4 Process Optimization for Improved Whole-Cell Biocatalysis

Mass transfer resistance is mainly caused by the membrane permeation of substrate and product. GlcNAc, the precursor of Neu5Ac, is transported mainly by the phosphotransferase system (PTS). To enhance the transportation of GlcNAc into cells, surfactants [such as Triton X-100, Tween 80, and Hexadecyltrimethy ammonium bromide (CTAB)] or organic solvents (such as xylene) were used as supplements in the reaction mixture [30, 34, 36]. Addition of 0.03 % CTAB increased the Neu5Ac yield by 1.54 times over the control without the surfactant [30]. With 0.2 % Triton X-100, Neu5Ac was 1.65-fold higher than that without [34]. These results demonstrated that detergents promoted substrate uptake resulting in an improved yield of Neu5Ac [30, 34, 36]. In the other hand, GlcNAc was transported by GlcNAc-specific PTS transport GlcNAc into cells as GlcNAc-6-P, followed by entering the GlcNAc metabolic pathway to be utilized as a source of carbon and nitrogen. The elimination of GlcNAc-specific PTS reduced GlcNAc catabolism, thus increased Neu5Ac production by 1.28-fold [30].

Equilibrium studies revealed that the epimerization of GlcNAc to ManNAc catalyzed by AGE was reversible and *E. coli* Neu5Ac aldolase favored the cleavage of Neu5Ac to ManNAc and pyruvate [43]. The effects of substrate concentration on Neu5Ac production demonstrated that the amount of Neu5Ac increased with increasing concentrations of GlcNAc and pyruvate in a considerable range [34]. A high concentration of GlcNAc could lead to a high concentration of ManNAc and shifted the reaction to Neu5Ac synthesis, resulting in a higher titer of Neu5Ac. The concentration of pyruvate played a more important role than that of GlcNAc in driving the reaction to Neu5Ac synthesis [34]. To achieve the maximal conversion rate, initial concentrations of 0.6 M GlcNAc/0.8 M pyruvate were used in Lin's process and 1.0MGlcNAc/0.5 M pyruvate were used in Tao's process [30, 34]. The activity of GlcNAc 2-epimerase was inhibited by excess pyruvate which in addition increased the complexity of downstream processing. Therefore, a continuous pyruvate feed other than the high initial concentration of pyruvate during the process may be employed in large-scale production. By promoting substrate transport and optimizing concentrations of substrates, performance of the system was further improved.

The whole-cell biocatalysis processes are developed for efficient production of Neu5Ac and the results are summarized in Table 11.1. The maximum titer of 240 mM Neu5Ac (74.2 g/L) was reported with a volumetric productivity of 6.2 g Neu5Ac/L/h [34]. The high productivity of whole-cell biocatalytic process is anticipated to make it a promising cost-effective resource for large-scale industrial production of Neu5Ac.

11.3 Polysialic Acid (PSA)

11.3.1 Biosynthesis and Metabolism of Polysialic Acid in E. coli

The biosynthesis of PSA in *E. coli* involves the synthesis of sialic acid monomers, polymerization of PSA from sialic acid, and transportation of PSA to the cell surface [28]. In *E. coli,* the *kps* gene cluster is involved in PSA biosynthesis, modification, and transport of the bacterial PSA chains (Fig. 11.2) [28, 44]. The *neuDBACES* region encodes the proteins that are involved in the biosynthesis, activation, and polymerization of Neu5Ac [45]. NeuD, a Neu5Ac 7-O (or 9-O)-acetyltransferase, acetylates Neu5Ac residues at carbon position 7 or 9 [46, 47]. NeuA, a bifunctional enzyme with both CMP-Neu5Ac synthetase [48, 49] and O-acetylesterase activity [50], catalyzes a reaction with free sialic acid to form cytidine 5′-monophosphate-sialic acid, thus generating the candidate donor form of sialic acid for all known sialyltransferases. NeuA also converts most of the CMP-O-acetyl-Neu5Ac to CMP-Neu5Ac before its incorporation into the polymers, and only a small amount of CMP-O-acetyl-Neu5Ac is incorporated into the

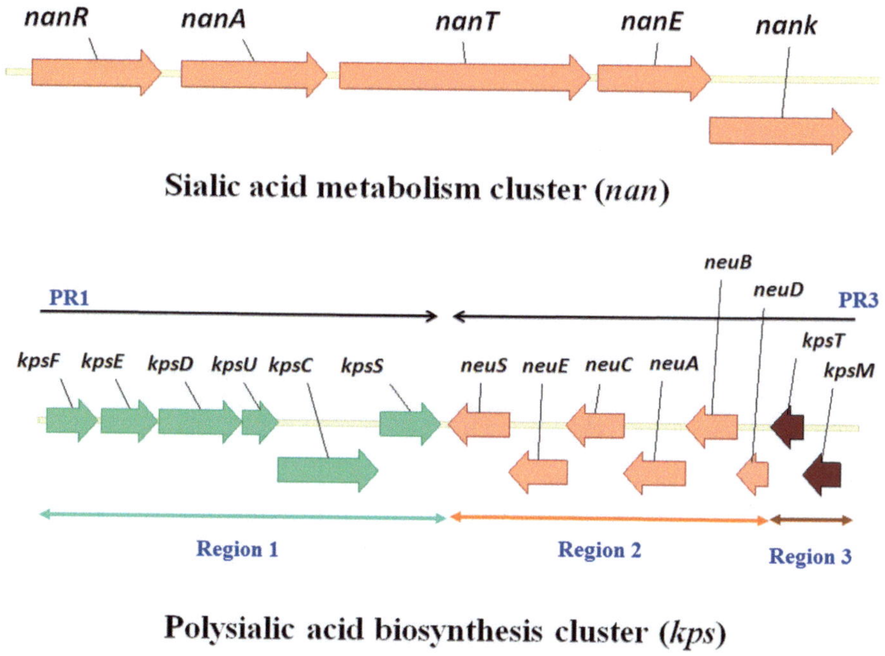

Fig. 11.2 Organization of gene clusters for the metabolism of polysialic acids: the sialic acid and polysialic acid biosynthesis cluster (*kps* cluster) and sialic acid catabolism operon (*nan* operon). Reprinted from Ref. [58], with permission from Springer Science+Business Media.

polymers [51]. NeuS, a α-Neu5Ac α-2,8-sialyltransferase, is responsible for polymerization of a homopolymer of α-2,8-linked Neu5Ac to form PSA [28, 47, 51]. The *kps* region 1 (*kpsFEDUCS*) and region 3 (*kpsTM*) genes participate in translocation of the polysaccharide across the periplasmic space and onto the cell surface [46, 52]. It was reported that KpsS could increase polysaccharide production by several fold (approximately 10–15 times) [53].

11.3.2 Metabolic Engineering for PSA Production

11.3.2.1 Enhancement of the Biosynthetic Pathway of PSA from Neu5Ac

Metabolic engineering strategies have been applied for improving PSA production (Fig. 11.3). For the biosynthesis of PSA, NeuD, NeuA, and NeuS are required to form PSA from free Neu5Ac [51]. To improve the efficiency of PSA production, the key enzymes of the PSA synthetic pathway were overexpressed separately or in combination. The strain that overexpressed NeuD (SA9/pDB1S-D) produced

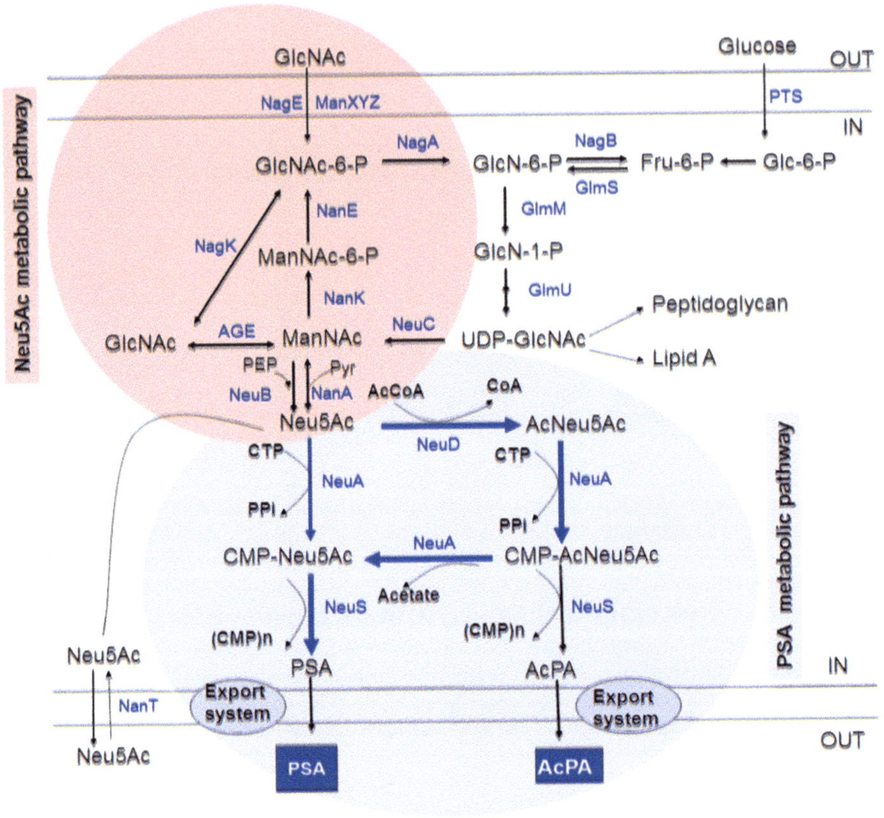

Fig. 11.3 Schematic presentation of the polysialic acid metabolic pathway in *E. coli*. Reprinted from Ref. [58], with permission from Springer Science+Business Media. AGE, GlcNAc 2-epimerase; NanA, Neu5Ac aldolase; NanT, sialic acid transporter; NanK, ManNAc kinase; NanE, ManNAc-6-P epimerase; Pgi, glucose-6-phosphate isomerase; GlmS, L-glutamine:D-fructose-6-phosphate aminotransferase; GlmM, phosphoglucosamine mutase; GlmU, bifunctional UDP-N-acetylglucosamine pyrophosphorylase/Glucosamine-1-phosphate N-acetyltransferase; NagA, N-acetylglucosamine-6-phosphate deacetylase; NagB, glucosamine-6-phosphate deaminase; NagK, GlcNAc kinase; NagE, GlcNAc-specific transporter; ManXYZ: mannose transporter; NeuA, CMP-Neu5Ac synthetase; NeuS, α-Neu5Ac α-2,8-sialyltransferase; NeuB, Neu5Ac synthase; NeuC, UDP-GlcNAc-2-epimerase; NeuD, Neu5Ac 7-O (or 9-O)-acetyltransferase

threefold more PSA (0.78 ± 0.04 g/L) than that of the wild-type strain (*E. coli* SA9: 0.24 ± 0.10 g/L) [54]. Whereas, only a slight increase in PSA production was observed in those strains with overexpressed NeuA or/and NeuS. On the other hand, co-expression of NeuD and NeuA in the strain harboring pDB1S-DA produced the most PSA (0.97 ± 0.06 g/L) [54]. In terms of the efficiency, NeuD was considered to be the most important factor. NeuA O-acetylesterase together with NeuD are deemed the most critical components for the production of PSA from Neu5Ac in *E. coli.*

11.3.2.2 Blocking the Competing Pathway

Neu5Ac is the precursor of PSA. In *E. coli*, the reactions catalyzed by NanA are reversible. The Neu5Ac cleavage activity of NanA is much greater than the Neu5Ac synthesis activity [34]. Knocking out the *nanA* gene blocked the competing catabolic pathway of Neu5Ac. The *nanA* gene was knocked out in *E. coli* SA9. A high yield of PSA (1.09 ± 0.05 g/L) was obtained with *E. coli* SA9Δ*nanA*/pbla-DA, with 22.5 % improvement compared to *E. coli* SA9/pbla-DA (0.89 ± 0.05 g/L) [54]. Blocking the Neu5Ac catabolic pathway by deleting NanA resulted in an enhanced production of PSA [54].

11.3.2.3 Fermentation of the Engineered Strain

The metabolically engineered *E. coli* SA9Δ*nanA*/pDB1S-DA strain was constructed by overexpressing NeuD and NeuA in the *nanA* knockout strain and was then used for PSA fermentation [54]. In the fed-batch cultivation, glucose limitation was avoided by maintaining a constant glucose concentration of 1 g/L in the culture broth. Feeding was started immediately after inoculation. The biomass of the strains was approximately 20 g/L DCW. The production of PSA by *E. coli* SA9Δ*nanA*/pDB1S-DA was increased by 85 % compared with the original strain, and 16.15 g/L PSA was obtained with a molecular mass of 113 kDa [54]. A number of reports have described the production of PSA using the wild-type strains [23, 55–57] and a maximal amount, 5.56 g/L, of PSA with molecular mass of 260 kDa was obtained [57]. The results of PSA production by fermentation are summarized in Table 11.2. The high productivity of the metabolically engineered strain provides evidence for improvement of PSA production by metabolic engineering and process engineering and makes it a promising cost-effective resource for PSA production.

11.4 Sialylated Oligosaccharides (3′ Sialyllactose)

Due to their important biological functions, sialylated structures have attracted considerable interest and many methods have been developed for the synthesis of sialylated oligosaccharides. Neu5Ac is the precursor for the synthesis of sialylated oligosaccharide. CMP-Neu5Ac synthetase catalyzes CTP and Neu5Ac to form CMP-Neu5Ac. Sialyltransferase adds sialic acid to oligosaccharide to form sialylated oligosaccharides. The enzymatic synthesis of sialylated oligosaccharides using

Table 11.2 Production of polysialic acids by process engineering and metabolic engineering in *E. coli*

Strains	Description	Molecular mass (kDa)	Chain length	PSA yields (g/L)	References
E. coli K-235	Incubated in 250-mL shake flasks. D-xylose and L-proline were used as sole carbon and nitrogen source	–	–	1.35	[22]
E. coli K235-WXJ4	pH control during fed-batch fermentation	–	–	2.61	[23]
E. coli K1	Multiple fed-batch cultivation	–	>130	0.95	[55, 59]
E. coli CCTCCM 208088	Lower level of initial phosphate and pH control during fed-batch fermentation	16–50	–	5.2	[55]
E. coli K1	Fed-batch cultivation at a constant glucose concentration of 50 mg/L	–	–	1.35	[60]
E. coli CCTCCM208088	Dual-stage pH control and fed-batch fermentation	260	890	5.65	[57, 61]
E. coli SA9	Strengthen the biosynthetic pathway of PSA	113	–	16.15	[54]

Reproduced from Ref. [58], with permission from Springer Science+Business Media

CMP-Neu5Ac synthetase and sialyltransferase fusion protein was developed, and sialyllactose at 100 g scale was obtained starting with lactose, sialic acid, PEP, and catalytic amounts of ATP and CMP [62]. Sialic acid, ATP and CMP are expensive and not readily available in large quantities. This enzymatic process was not suitable for the cost-effective production of sialylated oligosaccharides.

Endo et al. described a system for the production of 3′ sialyllactose by combining four bacterial strains [63]. Two recombinant *E. coli* strains overexpressing the genes for CMP-Neu5Ac synthetase and CTP synthetase, respectively, were responsible for CMP-Neu5Ac production. *C. ammoniagenes* was used to produce UTP from orotic acid. The fourth strain used was *E. coli* overexpressing the α-(2,3)-sialyltransferase from *Neisseria gonorrhoeae.* With this system, 3′-sialyllactose accumulated to a concentration of 52 mM (33 g/L) after an 11-h reaction that began with orotic acid, Neu5Ac, and lactose [63].

To simplify the system, Priem et al. (2002) assembled the biosynthetic pathway in one *E. coli* strain by overexpressing the genes for α-2,3 sialyltransferase and CMP-Neu5Ac synthase from *N. meningitides* to promote the production of sialyllactose from the exogenous added Neu5Ac and lactose [64]. The *nanA* gene was knocked out in order to accumulate Neu5Ac. The β-galactosidase gene (*lacZ*) was knocked out to allow lactose to accumulate intracellularly upon induction of *lacY* gene for β-galactoside permease. The engineered strain ($lacY^{+}$, lac^{-}, $nanT^{+}$, $nanA^{-}$)

that was transformed with two plasmids expressing α-2,3 sialyltransferase and the CMP-Neu5Ac synthase was cultivated on glycerol and the culture was supplemented with lactose and sialic acid at the fed-batch phase. After 22 h, 2.6 g/L of sialyllactose was produced (1.5 g/L in the intracellular and 1.1 g/L in the extracellular fraction) [64]. In the microbial process described by Endo et al. [63] and Priem et al. [64], the activated sialic acid donor (CMP-Neu5Ac) was generated from the exogenous sialic acid. Because Neu5Ac is expensive, a requirement for exogenous Neu5Ac represents a major constraint for optimizing the microbial production of sialyllactose.

Fierfort and Samain [42] circumvented the above problem by engineering an *E. coli* K12 strain that was capable of generating CMP-Neu5Ac using its own internal metabolism, a scheme shown in Fig. 11.4. Mutant strains devoid of Neu5Ac aldolase and of ManNAc kinase were created and shown to efficiently produce 3′ sialyllactose by co-expressing the α-2,3-sialyltransferase gene from *Neisseria meningitidis* with the genes *neuC*, *neuB,* and *neuA* from *Campylobacter jejuni* encoding GlcNAc-6-phosphate epimerase, sialic acid synthase and CMP-Neu5Ac synthetase, respectively. These results demonstrated that the over-expression of the *neuABC* genes confers on *E. coli* K12 the CMP-Neu5Ac biosynthetic capacity to efficiently produce sialylated oligosaccharides, such as 3′

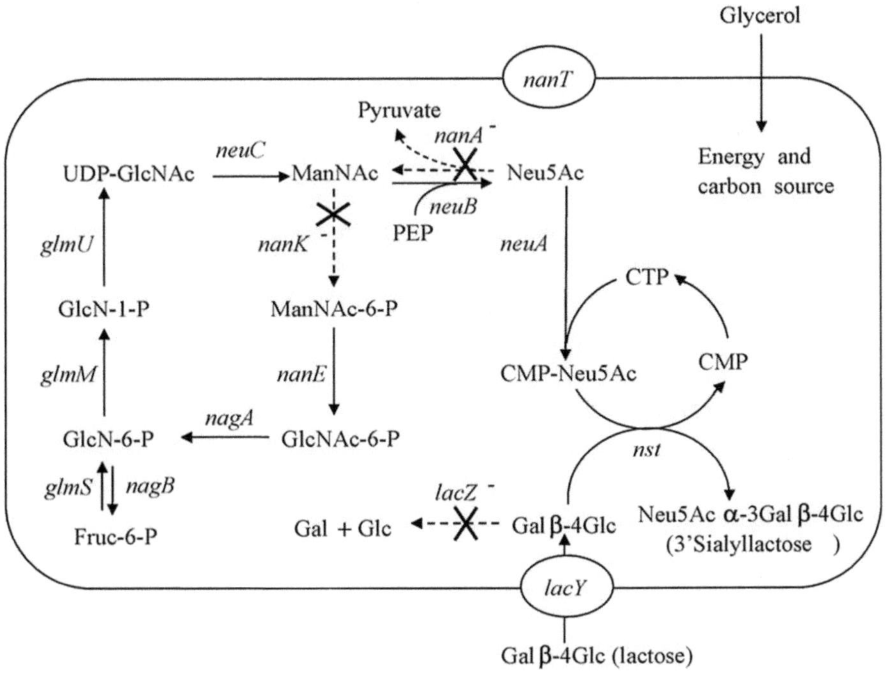

Fig. 11.4 Engineered metabolic pathways for the production of 3′ sialyllactose. Reproduced from Ref. [42], Copyright 2008, with permission from Elsevier

Table 11.3 Production of 3′ sialyllactose by metabolic engineering in *E. coli*

Strains	Approaches (enzymes overexpression)	Pathway (s) blocked	Substrate	Titer (g/L)	References
E. coli	α-2,3 sialyltransferase and CMP-Neu5Ac synthase from *N. meningitides*	*lac*$^-$ and *nanA*$^-$	Neu5Ac and lactose	2.6	[64]
Three *E. coli* strains *and C. ammoniagenes*	CMP-Neu5Ac synthetase, CTP synthetase, and α-(2,3)-sialyltransferase from *Neisseria gonorrhoeae*		Orotic acid, NeuAc, lactose	33	[63]
E. coli	Coexpress α-2,3-sialyltransferase from *Neisseria meningitidis* with the genes *neuC*, *neuB* and *neuA* from *Campylobacter jejuni*	*nanK*$^-$and *nanA*$^-$	Glycerol, lactose	25	[42]

sialyllactose. The knockout of the *nanK* and *nanA* genes improves the efficiency of sialylation considerably in the system by preventing ManNAc and Neu5Ac from being diverted from the biosynthesis of the CMP-Neu5Ac pathway and by avoiding the formation of energetically deleterious futile cycles [42]. The 3′ sialyllactose production yield of 25 g/L was obtained after 70 h in culture with a continuous lactose feed [42]. The results of the microbial production of 3′ sialyllactose are summarized in Table 11.3, the highest yield being an order of magnitude higher than that obtained by using an exogenous supply of Neu5Ac reported by Priem and coworkers [42, 63, 64].

This high yield and the low cost of fermentation medium should facilitate the use of sialylated oligosaccharides in new fields such as the food industry [42]. CMP-Neu5Ac synthetase can also be used to produce CMP derivatives of sialic acid analogs. This economical and efficient system of sialyllactose production can be extended to the production of a large number of sialylated oligosaccharides of biological interest.

11.5 Perspectives

E. coli provides a perfect platform for metabolic engineering due to its well studied genetic background, and highly advanced genetic engineering technologies for this microorganism as well as a relatively well-developed fermentation process with low-cost raw materials. Metabolic engineering approaches have contributed to the efficient production of sialic acid and its derivatives in *E. coli*. Based on the current results, we suggest the following focus areas of research for further improvement of sialic acids production by microorganisms:

(1) Neu5Ac: Neu5Ac is the precursor of PSA and sialylated oligosaccharides. Enhancing the Neu5Ac biosynthetic module should contribute to the improved production of PSA and sialylated oligosaccharides. NanA is the rate-controlling enzyme of Neu5Ac biosynthesis. Exploiting NanA from different species with high aldolase activities and low lyase activities would improve the efficiency of bioconversion from GlcNAc to Neu5Ac. Further studies should be conducted to improve the performance of the process by NanA substitution or modification, deletion of PTS, and blockage of GlcNAc catabolism, and improvement of pyruvate accumulation by reducing byproduct formation.
(2) PSA: The combination of metabolic engineering and process engineering strategies, including enhancing the precursor (Neu5Ac) supply, promoting capsule polysaccharide exportation, and optimizing the fermentation process, will lead to new approaches to further improve the production of PSA. These strategies will make the bioprocess a promising cost-effective resource for PSA production with higher titers, yields, and productivity, as well as with regulated PSA chain lengths.
(3) Sialylated oligosaccharides: By extending the donor from sialic acid to its analogs and the acceptor from lactose to other saccharides, a diverse range of sialylated oligosaccharide analogs can be produced.

Conflict of interest There is no conflict of interest.

References

1. Schauer R (2004) Sialic acids: fascinating sugars in higher animals and man. Zoology 107 (1):49–64
2. Almagro-Moreno S, Boyd EF (2009) Insights into the evolution of sialic acid catabolism among bacteria. BMC Evol Biol 9(1):118
3. Chen X, Varki A (2010) Advances in the biology and chemistry of sialic acids. ACS Chem Biol 5(2):163–176
4. Schauer R (2000) Achievements and challenges of sialic acid research. Glycoconj J 17 (7):485–499
5. Varki NM, Varki A (2007) Diversity in cell surface sialic acid presentations: implications for biology and disease. Lab Invest 87(9):851–857
6. Maru I, Ohnishi J, Ohta Y, Tsukada Y (2002) Why is sialic acid attracting interest now? Complete enzymatic synthesis of sialic acid with N-acylglucosamine 2-epimerase. J Biosci Bioeng 93(3):258–265
7. Lundgren BR, Boddy CN (2007) Sialic acid and N-acyl sialic acid analog production by fermentation of metabolically and genetically engineered *Escherichia coli*. Org Biomol Chem 5(12):1903–1909
8. Tanner ME (2005) The enzymes of sialic acid biosynthesis. Bioorg Chem 33(3):216–228
9. Wu W, Air GM (2004) Binding of influenza viruses to sialic acids: reassortant viruses with A/NWS/33 hemagglutinin bind to α2, 8-linked sialic acid. Virology 325(2):340–350

10. Kawai N, Ikematsu H, Iwaki N, Maeda T, Kawashima T, Hirotsu N, Kashiwagi S (2009) Comparison of the effectiveness of zanamivir and oseltamivir against influenza A/H1N1, A/H3N2, and B. Clin Infect Dis 48(7):996–997
11. von Itzstein M, Dyason JC, Oliver SW, White HF, Wu W-Y, Kok GB, Pegg MS (1996) A study of the active site of influenza virus sialidase: an approach to the rational design of novel anti-influenza drugs. J Med Chem 39(2):388–391
12. Zhang X, Kiechle FL (2004) Glycosphingolipids in health and disease. Ann Clin Lab Sci 34 (1):3–13
13. Sprenger N, Julita M, Donnicola D, Jann A (2009) Sialic acid feeding aged rats rejuvenates stimulated salivation and colon enteric neuron chemotypes. Glycobiology 19(12):1492–1502
14. Wang B, McVeagh P, Petocz P, Brand-Miller J (2003) Brain ganglioside and glycoprotein sialic acid in breastfed compared with formula-fed infants. Am J Clin Nutr 78(5):1024–1029
15. Wang B, Yu B, Karim M, Hu H, Sun Y, McGreevy P, Petocz P, Held S, Brand-Miller J (2007) Dietary sialic acid supplementation improves learning and memory in piglets. Am J Clin Nutr 85(2):561–569
16. Steinhaus S, Stark Y, Bruns S, Haile Y, Scheper T, Grothe C, Behrens P (2010) Polysialic acid immobilized on silanized glass surfaces: a test case for its use as a biomaterial for nerve regeneration. J Mater Sci Mater Med 21(4):1371–1378
17. Sprenger N, Duncan PI (2012) Sialic acid utilization. Adv Nutr Int Rev J 3(3):392S–397S
18. Bode L (2006) Recent advances on structure, metabolism, and function of human milk oligosaccharid*es*. J Nutr 136(8):2127–2130
19. Martin JE, Tanenbaum SW, Flashner M (1977) A facile procedure for the isolation of N-acetylneuramic acid from edible bird's-nest. Carbohydr Res 56(2):423
20. Koketsu M, Juneja LR, Kawanami H, Kim M, Yamamoto T (1992) Preparation of N-acetylneuraminic acid from delipidated egg yolk. Glycoconj J 9(2):70–74
21. Ferrero MA, Reglero A, Fernandez-Lopez M, Ordas R, Rodriguez-Aparicio LB (1996) N-acetyl-D-neuraminic acid lyase generates the sialic acid for colominic acid biosynthesis in *Escherichia coli* K1. Biochem J 317(Pt 1):157–165
22. Rodriguez-Aparicio L, Reglero A, Ortiz A, Luengo J (1988) Effect of physical and chemical conditions on the production of colominic acid by *Escherichia coli* in a defined medium. Appl Microbiol Biotechnol 27(5):474–483
23. Zhan X, Zhu L, Wu J, Zhen Z, Jia W (2002) Production of polysialic acid from fed-batch fermentation with pH control. Biochem Eng J 11(2):201–204
24. Wang TH, Chen YY, Pan HH, Wang FP, Cheng CH, Lee WC (2009) Production of N-acetyl-D-neuraminic acid using two sequential enzymes overexpressed as double-tagged fusion proteins. BMC Biotechnol 9:63
25. Hu S, Chen J, Yang Z, Shao L, Bai H, Luo J, Jiang W, Yang Y (2010) Coupled bioconversion for preparation of N-acetyl-D-neuraminic acid using immobilized N-acetyl-D-glucosamine-2-epimerase and N-acetyl-D-neuraminic acid lyase. Appl Microbiol Biotechnol 85(5):1383–1391
26. Vimr ER, Kalivoda KA, Deszo EL, Steenbergen SM (2004) Diversity of microbial sialic acid metabolism. Microbiol Mol Biol Rev 68(1):132–153
27. Martinez J, Steenbergen S, Vimr E (1995) Derived structure of the putative sialic acid transporter from *Escherichia coli* predicts a novel sugar permease domain. J Bacteriol 177 (20):6005–6010
28. Ferrero MÁ, Aparicio LR (2010) Biosynthesis and production of polysialic acids in bacteria. Appl Microbiol Biotechnol 86(6):1621–1635
29. Kalivoda KA, Steenbergen SM, Vimr ER, Plumbridge J (2003) Regulation of sialic acid catabolism by the DNA binding protein NanR in *Escherichia coli*. J Bacteriol 185 (16):4806–4815
30. Tao F, Zhang Y, Ma C, Xu P (2011) One-pot bio-synthesis: N-acetyl-D-neuraminic acid production by a powerful engineered whole-cell catalyst. Sci Rep 1:1–7
31. Boddy CN, Lundgren BR (2009) Metabolically engineered *Escherichia coli* for enhanced production of sialic acid. US Patent (PCT/US2007/079496)

32. Ishikawa M, Koizumi S (2010) Microbial production of N-acetylneuraminic acid by genetically engineered *Escherichia coli*. Carbohydr Res 345(18):2605–2609
33. Kang J, Gu P, Wang Y, Li Y, Yang F, Wang Q, Qi Q (2012) Engineering of an N-acetylneuraminic acid synthetic pathway in *Escherichia coli*. Metab Eng 14(6):623–629
34. Lin B-X, Zhang Z-J, Liu W-F, Dong Z-Y, Tao Y (2013) Enhanced production of N-acetyl-D-neuraminic acid by multi-approach whole-cell biocatalyst. Appl Microbiol Biotechnol 97 (11):4775–4784
35. Lee YC, Chien HC, Hsu WH (2007) Production of N-acetyl-D-neuraminic acid by recombinant whole cells expressing *Anabaena* sp. CH1 N-acetyl-D-glucosamine 2-epimerase and *Escherichia coli* N-acetyl-D-neuraminic acid lyase. J Biotechnol 129 (3):453–460
36. Tabata K, Koizumi S, Endo T, Ozaki A (2002) Production of N-acetyl-D-neuraminic acid by coupling bacteria expressing N-acetyl-D-glucosamine 2-epimerase and N-acetyl-D-neuraminic acid synthetase. Enzyme Microb Technol 30(3):327–333
37. Zhang Y, Tao F, Du M, Ma C, Qiu J, Gu L, He X, Xu P (2010) An efficient method for N-acetyl-D-neuraminic acid production using coupled bacterial cells with a safe temperature-induced system. Appl Microbiol Biotechnol 86(2):481–489
38. Tao F, Zhang Y, Ma C, Xu P (2010) Biotechnological production and applications of N-acetyl-D-neuraminic acid: current state and perspectives. Appl Microbiol Biotechnol 87 (4):1281–1289
39. Xu X, Gao C, Zhang X, Che B, Ma C, Qiu J, Tao F, Xu P (2011) Production of N-acetyl-D-neuraminic acid by use of an efficient spore surface display system. Appl Environ Microbiol 77(10):3197–3201
40. Samain E (2008) High yield production of sialic acid (Neu5Ac) by fermentation. (WO 2008/040717)
41. Sun W, Ji W, Li N, Tong P, Cheng J, He Y, Chen Y, Chen X, Wu J, Ouyang P (2013) Construction and expression of a polycistronic plasmid encoding N-acetylglucosamine 2-epimerase and N-acetylneuraminic acid lyase simultaneously for production of N-acetylneuraminic acid. Bioresour Technol 130:23–29
42. Fierfort N, Samain E (2008) Genetic engineering of *Escherichia coli* for the economical production of sialylated oligosaccharides. J Biotechnol 134(3):261–265
43. Zimmermann V, Hennemann H-G, Daußmann T, Kragl U (2007) Modelling the reaction course of N-acetylneuraminic acid synthesis from N-acetyl-D-glucosamine—new strategies for the optimisation of neuraminic acid synthesis. Appl Microbiol Biotechnol 76(3):597–605
44. Barrett B, Ebah L, Roberts I (2002) Genomic structure of capsular determinants. In: Pathogenicity islands and the evolution of pathogenic microbes. Springer, pp 137–155
45. Daines DA, Silver RP (2000) Evidence for multimerization of Neu proteins involved in polysialic acid synthesis in *Escherichia coli* K1 using improved LexA-based vectors. J Bacteriol 182(18):5267–5270
46. Cress BF, Englaender JA, He W, Kasper D, Linhardt RJ, Koffas M (2014) Masquerading microbial pathogens: capsular polysaccharides mimic host-tissue molecules. FEMS Microbiol Rev 38:660–697
47. Steenbergen SM, Lee Y-C, Vann WF, Vionnet J, Wright LF, Vimr ER (2006) Separate pathways for O acetylation of polymeric and monomeric sialic acids and identification of sialyl O-acetyl esterase in *Escherichia coli* K1. J Bacteriol 188(17):6195–6206
48. Zapata G, Vann W, Aaronson W, Lewis M, Moos M (1989) Sequence of the cloned *Escherichia coli* K1 CMP-N-acetylneuraminic acid synthetase gene. J Biol Chem 264 (25):14769–14774
49. Vann W, Silver R, Abeijon C, Chang K, Aaronson W, Sutton A, Finn C, Lindner W, Kotsatos M (1987) Purification, properties, and genetic location of *Escherichia coli* cytidine 5'-monophosphate N-acetylneuraminic acid synthetase. J Biol Chem 262(36):17556–17562
50. Liu G, Jin C, Jin C (2004) CMP-N-acetylneuraminic acid synthetase from *Escherichia coli* K1 is a bifunctional enzyme: identification of minimal catalytic domain for synthetase activity

and novel functional domain for platelet-activating factor acetylhydrolase activity. J Biol Chem 279(17):17738–17749
51. Song L, Zhou H, Cai X, Li C, Liang J, Jin C (2011) NeuA O-acetylesterase activity is specific for CMP-activated O-acetyl sialic acid in *Streptococcus suis* serotype 2. Biochem Biophys Res Commun 410(2):212–217
52. Willis LM, Whitfield C (2013) Structure, biosynthesis, and function of bacterial capsular polysaccharides synthesized by ABC transporter-dependent pathways. Carbohydr Res 378:35–44
53. Andreishcheva EN, Vann WF (2006) Gene products required for de novo synthesis of polysialic acid in *Escherichia coli* K1. J Bacteriol 188(5):1786–1797
54. Chen F, Tao Y, Jin C, Xu Y, Lin B-X (2015) Enhanced production of polysialic acid by metabolic engineering of *Escherichia coli*. Appl Microbiol Biotechnol 99(6):2603–2611
55. Wu J-R, Liu J-L, Zhan X-B, Lin C-C, Zhao H (2010) Enhancement of polysialic acid yield by reducing initial phosphate and feeding ammonia water to *Escherichia coli* CCTCC M208088. Biotechnol Bioprocess Eng 15(4):657–663
56. Liu J-L, Zhan X-B, Wu J-R, Lin C-C, Yu D-F (2010) An efficient and large-scale preparation process for polysialic acid by *Escherichia coli* CCTCC M208088. Biochem Eng J 53(1):97–103
57. Zheng Z-Y, Wang S-Z, Li G-S, Zhan X-B, Lin C-C, Wu J-R, Zhu L (2013) A new polysialic acid production process based on dual-stage pH control and fed-batch fermentation for higher yield and resulting high molecular weight product. Appl Microbiol Biotechnol 97(6):2405–2412
58. Lin B-X, Qiao Y, Shi B, Tao Y (2016) Polysialic acid biosynthesis and production in *Escherichia coli*: current state and perspectives. Appl Microbiol Biotechnol 100:1–8
59. Rode B, Endres C, Ran C, Stahl F, Beutel S, Kasper C, Galuska S, Geyer R, Muhlenhoff M, Gerardy-Schahn R, Scheper T (2008) Large-scale production and homogenous purification of long chain polysialic acids from *E. coli* K1. J Biotechnol 135(2):202–209
60. Chen R, John J, Rode B, Hitzmann B, Gerardy-Schahn R, Kasper C, Scheper T (2011) Comparison of polysialic acid production in *Escherichia coli* K1 during batch cultivation and fed-batch cultivation applying two different control strategies. J Biotechnol 154(4):222–229
61. Zheng ZY, Wang SZ, Li GS, Zhan XB, Lin CC, Wu JR, Zhu L (2013) A new polysialic acid production process based on dual-stage pH control and fed-batch fermentation for higher yield and resulting high molecular weight product. Appl Microbiol Biotechnol 97(6):2405–2412
62. Gilbert M, Bayer R, Cunningham A-M, DeFrees S, Gao Y, Watson DC, Young NM, Wakarchuk WW (1998) The synthesis of sialylated oligosaccharides using a CMP-Neu5Ac synthetase/sialyltransferase fusion. Nat Biotechnol 16(8):769–772
63. Endo T, Koizumi S, Tabata K, Ozaki A (2000) Large-scale production of CMP-NeuAc and sialylated oligosaccharides through bacterial coupling. Appl Microbiol Biotechnol 53 (3):257–261
64. Priem B, Gilbert M, Wakarchuk WW, Heyraud A, Samain E (2002) A new fermentation process allows large-scale production of human milk oligosaccharides by metabolically engineered bacteria. Glycobiology 12(4):235–240

Chapter 12
Phenolics Value Chain and L-Lactic Acid Bioproduction from Agricultural Biomass

Krista L. Morley and Peter C.K. Lau

Abstract To make the quantum leap from conceptualization of an idea to possible commercialization of a product, at first it would be desirable to have the opportunity to showcase the potential of a given product stream, successes, and the challenges involved. Using agricultural biomass as model feedstock, we feature in this chapter the tasks that were carried out in our laboratory in the context of bioproduction of vinyl phenols from two phenolic acids—vinyl guaiacol from ferulic acid and vinyl syringol from sinapic acid—as value-added chemicals. As a proof of concept, these monomers were polymerized to produce new biopolymers with improved properties. Along a similar product line, direct fermentation by a *Rhizopus* fungus of a nonfood feedstock to produce an optically pure L-lactic acid, a precursor to the well-known polyester, polylactic acid (PLA) was also featured. Technological advances made in these studies include the discovery and characterization of new enzymes viz. feruloyl esterases that release phenolic acids from lignocellulosic biomass. Guided by the 3-D structure of a phenolic acid decarboxylase (PAD), a novel sinapic acid decarboxylase (SAD) was evolved to effect the biotransformation of sinapic acid to vinyl syringol. Throughout the development, process optimization of various kinds was explored that include in situ recovery of water-insoluble compounds exemplified by vinyl guaiacol in a biphasic aqueous-solvent bioreactor system. We commented on some of the recent advances made in related areas and the necessity to push the green chemistry and innovation agenda forward for a sustainable future.

Keywords Chemurgy · Aromatic acids · Phenolic acids · Antioxidants · Building blocks · Biopolymers · PLA · Biphasic bioreactor system · Biocatalysis · Biorefinery · Green chemistry · Agricultural feedstock

---

K.L. Morley
National Research Council Canada, Montreal, Canada

P.C.K. Lau (✉)
Tianjin Institute of Industrial Biotechnology Chinese Academy of Sciences,
Tianjin 300308, China
e-mail: Peter.Lau@tib.cas.cn; Peter.Lau@mcgill.ca

P.C.K. Lau
McGill University, Montreal, Canada

P.C.K. Lau (ed.), *Quality Living Through Chemurgy and Green Chemistry*,
Green Chemistry and Sustainable Technology,
DOI 10.1007/978-3-662-53704-6_12

12.1 Introduction

Worldwide, there is a heightened interest in the use of biomass to reduce the dependence on dwindling supplies of fossil fuels as well as to reduce greenhouse gas emissions [1, 2]. The switch to a biorefinery approach that uses biomass to produce energy and to generate chemicals to replace petroleum-based materials has entered center stage. In the framework of a sustainable biorefinery, an assortment of feedstocks and diverse arrays of downstream products that include green chemicals, building blocks, and intermediates are in development [3]. Creation of high-value coproducts can provide income to offset the overall processing costs. Chemocatalysis [4, 5], biocatalysis [6], chemoenzymatic processes [7], and the integration of innovative bioprocessing [8] technologies are critical requirements of this strategy. Through advances in numerous disciplines including biotechnology, molecular biology, and metabolic engineering, great progress has been made in the development and implementation of technologies for the preparation of small aliphatic carbon building blocks and production of bio-based products from carbohydrates, for example [9].

In our laboratory, a variety of agricultural feedstocks, triticale bran in particular, was explored to demonstrate the potential value chain of an aromatic platform to access chemicals as monomers for possible biopolymer formation (Fig. 12.1). In addition, triticale starch and flour were used as a nonfood feedstock for the production of optically pure L-lactic acid monomer of many applications.

12.2 Phenolic Compounds in Nature

Phenolic compounds are found in nature as esters, amides, and glycosides of hydroxycinnamic acids (HC), glycosylated flavonoids, proanthocyanidins, and their derivatives. They constitute a major class of the plant secondary metabolites that are synthesized by plants via the shikimate/phenylpropanoid pathway or the polyketide metabolic pathway [10]. Phenolic compounds are involved in plant defence

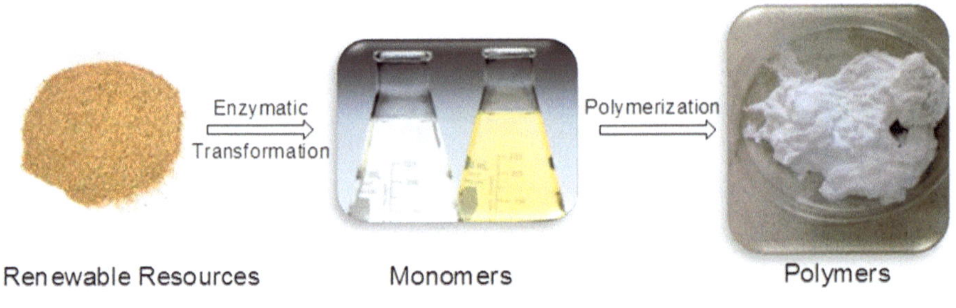

Fig. 12.1 A generalized scheme for value creation from agricultural biomass

mechanisms against pathogens and microbial infections [11], and function as antioxidants toward cancer and cardiovascular diseases [12]. Phenolic compounds are generally either in the free form or bound form. Agricultural lignocellulosic biomass is comprised of 40–50 % cellulose, 20–30 % hemicellulose, and 10–25 % lignin. Cellulose and hemicelluloses are polysaccharides consisting of hexoses and pentoses, and lignin is a polymer primarily made up of phenylpropane units derived from guaiacol, *p*-hydroxyphenol, and syringol. Suberin and pollen sporopollenin are other examples of phenolic-containing polymers.

12.2.1 Phenolic Acids/Hydroxycinnamic Acid Derivatives

Phenolic acids represent a significant fraction of the plant biomass and are considered as potential candidates for the replacement of aromatic compounds derived from fossil fuels. In a biorefinery setting, lignin is considered to be the primary source of aromatic chemicals because it is the only renewable feedstock that comprises aromatic rings. However, economical production of aromatic platform chemicals in acceptable purities and yields remains a challenge, mainly due to the heterogeneous nature of lignin [13]. Possible valorization of lignin to useful chemicals are discussed in two preceding chapters.

Phenolic acids which are esterified to polysaccharides and lignin represent an alternative and valuable source of renewable aromatic compounds from biomass feedstock [14]. Phenolic acids are divided into two main groups: benzoic acids, containing seven carbon atoms (C6-C1), and cinnamic acids, comprising nine carbon atoms (C6-C3). Examples of benzoic acids are salicyclic, gallic, *p*-hydroxybenzoic, protocatechuic, vanillic, and syringic acids. In plants such as soybean, barley, and rice, the level of salicylic acid ranges between 0.1 mg and 3.7 mg/100 g of biomass [15]. Gallic acid exists in plant material as the free acid, ester, catechin derivative, and hydrolyzable tannin. Significant amounts of *p*-hydroxybenzoic acid glycosides are found in raspberries (32–59 ppm, mg/kg of fresh fruit) [16]. Protocatechuic acid, vanillic acid, and syringic acids are active components of some traditional Chinese herbal medicines [17–19] and also occur in various berries [20].

Caffeic acid, ferulic acid (FA), *p*-coumaric acid (*p*-CA), and sinapic acid (SA) are the most common derivatives of cinnamic acid. The content of FA varies from 0.5 % (w/w) in wheat bran [21] to 3.1 % (w/w) in corn bran [22]. Although abundant in the plant biomass, FA is generally not found in free form, but covalently linked to polysaccharides by ester bonds [23]. Releasing FA from its bound, polysaccharides in the plant biomass can be carried out via chemical (sodium hydroxide) hydrolysis or using a feruloyl esterase (FAE, aka ferulic acid esterase, cinnamic acid or cinnamoyl esterases) as we will elaborate below. The many facets of FA as a “renewable chemical feedstock” and its biocatalytic transformation products had been the subject of a classical review by Rosazza et al. [24]. FA is widely used as a food preservative due to its antioxidant property and it is an active ingredient in sunscreens and also used as therapeutic agents [25, 26]. In addition, vinyl guaiacol (VG) and vanillin, both compounds of industrial importance in the food industry and as flavoring agents can be produced from FA [24, 27, 28]. Thus, it may not come to a surprise that FA was chosen as one of the top two aromatic acid building blocks from lignocellulosic materials and biomass [29, 30].

12.3 Feruloyl Esterases and Approaches to Extracting FA from Biomass

As key plant biomass degrading enzymes and having broader applications such as the synthesis of feruloylate glycosides, in animal feed and the pulp and paper industry, FAEs (EC 3.1.1.73) have been studied quite extensively and a topic of a great number of reviews [31–37]. FAEs are members of the superfamily of α/ß hydrolase-fold enzymes. Various classification schemes of FAEs have been proposed based on substrate preference, specificity for HC acid methyl esters and release of diferulic acid from model complex substrates as well as phylogenetic clustering of their amino acid sequences [38, 39].

Table 12.1 shows a summary of some key properties of three new recombinant FAEs with particular reference to the extraction of FA from triticale bran, wheat bran and an aqueous fraction of steam-exploded wheat straw. First, a side note about triticale. This is a nonfood crop, a man-made hybrid of wheat and rye grown worldwide mostly for its grain and forage, and used primarily as animal feed [40]. About the enzymes: TtFAE is a monomeric 36-kDa protein that originated from a thermophilic organism, *Thermoanaerobacter tencongensis* strain MB4^{T} isolated from a hot spring in China [41, 42]. The protein was found to be highly thermostable by virtue of its long half-life at high temperatures—at 75 °C, the enzyme retained at least 95 % of its original activity for over 80 min. Besides, TtFAE was solvent tolerant, e.g., 50 % v/v DMSO [42, 43]. ScFAE1 and ScFAE2 are the first of the kind of myxobacterium origin (*Sorangium cellulosum* So ce56) to be cloned and characterized [44]. One general conclusion drawn about these enzymes is that sequenced genomes provide a rich ground for potentially new biocatalysts. Finding the right one requires first the characterization of its properties.

TtFAE was shown to be an active biocatalyst for the conversion of feruloylated sugars to FA, the best yield was obtained after 5 h incubation of an autoclaved triticale bran suspension (5 % w/v) with partially purified TtFAE (40 mU/mL) and 10 U of *Trichoderma viride* xylanase at 37 °C. The released phenolic compounds were analyzed by HPLC. This represented 30 % of the total FA extractable by sodium hydroxide, established to be 227.8 ± 8.9 mg/100 g, and representing 0.23 % (w/w) of triticale bran. The triticale straw contained more *p*-CA (403.7 ± 19.3 mg/100 g) than FA (303.7 ± 12.7 mg/100 g).

ScFAE1 and ScFAE2 share only 30 % sequence identity with each other [42]. Comparison of their kinetic parameters indicated an apparent higher affinity of ScFAE2 than ScFAE1 toward a number of feruloyl substrates. This property was

Table 12.1 Properties of three recombinant FAEs

Enzyme [type]	pH optimum	Temp °C	Special properties	Feedstock	FA mg/100 g extracted[a]	Efficiency %	References
TtFAE [A]	8.0	80; Half life 50 min	Alkaline, thermostable and solvent stable	Triticale bran	686	30	[42, 43]
ScFAE1 [D]	7.0	50; Half life 68 min		Triticale bran-destarched	1223		[42]
				Wheat bran-destarched	1703		[42]
ScFAE2 [D]	6–8	55; Half life >800 min	More thermostable than ScFAE1	Triticale bran-destarched	1300	85	[42]
				Wheat bran-destarched	1721		[42]
				Steam-exploded wheat straw		89	[42]

[a]Fortified with xylanase; please refer to original paper [42] for details

reflected by the observation that ScFAE2 was capable of yielding up to 85 % of FA from destarched triticale bran, an improvement over TtFAE. In this study, we also used a steam-exploded wheat straw sample, where more than 85 % yield of FA or *p*-CA was obtained by ScFAE2. The steam-exploded sample came from a proprietary Vertical Linear Converter device, courtesy of BioVision Technology Inc.

An interesting and important highlight in this study came from the fact that useful chemicals, viz. 5-hydroxymethyl furfural and furfural were not decomposed as a result of enzymatic treatment but by alkali (Table 12.2). Lower concentrations of ScFAE2 were also used that led to the same conclusion [42].

We next explored the use of whole cell biocatalyst for the release of FA from wheat and triticale brans using mixtures of extracellular enzymes produced in culture by a dark mold *Alternaria alternata* strain FC007, originally isolated from Canadian wood log [45]. Conditions for the release of FA and reducing sugars from the two brans were evaluated. The highest level of FAE activity (89 ± 7 mU/mL fermentation culture) was obtained on the fifth day of fermentation on wheat bran as growth substrate. Depending on biomass and processing condition, up to 91.2 or 72.3 % of FA was released from wheat bran and triticale bran, respectively, indicating the proficiency of *A. alternata* extracellular enzymes in plant cell wall deconstruction. The apparent high extraction of FA from wheat and triticale brans signifies an advantage of using a whole fungal cell enzyme complement as previously seen in other systems, e.g., the filamentous fungus *Neosartorya spinosa* NRRL185 was able to release 99.5 % FA from corn fibers [46], and the crude enzyme extract from *Penicillium brasilianum* released over 70 % of the available phenolic acids in a number of agricultural substrates [47]. The use of whole cell enzyme complement versus an artificial assembly of enzymes likely offers the advantage of low-cost processing.

The suboptimal enzymatic hydrolysis in the release of FA has been encountered in other agricultural feedstocks, e.g., some 35 % recovery of FA from corn bran [46] despite corn residues (bran and fibers) being the best possible source of FA

Table 12.2 Chemical content of initial aqueous steam-exploded wheat straw fraction: enzymatic versus chemical treatment

Chemical	Condensate 1 (ppm)	NaOH-treated 55 °C, 5 h (ppm)	ScFAE2 90 mU/mL 24 h (ppm)
5-Hydroxymethylfurfural	77.5 (100 %)	n.d.	71.9 (93 %)
Furfural	342.4 (100 %)	2.4	270.6 (79 %)
Vanillin	9.5	19.1 (100 %)	9.6 (1 %)
Syringic acid	n.d.	n.d.	n.d.
Syringaldehyde	n.d.	19.1	n.d.
p-Courmaric acid	n.d.	14.0 (100 %)	13.2 (94 %)
FA	9.2	40.9 (100 %)	37.5 (89 %)

n.d. not determined (below 2 ppm). Sinapic acid and cinnamic acid were n.d. throughout. Adapted from Ref. [42], with kind permission from Springer Science+Business Media (up to 3 % by weight). Factors limiting efficient extraction of FA by enzymatic action on various biomasses include the complexity of the plant materials—compositional and linkage differences in the arabinoxylans—so that the right choice of enzymes is important. Faulds et al. [48] reported that the source of xylanase is important as well as the type of FAE used. For instance, family 10 xylanases with specificity toward the substituted or decorated region of the arabinoxylan backbone in conjunction with *Aspergillus niger* FAEA were more effective in releasing the 5,5′ form of diferulic acid from arabinoxylan derived from brewers' spent grain. On the other hand, family 11 xylanases cleaving the unsubstituted regions of the arabionoxylan backbone preferentially released FA.

12.4 Vinyl Phenols Production and Its Value Chain

The decarboxylation of phenolic acids to the corresponding vinyl phenols is commercially important since vinyl guaiacol (VG; 2-methoxy-4-vinyl phenol) for example is a GRAS compound that has been extensively used as a flavoring agent besides the fragrance and perfume industry. As a styrene-type molecule, VG can be polymerized to biodegradable oligomer like poly(3-methoxy-4-hydroxystyrene) [49]. This aspect of the work is elaborated in Sect. 12.7.1. (Fig. 12.2). VG is priced some 40 times more than its precursor, FA, and it can be biotransformed further to acetovanillone, ethylguaiacol, and vanillin [24, 50].

Although chemical decarboxylation of FA is the most widely used method for preparing VG or other styrene-type compounds, a typical reaction involves heating under reflux at 200–300 °C for 4–5 h in quinoline in the presence of copper powder [51]. Microwave-assisted and base-catalyzed decarboxylation reactions of FA have also been used [52, 53].

12.4.1 Phenolic Acid Decarboxylase (PAD)

A cofactor-free ferulic acid decarboxylase (Fdc; generic name, PAD) that effected the biotransformation of VG from FA was first purified from *Bacillus pumilus*

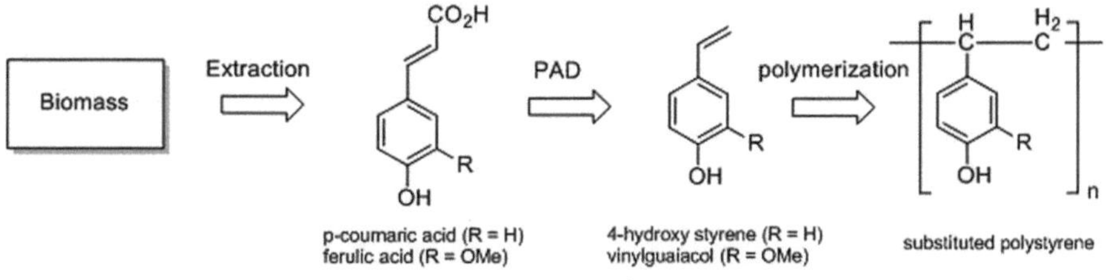

Fig. 12.2 Aromatic acid value chain. Modified from Ref. [61]

(formerly *Pseudomonas fluorescens*) strain UI-670 [24, 54]. The substrate specificity of PAD includes *p*-CA (4-hydroxycinnamic acid) but not other cinnamic acid derivatives without a free hydroxy group *para* to the unsaturated side chain. Sinapic acid (SA, 4-hydroxy-3,5-methoxycinnamic acid) was not a suitable substrate until protein engineering as discussed in a later section. For the decarboxylation of FA to VG, evidence for a non-oxidative mechanism involving an initial enzymatic isomerization of FA to a quinoid intermediate that spontaneously decarboxylates to the vinyl derivative was provided by deuterium-exchange experiments [24, 55].

We cloned the *B. pumilus* strain UI-670 PAD-encoding gene and subsequently the crystal structure of PAD was reported [56, 57]. The active enzyme is a dimer both in the crystal and in solution; each polypeptide consists of 161-residues. The PAD ß-barrel structure with two α-helices resemble those of the lipocalin-fold proteins, which are generally not catalytic but transport or binding proteins of hydrophobic ligands. When compared to structurally related proteins, a narrow cleft or tunnel containing several conserved amino acids was observed as a potential ligand-binding site for PAD. This structural feature implies that the substrate enters and the product leaves through the same opening. Amino acid residues near the surface of the cleft and those lining the bottom of PAD have been identified some of which were chosen for protein engineering as seen below.

12.4.2 Sinapic Acid Decarboxylase (SAD)

SAD is actually a happy enzyme in the sense that this new enzyme has not been found in nature according to the reported literature but now created in the laboratory. One amino acid substitution at position 85 of the PAD sequence, abbreviated as I85A, conferred the enzyme the new ability to act on SA, a substrate that has an extra methoxy group at the 5-position of the phenyl ring compared to FA. Ile85 was one of the five amino acids in the predicted active site targeted for saturation mutagenesis (Fig. 12.3). Ile85,Tyr19, Val38, Val70, and a few others were found to line the tunnel of the enzyme, whereas Phe87 and Asn15, and His92 (latter two, not shown) were found near the surface of the tunnel [57]. It is interesting that an I85G replacement did not give as high an activity as I85A even though Gly is the smallest amino acid and it keeps the micro-environment hydrophobic. Replacement of the branched side chain of Ile to charged amino acids of Ser and Thr resulted in one-third activity of I85A [58].

Table 12.3 shows that although the evolved SAD gained the ability to act on SA, its catalytic competence toward FA was compromised by about sevenfold compared to the native PAD, largely due to a reduction in Kcat. However, the affinity of the native PAD and SAD toward FA, and that of SAD toward SA, are very close, in the range from 2.7 to 3.2 mM.

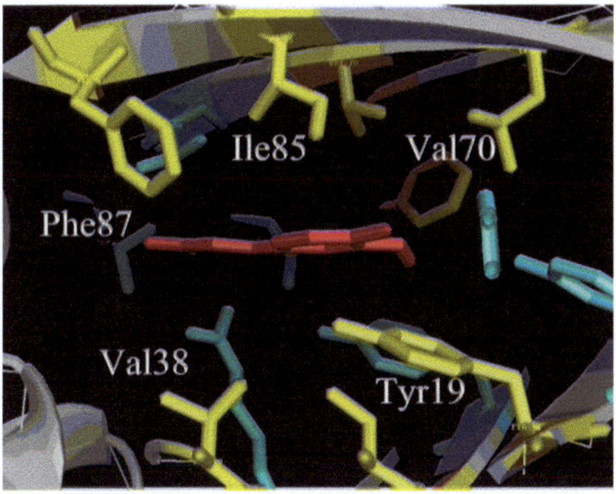

Fig. 12.3 Model of FA in the PAD active site. Reproduced from Ref. [58] with permission from The Royal Society of Chemistry

Table 12.3 Crude lysate activities of PAD and SAD and kinetic parameters of purified enzymes

Enzyme	Substrate	Specific activity (U/mg)	Vmax (U/mg)	Km (mM)	Kcat (s^{-1})	Kcat/Km (s^{-1} M^{-1})
PAD	FA	5.8	273.1	2.7 ± 0.8	182 ± 20	6.7×10^4
SAD	FA	2.3	46.4	3.2 ± 0.3	31 ± 8	9.7×10^3
SAD	SA	1.4	41.3	3 ± 1	28 ± 4	9.3×10^3

Less than 1 % conversion of SA was obtained with native PAD and so kinetic parameters could not be obtained. For details refer to [58]

12.5 Bioprocess Development for Enzymatic Decarboxylation of FA

Being water-insoluble and a light yellow clear liquid, VG is best isolated from a solvent system if not eluted off from an appropriate resin matrix. Lee et al. [59] carried out the decarboxylation of FA to VG by *B. pumilus* NRRL 14942 (DRV 52131) using hexane and a buffered phosphate system. The highest VG concentration achieved was about 10 g/L. But the product was not recovered. Yang et al. [56] investigated the in situ recovery of VG in a two-phase bioreactor system. Using octane as organic phase, a molar yield of 70 % (13.8 g VG of 98.4 % purity from 25 g of FA in a 2-L working volume) was achieved. In this two-phase system, the aqueous phase contains the growth media and the biocatalyst, natural, or recombinant whole cells [60].

Leisch et al. [61] revisited the use of the two-phase bioreactor system for the PAD-catalyzed decarboxylation of FA exploring a great variety of parameters—cell-free extract, whole cells, the use of resins (e.g., Amberlite XAD 4, Amberlite XAD 8, Lewatit VP OC 1064, Dowex Optipore L493, Sepabeads SP850

Table 12.4 Comparison of bioprocesses and their efficiencies in extracting VG from FA

Bioprocess	FA (g/L)	Time (h)	Yield (%)	Space time yield (g/L h)	Productivity/L cells OD $_{600}$ 35 (g/L)
Whole cells—biphasic (toluene)	12.5	2	70	3.39	13.8
Alginate immobolized cells—continuous reactor	8	96	49	0.03	40.3
Cell-free extract (2U/mL)	20	4	36	1.4	24.2
Cell-free extract (0.75 U/mL)—biphasic toluene (4 h)	25	4	93	4.83	224.1
Cell-free extract (2 U/mL)—resin	20	4	72	2.78	48.3

Refer to [61] for details

and Diaion HP20), alginate beads and type of solvents, etc. To maximize the source of starting material corn bran which contains the most amount of FA (3 % w/w) was used.

Table 12.4 shows that the use of cell-free extract and a biphasic system involving toluene was most productive among the various processes examined that included the use of resins in an attempt to circumvent the usage of organic solvents. In brief, nonionic resins such as Sepabeads SP850 gave the highest yield of VG (72 %), followed by Lewatit VPOC1163 of 68 %. Amberlite XAD 4 and 8 rendered between 51 % and 54 % yield (not shown). Nonetheless, among the processes examined, the use of resins as a possible green alternative to the use of toluene in a biphasic system came out second (48.3 g/L) in terms of productivity/L cells (Table 12.4).

12.6 Bioprocess Development for Enzymatic Decarboxylation of SA

Vinyl syringol (VS), the decarboxylated product of SA, aka canolol, has potent oxygen radical scavenging activity like its precursor or FA. The term canolol was coined by Wakamatsu et al. [63] when they first found it from roasting of canola seed. The optimum temperature for VS formation (720 μg/g) in this process was 160 °C [64]. Besides roasting there are chemical methods or microwave-induced decarboxylation in the presence of base. These methods include heating SA in quinoline/copper salt at 200 °C or thermal treatment in dimethylformamide at 130 °C using sodium acetate as a catalyst. Microwave methods include irradiation in the presence of the base 1,8-diazabicyclo[5.4.0]undec-7-ene (DBU) or irradiation in the ionic liquid [hmim]Br using sodium hydrogen carbonate as a mild base. VS has also been synthesized from syringaldehyde via Wittig olefination in hexane/tetrahydrofuran, or via simultaneous condensation–double decarboxylation of syringaldehyde and malonic acid under microwave irradiation. References cited in [58].

SA is the main phenolic acid in canola, a cultivar of rapeseed. In canola meal (by-product after oil is extracted), SA esters (sinapine and glucopyransol sinapate) make up 90–95 % of the total meal phenolics (10–12 mg/g), whereas the amount of free SA is less than 10 % (0.3–0.4 mg/g). Sinapine and glucopyranosyl sinapate, like SA all have significant antioxidative properties.

At first, preparative scale decarboxylation of SA with SAD in aqueous buffer (100 mM sodium phosphate buffer; pH 7.0) in the presence of 10.6 U/mL of crude lysate was carried out that resulted in 95 % conversion at a substrate concentration of 5 g/L (Table 12.5). The product VS was easily isolated in high yield (75 %) and purity (99 %) using ethyl acetate extraction of the aqueous medium at acidic pH, and a slightly basic buffer (pH 8) wash of the combined organic extracts to remove unreacted SA. The identity of VS was confirmed by 1H NMR spectrum in comparison to previously published spectra of VS.

In aqueous buffer, higher substrate concentrations (>5 g/L), or lower protein concentrations resulted in incomplete reactions. At 1 g/L VG was inhibiting the enzyme by about 66 %. In the case of VG even a lower concentration (0.25 g/L) impacted the decarboxylation of FA.

Since the addition of toluene as an immiscible organic solvent for in situ product removal significantly improved the efficiency of the FA to VG process, the decarboxylation of SA catalyzed by SAD in biphasic reactions was also investigated. Among the three solvents (Table 12.5) toluene was the best showing both high-reaction conversion and isolated yield. A 97 % conversion was observed after 16 h and an 81 % isolation yield in the case of 50 g/L SA catalyzed by 4.0 U/mL SAD. Although the biotransformation went to completion in ionic liquid (BMIM) PF6 (1-butyl-3-methylimidazol-3-ium hexafluorophosphate), we could not evaporate the solvent due to its viscosity. Also we were not able to isolate the product by solvent extraction due to its high solubility.

Table 12.5 Decarboxylation of SA to VS in buffer and biphasic media

SA (g/L)	SAD (U/mL)	Cosolvent	% Conversion	% Isolated yield
5	10.6	None	95	75
25	2.0	None	37	30
25	2.0	Toluene	99	83
50	2.0	Toluene	50	45
50	4.0	Toluene	97	81
25	2	Ethyl acetate	42	n.d.
25	2	Hexane	30	n.d.
25	2	(BMIM)PF_6	99.5	n.d.

Adapted from Ref. [58] with permission from The Royal Society of Chemistry
SAD was used as a crude lysate; Ratio of cosolvent to buffer was 1:1 (v/v); n.d. not determined. % Conversion was determined by HPLC

12.7 Scale-up Experiments

12.7.1 Extraction of FA and p-CA from Corn Bran

We wanted to showcase the utility of our technologies to the preparation of polyvinylphenol derivatives from biomass directly (Fig. 12.2). Due to the relatively high content of FA and *p*-CA compared to other feedstock, corn bran was chosen as a starting material [24, 50]. At first, we investigated the action of PAD in a biphasic reaction mixture on hydrolysates of corn bran prepared by alkaline hydrolysis followed by adjustment of the pH. This procedure was meant to avoid the isolation of phenolic acids thereby simplifying the reaction setup. But, it turned out that the vinyl derivatives were isolated in low purity due to contamination with fatty acids from the biomass. Hence, the purification of the desired vinyl phenols would be difficult. Therefore, we resorted to a more classical approach, whereby the biomass was treated with 2 N NaOH, centrifuged, extracted with ethyl acetate, acid–base washed and recrystallized from ethyl acetate and hexanes This gave a mixture of FA and *p*-CA in a ratio of 10:1 in high purity. Starting from 100 g of milled corn bran, we were able to isolate 1.90 g of a mixture of phenolic acids which accounts for 72 % of total alkali extractable phenolic acids (2.65 g/100 g corn bran). Decarboxylation of the mixture of phenolic acids was achieved by applying the optimized conditions for the enzymatic decarboxylation using a cell-free extract of PAD (0.75 U/mL) in combination with toluene for in situ product recovery. After a reaction time of 4 h, a mixture of VG and 4-hydroxy vinyl phenol in a ratio of 10:1 was recovered in good yield (0.99 g, 75 %) and in high purity (<95 %).

12.7.2 Extraction of SA from Canola Meal

Canola is the most profitable oilseed crop for Canadian farmers. The main canola products are oil for human consumption and meal for animal feed besides a wide variety of other by-products. Canada produces some 4 million tonnes of canola meal annually [65].

To explore the potential of producing VS as a value-added product from canola meal we investigated the enzymatic decarboxylation of canola meal extracts (Fig. 12.4). Alkaline treatment of 100 g of canola meal with 1 M NaOH for 18 h at 25 °C extracted 780 mg of SA (7.8 mg/g) as determined by HPLC. We predicted a maximum concentration of 8.8 mg/g based on previously reported concentrations of sinapine, sinapoyl glucose and SA in canola meal [66]. As in the case of VG production from corn bran an initial attempt using in situ enzymatic decarboxylation on the neutralized canola meal extracts rendered a low purity VS product that was heavily contaminated with fatty acids. To remedy this, we purified the canola meal extracts by defatting (washing) with hexane at pH 2 and then extracted the SA with diethyl ether–ethyl acetate. The isolated SA was used without further

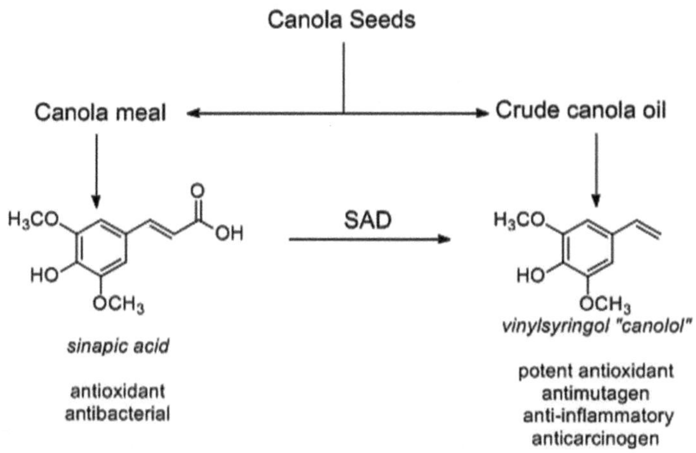

Fig. 12.4 Bioconversion of sinapic acid to canolol and their sources. Adapted from Ref. [58] with permission from The Royal Society of Chemistry

purification and decarboxylation with SAD in a biphasic reaction yielded 340 mg of VS in high purity (>90 %), which corresponded to an overall yield of 3.0 mg VS/g of canola meal. This represents 42 % of the available SA under the extraction conditions.

12.8 Polymerization of Vinyl Phenols to Polyvinylphenols and Their Solid State Properties

Using conventional chemistry, both cationic and radical polymerization conditions were set up to evaluate the possible polymers that could be obtained from VG and mixtures of VG and *p*-CA (Fig. 12.5) [61, 62]. The radical reaction made use of azobisisobutyronitrile (AIBN) and related compounds, whereas cationic polymerization used boron trifluoride diethyl etherate [BF3·(Et2O)2] as well as other derivatives.

Only the main results are summarized here. Boron trifluoride etherate gave a smooth conversion of VG to PVG and at low temperature (−40 °C) and it allowed the preparation of polymers with higher molecular weights (number average molecular weight, Mw = 12,077; and weight average molecular weight, Mn = 27,149).

Cationic polymerization at low temperature also yielded a mixed polymer consisting of VG and 4-vinylphenol (4-hydroxystyrene) in 97 % isolated yield. The ratio of monomer units in the polymer was determined by ^{31}P-NMR and was in close agreement with ratio of the applied mixture indicating that the copolymerization of VG and 4-hydroxystyrene proceeded smoothly. Size exclusion chromatography confirmed the existence of a polymer with a Mn of 10,543 and a polydispersity of 2.15.

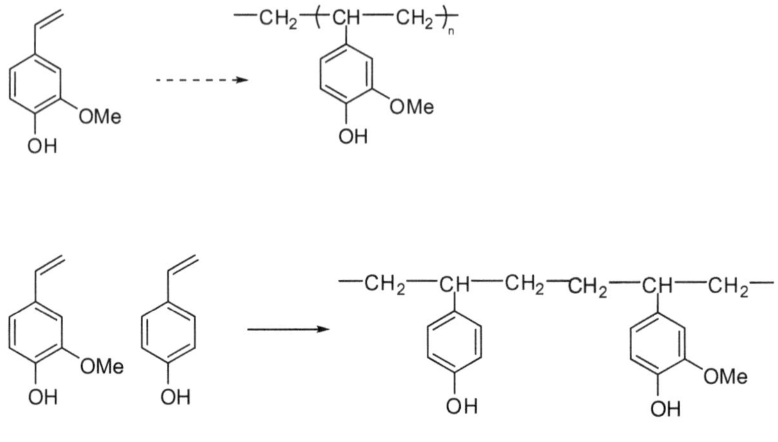

Fig. 12.5 Schematic of polyvinylguaiacol formation (*top*) and mixed polyvinylguaiacol (*bottom*)

Following this success, cationic conditions were also carried out to polymerize VS and explore the potential industrial applications of PVS. Treatment with boron trifluoride yielded a polymer with a Mn of 10,764 and a polydispersity of 1.4 [58].

12.8.1 Thermal Properties

Thermogravimetric analysis (TGA) using differential scanning calorimetry (DSC) observed that degradation of the two PVG polymers and the mixed PVG polymer began around 350 °C which is a good thermal characteristic but slightly lower than a commercial polystyrene (PS 1301; NOVA Chemicals) as a reference. The additional methoxy and hydroxy groups on the three PVG polymers may lend to the lower thermal stability since they show higher reactivity. DSC analysis showed that these polymers are amorphous with *Tg* of 86–92 °C for PVG_{rad}, 106–112 °C for PVG_{cat}, and 117–123 °C for PVG_{mixed}. This *Tg* temperature range is similar to that of the high molecular weight polystyrene and poly (methyl methacrylate), PMMA. The PVG_{cat} and PVG_{mixed} polymers are similar in Mn and polydispersity, however, the 9:1 ratio of guaiacol: 4-hydroxy phenol monomer units in the PVG_{mixed} polymer has contributed to a higher *Tg*.

Thermal analysis of the PVS polymer revealed a *Tg* range of 106–116 °C, determined by DSC, and a thermal stability of up to 350 °C. Hatakeyama and coauthors reported a *Tg* of 108 °C for a PVS polymer obtained via hydrolysis of poly(4-acetoxy-3,5-dimethoxystyrene) [49]. The Mw and polydispersity of the polymer were 9.9×10^4 and 3.7, respectively.

Table 12.6 Nanoindentation tests of PVG and PVS polymers

Polymer	Modulus, *E* (GPa)	Hardness, *H* (GPa)
PVG_{rad}	6.66 ± 0.05	0.421 ± 0.003
PVG_{cat}	6.75 ± 0.05	0.463 ± 0.006
PVG_{mixed}	6.5 ± 0.1	0.384 ± 0.012
PVS	5.22 ± 0.05	0.283 ± 0.003
PS 1301	4.15 ± 0.02	0.249 ± 0.001

Adapted from Ref. [58] with permission from The Royal Society of Chemistry

12.8.2 Indentation Hardness Tests

Nanoindentation characterization of the produced PVG and PVS polymers performed on polymer films compressed at 150 °C following the CSM (continuous stiffness measurement) is shown in Table 12.6. The two PVG's displayed similar hardness (*H*) and modulus (*E*) values that are significantly higher than that of the commercial polystyrene, PS 1301 that was analyzed simultaneously for comparison. The presence of the methoxy and hydroxyl groups appeared to be responsible for the improved performance. The mixed PVG polymer, containing a 9:1 ratio of guaiacol:4-hydroxy phenol monomer units, had slightly lower modulus and hardness than the two PVG polymers but significantly higher performance than the PS 1301 polymer.

The PVS polymer also afforded *H* and *E* values that were higher than the performance measured on the commercial polystyrene but lower than those of PVG polymers. It appeared that altering the number and location of ring substituents can significantly influence the polymer properties.

12.9 Miscellaneous Related Work

In this section some related work to the enzyme system and value chain described above or extension thereof are briefly discussed. This is not meant to be comprehensive and so we apologize to authors for any glaring omission.

From the known 3D-structure of a ferulic acid decarboxylase (FADase; PAD in our nomenclature) from *Enterobacter* sp. Px6-4, a triple mutant FADase (F95L/D112N/V151I) derived by random and site-directed mutagenesis was found to have some 34-fold enhanced activity toward FA for the production of VG [67, 68]. No product was isolated except by HPLC analysis. By docking study (for reason unknown the structure of a *p*-CA from *Lactobacillus plantarum* [69] was used as model), the increased activity of the triple mutant was postulated to be due to the formation of a more compact and stable binding mode in the active site compared to that of the wildtype enzyme [68]. From the crystal structure of FADase [67] and by site-directed mutagenesis it was found that during decarboxylation of

FA, amino acids W25 and Y27 were required for the entering and proper orientation of the substrate while E134 and N23 participated in proton transfer in the proposed mechanism that involves the formation of a quinoid intermediate [24].

In a non-mutagenic way, a clone of PAD from *Bacillus amyloliquefaciens* (BAPAD) in *E. coli* found most active toward *p*-CA by its conversion to *p*-hydroxystyrene (*p*-HS) was assessed in a biphasic bioreactor system [70]. Among various solvents and other parameters considered, 1-octanol:phosphate buffer (50 mM, pH 7.0) in a 3:1 ratio rendered the most yield (88.7 %), and gave a 1.34 g/h/g DCW productivity at 300 mM *p*-CA substrate concentration. The latter substrate concentration appeared to be the upper limit of solubility at the biphasic interface. It is noteworthy that this study confirmed our previous report of the use of 1-octanol as the best organic solvent for the extraction of VG from FA in a biphasic system that resulted in 70 % yield and a product of 98.4 % purity [56]. No extraction of *p*-HS was reported in [70]. Moreover, although BAPAD was also active toward FA (70 % relative to *p*-CA), its conversion to VG was not investigated.

Salgado et al. [71] reported on a 45 % bioconversion of *p*-CA to 4-VP (994 mg/L from 2223 mg/L) involving corn cob as starting material and an over-expressed PAD from *Lactobacillus plantarum* CECT 748^{T} as biocatalyst. In this particular case, the hydrolysate obtained after "pre-hydrolysis/alkaline hydrolysis" of corn cob was used as growth medium for the fermentation in a shaker format. Whereas in a 2 L bioreactor system, after 30 h of fermentation of the same type of hydrolysate, 878.4 mg 4-VP/L was achieved that translated to a 41.3 % conversion of the starting material. In relation to the conversion of FA to 4-VG this product was hardly produced although measurable. As in the previous study, the product of interest was not isolated.

Recently, Hu et al. [72] made use of a recombinant BLPAD, a PAD from *Bacillus licheniformis* CGMCC 7172, and reported high-conversion yields of 4-VP (97–near 100 %) from *p*-CA, and 4-VG (71 % – near 100 %) from FA at initial substrate concentrations of 200–500 mM in a biphasic system. In either case, an equal volume ratio of solvent to aqueous buffer (toluene for *p*-CA and cyclohexane for FA) was used. The organic solvent was chosen in comparison to a number of others of different partition coefficients that gave the highest relative activity against a control sample, i.e., without any added solvent. Similarly, the effect of various organic solvents on stability of BLPAD was reported at concentrations ranging from 20 to 50 %, v/v. This led to the purported conclusion that BLPAD is a highly solvent-tolerant enzyme. We reckon that one would consider the use of a counterpart "positive" control protein so as to provide a compelling piece of evidence for the current results. It is also not clear on what basis the original CGMCC strain was concluded as being solvent tolerant. One interesting notion of this work, however, was that beside *p*-CA, FA and caffeic acid as suitable substrates, the recombinant BLPAD was reported to be active toward SA with 0.3 % activity relative to that of *p*-CA.

Over the years there have been a series of publications that investigated the so-called "uphill" reaction of a number of PADs. This relates to the carboxylation side of the PAD reversible equation instead of the thermodynamically favored decarboxylation process. For details, readers are referred to a couple of representative reviews on carboxylation at large [73, 74]. To expand the scope of the usefulness of PAD, some recent progress on the subject are highlighted below.

In essence, in the presence of some 3 M potassium bicarbonate at pH 8.5 a number of PADs are capable of catalyzing the ß-carboxylation of the vinyl side chain of styrene derivative with yields of up to 40 % in the presence of 20 % acetonitrile [75] (Fig. 12.6). PADs from *Lactobacillus plantarum* (PAD_Lp) and *Bacillus amyloliquefaciens* (PAD_Ba) were the first two enzymes found capable of producing *p*-CA and FA from substrates *p*-VP and 4-VG, respectively, although the conversion rate was low and moderate (2–30 %) [76]. Subsequently, additional biocatalysts from *Methylobacterium* sp. (PAD_Ms), *Pantoea* sp. (PAD_Ps), *Lactoccocus lactis* (PAD_Ll) and PAD (FDC) from *Enterobacter* sp. (FDC_Es) but not *Mycobacterium colombiense* (PAD_Mc) were found to be able to hydrate a number of hydroxystyrene derivatives in an asymmetric and stereoselective manner [76] (Fig. 12.6).

An overarching significance of these studies is in the context of using carbon dioxide as C_1 carbon source and development of a "green" method for the synthesis of valuable organic compounds vis-à-vis the chemical Kolbe–Schmitt reaction that uses high temperature and pressure to produce aromatic hydroxy acid, viz., salicylic acid which also requires the use of sulfuric acid for its final production.

Kang et al. [77] assembled an artificial pathway in *E. coli* that introduced a *Bacillus amyloliquefaciens* PAD gene in an overexpression format to the endogenous L-tyrosine pathway, whereby the sequential phenolic acid intermediates viz. *p*-CA, caffeic acid and FA were transformed to 4-HS, 3,4-dihydroxystyrene, and 4-VG, respectively. The yields of these HSs were 355, 63, and 64 mg/L,

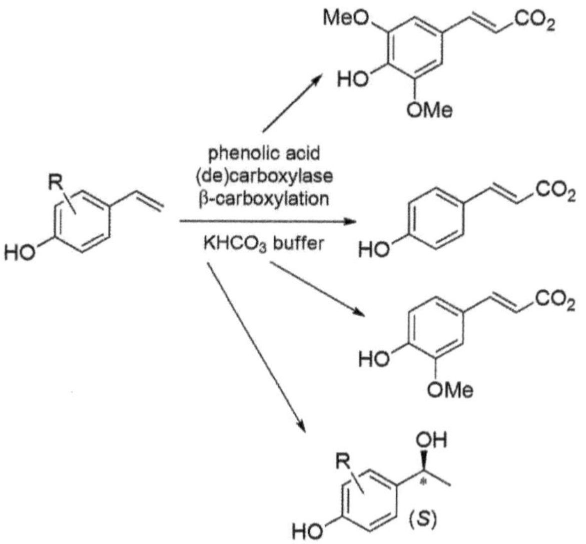

Fig. 12.6 Enzymatic carboxylation of hydroxystyrene derivatives using PAD. Selective data from Refs. [74, 76]

respectively, in shaking flasks after 36 h of cultivation. While this is interesting and the fact that the carbon source came from a simple glucose medium, it appears that the cytotoxicity of the products to the host that is not solvent tolerant needs to be addressed.

In another whole cell format, Furuya et al. [78] constructed a decarboxylase/oxygenase cascade system in an recombinant *E. coli* that involved a *Bacillus pumilus* PAD for the conversion of FA to VG followed by an oxygenase Cso2 from *Caulobacter segnis* to produce vanillin. Like PAD, Cso2 is coenzyme free. However, for an enhanced activity it was necessary to add 1 mM $FeCl_2$ to the cultivation medium since Cso2 requires iron as a prosthetic group. In an optimized system carried out in a 500 mL flask the *E. coli* cells expressing PAD was found to completely decarboxylate 75 mM of FA to 4-VG within 2 h at pH 9.0. Cell-free extract was then prepared by centrifugation, and following pH adjustment to the optimal pH of 10.5 for Cso2, the solution was subjected to oxygenation by Cso2 that effectively oxidized 4-VG to vanillin. In this enzyme cascade system, the concentration of vanillin was found to be 52 mM (7.8 g/L) in 24 h. To date this apparently was the best yield for the biotechnological production of vanillin using recombinant cells. It is noteworthy that a butyl acetate–water biphasic system was also investigated. However, under the various conditions tested, the best yield of vanillin was only 39 mM. These authors envisaged that an immobilized enzyme cascade system would possibly address the scale up issue.

Whereas VG can be polymerized to homo- or hetero-PVGs as seen in Sect. 12.8 beside the other VG-derived products, Zago et al. [79] recently described an epichlorohydrin-assisted glycidylation of 4-VG and canolol. Eugenol (4-allyl guaiacol) was another substrate. Cross metathesis coupling reaction of the glycidylated derivatives converted these phenolic compounds into a set of homo- and heterophenolic dimers in good yields and with a high diastereoselectivity. These diglycidylated diphenyl compounds that comprise two aromatic rings each bearing a methyl oxirane group, apparently are good candidates for the substitution of diglycidyl ether of bisphenol A, a notorious endocrine disruptor. A biobased alternative to such class of compound is an interesting proposition.

12.10 Optically Pure L-Lactic Acid from a Nonfood Crop

As a biodegradable aliphatic polyester, PLA (polylactic acid or polylactide), has spun considerable interest in the polymer industry and this momentum is increasing. The PLA market is projected to reach US $5.16 billion by 2020, growing at a compound annual growth rate (CAGR) of 20.9 %. The projected value of this monomeric building block, lactic acid (LA), is estimated at US $3.82 billion by 2020, growing at an equivalent CAGR of 18.6 % [80]. Besides PLA production, LA can be converted to commodity chemicals such as acetaldehyde, acrylic acid, and propionic acid [81, 82].

One of the trends driving PLA growth is an increased emphasis on sustainability, especially for consumer packaging, and a desire for renewably sourced raw materials. PLA is currently made from LA mostly derived from corn. Making PLA requires some 30–50 % less fossil fuel than polymers synthesized from hydrocarbons. As the price of oil has increased, renewably sourced polymers that use less fossil fuel have become economically attractive.

The growth of PLA shows that the market is ready for sustainable, economically attractive polymers. There is opportunity for new sources of LA that are from nonfood sources and have attractive costs to diversify the resource base and there is a market opportunity as well for additional new renewably sourced polymers with similar advantages of good physical properties, sustainability, and good economics.

12.10.1 L-Lactic Acid from Triticale Starch

Like its parent cereals wheat and rye, triticale is characterized by high starch and other carbohydrate content. Triticale grain dry matter contains 62.4–70.9 % starch, a content that is dependent upon cultivar and year. Triticale starch has the apparent advantage of low phenolic acids content that otherwise negatively impacts starch hydrolysis. In addition it has a lower gelatinization temperature, 65–68 °C, compared to those of barley, wheat, and corn, in the range of 72–75 °C [83 and references therein]. Triticale starch of high purity (<2.0 % w/w protein contamination) was kindly provided by Alberta Agriculture and Rural Development and University of Alberta. It was prepared from ground triticale grain flour using a wet-milling procedure that was initially developed for gluten-containing wheat grains. The isolation involved a "dough and wash" technology—preparing a dough or slurry with deionized water, followed by dilute alkaline washing and separation of the protein-enriched fiber fraction from starch milk by centrifugation, and then water washing [84].

Fermentative production of LA by microbes has several advantages, most importantly the production of optically pure L(+) or D(−)-LA instead of a racemic mixture obtained by chemical route. This subject has been adequately reviewed [81, 85–88]. The L(+)-isomer is the biologically important isomer and is the preferred substrate in most industrial applications [89]. Enantiomeric impurities even in minute quantity can drastically change the properties of the polymer, hence the quality of the monomers produced in the first place should be carefully controlled [89].

We explored the feasibility of LA production from triticale starch using a filamentous fungus, *Rhizopus oryzae* sb NRRL 29086, originally isolated from a dew-retted flax [90]. The *Rhizopus* genus together with *Lactobacillus* and *Bacillus* genera are considered commercially viable strains although each has its advantages and shortcomings. *Rhizopus* strains are known to generate L(+)-LA as a sole isomer of LA which is a big bonus. Other advantages include: the innate production of amylolytic enzymes for starch saccharification; requirement of few complex nutrients and only a small amount of inorganic salts for growth; tolerance to low pHs; and simpler downstream processing due to ease of separation of fungal biomass from the fermentation broth compared to bacterial biomass.

The inoculum size (1×10^4 spores/mL), temperature (28 °C), and correct timing and dosage of the neutralizing or titrating agent, calcium carbonate, to be added to maintain the pH of the medium at around 5 were all deemed important for the production of LA in batch fermentations. The best yield of 0.74 g LA/g starch was obtained in this direct fermentation. Under a set of optimized conditions in an automated control system of the DASGIP parallel bioreactor, the production of LA at a 2 L scale increased to 0.87/g triticale starch. Both yields are competitive with the direct fermentation of other agricultural materials, e.g., sweet potato starch and corn starch, where mycelia were used as inoculum and fermentations carried out in airlift or stirred bioreactors [83].

12.10.2 L-Lactic Acid from Triticale Flour

As starch is prepared from flour, converting the latter directly to LA would eliminate the steps involved in the extractions as outlined in Sect. 12.10.1 above. This also eliminates the use of hazardous or caustic chemical such as sodium hydroxide. Combined with *R. oryzae* in a single-stage simultaneous saccharification and fermentation process this feedstock is expected to be more economical than starch based. An example of economic analysis of LA production from wheat flour is provided by [91].

In a preliminary study, we conducted a 10 L working volume in a 14 L New Brunswick Bioreactor (Bioflow 110 system) that contained Vogel's medium at pH 5.0 and 6 % of triticale flour that was previously dissolved in hot water at 50 °C. To avoid foaming, 2 mL of antifoam Mazu 204 was added to the reactor before sterilization. Of special note: it appeared that foaming was a bigger problem when flour was used instead of starch due to the protein content. The inoculum was prepared with spores harvested from several plates and suspended in 5 mL sterile water. One 2.8 L Fernbach flask containing 900 mL of Vogel's medium with 3 % starch was inoculated with 1×10^4 *R. oryzae* spores/L (final concentration). The flask was incubated in a shaker for 43 h at 30 °C and 200 rpm. From this flask, 500 mL was used to inoculate the bioreactor.

The fermentation was performed at 28 °C. The dissolved oxygen was controlled by fixing the agitation rate at 250 rpm and the inlet-gas flow rate of 0.1 vvm (1 L/min). The pH was maintained at 5.0 ± 0.1 by addition of 28 % NH_4OH. Samples were taken twice a day, in the morning and at the end of the day for 72 h. The LA concentration was measured by HPLC.

LA was isolated via solvent extraction under acidic conditions using ethyl acetate. In total, 9 L of fermentation broth was extracted to afford 180 g of LA. A chiral analysis was performed via GC to determine enantiopurity. LA from the fermentation (as well as D, L-LA and L-LA standards from Sigma) were first derivatized with silylating agent, *N,O*-Bis(trimethylsilyl)trifluoroacetamide with

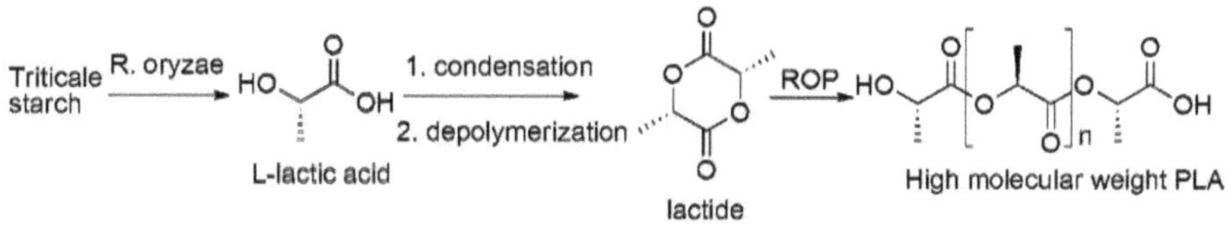

Fig. 12.7 Proposed L-lactic acid value chain from triticale starch

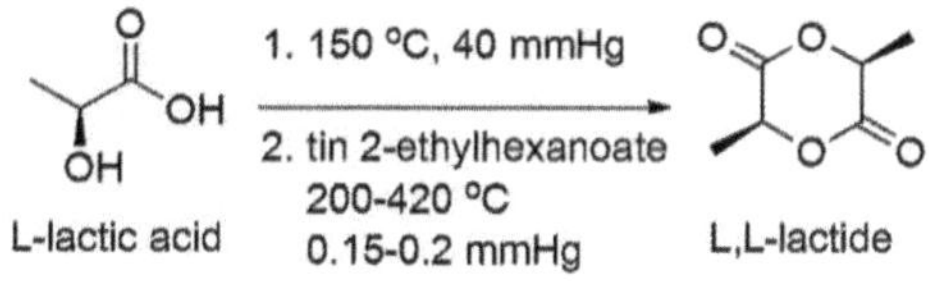

Fig. 12.8 Reaction conditions of L, L-lactide formation

trimethylchlorosilane (BSTFA + TMCS). Based on the chiral GC analysis, the LA produced was confirmed to be the L-enantiomer with >99.5 % ee purity.

A proposed L-lactic acid value chain from triticale starch or flour is shown in Fig. 12.7.

12.10.3 Synthesis of L, L-Lactide

Noda and Okuyama's procedure [92] for thermal depolymerization of L-PLA oligomer with tin (II) ethylhexanoate was followed (Fig. 12.8). Some experimental details are described below.

Solid L-LA (82.7 g, 0.918 mol) was slowly melted in a round bottom flask with attached reflux condenser at a pressure of 40 mmHg. The material was heated further to 150 °C for 8 h prior to cooling to room temperature overnight under a blanket of argon. Tin 2-ethylhexanoate (909 mg, 2.24 mmol, 1.10 wt%) was then added to this dark brown, viscous mixture, and with stirring the mixture was subjected to reduced pressure (0.15–0.20 mm Hg). The reaction was slowly heated and then the material was fractionally distilled. Fraction one was a clear and colorless oil (9.63 g, 73–80 °C) and fraction two (L, L-lactide, 45.4 g) was collected over a wide range of temperatures (200–420 °C) as a clear and colorless oil that crystallized upon cooling in the collection flask. The crude lactide (45.4 g) was recrystallized from warm ethyl acetate to yield 32.22 g (58 %) of white crystals: mp 84–86 °C (uncorrected, ethyl acetate); ^{1}H NMR ($CDCl_3$, 300 MHz) 1.64 (s, 3H), 1.67 (s, 3H), 5.04 (q, $J = 6.64$ Hz, 2H); > 99.5 % *ee* by chiral GC analysis: Varian Chirasil-Dex CB column (25 m × 0.25 mm × 0.25 μm); 120 °C for 5 min; 3 °C/min to 150 °C, 1 min at 150 °C, 10 °C/min to 200 °C, 2.5 min at 200 ° C: L-lactide = 12.2 min, D-lactide = 13.3 min.

The above reaction was carried out in batchwise, the best yield being 60 % that was considerably lower than the 90 % reported in the literature. However, analysis

Fig. 12.9 Optically pure L, L-lactide from triticale flour

of the pooled and recrystallized L, L-lactide by chiral GC showed >99.5 % ee. We have not carried out ring-opening polymerization (ROP) of the produced L, L-lactide (Fig. 12.9). But with such optical purity the quality of the resulting biopolymer is unlikely going to be disappointing [89, 93]. The challenge at hand is to optimize the production of the optically pure L-LA, a monomer of diverse applications other than the polymer industry [81].

Readers are referred to [94] for an analysis of "cradle-to-polymer-factory-gate life cycle" of Ingeo® polylactides, a range of PLA biopolymers that are made entirely from renewable resources, and produced since 2001 by NatureWorks LLC. In addition, a recent review of PLA production, processing techniques and applications, methods to tailor PLA properties, analysis of end-of-life scenarios and the environmental footprint is provided by Castro-Aguirre et al. [95].

12.11 Conclusion and Outlook

The type of value chain and creation presented in this chapter is just an example of myriad of products and opportunities that can be harnessed from renewal feedstock through the power and utility of biological methods that range from the use of simple naturally occurring enzymes, whole cells included, to genetically evolved ones even though by as few as one amino acid change to the polypeptide through rational design, and finally to the development of fermentation processes as a form of feasibility study. To make a more compelling case, isolation of the product and its characterization are essential instead of mere evidence of detection by analytical means.

The same approach applies to a given enzyme system. A priori in-depth knowledge of an enzyme via structure determination opens the door to its possible genetic improvement although directed evolution using approaches such as error-prone PCR or iterative mutagenesis have been shown to fare very well [96, 97]. Is it possible that nature has already provided a biocatalyst of interest without human intervention? Indeed this was the case when a naturally occurring thermostable biocatalyst obtained via genome mining was found to outperform experimentally evolved homologues of an enantioselective esterase that carried out the same hydrolysis. The former candidate has a melting temperature (T_m) increase of 10–12 °C compared to that of the mesophilic counterpart protein [98]. From that study there are at least two other lessons that are useful to this day: (i) judicious genome mining remains a fruitful strategy to discover new enzymes as illustrated by TtFAE [42] and ScFAE2 [44] for example, in the extraction of FA from biomass, and (ii) like the multistep synthetic pathway of the antibiotic levofloxacin [98], a combination of chemoenzymatic approach would appear attractive in the value chain of FA and SA extraction and transformation since this strategy combines proven economics of some unavoidable chemical processes with environmental care through biocatalysis. This echoes the call to "uninitiated" synthetic chemists to take up and combine biology and organic synthesis in their design of target compounds that would lead to increased efficiency in the overall process [99].

In a recent review of the use of FAE for biorefining of lignocellulosic biomass, a number of agricultural by-products such as brewers spent grain, sugarcane bagasse, rice bran, jojoba meal, etc., are listed as having significant potential for extraction of phenolic acids [37]. However, availability of feedstock is not a limitation. Rather, the bottleneck lies in the efficiency of a given method or combination of approaches to enable an overall economical process. Clearly there are many research gaps that need to be addressed by a multidisciplinary team and effort involving state-of-the-art technologies and engineering of all kind. Process engineering and system optimization at various stages of the value chain are key elements. By way of biocatalyst development, more robust characteristics and recyclability property of the biological entities need to be realized by immobilization technology, etc. A systems biology approach involving metabolic engineering for the production of various chemicals of interest is another consideration [100].

Whereas monomers like VG and SG are value-added chemicals in their own rights, for the first time, we showcased the production of polyvinylphenols (PVG, PVG_{mixed} and PVS) primarily derived from a biobased feedstock and they displayed improved performance compared to a commercial polystyrene [58, 61]. The various 4-HS derivatives may eventually be used for substituting significant amounts of styrene in the current polymer industries, similar to the first and second generation bioethanol to displace oil-derived fuels. The production of both VG and VS monomers and their polymerized products is anticipated to add significant value to a biorefinery setting by providing a biobased platform as well as addressing the engaging sustainability issue [101, 102].

Finally, there is a great lesson to be learned from the first commercial success of a "green" platform chemical, 1,3-propanediol (PDO) produced from glucose by metabolic engineering [103]. Besides being a building block for polyethers and polyurethanes, the best known PDO-based polyester is PTT (polytrimethylene terephthalate) of tradename Sorona 3GT. It is known that the commercially attractive titer of PDO production in *E. coli* took some seven years to achieve. It is little wonder that in a Perspective article by the Swiss Industrial Biocatalysis Consortium (SIBC) the authors quipped in its opening sentence: "It takes 20 years to become an overnight success." The authors were referring to the slow uptake of biotechnology in organic chemistry in general and in particular the use of biocatalysis in industrial organic synthesis considering the fact that enzymes have been used since time immemorial in cheese making for example [104].

Timeframe aside, the seeds of green chemistry have been sown and it is a matter of nourishing and nurturing in an appropriate field that will bear abundant fruits. The quality of life in a chemurgic and environmentally conscious framework is ensured.

Acknowledgments P.C.K.L. would like to thank the past members of his lab at the National Research Council Canada (NRCC) for their invaluable contributions to the work described here. Their names appeared in the cited references as well as those of numerous collaborators who provided samples or helped with analyses [42, 44, 45, 56–58, 61, 83]. Funding of various aspects of the work was provided by the Canadian Triticale Biorefinery Initiative (CTBI) of the Agricultural Bioproducts Innovation Program of Agriculture and Agri-Food Canada (AAFC), the National Bioproducts Program of NRCC-AFCC-NRCan (Natural Resource Canada), and NRC internal research programs. We also thank the Climate Change Technology and Innovation Biotechnology Program of the Canadian Biomass Innovation Network (CBIN) of NRCan for initial financial support. Current support by Tianjin "Thousand Talents" Program for Senior International Scientists to P.C.K.L. is gratefully acknowledged. The Fonds québécois de la recherche sur la nature et les technologies (FQRNT) Center for Green Chemistry and Catalysis is also acknowledged for partial support.

References

Note

For lack of reproducibility of results, the authors of reference 68 had made a retraction of their article in Appl Microbiol Biotechnol (2016) 100 (22):9807.

1. Carr M, Davies S, Locke B (2010) Job creation and market opportunities for biobased chemicals and products. Ind Biotechnol 6:74–77
2. Christensen CH, Rass-Hansen J, Marsden CC, Taaring E, Egeblad K (2008) The renewable chemicals industry. ChemSusChem 1:283–289
3. Gallezot P (2012) Conversion of biomass to selected chemical products. Chem Soc Rev 41:1538–1558
4. Zakzeski J, Bruinincx PCA, Jongerius AL, Weckhuysen BM (2010) The catalytic valorization of lignin for the production of renewable chemicals. Chem Rev 110:3552–3599
5. Besson M, Gallezot P, Pinel C (2014) Conversion of biomass into chemicals over metal catalysts. Chem Rev 114:1827–1870
6. Straathof AJJ (2014) Transformation of biomass into commodity chemicals using enzymes or cells. Chem Rev 114:1871–1908
7. Schwartz TJ, Shanks BH, Dumesic JA (2016) Coupling chemical and biological catalysis: a flexible paradigm for producing biobased chemicals. Curr Opin Biotechnol 38:54–62
8. Kawaguchi H, Hasunuma T, Ogino C, Kondo A (2016) Bioprocessing of bio-based chemicals produced from lignocellulosic feedstocks. Curr Opin Biotechnol 42:30–39

9. Chen Y, Nielsen J (2016) Biobased organic acids production by metabolically engineered microorganisms. Curr Opin Biotechnol 37:165–172
10. Lattanzio V (2013) Phenolic compounds: introduction. In: Ramawat KG, Merillon JM (eds) Natural products. Springer, Berlin, pp 1543–1580
11. Bhattacharya A, Sood P, Citovsky V (2010) The roles of plant phenolics in defence and communication during *Agrobacterium* and *Rhizobium* infection. Mol Plant Pathol 11:705–719
12. Middleton E, Kandaswami C, Theoharides TC (2000) The effects of plant flavonoids on mammalian cells: implications for inflammation, heart disease and cancer. Pharmacol Rev 52:673–839
13. Li C, Zhao X, Wang A, Huber GW, Zhang T (2015) Catalytic transformation of lignin for the production of chemicals and fuels. Chem Rev 115:11559–11624
14. Isikgor FH, Becer CR (2015) Lignocellulosic biomass: a sustainable platform for the production of bio-based chemicals and polymers. Polym Chem 6:4497–4559
15. Raskin I, Skubatz H, Tang W, Meeuse BJD (1990) Salicylic acid levels in thermogenic and nonthermogenic plants. Ann Bot 66:376–383
16. Rommel A, Wrolstad RE (1993) Influence of acid and base hydrolysis on the phenolic composition of red raspberry juice. J Agric Food Chem 41:1237–1241
17. Kakkar S, Bais S (2014) A review on protocatechuic acid and its pharmacological potential. ISRN Pharmacol. Article ID 952943
18. Chiu CS, Deng JS, Chang HY, Chen YC, Lee MM, Hou WC, Lee CY, Huang SS, Huang GJ (2013) Antioxidant and anti-inflammatory properties of Taiwanese yam [*Dioscorea japonica* Thunb. var. *pseudojaponica* (Hayata) Yamam.] and its reference compounds. Food Chem 141:1087–1096
19. Wen D, Li C, Di H, Liao Y, Liu H (2005) A universal HPLC method for the determination of phenolic acids in compound herbal medicines. J Agric Food Chem 53:6624–6629
20. Zhao, Y (ed) (2007) Chapter 3: Berry fruit Phytochemicals. In: Berry fruit: value-added products for health promotion. CRC Press, Boca Raton, pp 73–99
21. Barberousse H, Roiseux O, Robert C, Paquot M, Deroanne C, Blecker C (2008) Analytical methodologies for quantification of ferulic acid and its oligomers. J Sci Food Agric 88:1494–1511
22. Zhao Z, Egashira Y, Sanada H (2005) Phenolic antioxidants richly contained in corn bran are slightly bioavailable in rats. J Agric Food Chem 53:5030–5035
23. Liyama K, Lam Thi B-T, Stone BA (1994) Covalent cross-links in the cell wall. Plant Physiol 104:315–320
24. Rosazza JPN, Huang Z, Dostal L, Volm T, Rosseau B (1995) Review: biocatalytic transformations of ferulic acid: an abundant aromatic natural product. J Ind Microbiol 15:457–471
25. Graf E (1992) Antioxidant potential of ferulic acid. Free Radic Biol Med 13:435–448
26. Srinivasan M, Sudheer AR, Menon VP (2007) Ferulic acid: therapeutic potential through its antioxidant property. J Clin Biochem Nutr 40:92–100
27. Priefert H, Rabenhorst J, Steinbuchel A (2001) Biotechnological production of vanillin. Appl Microbiol Biotechnol 56:296–314
28. Kaur B, Chakraborty D (2013) Biotechnological and molecular approaches for vanillin production: a review. Appl Biochem Biotechol 169:1353–1372
29. Werpy T, Peterson G (2004) Top value added chemicals from biomass, volume I: results of screening for potential candidate from sugars and synthesis gas. US Department of Energy, Washington, DC. http://www.nrel.gov/docs/fy04osti/35523.pdf
30. Dodd DR, Gross RA (2007) Chemicals from biomass. Science 318:1250–1251
31. Faulds CB, Bartolomé B, Williamson G (1997) Novel biotransformations of agro-industrial cereal waste by ferulic acid esterases. Ind Crops Prod 6:367–374
32. Wong DWS (2006) Feruloyl esterase: a key enzyme in biomass degradation. Appl Biochem Biotechnol 133:87–112

33. Benoit I, Navarro D, Marnet N, Rakotomanomana N, Leasage-Messen L, Sigoilot JC, Asther M, Asther M (2006) Feruloyl esterases as a tool for the release of phenolic compounds. Carbohydr Res 341:1820–1827
34. Fazary AE, Ju YH (2007) Feruloyl esterases as biotechnological tools: current and future perspectives. Acta Biochim Biophys Sin 39:811–828
35. Koseki T, Fushinobu S, Ardiansyah SH, Komai M (2009) Occurrences, properties, and applications of feruloyl esterases. Appl Microbiol Biotechnol 84:803–810
36. Topakas E, Vafiadi C, Christakopoulos P (2007) Microbial production, characterization and applications of feruloyl esterases. Process Biochem 42:497–509
37. Gopalan N, Rodríguez-Duran LV, Saucedo-Castaneda G, Nampoothiri KM (2015) Review on technological and scientific aspects of feruloyl esterases: a versatile enzyme for biorefining of biomass. Bioresour Technol 193:534–544
38. Crepin VF, Faulds CB, Connerton IF (2004) Functional classification of the microbial feruloyl esterases. Appl Microbiol Biotechnol 63:647–653
39. Benoit I, Danchin EGJ, Bleichrodt RJ, de Vries RP (2008) Biotechnological applications and potential of fungal feruloyl esterases based on prevalence, classification and biochemical diversity. Biotechnol Lett 30:387–396
40. Oettler G (2005) The fortune of a botanical curiosity—triticale, past, present, and future. J Agric Sci 143:329–346
41. Bao Q, Tian Y, Li W, Xu Z, Xuan Z, Hu S, Dong W, Yang J, Chen Y, Xue Y, Xu Y, Lai X, Huang L, Dong X, Ma Y, Ling L, Tan H, Chen R, Yang H (2002) A complete sequence of the *T. tengcongensis* genome. Genome Res 12:689–700
42. Abokitse K, Wu M, Bergeron H, Grosse S, Lau PCK (2010) Thermostable feruloyl esterase for the bioproduction of ferulic acid from triticale bran. Appl Microbiol Biotechnol 87:195–203
43. Royter M, Schmidt M, Elend C, Höbenreich H, Schäfer T, Bornscheuer UT, Antranikian G (2009) Thermostable lipases from the extreme thermophilic anaerobic bacteria *Thermoanaerobacter thermohydrosulfuricus* SOL1 and *Caldanaerobacter subterraneus* subsp. *tengcongensis*. Extremophiles 13:769–783
44. Wu M, Abokitse K, Grosse S, Leisch H, Lau PCK (2012) New feruloyl esterases to access phenolic acids from grass biomass. Appl Biochem Biotechnol 168:129–143
45. Xiao Z, Bergeron H, Lau PCK (2012) *Alternaria alternata* as a new fungal enzyme system for the release of phenolic acids from wheat and triticale brans. Antonie Van Leeuwenhoek 101:837–844
46. Shin HD, McClendon S, Le T, Taylor F, Chen RR (2006) A complete enzymatic recovery of ferulic acid from corn residues with extracellular enzymes from *Neosartorya spinosa* NRRL185. Biotechnol Bioeng 95:1108–1115
47. Panagiotou G, Olavarria R, Olsson L (2007) *Penicillium brasilianum* as an enzyme factory; the essential role of feruloyl esterases for the hydrolysis of the plant cell wall. J Biotechnol 130:219–228
48. Faulds CB, Mandalari G, Lo Curto RB, Bisignano G, Christakopoulos P, Waldron KW (2006) Synergy between xylanases from glycoside hydrolase family 10 and family 11 and a feruloyl esterase in the release of phenolic acids from cereal arabinoxylan. Appl Microbiol Biotechnol 71:622–629
49. Hatakeyama H, Hayashi E, Haraguchi T (1977) Biodegradation of poly (3-methoxy-4-hydroxystyrene). Polymer 18:759–763
50. Mathew S, Abraham TE (2006) Bioconversions of ferulic acid, a hydroxycinnamic acid. Crit Rev Microbiol 33:115–125
51. Griengl H, Appenroth M, Dax K, Schwarz H (1969) Zur Umsetzung von Formaldehyd mit Hydroxystyrolen in alkalischem Medium. Monatsh Chem 100:316–327
52. Nomura E, Hosoda A, Mori H, Taniguchi H (2005) Rapid base catalyzed decarboxylation and amide-forming reaction of substituted cinnamic acids via microwave heating. Appl Microbiol Biotechnol 7:863–866

53. Bernini R, Mincione E, Barontini M, Provenzano G, Setti L (2007) Obtaining 4-vinylphenols by decarboxylation of natural 4-hydroxycinnamic acids under microwave irradiation. Tetrahedron 63:9663–9667
54. Huang Z, Dostal L, Rosazza JPN (1994) Purification and characterization of a ferulic acid decarboxylase from *Pseudomonas fluorescens*. J Bacteriol 1176:5912–5918
55. Huang Z, Dostal L, Rosazza JPN (1993) Microbial transformations of ferulic acid by *Saccharomyces cerevisiae* and *Pseudomonas fluorescens*. Appl Environ Microbiol 59:2244–2250
56. Yang J, Wang S, Lorrain M-J, Rho D, Abokitse K, Lau PCK (2009) Bioproduction of lauryl lactone and 4-vinyl guaiacol as value-added chemicals in two-phase biotransformation systems. Appl Microbiol Biotechnol 84:867–876
57. Matte A, Grosse S, Bergeron H, Abokitse K, Lau PCK (2010) Structural analysis of *Bacillus pumilus* phenolic acid decarboxylase, a lipocalin-fold enzyme. Acta Cryst F66:1407–1414
58. Morley KL, Grosse S, Leisch H, Lau PCK (2013) Antioxidant canolol production from a renewable feedstock via an engineered decarboxylase. Green Chem 15:3312–3317
59. Lee IY, Volm TG, Rosazza JPN (1998) Decarboxylation of ferulic acid to 4-vinylguaiacol by *Bacillus pumilus* in aqueous-organic solvent two-phase systems. Enzyme Microb Technol 23:261–266
60. Daugulis AJ (2001) Two-phase partitioning bioreactors: a new technology platform for destroying xenobiotics. Trends Biotechnol 19:457–462
61. Leisch H, Grosse S, Morley K, Abokitse K, Perrin F, Denault J, Lau PCK (2013) Chemicals from agricultural biomass: chemoenzymatic approach for production of vinylphenols and polyvinylphenols from phenolic acids. Green Process Synth 2:7–17
62. Kodaira K, Onishi Y, Ito K (1980) An oligomerization of 2-methoxy-4-vinyl phenol. Makromol Chem Rapid Commun 1:427–431
63. Wakamatsu D, Morimura S, Sawa T, Kida K, Nakai C, Maeda H (2005) Isolation, identification, and structure of a potent alkyl-peroxyl radical scavenger in crude canola oil, canolol. Biosci Biotechnol Biochem 69:1568–1574
64. Spielmeyer A, Wagner A, Jahreis G (2009) Influence of thermal treatment of rapeseed on the canolol content. Food Chem 112:944–948
65. http://www.canolacouncil.org/markets-stats/industry-overview/
66. Krygier K, Sosulski F, Hogge L (1982) Free, esterified, and insoluble-bound phenolic acids. 1. Extraction and purification procedure. J Agric Food Chem 30:330–334
67. Gu W, Yang J, Lou Z, Liang J, Sun Y, Huang J, Li X, Cao Y, Meng Z, Zhang KQ (2011) Structural basis of enzymatic activity for the ferulic acid decarboxylase (FADase) from *Enterobacter* sp. P6–4. PLoS ONE 6:e16262
68. Lee H, Park J, Jung C, Han D, Seo J, Ahn J-H, Chong Y, Hur H-G (2015) Enhancement of the catalytic activity of ferulic acid decarboxylase from *Enterobacter* sp. P6–4 through random and site-directed mutagenesis. Appl Microbiol Biotechnol 99:9473–9481
69. Rodriguez H, Angulo I, De Las RB, Campillo N, Paez JA, Munoz R, Mancheno JM (2010) *p*-Coumaric acid decarboxylase from *Lactobacillus plantarum*: structural insights into the active site and decarboxylation catalytic mechanism. Proteins 78:1662–1676
70. Jung D-H, Choi W, Choi KY, Jung E, Yun H, Kazlauskas RJ, Kim B-G (2013) Bioconversion of *p*-coumaric acid to p-hydroxystyrene using phenolic acid decarboxylase from *B. amyloliquefaciens* in biphasic reaction system. Appl Microbiol Biotechnol 97:1501–1511
71. Salgado JM, Rodríguez-Solana R, Curiel JA, de las Rivas B, Muñoz R, Domínguez JM (2012) Production of vinyl derivatives from alkaline hydrolysates of corn cobs by recombinant *Escherichia coli* containing the phenolic acid decarboxylase from *Lactobacillus plantarum* CECT 748T. Bioresour Technol 117:274–285
72. Hu H, Li L, Ding S (2015) An organic solvent-tolerant phenolic acid decarboxylase from *Bacillus licheniformis* for the efficient bioconversion of hydroxycinnamic acids to vinyl phenol derivatives. Appl Microbiol Biotechnol 99:5071–5081

73. Glueck SM, Guemues S, Fabian WMF, Faber K (2010) Biocatalytic carboxylation. Chem Soc Rev 39:313–328
74. von Borzyskowski LS, Rosenthal RG, Erb TJ (2013) Evolutionary history and biotechnological future of carboxylases. J Biotechnol 168:243–251
75. Wuensch C, Pavkov-Keller T, Steinkellner G, Gross J, Fuchs M, Hromic A, Lyskowski A, Fauland K, Gruber K, Glueck SM, Faber K (2015) Regioselective enzymatic ß-carboxylation of *para*-hydroxystyrene derivatives catalyzed by phenolic acid decarboxylases. Adv Synth Catal 357:1909–1918
76. Wuensch C, Gross J, Steinkellner G, Gruber K, Glueck SM, Faber K (2013) Asymmetric enzymatic hydration of hydroxystyrene derivatives. Angew Chem Int Ed 52:2293–2297
77. Kang S-Y, Choi O, Lee JK, Ahn J-O, Ahn JS, Hwang BY, Hong Y-S (2015) Artificial de novo biosynthesis of hydroxystyrene derivatives in a tyrosine overproducing *Escherichia coli* strain. Microb Cell Fact 14:78
78. Furuya T, Miura M, Kuroiwa M, Kino K (2015) High-yield production of vanillin from ferulic acid by a coenzyme-independent decarboxylase/oxygenase two-stage process. New Biotechnol 32:335–339
79. Zago E, Dubreucq E, Lecomte J, Villeneuve P, Fine F, Fulcrand H, Aouf C (2016) Synthesis of bio-based epoxy monomers from natural allyl- and vinyl phenols and the estimation of their affinity to the estrogen receptor α by molecular docking. New J Chem 40:7701–7710
80. http://www.marketsandmarkets.com/Market-Reports/polylacticacid-387.html
81. Wee YJ, Kim JN, Ryu HW (2006) Biotechnological production of lactic acid and its recent applications. Food Tech Biotechnol 44:163–172
82. Katryniok B, Paul S, Dumeignil F (2010) Highly efficient catalyst for the decarbonylation of lactic acid to acetaldehyde. Green Chem 12:1910–1913
83. Xiao Z, Wu M, Beauchemin M, Groleau D, Lau PCK (2011) Direct fermentation of triticale starch to lactic acid by *Rhizopus oryzae*. Ind Biotechnol 7:129–135
84. Kandil A, Li J, Vasanthan V, Bressler DC, Tyler RT (2011) Compositional changes in whole grain flours as a result of solvent washing and their effect on starch amylolysis. Food Res Int 44:167–173
85. Nampoothiri KM, Nair NR, John RP (2010) An overview of the recent developments in polylactide (PLA) research. Bioresour Technol 101:8493–8501
86. Okano K, Tanaka T, Ogino C, Fukuda H, Kondo A (2010) Biotechnological production of enantiomeric pure lactic acid from renewable resources: recent achievements, perspectives, and limits. Appl Microbiol Biotechnol 85:413–423
87. John RP, Anisha GS, Nampoothiri KM, Pandey A (2009) Direct lactic acid fermentation: focus on simultaneous saccharification and lactic acid production. Biotechnol Adv 27:145–152
88. Abdel-Rahman MA, Tashiro Y, Sonomoto K (2011) Lactic acid production from lignocellulose-derived sugars using lactic acid bacteria: overview and limits. J Biotechnol 156:286–301
89. Inkinen S, Hakkarainen M, Albertsson A-C, Sodergard A (2011) From lactic acid to poly (lactic acid) (PLA): characterization and analysis of PLA and its precursors. Biomacromolecules 12:523–532
90. Henriksson G, Akin DE, Hanlin RT, Rodriguez C, Archibald DD, Rigsby LL, Eriksson KL (1997) Identification and retting efficiencies of fungi isolated from dew-retted flax in the United States and Europe. Appl Environ Microbiol 63:3950–3956
91. Akerberg C, Zacchi G (2000) An economic evaluation of the fermentative production of lactic acid from wheat flour. Bioresour Technol 75:119–126
92. Noda M, Okuyama H (1999) Thermal catalytic depolymerization of poly(L-lactic acid) oligomer into LL-lactide: Effects of Al, Ti, Zn and Zr compounds as catalysts. Chem Pharm Bull 47:467–471
93. Pang X, Zhuang X, Tang Z, Chen X (2010) Polylactic acid (PLA): research, development and industrialization. Biotechnol J 5:1125–1136

94. Vink ETH, Davies S, Kolstad JJ (2010) The eco-profile for current Ingeo® polylactide production. Ind Biotechnol 6:212–224
95. Castro-Aquirre E, Iniquez-Franco F, Samsudin H, Fang X, Auras R (2016) Poly(lactic acid) —mass production, processing, industrial applications, and end of life. Adv Drug Deliv Rev. doi:10.1016/j.addr.2016.03.010
96. Wang M, Si T, Zhao H (2012) Biocatalyst development by directed evolution. Bioresour Technol 115:117–125
97. Roiban GD, Reetz MT (2015) Expanding the toolbox of organic chemists: directed evolution of P450 monooxygenases as catalysts in regio- and stereoselective oxidative hydroxylation. Chem Commun (Camb) 51:2208–2224
98. Grosse S, Bergeron H, Imura A, Boyd J, Wang S, Kubota K, Miyadera A, Sulea T, Lau PCK (2010) Nature versus nurture in two highly enantioselective esterases from *Bacillus cereus* and *Thermoanaerobacter tengcongensis*. Microb Biotechnol 3:65–73
99. Hudlicky T, Reed JW (2009) Applications of biotransformations and biocatalysis to complexity generation in organic synthesis. Chem Soc Rev 38:3117–3132
100. Vargas-Tah A, Gosset G (2015) Production of cinnamic and *p*-hydroxycinnamic acids in engineered microbes. Front Bioeng Biotechnol 3:116
101. Tuck CO, Perez E, Horvath IT, Sheldon RA, Poliakoff M (2012) Valorization of biomass: deriving more value from waste. Science 337:695–699
102. Llevot A, Dannecker PK, von Czapiewski M, Over LC, Soyler Z, Meier MA (2016) Renewability is not enough: recent advances in the sustainable synthesis of biomass-derived monomers and polymers. Chem Eur J 22:11510–11521
103. Nakamura CE, Whited GM (2003) Metabolic engineering for the microbial production of 1,3-propanediol. Curr Opin Biotechnol 14:454–459
104. Meyer H-P, Eichhorn E, Hanlon S, Lutz S, Schurmann M, Wohlgemuth R, Coppolecchia R (2013) The use of enzymes in organic synthesis and the life sciences: perspectives from the Swiss Industrial Biocatalysis Consortium (SIBC). Catal Sci Technol 3:29–40

Chapter 13
Regional Pillars of Competitiveness in Chemurgy and Green Chemistry

Manfred Kircher

Abstract The change from fossil- to bio-based chemical processing and value chains is a tremendous challenge and if done right it provides a great opportunity for established chemical clusters as well as biomass-producing regions. Both need to adapt or develop pillars of competence which determine their regional competitive position: Infrastructure, industries, science and education, public administration, public acceptance and cluster-building. While biomass regions benefit from feedstock availability, established clusters are especially strong in further processing along the value chain. Based on the example of Europe's leading chemical cluster Antwerp–Rotterdam–Rhine–Ruhr (ARRR), a bottom-up approach into chemurgy and green chemistry is presented which combines strategies and activities on county, state, and multi-national level.

Keywords Bio-based fuel · Bio-based chemicals · Bio-industry infrastructure · Bio-based industries · Bio-based industry research and education · Bio-based industry acceptance · Bio-based industry cluster · Bio-based industry region

13.1 Introduction

It is well known that our current way of living and our wealth is based on fossil resources. Generating heat, power, and fuel for living and industrial purposes depends very much on fossil carbon sources. And the very same is true for chemicals starting from bulk platform chemicals such as ethylene up to lower

M. Kircher (✉)
KADIB Kircher Advice in Bioeconomy,
Kurhessenstr. 63, 60431 Frankfurt, Germany
e-mail: kircher@kadib.de; kircher@clib2021.de
URL: http://www.KADIB.de; http://www.CLIB2021.de

M. Kircher
CLIB2021, Völklingerstr. 4, 40219 Düsseldorf, Germany

P.C.K. Lau (ed.), *Quality Living Through Chemurgy and Green Chemistry*,
Green Chemistry and Sustainable Technology,
DOI 10.1007/978-3-662-53704-6_13

volume specialty (food- and feed additives, agro-chemicals, adhesives, lubricants, detergents, etc.) and fine chemicals (skin care actives, pharmaceuticals, etc.).

The processing chains leading from fossil resources to these product categories have become so much integrated into public infrastructure and private industries, science, education, and administration that their impact on these pillars has to be brought to consciousness again. Their quality defines the position of a region in attracting industries, generating jobs, and private as well as public wealth.

In turn, this means that if processing chains change fundamentally, specific pillars of competitiveness need to be adapted. Otherwise regions lose their competitive edge and new industries will generate jobs and wealth elsewhere.

This chapter presents regional pillars of competitiveness in today's fossil-based economy and discusses how to prepare and transform regions for chemurgy and green chemistry in the coming bio- and circular economy.

13.2 What Are Regional Pillars of Competitiveness

Before entering the topics specific for chemurgy and green chemistry, the current setting in the fossil-based economy should be discussed. Its basic pillars have been developed and optimized since the beginning of the "fossil era" in mid-nineteenth century, built its basis in the industrialized regions of the world and determined their ranking and competitiveness.

13.2.1 Infrastructure

The logistic infrastructure to feed an industry very much depends on the characteristics of the material to transport and store. Whether it is solid (coal), liquid (oil) or gaseous (natural gas) most important are

- composition,
- energy density,
- carbon density,
- storage suitability.

All fossil energy and carbon sources are quite homogenous and consist essentially of carbon and hydrogen (Table 13.1). The energy and carbon density is quite high (Table 13.2) and all these materials are stable to store under regular transport conditions. Therefore, it is technically easy and cost-efficient to use railway (coal), freighter and tanker (coal, oil) and pipelines (oil, gas) to supply industries over large distances. A cost-efficient global infrastructure using all these freight possibilities has been built over the past 100 years. Being integrated into this system is a decisive regional pillar of competitiveness.

Table 13.1 Composition of fossil feedstock (%)

	C	H	N	O	S
Natural gas	75–85	9–24	Traces	Traces	Traces
Mineral oil	83–87	10–14	0.1–2	0.5–6	0.5–6
Black coal	60–75	6	Traces	17–34	0.5–3
Lignite	58–73	4.5–8.5	Traces	21–36	3

Table 13.2 Material, carbon and energy density of fossil resources (ton/m^3)

	Density	Bulk density	Carbon	Energy (MJ/kg)
Natural gas				32–45
Mineral oil	0.8–0.9		0.68–0.77	43
Coal		0.4–0.8	0.32–0.64	25–33

13.2.2 Industries

The combination of good transport, storage characteristics and a technically easy and cost-efficient global logistics network are critical factors for processing industries to decide where to settle. For example, crude oil refineries and chemical plants prefer logistics centers such as sea ports and along rivers to reach impressing capacities of 10–20 million tons of oil per year in order to fully play out the economy of scale. These plants fractionate crude oil (Table 13.3) into its components which either are commercialized directly (e.g., gasoline) or further processed by subsequent industries.

The chemical industry, very much depending on naphtha [1] (Fig. 13.1), is an integral element of these processing chains. Therefore, the quality of industrial sites linking logistics and processing is another crucial pillar of competitiveness.

13.2.3 Science and Education

Since the first industrial use of crude oil in the late eighteenth century scientists worked on technologies to produce and process coal, gas, and oil. Establishing these materials as industrial feedstock even gave birth to what was known at the

Table 13.3 Oil refinery output (depending on refining temperature) and use

25 °C > > > > 350 °C						
Refinery gas	Gasoline	Naphtha	Kerosene	Diesel oil	Fuel oil	Residue
Bottled gas	Automotive fuel	Chemical feedstock	Aircraft-fuel	Truck-fuel, bus-fuel	Ship-fuel, power generation	Bitumen for road construction

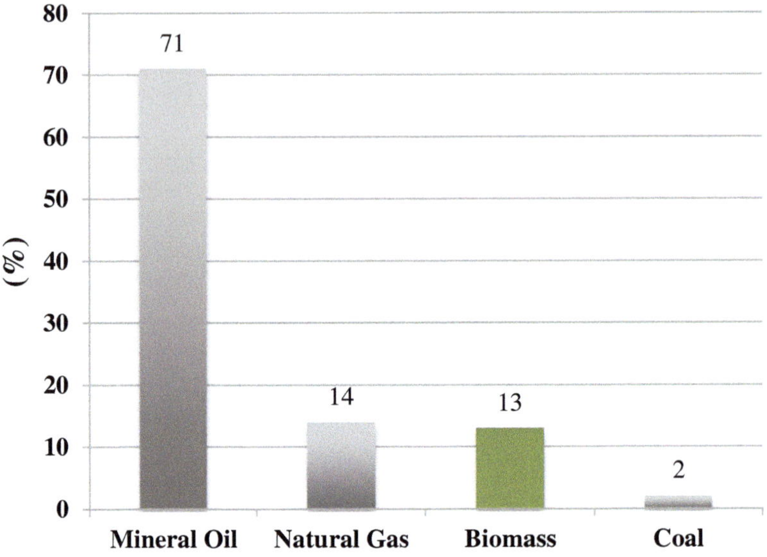

Fig. 13.1 Feedstock mix in chemical industries (Germany 2011)

time "science of chemistry" that started an unprecedented innovation cycle in dyes, polymers, and pharmaceuticals to name just a few categories. Since then science continued to add value and functionalized chemical compounds for nearly all needs in our daily life. Consequently, there is a direct correlation of science and know-how invested in chemical functionalization and its commercial value (Fig. 13.2). Successful regions are characterized by top research and development facilities in chemistry and chemical engineering. In 2013 the share of total chemical sales (2 billion $) in world trade (18.3 billion $) [2] has reached 11 %, thus proving the central function of this sector in world economics.

The more sophisticated industrial processes and products become, the more important it is to have an adequate staff of well-trained technical and academic personnel. Excellent institutions for vocational training and education in chemistry and chemical process engineering are added mandatory pillars of competitiveness.

13.2.4 Public Administration

Infrastructure, industries, science and education depend on public facilities like transportation systems, production sites and research and educational institutions. In addition, industrial activities interact with public goods and externalities. A prime example is nature and climate protection that necessitate an efficient environment policy by governments and its implementation by a properly functioning civil service. Both, well-balanced policies and an efficient administration are strong factors in regional competitiveness.

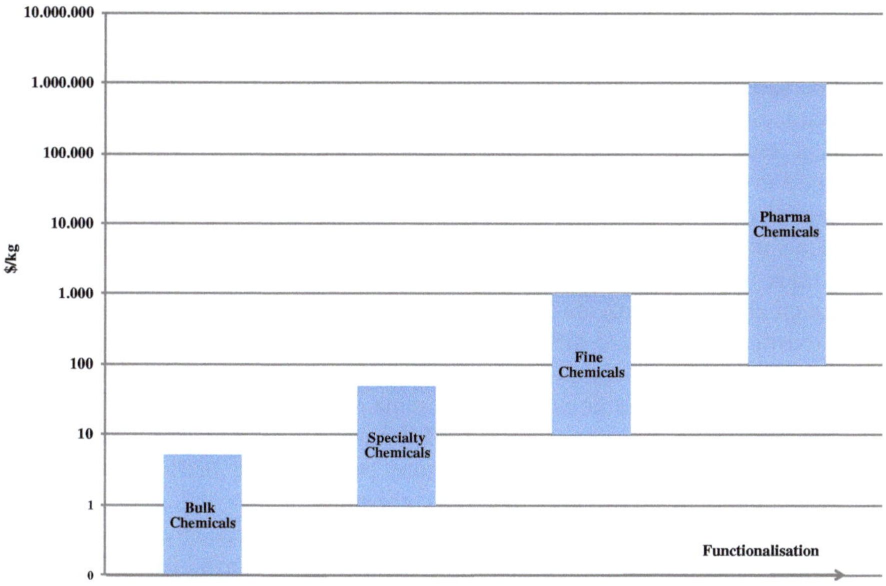

Fig. 13.2 Correlation of chemical functionalization and commercial value in chemical categories

13.2.5 Public Acceptance

Public acceptance is a very basic requirement. If social expectations (employment, salaries), ecological requirements (nature and climate protection) and economical output (profit, taxes) are not well-balanced an industrial sector or a specific technology will lose acceptance and finally its regional base.

Public acceptance is a real soft factor. It cannot be engineered and cannot be achieved by administrative measures. However, it is the result of public communication with all stakeholders represented by academic and industrial associations, political parties, unions, NGOs and more representatives of the civil society. The specific culture of consensus finding with and among the public stakeholders is therefore another pillar of competitiveness which incidentally is strongly connected with education.

13.2.6 Cluster

"Clusters are geographic concentrations of interconnected companies, specialized suppliers and service providers, firms in related industries, and associated institutions (e.g., universities, standards agencies, and trade associations) in particular fields that compete but also cooperate [3]." In order to attract top companies, investors and specialists, cluster regions need to organize their infrastructure in logistics, in science and education, their governmental policies and administrative measures for the benefit of a specific industrial sector. While cooperation is an essential success factor, dynamic competition stimulating continuous advancement in technologies and business models and the entry of new firms is another [4]. In the chemical sector, the 5 leading global clusters are: Houston (USA), Shanghai (China), Jurong (Singapore), Jubail (Saudi Arabia) and ARRR (EU; see below). The crucial role of clusters is underlined by the fact that chemical clusters (producing about 50 % of global chemistry value) generate 5 % (1 billion $) of total world trade (Fig. 13.3).

Though all clusters provide the very same set of pillars of competitiveness, their characteristics can be quite different. For example, the Near East cluster is strong because it has direct and cost-efficient access to oil and gas wells thus offering extremely beneficial conditions for bulk volume chemical production in the million tons per year range. Seventeen percent of the world's biggest platform chemical ethylene (world production 140 million tons per year) are produced here. This cluster is dominated by big multinational companies.

Another model is presented by the most important European chemical cluster covering Flanders (Belgium), the Netherlands and North Rhine-Wesphalia (Germany). This region is characterized by providing a multitude of fine and specialty chemicals in the 100,000 tons per year range by big players like Solvay, DSM and Evonik. Hundreds of small- and medium-sized enterprises (SME) further process these intermediates to consumer products such as plastic casings, adhesives, skin care products, pharmaceuticals and more. Obviously such a cluster very much depends on an excellent science and education landscape.

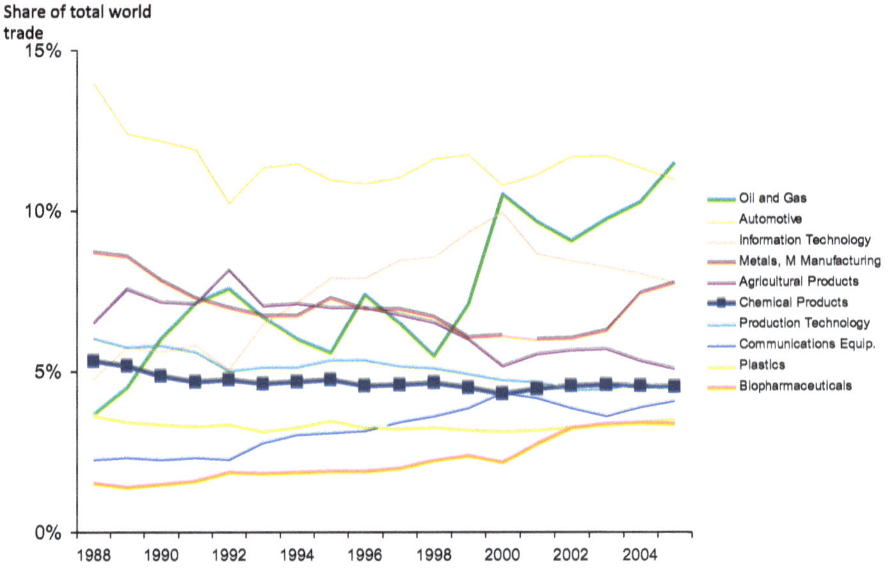

Fig. 13.3 Share of cluster categories in world trade. Reprinted from Ref. [5] with permission from author

13.3 What Makes a Region Competitive in Chemurgy and Green Chemistry

The very same pillars of competence making a successful fossil-based chemical region are key to regions targeting bio-based prosperity. However, different characteristics of the transforming regions would require adjustment or adaptation of those pillars.

13.3.1 Infrastructure

As with fossil resources, composition, energy and carbon density as well as storage suitability determine the logistic requirements. Bio-based industrial feedstock is principally available from agricultural, forestry and marine sources. Plant biomass is most relevant and commercially available in different processing states:

- crude biomass (wood, straw, empty fruit bunches, corn and rice husks)
- purified fruits and tubers (cereals, oleaginous fruits, potatoes),
- fractionated ingredients (sugar, starch, vegetable oil).

Today the chemical industry uses about 10–13 % bio-based feedstock, most of it fractionated and purified from vegetable sources. The resulting products spread (except bio-ethanol) mainly across low-volume specialty and fine chemicals (Fig. 13.4) [6, 7].

Plant materials and most plant products are solid and need to be transported in (compared to fossil carbon sources) relatively small batches by trucks, railways or

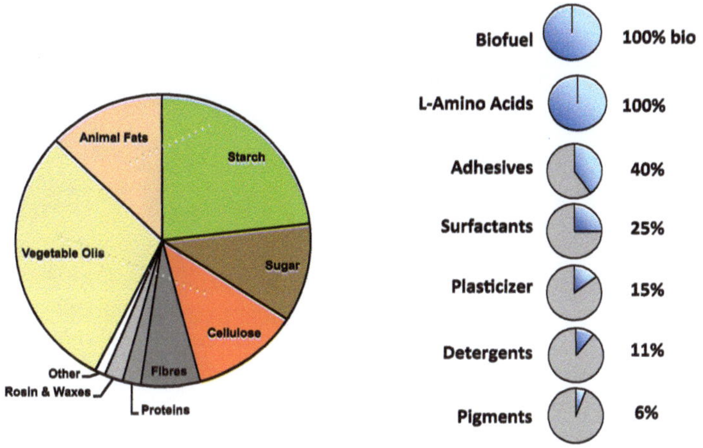

Fig. 13.4 Bio-based chemical feedstock, products groups and bio-based share. Reprinted from Ref. [7], Copyright 2012, with permission from Elsevier

ships. Only plant oil is liquid but the commercial volume does not justify cheaper transport in pipelines. As agriculture and forestry areas are extensive bio-feedstock regions need a close meshed logistic infrastructure and facilities to load and distribute solid materials.

The carbon content as well as carbon- and energy density of plant materials is significantly lower than that of fossil resources (Tables 13.4, 13.5). Consequently, a much bigger volume needs to be handled which has a direct impact on the cost of transport.

Nevertheless, sugar, plant oil, cereals, fruits are shipped globally in bulk volumes especially to supply food and feed markets where the chemical complexity of vegetable protein, sugar and oil defines the nutritional value, thus satisfying the demand of the market. If used in future bio-based processes to large volume hydrocarbons and derivatives, however, only carbon is really relevant; the rest (especially oxygen) is a byproduct.

This conflict becomes even more evident when it comes to transporting agricultural and forestry biomass such as straw, corn husks, empty oil palm bunches, or sugar cane bagasse. These materials are extremely bulky and demand relatively wide shipping volumes. Extra-cost for transporting feedstock components which are unsuitable in bio-based chemical operation need to be balanced by recycling and generating revenue in other markets. Bio-feedstock producing regions should

Table 13.4 Composition (%) and heat value (MJ/kg) of vegetable biomass and biomass compounds

	C	H	N	O	S
Vegetable biomass (average)	45	6	2	42	Traces
Linoleic acid	77	12	0	11	0
Glucose	40	7	0	53	0
Sucrose	43	6	0	51	0
Starch	44	6	0	50	0
Lignocellulose	44	6	0	50	0
Wood	50	6	3	41	Traces

Table 13.5 Energy (MJ/kg), bulk and carbon density (t/m^3) of biomass

	Bulk density	Carbon density	Energy density
Vegetable biomass (average)			6.8
Corn (whole biomass)			10.8
Linoleic acid			39.1
Glucose			15.6
Starch			17.5
Lignocellulose			10–25
Wood	0.2–0.6	0.1–0.3	14.4–15.8
Straw bale	0.05–0.11	0.03–0.05	
Cereal	0.4–0.48	0.2–0.24	

therefore not only provide facilities for transport but also for (pre-)processing in order to reduce biomass volume and complexity.

In summary, whether a region is positioned as a bio-feedstock provider or as a processor it needs to build up its logistics infrastructure.

13.3.2 *Industries*

The characteristics of bio-feedstock strongly influence the optimal capacity of a biorefinery. To keep the cost of feedstock transport down it is located in the biomass region and the capacity is adequate to the feedstock volume produced in an economically accessible radius. An example is given by the world's largest bio-ethanol plant which has been put into operation in November 2015 by DuPont in Iowa (USA) [8]. It is located in the midst of an agricultural area and feedstock supply of 375,000 dry tons of corn stover (containing around 45 % carbon) secured by contracts with 500 farmers in a 30-mile radius. The plant has a production capacity of 90,000 tons of bioethanol which is equivalent to 47,000 tons of ethanol-carbon. In contrast, e.g., the largest German fossil-oil refinery has a capacity of 13.9 million tons of oil-carbon (in 16.9 million tons of oil) which is equivalent to bio-carbon fixed in 1 Brazilian sugar harvest [9]. Balancing the much smaller economy of scale of biorefineries by other cost benefits is therefore a crucial pillar of competitiveness in rural areas.

In the long range such biorefineries will put biomass-producing regions in the position of evolving industry chains for the delivery of the product portfolio of today's fossil-based industries. A crucial prerequisite is building and attracting companies active in versatile processing chains to various markets as it was successfully demonstrated by the leading European integrated biorefinery, Bazancourt-Pomacle (France) [10]. Biorefinery products begetting more products are waiting for realization [11]. For the time being further processing is still the special strong pillar and competitive advantage of established chemical clusters.

As discussed earlier, established fossil-based industrial clusters are characterized by large-volume capacities. Although, for example, the European chemical cluster mentioned before has significant plant-based feedstock within its region it is not enough to feed large-scale plants and export-oriented industries. Importing pre-processed bio-based feedstock, platform chemicals or intermediates will connect biomass- and industrial regions.

Another alternative is using regional material flows in a cascade mode, thus linking cross-sectorial carbon chains in a new way. This potentially improves the carbon- and energy efficiency significantly and is therefore a competitive factor for individual companies and also on the regional level. An example is the industrial site of Frankfurt-Hoechst (Germany) where bio-based pharmaceuticals, chemicals and fuel are produced and residue flows from private and public providers feed Europe's biggest biogas plant (20 MW) which delivers power to the Hoechst-site and in parallel biogas into the public grid [12]. This scenario serves as a model for all graduations from biomass to industrial regions as to how to act beyond an individual plant and benefit regional stakeholders.

13.3.3 Science and Education

Also in bio-based industries, chemistry remains a key discipline because most bio-based production chains include chemical steps. For example, bioplastics start with a bio-based monomer produced by biotechnology but polymerization and further steps are pure chemistry (Fig. 13.5). In addition, other disciplines like microbiology, enzymology, biochemistry, and bioprocessing that include fermentation and enzymatic catalysis as well as mechanical and plant engineering are mandatory. Combining and integrating biotechnological and chemical processes will be a key factor in reducing investment and operating cost. Also, the growing share of volatile renewable energy (e.g., hydrogen) needs to be considered when developing bio-based processes. Especially the interfaces of traditional disciplines and those gaining relevance in the bioeconomy become important and it goes without saying that all these topics require a well-educated technical and academic staff.

13.3.4 Public Administration

Readiness to adapt to a changing environment remains a key pillar of public administration that is applicable to chemurgy. The need to prepare the logistics infrastructure for more solid feedstock supply, recycling residue material flows and balancing volatile renewable energy sources have already been mentioned. Another

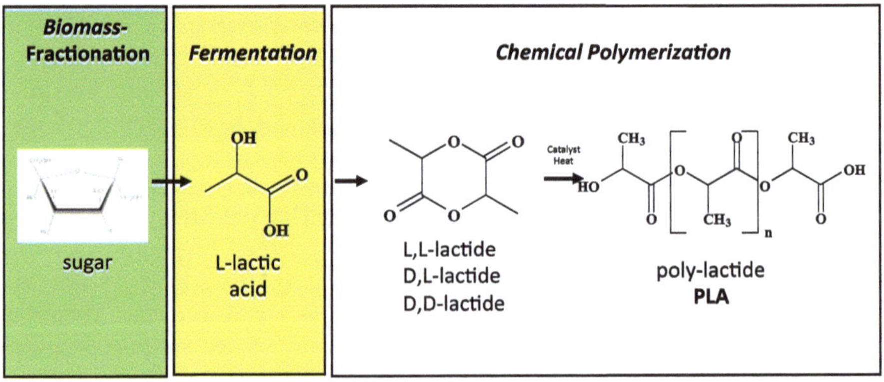

Fig. 13.5 Combined bio- and chemical processing in producing the polymer poly-lactide (PLA)

task is how to integrate the public sector into industrial material flows. Examples are the cascade use of industrial and municipal waste as well as carbon emission that become increasingly important as recyclable carbon sources. A public administration realizing the innovative potential and preparing an administrative framework for such options will become a decisive factor in green chemistry.

13.3.5 Public Acceptance

In general, bio-based products are well accepted especially by those brand-owners who win customers by emphasizing the bio-based nature of their products. A model example is Coca Cola's PlantBottle® (partially bio-based PET) in the leading bioplastics packaging sector (Fig. 13.6).

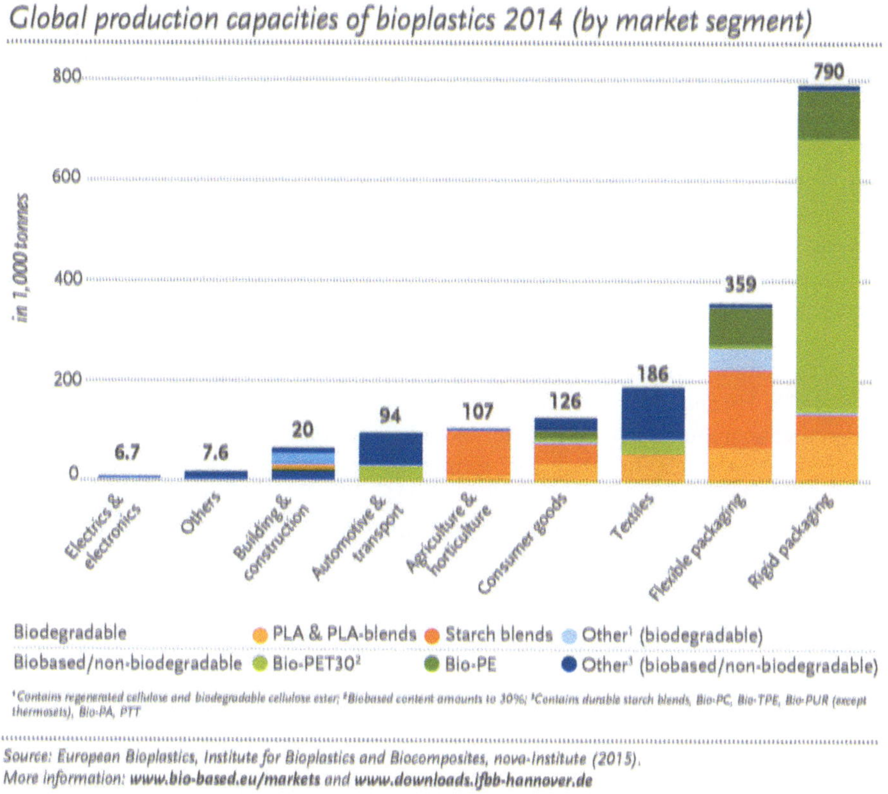

Fig. 13.6 Capacities and applications of bioplastics (2014). Reprinted from Ref. [13] with permission from European Bioplastics

Nevertheless, bioeconomical production lines are not without conflicts, especially bio-sourced fuels and chemicals are perceived not sustainable per se [14]. When it comes to plant-based raw materials the "food versus tank" topic has been discussed all over the world. The acceptance of genetically modified (GM) crops is a regional phenomenon. It is low in most of Europe but high in the Americas and Asia. Also, the expansion of sustainable resources of energy is not without conflicts as it comes with small scale power generating facilities in rural regions and unspoilt countryside. Regions that are able to adequately shape the public opinion fulfil one of the crucial preconditions for pioneering bioeconomy.

13.3.6 Cluster

There are 2 types of bioeconomy clusters:

- those evolving from established fossil-based clusters, and
- new clusters starting from bio-feedstock supply.

Established clusters on the one hand benefit from a proven industry, infrastructure, science landscape and administration, but on the other hand, running structures and systems tend to come along with strong inertia. Such clusters already comprise multi-faced industries for further processing which can build on established competences whether their raw stock or semi-finished materials are fossil- or bio-based. As it takes years and decades to evolve such a complex industrial landscape this is a competitive advantage not to be underestimated.

Biomass regions have a clear advantage when feedstocks take a high share of operating cost. An example is the Brazilian bio-ethanol industry which delivers 25 % of global bio-ethanol due its superior position in cultivating sugar cane. More than 400 "usinas" (sugar-cane refineries) produced 23.4 billion liters of bio-ethanol in 2014 [15, 16] and there are plans to expand the volume and product spectrum [17].

Another accumulation of biorefineries and ethanol fermentation is seen in the US corn producing states. In 2013, 210 fermentation plants produced 40 million tons (56 % of global production) of bioethanol [18, 19]. Together, Brazil and the US account for more than 80 % of the global bioethanol capacity.

In Indonesia and Malaysia, which account for 85 % (60 million tons) of the global palm oil production, hundreds of oil mills are running with further goals to capture added value chains in the energy as well as the chemical fields [20–22].

All these examples show evolving clusters comprising feedstock production and processing. Additional competencies in crop breeding and cultivation as well as plant and equipment engineering are being developed in parallel. In the foreseeable future both types of clusters offer good chances, but while the bioeconomy unfolds globally, established industrial clusters will find a serious competitor in feedstock-based clusters. Sustainability of these clusters will depend very much on science, education and entrepreneurship that is driving innovation.

13.4 Ways into Chemurgy and Green Chemistry Regions

There is no single roadmap to the fledgling bioeconomy. It depends on the starting conditions in feedstock supply, existing industries, target sector, infrastructure and other pillars as to how to transform a region into chemurgy and green chemistry.

For example, Malaysia is developing a renewable energy and chemical sector based on its strong position in palm oil production and has published the Malaysian Biomass Strategy in 2011 [23]. A special topic in this program is building industries in the framework of the so far neglected resources such as empty fruit bunches and palm oil effluents in the producing regions.

The US is using its leading position in corn production to build a renewable energy and chemical sector with a special focus on 2nd generation fuel from lignocellulosic sources. This strategy follows the National Bio-Economy Blueprint published in 2011 [24].

Europe also produces a significant amount of biomass resources. It published its bioeconomy vision back in 2007 [25], which was complemented by the European Vanguard Initiative [26] aiming at regional competitiveness in general. Since 2014, several developed nations, from middle- to south-east Europe along the river Danube (among them Austria, Croatia, Czech Republic, Hungary, Slovakia, Slovenia, Germany), run the project Danube-INCO.NET [27] which works on the topic of bioeconomy in these regions beside others. The latest example is the 3BI cluster project (Brokering Bio-Based Innovation) [28] where organizations from The Netherlands, France, UK and Germany joined forces in 2015.

In Europe, key regions that can be transformed towards chemurgy and green chemistry are in north-western Europe the directly adjacent regions of the Belgium province of Flanders, The Netherlands and the German state of North Rhine-Westphalia (NRW). This cross-border region starts from the sea ports of Antwerp and Rotterdam and reaches south along Europe's strongest logistics line along the rivers Rhine and Ruhr. Because of these anchor points the region is called the ARRR cluster; one of the 5 globally leading chemistry clusters generating 30 % of European chemical sales. It owes its excellent global rank and reputation from well-established and outstanding pillars in infrastructure, industries, science, education and public administration supported by understanding civil societies. The relevance of the chemical sector in this border-crossing mega-cluster is exemplified by the world's highest sales volume of chemicals per capita (Fig. 13.7).

Not only the essential role of chemistry in all three regions but also comparable economical performance based on adequate regional pillars of competence make this border-crossing chemistry cluster a model for well-balanced cooperation (Table 13.6).

ARRR's way into chemurgy and green chemistry follows the bioeconomy strategy of the EU, Flanders/Belgium, The Netherlands and NRW/Germany [31–35]. For many years bioeconomy initiatives are working separately at the national and regional levels. In Germany, biotechnology initiatives started in the 1980's [36] and in 2007 the Cluster Industrial Biotechnology e.V. (CLIB2021) was

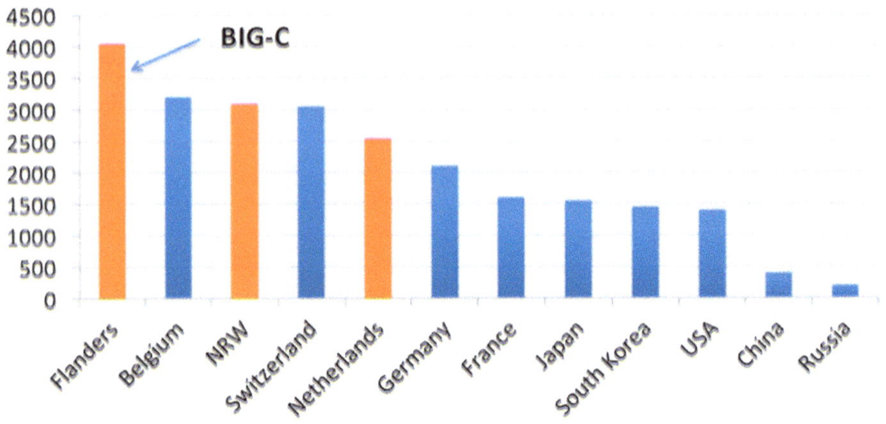

Fig. 13.7 Chemical sales volume per capita in different countries (2010; minus life sciences). Reprinted from Ref. [29] with permission from author

Table 13.6 Key data of the ARRR mega-cluster and its regions [30]

	ARRR	Flanders (Belgium)	The Netherlands	NRW (Germany)
Population (mio)	40.9	6.35	17	17.5
GDP (bn EUR)	1382	221	579	582
GDP/capita (EUR)	33,572	33,400	34,059	33,257
Industry (% of GDP)	25	19.5	29	25.4
Chemical ind. (bn EUR)	168	43	60	65

initiated by top-down official cluster policies [37] primarily to address bio-based chemicals. Since then CLIB evolved to become a bottom-up cluster championed by about 100 members from big industries (>1000 full-time employees), SME (50 % of membership), academia, investors, industrial sites and associations. Thirty percent are non-German members from all over the world (Fig. 13.8). These membership categories give CLIB both a critical mass and a well-balanced industrial, academic, regional and cross-regional competence profile.

The special feature of CLIB's strategy is to work with and connect world leading technology owners, industries and bioeconomy regions by providing open innovation platforms and actively acting as a value chain broker. Member industries are especially interested in greening supply chains and adding functional value to products. To strengthen the science base CLIB has run an academic technology program at 4 academic institutions in NRW from 2009 to 2013 which stably established the key disciplines of polyomics (genome and metagenome analysis), expression, biocatalysis and DSP (downstream processing). Since 2009, the very same institutions implemented Europe's biggest graduate program for biotechnologists and (bio-)chemical engineers (CLIB GC) from all over the world [39]. So far

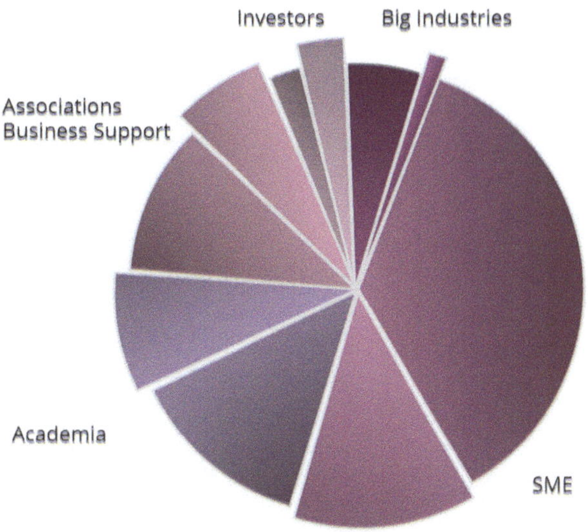

Fig. 13.8 Membership of CLIB2021 (total 100; non-German members highlighted). Reprinted from Ref. [38] with permission from CLIB2021

60 alumni entered academic or industrial careers. As CLIB's home region NRW is producing 30 % of Germany's carbon emission especially from power generation and heavy industries, the development of C_1-based biocatalyzed value chains is of special interest (Fig. 13.9).

The C_1-based value chain is one of the topics in "RIN (Regional Innovation Network) innovative and sustainable Material Flow" [41], a CLIB-project which analyses and evaluates the regional green chemistry potential concerning feedstock base, logistics, private and public stakeholders, administration, cross-sectorial options in a specific county [Rheinland (NRW); Fig. 13.10] and looks for exchange with the local population.

The next level of CLIB's regional activities is partnering with its ARRR-neighbors, the Netherlands and Flanders. Each of these has a strong foundation. In the Netherlands, the Biotechnology Based Ecologically Balanced Sustainable Industrial Consortium (BE-BASIC Foundation) has been working since 2010; and in Flanders essencia, the Belgian Association of the Chemical Industry partnered with the Flanders Innovation Hub for Sustainable Chemistry FI-SCH. Promoting cluster growth by exchange with other clusters [4] is a key element of CLIB, BE-BASIC and FI-SCH. In a bottom-up initiative in 2013 [42], these and 11 more national and regional organizations took the initiative to form the border-crossing Bio-Innovation Growth mega-cluster (BIG-C) with a clear focus on transforming the ARRR mega-cluster into Europe's leading chemurgy and green chemistry region [42]. This initiative is backed by a critical mass of strong industries, a crucial factor in cluster development [37]. Besides, the governments of Flanders, the Netherlands and NRW/Germany have been involved from the very beginning and supported the BIG-C's strategy. According to the German support strategy to "strengthen strengths" [43] the German Federal Ministry of Education

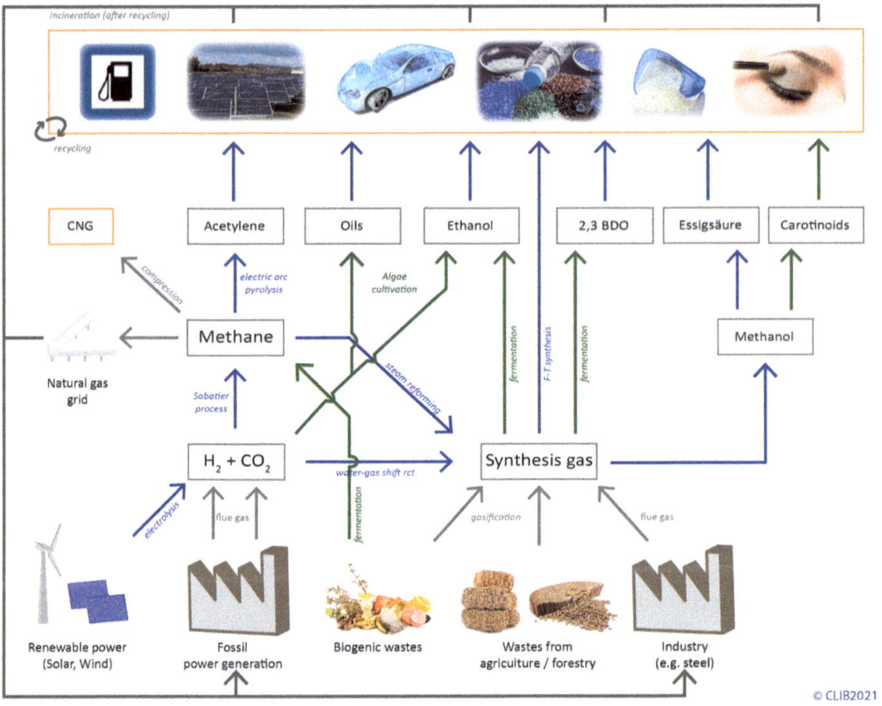

Fig. 13.9 Potential C_1-based value chains. Reprinted from Ref. [40] with permission from CLIB2021

and Research (BMBF) announced recently its support of CLIB's BIG-C activities [44].

The BIG-C's strategy meets the European strategy "Towards a Circular Economy: A Zero Waste Program for Europe" [45] for making better use of resources for the economic and environmental benefit of Europe and makes further link with the Europe 2020 Strategy [46, 47] for smart, sustainable and inclusive growth. Processing chains should valorize all industrial material flows, thus:

- increasing feedstock efficiency,
- reducing side streams, and
- climate damaging emission.

Specifically BIG-C aspires to develop and realize four complete value chains from feedstock through intermediates and components up to consumer product (F2P) [48]:

- F2P woody biomass (lignocellulose)
- F2P agro-based biomass
- F2P organic side streams
- F2P CO/CO_2-containing gas

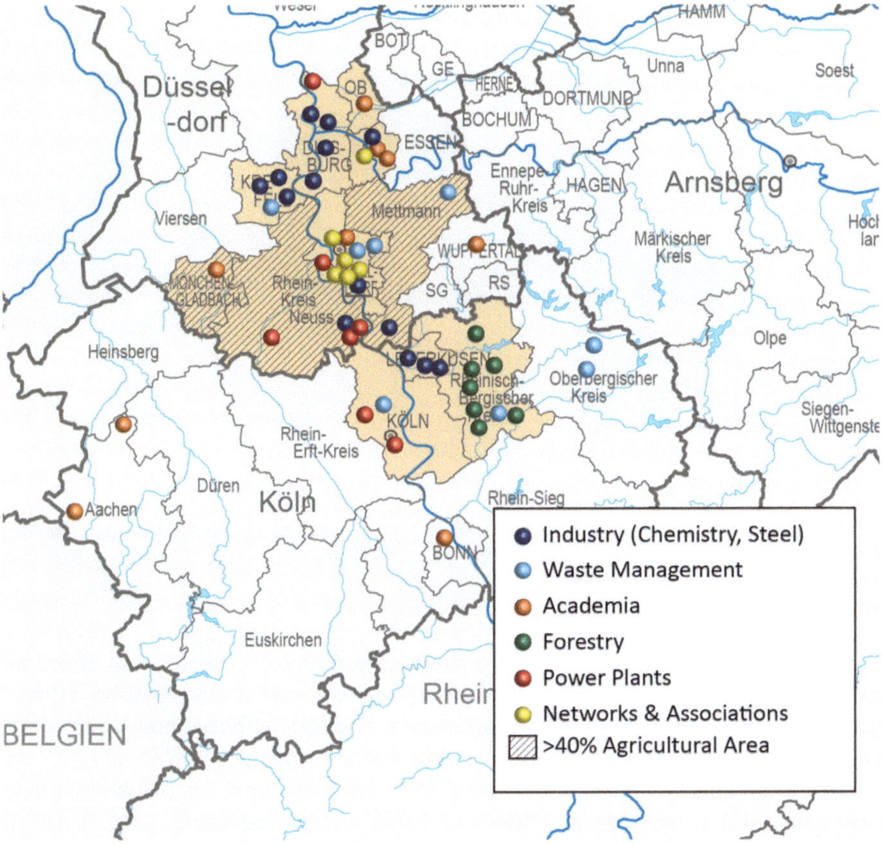

Fig. 13.10 The Rheinland county in NRW (Germany) is evaluated concerning pillars of competence in bioeconomy. Reprinted from Ref. [40] with permission from CLIB2021

This ambitious and long-termed vision (Fig. 13.11) will rely on a largely integrated infrastructure and a world-class complementary industrial as well as academic landscape in Europe's economical center with two multi-purpose pilot facilities (Bio Base Europe-Pilot Plant in Gent, Belgium and Bioprocess Pilot Facility in Delft, Netherlands).

BIG-C's work program (2014/2015) addresses the regional pillars of competence as discussed earlier. Logistics and raw materials potential are analyzed on county level in the project Regional Innovation Network (RIN) [41]. In Gent (Flanders) the connection of an industrial gaseous carbon source and gas fermentation is investigated on production site level [49] and at the Port of Rotterdam (The Netherlands) SkyNRG [50] works on producing biobased aviation fuel. Corresponding to the industrial demand of biobased supply chains market-oriented areas of work have been defined (biobased aromatics and aviation fuel, use of

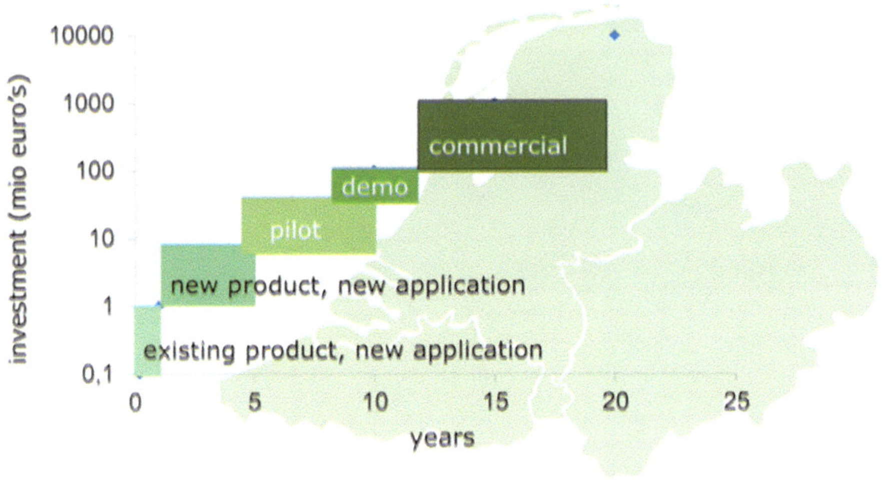

Fig. 13.11 Time and investment frame of BIG-C from R&D to commercialization. The map shows BIG-C's regions Flanders (Belgium), The Netherlands and NRW (Germany). Reprinted from Ref. [29] with permission from author

C_1-feedstock). These topics are also addressed in BIG-C's research and education institutions in order to provide the necessary technologies and specialists. How to deal with public expectations and adjustment needs is a horizontal topic which includes communicating with private and public stakeholders.

Green chemistry will form regional, cross-regional and international/ intercontinental value chains. BIG-C seeks to address all levels and strives to cooperate with global regions. Since 2009, CLIB has worked beyond Europe with North-America, Brazil, Russia, Malaysia and China, sharing valuable contacts and experiences with its BIG-C partners who are equally active especially in Brazil and India.

The 2015 G7 Summit (Elmau; Germany) and the 2015 Climate Change Conference (Paris; France) strongly called for decarbonization by the end of this century which means the displacement of fossil carbon sources completely. Accordingly, chemical industries, other than the energy sector that absolutely depend on carbon sources, are most vulnerable. The future lies in bio- and circular economy whose urgency has been confirmed by the 1st Global Bioeconomy Summit in November 2015 in Berlin (Germany) [51]. Up to now more than 40 countries have adopted specific bioeconomy strategies (Fig. 13.12), all have been developed along national and regional requirements [52, 53]. It is high time for implementation of future chemurgy and green chemistry regions. In doing so it is important to prepare the regional pillars of competitiveness the right way. Especially in the coming transition phase from the fossil- to the bio-based economy this is complex and difficult task. Nevertheless, those early receptors of the challenge and ability to adapt fast will obtain a real competitive advantage.

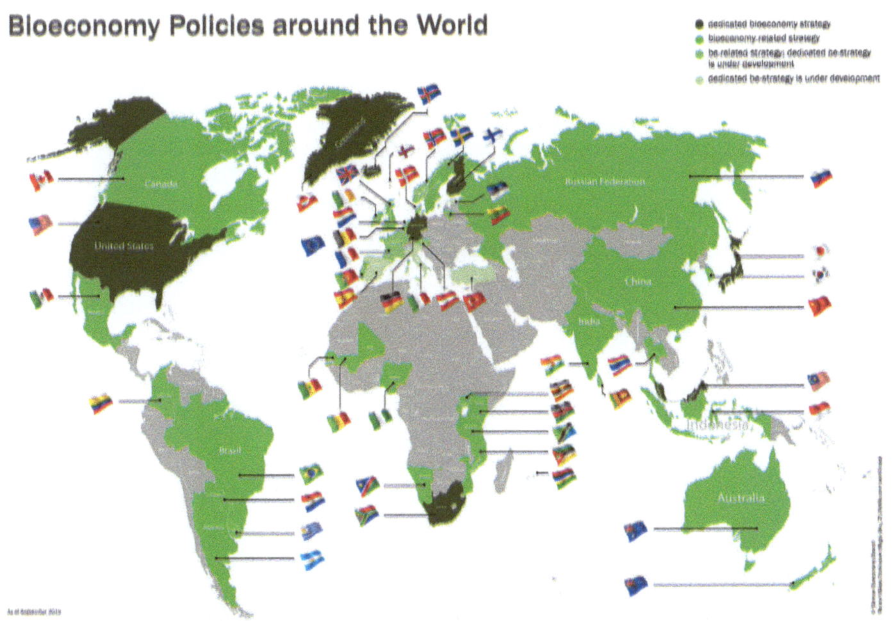

Fig. 13.12 Countries that have adopted bioeconomy strategies (2015). Reprinted from Ref. [54] with permission from German Bioeconomy Council

References

1. VCI (2015) Rohstoffbasis der chemischen Industrie. https://www.vci.de/vci/downloads-vci/top-thema/daten-fakten-rohstoffbasis-der-chemischen-industrie-de.pdf. Accessed 30 Dec 2015
2. WTO (2014) International trade statistics 2014. https://www.wto.org/english/res_e/statis_e/its2014_e/its2014_e.pdf. Accessed 30 Dec 2015
3. Porter M (2000) Locations clusters and company strategy. In: Clark G, Feldman M, Gertler M (eds) Oxford handbook of economic geography. Oxford University Press, Oxford
4. Sölvell Ö, Lindqvist G, Keltels C (2003) The cluster initiative greenbook. Ivory Tower AB, Stockholm
5. Ketels C (2007) The role of clusters in the chemical industry. EPCA European Petrochemical Association, Budapest
6. Hill K, Höfer R (2009) Biomass for green chemistry. In: Höfer R (ed) Sustainable solutions for modern economies. RSC Publishing, London
7. Kircher M (2011) Biotechnology pushes greening the chemical industry—biomass transformation the most relevant driver of industrial biotechnology. Ind Biotechnol 7(2):38–42
8. DuPont (2015) DuPont opens world's largest cellulosic ethanol plant in Iowa. http://www.biofuelsdigest.com/bdigest/2015/11/01/dupont-opens-worlds-largest-celulosic-ethanol-plant-in-iowa-the-full-story-in-pictures/. Accessed 30 Dec 2015
9. Kircher M (2012) The transition to a bio-economy: emerging from the oil age. Biofuels Bioprod Biorefin 6:369–375

10. Schieb PA, Lescieux-Katir H, Thenot M et al (2015) Biorefinery 2030 Future prospects for the bioeconomy. Springer, New York
11. Pandey R, Höfer M, Taherzadeh K, Nampoothiri M, Larroche C (eds) (2015) Industrial biorefineries and white biotechnology. Elsevier, Amsterdam
12. Infraserv Hoechst (2015) Biogas plant—cutting-edge energy recovery from biowaste. http://www.industriepark-hoechst.com/en/index/industriepark/investitionsprojekte/energiesicherheit/co-fermentationsanlage.htm. Accessed 30 Dec 2015
13. European Bioplastics (2014) Global production capacities of bioplastics 2014. http://en.european-bioplastics.org/market/applications/. Accessed 30 Dec 2015
14. Kircher M (2015) Sustainability of biofuels and renewable chemicals production from biomass. Curr Opin Chem Biol 29:26–31
15. Renewable Fuels Association (RFA) (2015) Going global—2015 ethanol industry outlook. http://www.ethanolrfa.org/wp-content/uploads/2015/09/c5088b8e8e6b427bb3_cwm626ws2.pdf. Accessed 30 Dec 2015
16. Alternative Fuels Data Center (2014) Global ethanol production 2007–2014. http://www.afdc.energy.gov/data/10331. Accessed 30 Dec 2015
17. Souza MG, Reynaldo LV, Joly CA, Verdade LM (eds) (2014) Bioenergy & sustainability: bridging the gaps. SCOPE, Sao Paulo
18. US-EIA (2015) U.S. fuel ethanol plant production capacity. http://www.eia.gov/petroleum/ethanolcapacity/. Accessed 30 Dec 2015
19. Bundesverband der deutschen Bioethanolwirtschaft (2015) Bioethanol weltweit. http://www.bdbe.de/daten/bioethanol-weltweit. Accessed 30 Dec 2015
20. Kuoy A (2014) New wealth in oil palm. http://www.thestar.com.my/news/environment/2014/03/24/green-wealth-in-oil-palm/March-24-2014. Accessed 30 Dec 2015
21. Malaysian Palm Oil Council (MPOC). Malaysian palm oil industry. http://www.mpoc.org.my/Malaysian_Palm_Oil_Industry.aspx. Accessed 30 Dec 2015
22. Sime Darby (2015) Facts and figures. http://www.simedarby.com/upload/Palm_Oil_Facts_and_Figures.pdf. Accessed 30 Dec 2015
23. AgansiInovasi Malaysia (2011) National biomass strategy 2020: new wealth creation for Malaysia's palm oil industry. http://innovation.my/wpcontent/downloadabels/National/20Biomass/20Strategy/20Nov/202011/20FINAL.pdf. Accessed 30 Dec 2015
24. US White House (2011) Building a 20th century bioeconomy. http://www.whitehouse.gov/the-press-office/2011/09/16/president-obama-signs-america-invents-act-overhauling-patent-system-stim. Accessed 30 Dec 2015
25. EU (2007) En route to the knowledge-based bio-economy (KBBE). http://www.bio-economy.net/reports/files/koln_paper.pdf. Accessed 30 Dec 2015
26. EU (2014) Vanguard initiative. New growth through smart specialization. http://www.s3vanguardinitiative.eu. Accessed 30 Dec 2015
27. INCO.Net (2015) Inco-net energy and Bioeconomy. https://danube-inco.net/ourpillars/energy_and_bio_economy. Accessed 30 Dec 2015
28. BI (2015) 3BI Europes Bioeconomy intercluster. http://www.3bi-intercluster.org/home/. Accessed 30 Dec 2015
29. van Breugel J, van der Wielen LAM (2013) 17th ICABR international consortium on applied bioeconomy research conference, Ravello, Italy
30. BIG-C (2015) Agenda 2014/2015. 1. BIG-C stakeholder meeting, Düsseldorf, 19.5.2014
31. EU (2012) The bioeconomy strategy. http://ec.europa.eu/research/bioeconomy/index.cfm?pg=policy&lib=strategy. Accessed 30 Dec 2015
32. Vlaanderen Dep Leefmilieu Natuur en Energie (2013) Bioeconomy in Flanders. The vision and strategy of the government of Flanders for a sustainable and competitive bioeconomy in 2030. https://www.vlaanderen.be/nl/publicaties/detail/bioeconomy-in-flanders. Accessed 30 Dec 2015
33. Wetenschappelijke en Technologische Commissie voor de biobased economy (2013) Strategie voor een groene samenleving. https://biobs.jrc.ec.europa.eu/country/netherlands. Accessed 30 Dec 2015

34. BMBF (2010) Nationale Forschungsstrategie Bio Ökonomie 2030. https://www.bmbf.de/pub/biooekonimie.pdf. Accessed 30 Dec 2015
35. MIWF NRW (2013) Eckpunkte einer Bioökonomiestrategie für Nordrhein-Westfalen. http://www.wissenschaft.nrw.de/forschung/fortschritt-nrw/grosse-gesellschaftliche-herausforderungen-bewaeltigen/multitalent-biotechnologie/biooekonomie-in-nrw/. Accessed 30 Dec 2015
36. Cooke P (2002) Biotechnology clusters as regional, sectorial innovation systems. Int Reg Sci Rev 25:8–37
37. Fromhold-Eisebith M, Eisebith G (2004) How to institutionalize innovative clusters? Comparing explicit top-down and implicit bottom-up approaches. Conference regionalization of innovation policy—options and experiences, Berlin, 4–5 June
38. CLIB2021 (2014). Annual brochure. online. http://www.clib2021.de/sites/default/files/pdf/Broschüre_2014_lq_neu.pdf. Accessed 9 Feb 2016
39. CLIB2021 (2015) Graduate Cluster. http://www.clib2021.de/en/graduierten-cluster. Accessed 30 Dec 2015
40. Köster F, Schulze-Beusingsen L (2014) RIN Stoffströme—Regionale Umsetzung der Bioökonomie. online. http://www.energieagentur.nrw.de/content/anlagen/Koester.pdf. Accessed 30 Dec 2015
41. RIN Stoffströme (2015) RIN Stoffströme. http://www.stoffstroeme.de. Accessed 30 Dec 2015
42. BE-Basic (2014) Bio innovation growth mega-cluster BIG-C launched in Düsseldorf May 19, 2014. http://www.be-basic.org/news-center/news/bio-innovation-growth-megacluster-big-c-launched-in-duesseldorf.html. Accessed 30 Dec 2015
43. Staehler D, Dohse P, Cooke P (2007) Evaluation der FördermaßnahmenBioRegio und BioProfile, BMBF FKZ 0313708
44. CLIB2021 (2015) BIG-C wins German Federal Research Ministry competition—CLIB2021 promotes international network between NRW, Flanders and The Netherlands. http://www.clib2021.de/en/news/news-clib2021?id=275. Accessed 30 Dec 2015
45. EU (2014) COM(2014)398 Towards a circular economy: a zero waste program for Europe. http://eur-lex.europa.eu/legal-content/EN/TXT/?uri=CELEX:3A52014DC0398. Accessed 30 Dec 2015
46. EU (2010) COM(2010) Europe 2020—a strategy for smart, sustainable and inclusive growth. http://eur-lex.europa.eu/legal-content/en/ALL/?uri=CELEX:52010DC2020. Accessed 30 Dec 2015
47. EU (2011) COM(2011)21, Communication from the commission to the European Parliament, the Council, the European Economic and social committee and the committee of the regions making raw materials available for Europe's future well-being. Proposal for a European innovation partnership on raw materials. http://eurlex.europa.eu/legalcontent/AUTO/?uri=CELEX:52012DC0082&qid=1451586715092&rid=7. Accessed 30 Dec 2015
48. Wielen van der L (2014) 1. BIG-C stakeholder meeting, 19.5.2014 Düsseldorf
49. ArcelorMittal (2015) ArcelorMittal, LanzaTech and Primetals Technologies announce partnership to construct breakthrough €87m biofuel production facility. http://corporate.arcelormittal.com/news-and-media/news/2015/july/13-07-2015. Accessed 30 Dec 2015
50. SkyNRG (2015) We supply aviation with sustainable jet fuel. http://skynrg.com. Accessed 30 Dec 2015
51. Global Bioeconomy Summit (2015) Communiqué. http://gbs2015.com/fileadmin/gbs2015/Downloads/Communique_final_neu.pdf. Accessed 30 Dec 2015
52. German Bioeconomy Council (2014) Bioeconomy policy—synopsis and analysis of strategies in the G7. BIOCOM AG, Berlin
53. de Besi M, McCormick K (2015) Towards a bioeconomy in Europe: national, regional and industrial strategies. Sustainability 7(8):10461–10478
54. Bioökonomierat (2015) Global bioeconomy policies. http://gbs2015.com/fileadmin/gbs2015/Downloads/GBS2015_05_Bioeconomy_Around_World_Map.pdf. Accessed 30 Dec 2015

MIX
Papier aus verantwortungsvollen Quellen
Paper from responsible sources
FSC® C105338

If you have any concerns about our products,
you can contact us on
ProductSafety@springernature.com

In case Publisher is established outside the EU,
the EU authorized representative is:
Springer Nature Customer Service Center GmbH
Europaplatz 3, 69115 Heidelberg, Germany

Printed by Libri Plureos GmbH
in Hamburg, Germany